Lecture Notes in Computer Science 519

Edited by G. Goos and J. Hartmanis

Advisory Board: W. Brauer   D. Gries   J. Stoer

F. Dehne, J.-R. Sack,
N. Santoro (Eds.)

Algorithms and
Data Structures

2nd Workshop WADS '91
Ottawa, Canada, August 14-16, 1991
Proceedings

Springer-Verlag
Berlin Heidelberg New York
London Paris Tokyo
Hong Kong Barcelona
Budapest

F. Dehne   J.-R. Sack
N. Santoro (Eds.)

Algorithms and Data Structures

2nd Workshop, WADS '91
Ottawa, Canada, August 14-16, 1991
Proceedings

Springer-Verlag

Berlin Heidelberg New York
London Paris Tokyo
Hong Kong Barcelona
Budapest

Series Editors

Gerhard Goos
GMD Forschungsstelle
Universität Karlsruhe
Vincenz-Priessnitz-Straße 1
W-7500 Karlsruhe, FRG

Juris Hartmanis
Department of Computer Science
Cornell University
Upson Hall
Ithaca, NY 14853, USA

Volume Editors

Frank Dehne
Jörg-Rüdiger Sack
Nicola Santoro
School of Computer Science, Carleton University
Ottawa, Canada K1S 5B6

CR Subject Classification (1991): F.1-2, G.2-3, H.3, I.3.5

ISBN 3-540-54343-0 Springer-Verlag Berlin Heidelberg New York
ISBN 0-387-54343-0 Springer-Verlag New York Berlin Heidelberg

Printing and binding: Druckhaus Beltz, Hemsbach/Bergstr.
2145/3140-543210 - Printed on acid-free paper

PREFACE

The papers in this volume were presented at the Second Workshop on Algorithms and Data Structures (WADS'91). The workshop took place August 14 - 16, 1991, at Carleton University in Ottawa and was organized by the School of Computer Science at Carleton University (Ottawa, Ont). The workshop alternates with the Scandinavian Workshop on Algorithm Theory (SWAT) continuing the tradition of SWAT'88, WADS'89, and SWAT'90.

In response to the program committee's call for papers, 107 papers were submitted. From these submissions, the program committee selected 38 for presentation at the workshop. In addition to these papers, the workshop included five invited presentations.

August 1991

Frank Dehne
Jörg-Rüdiger Sack
Nicola Santoro

PROGRAM COMMITTEE:
A. Apostolico (Purdue U./Aquila U.)
M. Atallah (Purdue U.)
J.-D. Boissonat (INRIA, Sophia Antipolis)
S. Carlson (Lund U.)
K. Clarkson (AT&T Bell Labs.)
F. Dehne (Carleton U.)
J. Gilbert (XEROX, Palo Alto)
M. Goodrich (Johns Hopkins U.)
S. Hambrusch (Purdue U.)
M. Keil (U. of Saskatchewan, Saskatoon)
M. C. Loui (NSF)
F. Meyer auf der Heide (Paderborn)
J.-R. Sack (Carleton U.)
N. Santoro (Carleton U.)
R. Seidel (U. of Berkeley)
R. Tamassia (Brown U.)
N. M. Thalmann (U. of Geneva)
J. Urrutia (Ottawa U.)
J. van Leeuwen (U. of Utrecht)
C.K. Wong (IBM, Yorktown Heights)
D. Wood (U. of Waterloo)

ORGANIZING COMMITTEE:
F. Fiala (Carleton U., Chair)
R. Carter (Carleton U.)
E. Otoo (Carleton U.)
R. Probert (Ottawa U.)

SPONSORED BY
NSERC, ITRC, OCRI, TRIO,
Carleton U. and Queen's U.

TABLE OF CONTENTS

A Case Study in Comparison Based Complexity: Finding the Nearest Value(s)

Walter Cunto

Centro Científico IBM de Venezuela
A.P. 64778, Caracas, Venezuela

J. Ian Munro

Department of Computer Science, University of Waterloo
Waterloo Ont., N2L 3G1, Canada

Patricio V. Poblete

Departamento de Ciencias de la Computación, Universidad de Chile
Blanco Encalada 2120, Casilla 2777, Santiago, Chile

Abstract. It is shown that $5n/4$ plus-minus lower order terms comparisons on average are necessary and sufficient to solve the problem of finding the values of ranks immediately above and below a specified element x in a set X of size $n > 1$. When x turns out to be the median of X, $1.5n + \sqrt{\pi n/8} + O(\lg n)$ comparisons are proven to be sufficient. $n + \min(k, n - k) + 3 \ln n + O(1)$ comparisons are sufficient if k, the rank of x in X, differs from $n/2$ by $\Theta(n)$.

1 Introduction

An interesting although surprisingly-little studied problem in selection is that of determining the nearest value in an unordered array to a given value under a pure comparison based model of computation. We address the average case complexity of this problem more formally given as:

Problem 1. *Given a set X of $n > 1$ elements, including a designated $x \in X$, find the elements of ranks one above and one below x; or report the absence of one of these.*

Two variants of the previous problem are also useful. The *left neighbor* problem is that of finding the largest element in X that is less than x if such element exists; otherwise, reporting its absence. The *right neighbor* problem is defined symmetrically. The worst case complexity of this problem has been fully studied [2]. A simple algorithm making at most $2n - 3$ comparisons has shown to be optimal in that case. Worst case optimality is proven by designing a simple adversary which resembles the one given in [1] for the worst case selection problem.

Assuming all input permutations are equally likely, a somewhat faster method, on average, can be demonstrated. This algorithm performs $1.5n + \sqrt{\pi n/8} + O(\lg n)$ comparisons if x turns out to be (virtually) the median of X; otherwise, it performs $n + \min(k, n - k) + 3 \ln n + O(1)$ comparisons where k denotes the rank of x in X. If any of the possible ranks of x in X is also equally likely, an average of $5n/4 + H_n/2 + 11 H_{\lfloor n/2 \rfloor}/4 + O(1)$ comparisons are performed by the algorithm.

These estimates of runtime are derived with the help of a Markovian graph model wherein nodes represent computational states and edges represent transitions among states performed during the computation of any problem instance. Computations are traced by traversing paths in the graph and average performances are obtained by counting average costs (comparisons) associated either to the nodes or to the edges of the graph. Edge oriented counts along the traversals were used to derive the performance of our algorithm conditioned to the (previously unknown) rank of x in X. Node oriented counts were used to derive a closed formula for the average performance of our algorithm. Finally, $5n/4 - \Omega(1)$ comparisons are shown to be required by any algorithm that solves the problem with a technique slightly different than that discussed in [3] which counts different types of comparisons along the computation of the solution.

2 The algorithm

The algorithm keeps track of closest neighbors found thus far on either side of x, together with a count on the number elements seen on each side. More formally:

i) Compare the first element with x, making it a neighbor candidate on the appropriate side.

ii) Process each remaining element by comparing it with the current neighbor on the more populous side of x. In case of equal population, the neighbor is randomly chosen.

iii) If necessary, compare the new element with the other neighbor.

iv) If the new element falls between the current neighbors, compare it with x and replace the appropriate neighbor candidate with the new value.

This algorithm, which is also suitable for on-line applications, performs at most $3n-6$ comparisons; but, as we shall see, its average case behavior is more interesting.

3 A Markovian graph model

At each step, the algorithm determines whether the next element is larger or smaller than x. This process is modeled as a Markovian graph $G = (V, E)$. V includes all states $\langle p, q \rangle$ with $p \geq 0$, $q \geq 0$ and $p + q \leq n - 1$ such that p and q are the numbers of elements smaller and larger than x respectively after $p + q$ steps. E contains all possible state transitions. The collection of subsets $V_t = \{\langle p, q \rangle \mid p + q = t - 1\}$, $1 \leq t \leq n$, partitions the set V. Clearly, $|V_t| = t$ and $|V| = \binom{n+1}{2}$ as transitions occur only from nodes in V_{t-1} to nodes in V_t, $2 \leq t \leq n$. The computation starts at $s = \langle 0, 0 \rangle$ and finishes at any of the n states $\langle p, q \rangle$ such that $p + q = n - 1$ with $p, q \geq 0$.

A directed edge is denoted by (v, w, j) where $v \in V_{t-1}$, $w \in V_t$, $2 \leq t \leq n$, and $j \in E(v, w)$ is the label of one of the transitions from v to w. Each edge $(v, w, j) \in E$ specifies the number of comparisons $c(v, w, j)$ to be executed and its transition probability $pr(v, w, j)$. Note that G is nonsimple.

Since G is a Markovian graph, the sum of probabilities associated with transitions starting from the same node must equal 1, that is

$$\forall v \in V_{t-1}, \ 2 \leq t \leq n, \ \sum_{w \in V_t} \sum_{j \in E(v,w)} pr(v, w, j) = 1 \ . \tag{1}$$

Also, transitions in G are symmetric with respect to central nodes (those states $\langle p, q \rangle$ such that $|p - q| \leq 1$). Figure 1 summarizes the number of comparisons and the transition

First type of transitions, $p = q = 0$,

$$c(\langle 0,0\rangle, \langle 0,1\rangle, 1) \quad = \quad c(\langle 0,0\rangle, \langle 1,0\rangle, 1) \quad = \quad 1$$

$$pr(\langle 0,0\rangle, \langle 0,1\rangle, 1) \quad = \quad pr(\langle 0,0\rangle, \langle 1,0\rangle, 1) \quad = \quad \tfrac{1}{2}$$

Second type of transitions, $\min(p,q) = 0$ and $\max(p,q) > 0$

$$
\begin{aligned}
c(\langle p,q\rangle, \langle p,q+1\rangle, 1) &= c(\langle q,p\rangle, \langle q+1,p\rangle, 1) &=& \quad 1 \\
c(\langle p,q\rangle, \langle p,q+1\rangle, 2) &= c(\langle q,p\rangle, \langle q+1,p\rangle, 2) &=& \quad 2 \\
c(\langle p,q\rangle, \langle p+1,q\rangle, 1) &= c(\langle q,p\rangle, \langle q,p+1\rangle, 1) &=& \quad 2
\end{aligned}
$$

$$
\begin{aligned}
pr(\langle p,q\rangle, \langle p,q+1\rangle, 1) &= pr(\langle q,p\rangle, \langle q+1,p\rangle, 1) &=& \quad \tfrac{p+q}{p+q+2} \\
pr(\langle p,q\rangle, \langle p,q+1\rangle, 2) &= pr(\langle q,p\rangle, \langle q+1,p\rangle, 2) &=& \quad \tfrac{1}{p+q+2} \\
pr(\langle p,q\rangle, \langle p+1,q\rangle, 1) &= pr(\langle q,p\rangle, \langle q,p+1\rangle, 1) &=& \quad \tfrac{1}{p+q+2}
\end{aligned}
$$

Third type of transitions, $p, q > 0$ and $p \neq q$,

$$
\begin{aligned}
c(\langle p,q\rangle, \langle p,q+1\rangle, 1) &= c(\langle q,p\rangle, \langle q+1,p\rangle, 1) &=& \quad 1 \\
c(\langle p,q\rangle, \langle p,q+1\rangle, 2) &= c(\langle q,p\rangle, \langle q+1,p\rangle, 2) &=& \quad 3 \\
c(\langle p,q\rangle, \langle p+1,q\rangle, 1) &= c(\langle q,p\rangle, \langle q,p+1\rangle, 1) &=& \quad 2 \\
c(\langle p,q\rangle, \langle p+1,q\rangle, 2) &= c(\langle q,p\rangle, \langle q,p+1\rangle, 2) &=& \quad 3
\end{aligned}
$$

$$
\begin{aligned}
pr(\langle p,q\rangle, \langle p,q+1\rangle, 1) &= pr(\langle q,p\rangle, \langle q+1,p\rangle, 1) &=& \quad \tfrac{\max(p,q)}{p+q+2} \\
pr(\langle p,q\rangle, \langle p,q+1\rangle, 2) &= pr(\langle q,p\rangle, \langle q+1,p\rangle, 2) &=& \quad \tfrac{1}{p+q+2} \\
pr(\langle p,q\rangle, \langle p+1,q\rangle, 1) &= pr(\langle q,p\rangle, \langle q,p+1\rangle, 1) &=& \quad \tfrac{\min(p,q)}{p+q+2} \\
pr(\langle p,q\rangle, \langle p+1,q\rangle, 2) &= pr(\langle q,p\rangle, \langle q,p+1\rangle, 2) &=& \quad \tfrac{1}{p+q+2}
\end{aligned}
$$

Fourth type of transitions, $p = q > 0$,

$$
\begin{aligned}
c(\langle p,p\rangle, \langle p,p+1\rangle, 1) &= c(\langle p,p\rangle, \langle p+1,p\rangle, 1) &=& \quad 1 \\
c(\langle p,p\rangle, \langle p,p+1\rangle, 2) &= c(\langle p,p\rangle, \langle p+1,p\rangle, 2) &=& \quad 2 \\
c(\langle p,p\rangle, \langle p,p+1\rangle, 3) &= c(\langle p,p\rangle, \langle p+1,p\rangle, 3) &=& \quad 3
\end{aligned}
$$

$$
\begin{aligned}
pr(\langle p,p\rangle, \langle p,p+1\rangle, 1) &= pr(\langle p,p\rangle, \langle p+1,p\rangle, 1) &=& \quad \tfrac{p/2}{2p+2} \\
pr(\langle p,p\rangle, \langle p,p+1\rangle, 2) &= pr(\langle p,p\rangle, \langle p+1,p\rangle, 2) &=& \quad \tfrac{p/2}{2p+2} \\
pr(\langle p,p\rangle, \langle p,p+1\rangle, 3) &= pr(\langle p,p\rangle, \langle p+1,p\rangle, 3) &=& \quad \tfrac{1}{2p+2}
\end{aligned}
$$

Fig. 1. Summary of comparisons and probabilities per type of transition

probabilities associated with edges in the graph. The probability value of each transition follows from the assumption that any permutation of the input data is equally likely.

The computation of any given input instance is traced by a path starting from s and ending at one of the nodes in V_n. Transitions in the path follow an increasing sequence according to the partition of V and different instances may follow the same path. For a given instance, the number of comparisons performed is the sum of comparisons of each edge along the path followed. The probability of traversing any path is the product of probabilities of each edge in it.

Let $v \to w$ denote any possible transition between two designated nodes and $s \xrightarrow{*} w$, any path from the initial node to a node w. The average number of comparisons performed by the algorithm with an input of size n is

$$C_n = \sum_{\substack{w \in V_n \\ s \xrightarrow{*} w}} c(s \xrightarrow{*} w) pr(s \xrightarrow{*} w) \, . \tag{2}$$

The probability of reaching a node $w \in V$ is

$$pr(w) = \sum_{s \xrightarrow{*} w} pr(s \xrightarrow{*} w) \tag{3}$$

and the probability that the algorithm performs a transition in $E(v,w)$ is given by

$$pr(v,w) = pr(v) \left(\sum_{j \in E(v,w)} pr(v,w,j) \right) \, . \tag{4}$$

Equations (3) and (4) are related. A simple induction on the path sequence shows that

$$pr(w) = \sum_{v \in V} pr(v,w) \, . \tag{5}$$

The average cost associated with each vertex $v \in V$ and the average cost associated with each set of transitions $E(v,w)$ are defined respectively as

$$\bar{c}(v) = \sum_{w \in V} \sum_{j \in E(v,w)} c(v,w,j) pr(v,w,j) \tag{6}$$

and

$$\bar{c}(v,w) = \frac{\sum_{j \in E(v,w)} c(v,w,j) pr(v,w,j)}{\sum_{j \in E(v,w)} pr(v,w,j)} \, . \tag{7}$$

The following lemma presents two methods for computing the average number of comparisons C_n. The first one is *node-oriented* while the second is *edge-oriented*. In addition, both methods can be adapted to any dynamic process described by an acyclic Markovian graph with transition costs.

Lemma 1. *The average number of comparisons C_n performed by the algorithm can is*

$$C_n = \sum_{v \in V} \bar{c}(v) pr(v) = \sum_{v,w \in V} \bar{c}(v,w) pr(v,w) \, .$$

Proof. The proof is by induction on n which trivially holds for $n = 1$. When $n > 1$ and transitions from V_{n-1} to V_n are fixed, equation (2) can be rewritten as

$$C_n = \sum_{v \in V_{n-1}} \sum_{w \in V_n} \sum_{s \xrightarrow{*} v \to w} c(s \xrightarrow{*} v \to w) pr(s \xrightarrow{*} v \to w) \, .$$

Grouping all possible transitions from v to w gives

$$C_n = \sum_{v \in V_{n-1}} \sum_{s \xrightarrow{*} v} \sum_{w \in V_n} \sum_{j \in E(v,w)} (c(s \xrightarrow{*} v) + c(v,w,j)) pr(s \xrightarrow{*} v) pr(v,w,j) \, . \tag{8}$$

$$\bar{c}(\langle p,q\rangle) = \begin{cases} 1 & \text{if } p=q=0, \\ 1+\frac{2}{p+q+2} & \text{if } \min(p,q)=0 \text{ and } \max(p,q)>0, \\ 1+\frac{\min(p,q)+4}{p+q+2} & \text{if } p,q>0 \text{ and } p\neq q, \\ \frac{3}{2}+\frac{3}{2p+2} & \text{if } p=q>0. \end{cases}$$

Fig. 2. Summary of average cost by type of state

From (1), equation (8) becomes

$$C_n = \sum_{\substack{v\in V_{n-1} \\ s\xrightarrow{*}v}} c(s\xrightarrow{*}v)pr(s\xrightarrow{*}v)$$

$$+ \sum_{v\in V_{n-1}}\left(\sum_{s\xrightarrow{*}v}pr(s\xrightarrow{*}v)\right)\left(\sum_{w\in V_n}\sum_{j\in E(v,w)}c(v,w,j)pr(v,w,j)\right). \tag{9}$$

When (2), (3) and (6) are taken into consideration,

$$C_n = C_{n-1} + \sum_{v\in V_{n-1}}pr(v)\bar{c}(v). \tag{10}$$

Otherwise, if (2), (4) and (7) are substituted into (9),

$$C_n = C_{n-1} + \sum_{\substack{v\in V_{n-1} \\ w\in V_n}}pr(v,w)\bar{c}(v,w). \tag{11}$$

The lemma follows by carrying forward the inductive hypothesis. ∎

Since equations (10) and (11) are recurrent, the average number of comparisons C_n can be easily computed.

4 Average case upper bounds

When the execution of an instance is traced with a Markovian graph described above, any of the nodes in the same partition subset is equally likely to be reached by the algorithm. This property is stated in the next lemma.

Lemma 2. $\forall w\in V_t$, $1\le t\le n$, $pr(w)=1/t$.

Sketch of proof. From (4) and (5), the probability of reaching any node in V_t can be inductively defined as

$$pr(w) = \sum_{v\in V_{t-1}}pr(v)\sum_{j\in E(v,w)}pr(v,w,j).$$

A proof by cases with the cases presented in Figure 1 completes the proof of the lemma. ∎

Corollary 3.

$$C_n = \sum_{1\le t\le n-1}\frac{1}{t}\sum_{w\in V_t}\bar{c}(w). \tag{12}$$

Figure 2 displays the average cost for type of nodes in the graph.

Theorem 4. *If $|X| = n \geq 1$, the average number of comparisons performed by the algorithm to find both neighbors of $x \in X$ is*

$$C_n = \begin{cases} 0 & \text{if } n = 0, \\ 1 & \text{if } n = 1, \\ \frac{8}{3} & \text{if } n = 3, \text{ and} \\ \frac{5}{4}n + \frac{1}{2}H_n + \frac{11}{4}\left(H_{\lfloor n/2 \rfloor} - \frac{n \bmod 2}{2}\right) - \frac{27}{4} + \frac{4}{n} & \text{if } n \geq 4. \end{cases}$$

where $H_n = \sum_{i=1}^n 1/i = \ln n + O(1)$.

Proof. Regrouping equation (12) by type of nodes,

i) $n \geq 5$.

$$C_n = C_{n-1} + \frac{2}{n-1}\left(\frac{n+2}{n} + \sum_{1 \leq p \leq \lfloor n/2 \rfloor - 2} \frac{n+p+4}{n} + \frac{3n+6}{4n}((n+1) \bmod 2)\right)$$

ii)

$$C_1 = 0, \ C_2 = 1, \ C_3 = \frac{8}{3} \text{ and } C_4 = \frac{25}{6} .$$

Algebraic manipulation of the previous equation leads us to

i) odd $n \geq 7$

$$C_n = C_{n-2} + \frac{5}{2} + \frac{29n}{4} + \frac{6}{n-1} - \frac{27}{4(n-2)} ,$$

ii) even $n \geq 6$

$$C_n = C_{n-2} + \frac{5}{2} + \frac{10}{n} + \frac{1}{2(n-1)} - \frac{4}{n-2} .$$

The desired result stated above is obtained by recurring on n. ∎

An interesting question is how many comparisons are performed on the average if the rank of x in X turns out to be k. In this case, we will show that $n + \min(k, n - k) + o(n)$ comparisons suffice. Moreover, when x happens to be the median of X, the lower order term becomes $O(\sqrt{n})$, as the algorithm is essentially betting the new element will fall on the less likely side of x.

Let us consider the algorithm starting at state s and stopping when some predefined state $\langle p, q \rangle$ is reached. As explained before, each possible execution of the program determines a path from $\langle 0, 0 \rangle$ to $\langle p, q \rangle$ and since we are sampling without replacement, each one of the $(p, q) = \binom{p+q}{p}$ paths is equally likely.

Conditioned to the fact that $\langle p, q \rangle$ is the final state, the probability that any of the possible transitions between two states is traversed by the algorithm is

$$pr(\langle i, j \rangle, \langle i+1, j \rangle) = \frac{(i,j)(p-i-1, q-j)}{(p,q)} \text{ and}$$

$$pr(\langle i, j \rangle, \langle i, j+1 \rangle) = \frac{(i,j)(p-i, q-j-1)}{(p,q)} .$$

Such probabilities are zero if the corresponding transitions are not included in any of the paths between the initial and the fixed final state.

$$\bar{c}(\langle i,j \rangle \rightarrow \langle i,j+1 \rangle) = \begin{cases} 1 & \text{if } i = j = 0, \\ 1 + \frac{1}{j+1} & \text{if } i = 0 \text{ and } j > 0, \\ 2 & \text{if } i > 0 \text{ and } j = 0, \\ 2 + \frac{1}{j+1} & \text{if } i > 0 \text{ and } 0 < j < i, \\ \frac{3}{2} + \frac{3/2}{j+1} & \text{if } i > 0 \text{ and } j = i, \\ 1 + \frac{2}{j+1} & \text{if } i > 0 \text{ and } j > i. \end{cases}$$

Fig. 3. Summary of average cost per type of grouped transitions

The average performance $C(p,q)$ will be computed with equation (11) adapted to this specific context. Observe that the subgraph associated with the execution of the algorithm will be confined within states $\langle i,j \rangle$ such that $0 \le i \le p$ and $0 \le j \le q$. The average cost per type of transition is given in Figure 3. Such average costs are symmetric, that is

$$\bar{c}(\langle i,j \rangle \rightarrow \langle i+1,j \rangle) = \bar{c}(\langle j,i \rangle \rightarrow \langle j,i+1 \rangle) .$$

Theorem 5. *The average number of comparisons performed by the algorithm conditioned to the fact that it stops at state $\langle p,q \rangle$ is*

$$\begin{aligned} C(p,q) = {} & \max(p,q) + 2\min(p,q) + 2H_{\max(p,q)} + H_{\min(p,q)} - 2H_{p+q} + H_p + H_q - 2 \\ & + \delta_{p,0} + \delta_{q,0} - \frac{1}{2}\left(1 - \frac{1}{p+1}\right)\delta_{p,q} \\ & + \frac{1}{2}\sum_{0 \le j \le \min(p,q)}\left(1 - \frac{1}{j+1}\right)\frac{(j,j)(p-j,q-j)}{(p,q)} . \end{aligned} \tag{13}$$

Proof.

$$\begin{aligned} C(p,q) = {} & \sum_{j \ge 0}\left(1 + \frac{2}{j+1}\right)\sum_{0 \le i \le j}pr(\langle i,j \rangle, \langle i,j+1 \rangle) \\ & + \sum_{i \ge 0}\left(1 + \frac{2}{i+1}\right)\sum_{0 \le j \le i}pr(\langle i,j \rangle, \langle i+1,j \rangle) \\ & + \sum_{j \ge 0}\left(2 + \frac{1}{j+1}\right)\sum_{i > j}pr(\langle i,j \rangle, \langle i,j+1 \rangle) + \sum_{i \ge 0}\left(2 + \frac{1}{i+1}\right)\sum_{j > i}pr(\langle i,j \rangle, \langle i+1,j \rangle) \\ & + \frac{1}{2}\sum_{j \ge 0}\left(1 - \frac{1}{j+1}\right)(pr(\langle j,j \rangle, \langle j,j+1 \rangle) + pr(\langle j,j \rangle, \langle j+1,j \rangle)) \\ & - \sum_{j \ge 0}\frac{1}{j+1}pr(\langle 0,j \rangle, \langle 0,j+1 \rangle) - \sum_{i \ge 0}\frac{1}{i+1}pr(\langle i,0 \rangle, \langle i+1,0 \rangle) \\ & - \sum_{j \ge 0}pr(\langle 0,j \rangle, \langle 1,j \rangle) - \sum_{i \ge 0}pr(\langle i,0 \rangle, \langle i,1 \rangle) . \end{aligned}$$

Simplification of the inner summations gives

$$C(p,q) = \sum_{0 \le j \le \max(p,q)} \left(1 + \frac{2}{j+1}\right) + \sum_{0 \le j \le \min(p,q)} \left(2 + \frac{1}{j+1}\right)$$

$$+ \frac{1}{2} \sum_{\substack{0 \le j \le \min(p,q) \\ j < \max(p,q)}} \left(1 - \frac{1}{j+1}\right) \frac{(j,j)(p-j,q-j)}{(p,q)} - \sum_{0 \le j < q} \frac{1}{j+1} \frac{(p,q-j-1)}{(p,q)}$$

$$- \sum_{0 \le i < p} \frac{1}{i+1} \frac{(p-i-1,q)}{(p,q)} - \sum_{0 \le j \le q} \frac{(p-1,q-j)}{(p,q)} (1 - \delta_{p,0})$$

$$- \sum_{0 \le i \le p} \frac{(p-i,q-1)}{(p,q)} (1 - \delta_{q,0}) .$$

Expression (13) is obtained by using identities A1 and A2 from the Appendix in the previous expression. ∎

Theorems 4 and 5 are related since it is not difficult to realize that

$$C_n = 2/n \sum_{0 \le j \le \lceil n/2 \rceil - 1} C(j, n-1-j) .$$

From Theorem 5, two particular cases are considered:
i) $p = q = (n+1)/2$ with odd n,
ii) $p = \alpha n$, $q = (1-\alpha)n$ for a fixed $\alpha \in (0, \frac{1}{2})$.
The case $\alpha \in (\frac{1}{2}, 1)$ is symmetric to *ii*.

Theorem 6. *The asymptotic average number of comparisons performed by the algorithm when x is the median of X (and n is odd) is*

$$C\left(\frac{n+1}{2}, \frac{n+1}{2}\right) = \frac{3}{2}n + \frac{1}{2}\sqrt{\frac{\pi n}{2}} + 3\ln n + O(1) .$$

Proof. If p is substituted for q into equation (13),

$$C(p,p) = 3p + 5H_p - 2H_{2p} - \frac{5}{2} + \frac{1}{2(p+1)} + 2\delta_{p,0} + \frac{1}{2} \sum_{0 \le j \le p} \left(1 - \frac{1}{j+1}\right) \frac{(j,j)(p-j,p-j)}{(p,p)} .$$

Further simplification of the previous expression is obtained with identities A3 and A4 from the Appendix, that is

$$C(p,p) = 3p + 5H_p - 2H_{2p} - \frac{5}{2} + 2\delta_{p,0} - \frac{p}{1+p} + \frac{1}{2} \frac{4^p}{(p,p)} .$$

The previous equation is then asymptotically expanded with identities A6 and A8 from the Appendix getting

$$C(p,p) = 3p + \frac{1}{2}\sqrt{\pi p} + 3\ln n + 3\gamma - \frac{7}{2} - 2\ln 2 + \frac{1}{16}\sqrt{\frac{\pi}{p}} + \frac{3}{p} + \frac{1}{256}\sqrt{\frac{\pi}{p^3}} - \frac{11}{8}\frac{1}{p^2} + \cdots$$

A simple replacement of p by $(n+1)/2$ proves the lemma. ∎

Theorem 7. *The average number of comparisons performed by the algorithm when the rank of x in X is αn, for any fixed $\alpha < 1/2$, is*

$$C(\alpha n, (1 - \alpha n)) = (1 + \alpha)n + 3\ln n + O(1) \ .$$

Proof. If $p = \alpha n$ and $q = (1 - \alpha)n$ in equation (13) and identity A5 from the Appendix is applied, then

$$C(\alpha n, (1 - \alpha n)) = (1 + \alpha)n + 2H_{\alpha n} + 3H_{(1-\alpha)n} - 2H_n - 2$$

$$+ \frac{1}{2}\frac{\alpha n}{(1 - \alpha)n + 1}\sum_{j \geq 0}\frac{(\alpha n)^{\underline{j}}}{n^{\underline{j}}}2^k - \frac{1}{2}\frac{1}{(1 - \alpha)n + 1}\sum_{j \geq 0}\frac{(\alpha n)^{\underline{j}}}{n^{\underline{j}}}k2^k \ .$$

Identity A8 from the Appendix with $g(x) = 1/(1 - 2x)$ and $g(x) = 2x/(1 - 2x)$ solves the summations in the previous expressions. ∎

5 Average case lower bounds

Let us consider the following (easier) problem:

Problem 2. *X is a set of $n > 1$ numbers with two designated neighbors w and x, such that $w < x$, verify that w and x are indeed of consecutive ranks (ranks of w and x are unknown in advance).*

It is assumed that elements w and x are stored in registers, while the other $n - 2$ elements are in the array X. To simplify the discussion, we will assume the values in question constitute the distinct integers 1 through n with w and x of consecutive but unknown ranks; thus, there is no need to distinguish between the kth smallest element of X and the number k. Clearly, any lower bound for Problem 2 is also one for the more general Problem 1.

In developing a lower bound for the number of comparisons to be performed by any algorithm which solves Problem 2, three types of comparisons will be considered: *partition comparisons, straddle comparisons* and *closer comparisons*. Any solution of Problem 2 must identify the elements smaller and larger than x. Thus, for any element smaller than w, its *partition comparison* is the first comparison between it and either w or another element lying between these two. Symmetrically, if the element is greater than x, its partition comparison is the first comparison between it and either x or an intermediate element between both elements. It is not difficult to realize that $n - 2$ of such comparisons must be performed in order to get a consistent solution. It is expected, however, that some comparisons which are not partition comparisons will be performed, and in this case, we will focus only on *straddle comparisons*: those involving an element not greater than w with another not smaller than x.

Partition and straddle comparisons will be related through the concept of *closer comparison*. Let $\theta_\pi(w)$ and $\theta_\pi(x)$ denote the rank of w and x in X for a given the input permutation π respectively such that $\theta_\pi(x) = \theta_\pi(w) + 1$. The closer comparison of any element $k \in X$ for a given π is the first comparison between it and an element $l \in X$ subject to

i) $k = \theta_\pi(w) - i$, $i \in [1..\min\{\theta_\pi(w) - 1, n - \theta_\pi(w)\}]$, and $l \in [k + 1..\theta_\pi(w) + i]$, or

ii) $k = \theta_\pi(x) + i$, $i \in [1..\min\{\theta_\pi(x) - 1, n - \theta_\pi(x)\}]$, and $l \in [\theta_\pi(x) - i..k - 1]$.

The closer comparison of k with l for a given π straddles if

i) $k = \theta_\pi(w) - i$ and $l \in [\theta_\pi(w) + 1 .. \theta_\pi(w) + i]$, or

ii) $k = \theta_\pi(x) + i$ and $l \in [\theta_\pi(x) - i .. \theta_\pi(x) - 1]$.

This closer comparison is a partition comparison if the opposite verifies.

Note: If the permutation π is fixed, the closer comparison for k implies that such a comparison is not closer for any of the possible elements l which k may be compared.

Closer comparisons are similar to *close comparisons* defined in [3] to prove a tight lower bound for the average number of comparisons required to select the kth smallest. However, an important difference between the two concepts arises: when the kth smallest is selected, if the close comparison of the element $k - i$ happens to occur with the element $k + i$, this comparison is also close for the second element and thus classified as *double close*; that is not the case with closer comparisons in Problem 2.

Theorem 8. *On the average, $\frac{5}{4}n - \frac{9}{4} - \frac{(n+1) \bmod 2}{4(n-1)}$ comparisons are required by any algorithm that solves the closest neighbor(s) problem of a designated x in a set of n numbers.*

Proof. Let $\theta(x)$, $2 \le \theta(x) \le n$ denote a specific rank of x in X. For any fixed value of $\theta(x)$, there are $(n - 2)!$ permutations.

If π is a permutation such that $\theta_\pi(x) = \theta(x)$, and $\theta(x) + i$, $i \in [1 .. \min\{\theta_\pi(x) - 1, n - \theta_\pi(x)\}]$, is a given element, let us define $f_{\theta(x),i}\pi$ to be the permutation featuring

$$f_{\theta(x),i}\pi(\theta(x) + i) = \theta(w) - i,$$
$$f_{\theta(x),i}\pi(j) = j + 1, \quad j \in [\theta(x) - i, \theta(x) + i - 1],$$
$$f_{\theta(x),i}\pi(j) = j, \quad \text{otherwise.}$$

This permutation cyclically shifts to the left the subinterval $[\theta(x) - i, \theta(x) + i]$. Observe the transformation increases by one the ranks of w and x.

Symmetrically, for a given permutation π such that $\theta_\pi(w) = \theta(w)$, and a given element $\theta(w) - i$, $i \in [1 .. \min\{\theta_\pi(w) - 1, n - \theta_\pi(w)\}]$, let us define $g_{\theta(w),i}\pi$ to be permutation such that

$$g_{\theta(w),i}\pi(\theta(w) - i) = \theta(w) + i,$$
$$g_{\theta(w),i}\pi(j) = j - 1, \quad j \in [\theta(w) - i + 1, \theta(w) + i],$$
$$g_{\theta(w),i}\pi(j) = j, \quad \text{otherwise.}$$

In this case, g cyclically shifts to the left the subinterval $[\theta(w) - i, \theta(w) + i]$ and decreases by one the original ranks of w and x.

From the previous two definitions, if $f_{\theta(x),i}\pi = \pi'$ then

$$g_{\theta_{\pi'}(w),i}\pi' \circ f_{\theta(x),i}\pi = I$$

where I is the identity permutation function. Furthermore, up to the comparison in question:

i) $\theta_\pi(x) = \theta_{\pi'}(w)$,

ii) the algorithm will run identically on π and π',

iii) the comparison is the closer for $\theta_\pi(x) + i$ in π if and only if it is closer for $\theta_{\pi'}(w) - i$ in π', and

iv) this comparison is a straddle in π if and only if it is a partition comparison in π'.

Table 1. Closer comparison relationships

	Straddles
$[1, 2, 3] \leftrightarrow [2, 3, 1]$	1
$[2, 3, 4] \leftrightarrow [3, 4, 2]$ $[2, 3, 5] \leftrightarrow [3, 4, 1]$	2
$[3, 4, 5] \leftrightarrow [4, 5, 3]$ $[3, 4, 6] \leftrightarrow [4, 5, 2]$ $[3, 4, 7] \leftrightarrow [4, 5, 1]$	3
$[4, 5, 6] \leftrightarrow [5, 6, 4]$ $[4, 5, 7] \leftrightarrow [5, 6, 3]$	2
$[5, 6, 7] \leftrightarrow [6, 7, 5]$	1

We can now average over all ranks of x, all permutations, and all elements different than w and x, to obtain the average number of straddles required. When n is even, this amount is $\frac{1}{n-1} \sum_{i=1}^{n/2-1} i$. In the case of n odd, at least $\frac{1}{n-1} \sum_{i=1}^{(n-1)/2} i + \frac{1}{2}$ straddles are performed. The theorem follows by adding the $n - 2$ partition comparisons required. ∎

Table 1 indicates the correspondence between permutations through functions f and g when $n = 7$. Each entry shows the elements involved in a closer comparison and the consecutive ranks of w and x. The rightmost column counts the number of straddles.

6 Final Remarks

We have shown rather tight bounds on the number of comparisons necessary to determine the neighbor(s) of a given value. One of the nice features of our study has been the fact that the optimal algorithms are clean and easy to code. A natural generalization of our problem, to be discussed in a fuller paper, is to find the ith closest neighbor(s) of a designated element x.

References

1. M. Blum, R.W. Floyd, V.R. Pratt, R.L. Rivest and R.E. Tarjan, *Time Bounds for Selection*, J. Compt. and Sys. Sci., 7, 1973, pp. 448-461.
2. W. Cunto and J.I. Munro, *Closest Neighbor Problem*, Proceedings 22-nd Allerton Conference, Urbana, Illinois, 1984, pp. 510-515.
3. W. Cunto and J.I. Munro, *Average Case Selection*, JACM, 36, 1989, pp. 270-279.
4. R.W. Floyd and R.L. Rivest, *Expected Time Bounds for Selection*, CACM, 18, 1975, pp. 165-172.
5. R.W. Floyd and R.L. Rivest, *Algorithm 489: The Algorithm SELECT for Finding the ith Smallest of n elements*, CACM, 18, 1975, pp. 173.

Appendix

A1.

$$\sum_{0 \le j \le q} (p,j) = (p+1,q) \ .$$

A2.

$$\sum_{0 \le j < q} \frac{1}{j+1} \frac{(p,q-j-1)}{(p,q)} = H_{p+q} - H_p \ .$$

A3.

$$\sum_{0 \le j \le p} (j,j)(p-j,p-j) = 4^p \ .$$

A4.

$$\sum_{0 \le j \le p} \frac{(j,j)(p-j,p-j)}{j+1} = \frac{(p+1,p+1)}{2} \ .$$

A5.

$$\sum_{0 \le j \le p} \left(1 - \frac{1}{j+1}\right) \frac{(j,j)(p-j,q-j)}{(p,q)} = \frac{p}{q+1} \sum_{j \ge 0} \frac{p^j}{(p+q)^{\underline{j}}} 2^j - \frac{1}{q+1} \sum_{j \ge 0} \frac{p^j}{(p+q)^{\underline{j}}} j 2^j \ .$$

A6.

$$H_n = \gamma + \ln(n) + \frac{1}{2n} - \frac{1}{12n^2} + \frac{1}{120n^4} - \frac{1}{252n^6} + \frac{1}{240n^8} + \dots \ .$$

A7.

$$\frac{4^p}{(p,p)} = \sqrt{\pi p} \left(1 + \frac{1}{8p} + \frac{1}{128p^2} - \frac{5}{1024p^3} - \frac{21}{32768p^4} + \frac{399}{262144p^5} + \dots\right) \ .$$

A8.

$$\sum_{i \ge 0} a_i \frac{n^{\underline{i}}}{m^{\underline{i}}} = g(\alpha) + \frac{\alpha(a-\alpha)g''(\alpha)}{2m}$$

$$+ \frac{\alpha(1-2\alpha)}{24m^2} \left(3\alpha(1-\alpha)g^{iv}(\alpha) + 8(1-2\alpha)g'''(\alpha) - 12g''(\alpha)\right) + \dots$$

where $\alpha = n/m$ and $g(x) = \sum_{i \ge 0} a_i x_i$.

On the Zone of a Surface in a Hyperplane Arrangement[*]

Boris Aronov [†]   Micha Sharir[‡]

Abstract

Let H be a collection of n hyperplanes in $\mathbb{R}^d$, let $\mathcal{A}$ denote the arrangement of H, and let σ be a $(d-1)$-dimensional algebraic surface of low degree, or the boundary of a convex body in $\mathbb{R}^d$. The *zone* of σ in $\mathcal{A}$ is the collection of cells of $\mathcal{A}$ crossed by σ. We show that the total number of faces bounding the cells of the zone of σ is $O(n^{d-1} \log n)$.

1   Introduction

A set H of n hyperplanes in d-dimensional space $\mathbb{R}^d$ decomposes $\mathbb{R}^d$ into open *cells* of dimension d (also called *d-faces*) and into relatively open faces of dimension k, $0 \leq k < d$. These cells and faces define a cell complex which is commonly known as the *arrangement* $\mathcal{A} = \mathcal{A}(H)$ of H. We define the *complexity* of a cell in $\mathcal{A}$ to be the number of faces that are contained in the closure of the cell.

Let σ be an arbitrary subset of $\mathbb{R}^d$. We define $\mathcal{Z}_\sigma(H)$, the *zone* of σ in $\mathcal{A}$, to be the set of all cells in $\mathcal{A}$ that intersect σ; the *complexity* of a zone is the sum of the complexities of its constituent cells. In the following analysis, we will concentrate on the case where σ is either a $(d-1)$-dimensional algebraic surface of degree δ, where δ is a small constant, or the boundary of a convex body in $\mathbb{R}^d$. However, most of our analysis, with the notable exception of Lemma 2.1(2-3), holds for an arbitrary set σ.

A fundamental result on hyperplane arrangements is the Zone Theorem [4, 6], in which σ is assumed to be a hyperplane distinct from those in H. It asserts that the zone of a hyperplane in an arrangement of n hyperplanes in $\mathbb{R}^d$ has complexity $\Theta(n^{d-1})$. A recent proof of the Zone Theorem is given in [6].

In this paper we extend the Zone Theorem to cases where σ is a more general set, as described above. Specifically, we show:

Theorem 1.1 (Extended Zone Theorem) *The complexity of the zone of a surface σ, which is either a small-degree algebraic surface or the boundary of an arbitrary convex*

[*]Work on this paper by the second author has been supported by Office of Naval Research Grant N00014-90-J-1284, by National Science Foundation Grant CCR-89-01484, and by grants from the U.S.–Israeli Binational Science Foundation, the G.I.F. — the German Israeli Foundation for Scientific Research and Development, and the Fund for Basic Research administered by the Israeli Academy of Sciences.

[†]Department of Computer Science, Polytechnic University, Brooklyn, NY 11201 USA

[‡]Courant Institute of Mathematical Sciences, New York University, and School of Mathematical Sciences, Tel Aviv University, Tel Aviv, Israel

body, in an arrangement of n hyperplanes in $\mathbb{R}^d$ is $O(n^{d-1} \log n)$, where $d \geq 3$ and the constant of proportionality depends on d and on the degree δ of σ.

We note that when $d = 2$, a somewhat better bound of $O(n\alpha(n))$ is known for the complexity of such a zone [5], where $\alpha(n)$ is the inverse Ackermann function. We have so far been unable to obtain similarly improved bounds in higher dimensions. We do not even know if the bound $O(n\alpha(n))$ is tight in the worst case in the plane. The best known lower bound, in any dimension d, is $\Omega(n^{d-1})$, so there remains a small gap between the lower and upper bounds. Related results for planar arrangements have been obtained by Bern et al [3].

The main difference between the zone of a hyperplane and that of a more general surface σ is the behavior of what we call *popular facets*. These are facets f (i.e. $(d-1)$-dimensional faces) that bound two adjacent cells C, C' of $\mathcal{A}$ so that both C and C' belong to the zone of σ. Even though popular facets do exist in the zone of a hyperplane, there they must always meet the zone hyperplane, which is not necessarily the case for zones of other surfaces. A main step in our analysis is to obtain a sharp bound on the number and complexity of the popular facets in a zone.

2 Proof of the Extended Zone Theorem

For a d-polyhedron P let $f_k(P)$ denote the number of k-faces of P (i.e. faces of dimension k). For $0 \leq k \leq d$, let $z_k(\sigma; H)$ denote $\sum_{C \in \mathcal{Z}_\sigma(H)} f_k(\bar{C})$, where $\bar{C}$ denotes the topological closure of cell C. Finally, for $n > 0$, $d > 0$, and $0 \leq k \leq d$, let $z_k^{(\delta)}(n, d)$ denote the maximum of $z_k(\sigma; H)$ over all surfaces σ of degree at most δ and all sets H of n hyperplanes in $\mathbb{R}^d$; to reduce the proliferation of indices, we will omit the superscript (δ) in what follows.

As we are interested in the asymptotic behavior of z_k, we assume $n > d$ throughout the proof, unless stated otherwise.

First note that $z_k(\sigma; H)$ achieves its maximum when σ and H are in general position, i.e. every $j \leq d$ hyperplanes in H intersect in a $d - j$-flat, no $d + 1$ hyperplanes have a point in common, and σ is not tangent to any flat formed by the intersection of $j \leq d$ hyperplanes of H. This can be proved using a standard perturbation argument: displacing the hyperplanes of H slightly will put σ and H in general position, and can only increase the complexities of the cells in $\mathcal{Z}_\sigma(H)$, through vertex truncation or the actions dual to vertex pulling or pushing (see [7, pp. 78–83]).

Let H be a set of n hyperplanes in $\mathbb{R}^d$, and let σ be an algebraic surface as above, so that σ and H are in general position. A k-face f in $\mathcal{A}(H)$ now lies in exactly $d - k$ hyperplanes of H and is part of the boundary of 2^{d-k} cells of $\mathcal{A}(H)$. More than one of those cells can lie in $\mathcal{Z}_\sigma(H)$, and thus the contribution of f to $z_k(\sigma; H)$ can be larger than one. In order to have entities that contribute at most one to the count $z_k(\sigma; H)$ we define a k-*border* to be a pair (f, C), where f is a k-face in $\mathcal{A}(H)$ and C is a cell that has f on its boundary. Thus $z_k(\sigma; H)$ counts all borders of dimension k in $\mathcal{Z}_\sigma(H)$, i.e. k-borders (f, C) with $C \in \mathcal{Z}_\sigma(H)$. More generally, for $0 \leq k \leq i \leq d$, a (k, i)-*border* is a pair of faces (f, g) in $\mathcal{A}$ of dimension k and i respectively, with $f \subseteq \bar{g}$. We refer to a pair of faces, f, g, with $f \subseteq \bar{g}$, as *incident faces*. Note that k-borders defined above are simply (k, d)-borders.

We call a k-face f in $\mathcal{A}$ *popular* if all 2^{d-k} cells in $\mathcal{A}$ incident to f belong to $\mathcal{Z}_\sigma(H)$. In particular, a popular cell is simply a zone cell, i.e. a cell of $\mathcal{A}$ met by σ.

A (k, i)-border (f, g) is *popular* if g is a popular i-face. Let $\tau_k^{(i)}(\sigma; H)$ be the number of popular (k, i)-borders. Notice that the problem of bounding the complexity of the zone of σ in $\mathcal{A}$ reduces to bounding the quantities $\tau_k^{(d)}(\sigma; H)$, for all $0 \leq k \leq d$, as $z_k(\sigma; H) = \tau_k^{(d)}(\sigma; H)$. We will obtain such bounds by inductively estimating $\tau_k^{(i)}$, for all $0 \leq k \leq i \leq d$.

We begin by providing a bound on $\tau_k^{(k)}$, for all $0 \leq k \leq d$.

Lemma 2.1 *Let H be a collection of n hyperplanes in general position in $\mathbb{R}^d$. Then*
(1) for any set $X \subset \mathbb{R}^d$ and $0 \leq k \leq d$,

$$\tau_k^{(k)}(X; H) = \binom{d}{k} \tau_d^{(d)}(X; H) + O(n^{d-1});$$

(2) for an algebraic surface σ of codimension 1 and degree δ,

$$\tau_k^{(k)}(\sigma; H) = O(n^{d-1}), \quad 0 \leq k \leq d,$$

where the constant of proportionality depends on k, d, and δ; and
(3) $\tau_k^{(k)}(\sigma; H) = O(n^{d-1})$, whenever σ is the boundary of an arbitrary convex body.

Proof. Let $k < d$. Recall that $\tau_k^{(k)}(X; H)$ is simply the number of popular k-faces, i.e. k-faces f for which all 2^{d-k} incident cells belong to the zone of X. To show (1), it is sufficient to associate each such face with one of the incident cells, and argue that no zone cell gets charged more than $\binom{d}{k}$ times.

We set up the correspondence as follows: Rotate the arrangement in such a fashion that every face that has a lowest vertex has a unique lowest vertex, with the height measured in terms of the x_d coordinate. Observe that faces that do not have a lowest vertex are exactly those that meet the hyperplane $x_d = c$ for sufficiently large negative c—they are in 1-1 correspondence to the faces in the arrangement induced by H in this hyperplane. Thus their number is only $\Theta(n^{d-1})$ and we may restrict our attention to the faces which do have a lowest vertex. We claim that, since $\mathcal{A} = \mathcal{A}(H)$ is a simple arrangement, the lowest vertex v_f of a face f is the lowest vertex of exactly one of the cells incident to f. The way to see this is to observe that, among all cells incident to v_f, the unique cell that has v_f as its lowest vertex has the property that its bounding faces incident to v_f are exactly those faces of $\mathcal{A}$ incident to v_f and having v_f as their lowest vertex.

Now each popular k-face is assigned to a unique cell with which it shares its lowest vertex. No cell in $\mathcal{Z}_X(H)$ is charged more than $\binom{d}{k}$ times, as this is the total number of k-faces in $\mathcal{A}$ sharing its lowest vertex, since $\mathcal{A}$ is a simple arrangement. This completes the proof of (1).

(2) It is sufficient to show that $\tau_d^{(d)}(\sigma; H)$ is $O(n^{d-1})$, i.e. σ meets $O(n^{d-1})$ cells of $\mathcal{A}$. Let C be a cell of $\mathcal{Z}_\sigma(H)$ and let $0 < k \leq d$ be the smallest integer so that σ meets

a face f of $\bar{C}$ of dimension k ($k \neq 0$ because of our assumptions on general position). Thus $\sigma \cap f \neq \emptyset$ but σ does not meet the relative boundary of f. Consider the k-flat F containing f. Then f contains one or more connected components of $\sigma \cap F$. But $\sigma \cap F$ is an algebraic surface in F of degree at most δ, so it has a constant number (depending on δ) of connected components. Since the number of k-flats, for $1 \leq k \leq d$, formed by intersections of the hyperplanes of H is $O(n^{d-1})$, it follows that the number of faces f with the above property is $O(n^{d-1})$. Since every face of dimension k bounds at most 2^{d-k} cells of $\mathcal{Z}_\sigma(H)$, the claim follows.

(3) The claim is immediate by noting that the argument in (2) applies to the case of the boundary of a convex body as well, with the number of connected components of $\sigma \cap F$, for any flat F, being at most 1 (2 if F is a line). □

Note: The upper bounds in Lemma 2.1(2–3) are easily seen to be asymptotically tight— take σ to be a generic hyperplane and notice that every k-face of the arrangement induced by $\mathcal{A}$ in σ corresponds to a popular $(k+1)$-face in $\mathcal{A}$. A lower bound on $\tau_0^{(0)}$ can be obtained by a slight modification of this argument.

Corollary 2.2 *For any algebraic surface $\sigma \subset \mathbb{R}^d$ and a set H of n hyperplanes, $z_d(\sigma; H)$ is $O(n^{d-1})$ with the constant of proportionality depending on d and on the degree δ of σ. The assertion also holds for the boundary of an arbitrary convex body.*

Proof. Recall that $z_d(\sigma; H) = \tau_d^{(d)}(\sigma; H)$ by definition. □

We now proceed by induction on i, and derive a recurrence for $\tau_k^{(i)}(\sigma; H)$, for $0 \leq k < i$, using an approach similar to that used in [6] and also in [2]. In more detail, fix a hyperplane $h \in H$ and consider all popular (k, i)-borders (f_0, g_0) in $\mathcal{Z}_\sigma(H)$ with $f_0 \not\subset h$. When we remove h, the face g_0 becomes part of a possibly larger i-face g, which is clearly also popular. Moreover, f_0 is a part of some k-face contained in $\bar{g}$. So let (f, g) be a popular (k, i)-border in $\mathcal{Z}_\sigma(H \setminus \{h\})$, and consider what happens to it when h is reinserted into the arrangement. Let C_ℓ, $\ell = 1, \ldots, 2^{d-i}$, be the cells in $\mathcal{Z}_\sigma(H \setminus \{h\})$ incident to g. The following cases may occur:

$h \cap g = \emptyset$: In this case g may or may not be popular in $\mathcal{Z}_\sigma(H)$, but (f, g) contributes at most one popular (k, i)-border to this zone, namely itself.

$h \cap g \neq \emptyset$ and $h \cap f = \emptyset$: Again, (f, g) can contribute at most one popular (k, i)-border to $\mathcal{Z}_\sigma(H)$, namely (f, g^+), where g^+ is the portion of g lying to the same side of h as f.

$h \cap g \neq \emptyset$ and $h \cap f \neq \emptyset$: Let h^+, h^- denote the two open halfspaces bounded by h, and consider the two (k, i)-borders $(f \cap h^+, g \cap h^+)$ and $(f \cap h^-, g \cap h^-)$. We are only interested in cases where both of them become popular borders in $\mathcal{Z}_\sigma(H)$, for only then will our count go up. Let $C_\ell^+ = C_\ell \cap h^+$, $C_\ell^- = C_\ell \cap h^-$, for $\ell = 1, \ldots, 2^{d-i}$. Thus we are interested in situations where σ meets all 2^{d-i+1} cells C_ℓ^+, C_ℓ^-. Notice that all these cells are incident to $g \cap h$, an $(i-1)$-face in $\mathcal{A}$. Hence $g \cap h$ is a popular face and $(f \cap h, g \cap h)$ is a popular $(k-1, i-1)$-border in $\mathcal{Z}_\sigma(H)$.

To sum up, the number of popular (k, i)-borders in $\mathcal{Z}_\sigma(H)$ which are not contained in h is bounded by

$$\tau_k^{(i)}(\sigma; H \setminus \{h\}) + \rho_h,$$

where ρ_h is the number of popular $(k-1, i-1)$-borders (f', g') with $g' \subset h$. If we sum these bounds over all hyperplanes $h \in H$ and observe that every popular (k, i)-border in $Z_\sigma(H)$ is counted exactly $n - d + k$ times (it is not counted if and only if h is one of the $d - k$ hyperplanes containing the k-face of the border), we obtain, similar to [6]:

$$(n - d + k)\tau_k^{(i)}(\sigma; H) \leq \sum_{h \in H} \tau_k^{(i)}(\sigma; H \setminus \{h\}) + (d - i + 1)\tau_{k-1}^{(i-1)}(\sigma; H),$$

where the factor $(d - i + 1)$ comes from the fact that a popular $(k-1, i-1)$-border is be charged $d - i + 1$ times, once for each hyperplane h containing it.

For a fixed number δ, let us denote by $\tau_k^{(i)}(n, d)$ the maximum of $\tau_k^{(i)}(\sigma; H)$ over all choices of a set H of n hyperplanes in $\mathbb{R}^d$ and an algebraic surface σ of degree at most δ. We thus have

$$\tau_k^{(k)}(n, d) = O(n^{d-1}), \quad 0 \leq k \leq d \tag{1}$$

and

$$\tau_k^{(i)}(n, d) \leq \frac{n}{n - d + k}\tau_k^{(i)}(n - 1, d) + \frac{d - i + 1}{n - d + k}\tau_{k-1}^{(i-1)}(n, d), \quad 0 \leq k < i \leq d. \tag{2}$$

When $k = 0$, the rightmost term in (2) vanishes, but the recurrence solves to $O(n^d)$ (see [2, 6]), which is too large for our purposes. However, we observe that, trivially, $\tau_0^{(i)}(n, d) \leq 2\tau_1^{(i)}(n, d)$. Thus it suffices to analyze (2) only for $k \geq 1$.

We first transform the relation (2) into a simpler one, by substituting $\tau_k^{(i)}(n, d) = \binom{n}{d-k}\psi_k^{(i)}(n, d)$. (Recall that we have assumed that $n > d$.) This yields the following recurrences, as is easily verified:

$$\psi_k^{(k)}(n, d) = O(n^{k-1}), \quad 1 \leq k \leq d$$

and

$$\psi_k^{(i)}(n, d) \leq \psi_k^{(i)}(n - 1, d) + \frac{d - i + 1}{d - k + 1}\psi_{k-1}^{(i-1)}(n, d), \quad 1 \leq k < i \leq d. \tag{3}$$

Our goal is now to show that $\psi_k^{(i)}(n, d) = O(n^{k-1} \log n)$. We prove this by induction on i. The base case $i = 0$ only allows $k = 0$, and we have already shown that $\tau_0^{(0)}(n, d) = O(n^{d-1})$. Similarly, the case $i = 1$ has also been established in (1), since we only consider the case $k \geq 1$.

The case $i = 2$ is the most interesting one, since it is there where the $\log n$ factor enters our analysis. To be more precise, the interesting case is $i = 2$, $k = 1$, as the case $k = 2$ has already been dealt with in (1). In this special case, (3) becomes

$$\psi_1^{(2)}(n, d) \leq \psi_1^{(2)}(n - 1, d) + \frac{d - 1}{d}\psi_0^{(1)}(n, d) .$$

But we have already shown that

$$\psi_0^{(1)}(n, d) = \frac{1}{\binom{n}{d}}\tau_0^{(1)}(n, d) \leq \frac{2}{\binom{n}{d}}\tau_1^{(1)}(n, d) = O(\frac{1}{n}) .$$

Thus we obtain the recurrence

$$\psi_1^{(2)}(n, d) = \psi_1^{(2)}(n - 1, d) + O(\frac{1}{n}),$$

whose solution is $\psi_1^{(2)}(n, d) = O(\log n)$, as asserted.

For $i > 2$, we first ignore both cases $k = 0$ and $k = 1$. By induction hypothesis on i we obtain the following recurrence for $k < i$:

$$\psi_k^{(i)}(n, d) \leq \psi_k^{(i)}(n - 1, d) + An^{k-2}\log n,$$

where A is a constant depending on k, i, d, and δ. Since $k \geq 2$, this recurrence solves to $O(n^{k-1}\log n)$, yielding $\tau_k^{(i)}(n, d) = O(n^{d-1}\log n)$, with a constant of proportionality depending on i, k, d, δ, as claimed.

To complete the argument, we need to extend this bound to the case $k = 1$. For this we recall that $\tau_k^{(i)}(\sigma; H)$ is the number of popular (k, i)-borders, which is the same as the total number of k-faces of the popular i-faces in $\mathcal{Z}_\sigma(H)$. Each bounded i-face is a simple i-polytope, so the number of its faces of all dimensions is at most a constant multiple (depending on i) of the number of its $\lceil i/2 \rceil$-faces (see, for example, Problem 6.2 in [4], or [2]). (Unbounded faces can be handled by a variant of this argument, or one may note that the Zone Theorem for a hyperplane implies that the total number of features bounding all unbounded cells of $\mathcal{A}(H)$ is $O(n^{d-1})$. We omit the details of this argument.) Hence $\tau_1^{(i)}(\sigma; H) = O(\tau_{\lceil i/2 \rceil}^{(i)}(\sigma; H))$. But since $i > 2$ we have $\lceil i/2 \rceil > 1$, which implies that $\tau_1^{(i)}(n, d)$ is also $O(n^{d-1}\log n)$.

This completes the proof of the Extended Zone Theorem.   □.

3   Discussion

The following immediate application of the Extended Zone Theorem is obtained from an observation of Pellegrini [9, 10], as also discussed in [2]:

Theorem 3.1 *Given n triangles in 3-dimensional space and any $\epsilon > 0$, one can preprocess them in randomized expected time $O(n^{4+\epsilon})$ into a data structure of size $O(n^{4+\epsilon})$, so that for a query ray ρ, we can identify the first triangle met by ρ in time $O(\log n)$.*

This result improves the preprocessing and space complexity of the best previous solution, given in [2], by a factor of roughly $n^{1/2}$.

In terms of further extending the Zone Theorem, we plan to investigate how wide the class of surfaces is for which the complexity of the zone in an arrangement of n hyperplanes in $\mathbb{R}^d$ is close to $O(n^{d-1})$. One immediate observation is that any surface whose intersection with an arbitrary k-flat in $\mathbb{R}^d$ has a bounded number of components falls into this category. Another possible extension is to see if a stronger upper bound can be obtained for more restricted families of surfaces or other classes of sets $\sigma \subset \mathbb{R}^d$. Classes that have been recently investigated include k-flats[8] and algebraic surfaces of dimension $k < d - 1$[11]. Another intriguing and largely uninvestigated area is that of replacing hyperplane arrangements by arrangements of more general algebraic surfaces or some other classes of objects—we discuss one such situation in [1].

References

[1] B. Aronov and M. Sharir, On the complexity of a single cell in an arrangement of simplices in d-space, in preparation.

[2] B. Aronov, J. Matoušek and M. Sharir, On the sum of squares of cell complexities in hyperplane arrangements, *Proc. 7th Symp. on Computational Geometry*, 1991, to appear.

[3] M. Bern, D. Eppstein, P. Plassman, and F. Yao, Horizon theorems for lines and polygons, manuscript, 1990.

[4] H. Edelsbrunner, *Algorithms in Combinatorial Geometry*, Springer-Verlag, Heidelberg, 1987.

[5] H. Edelsbrunner, L. Guibas, J. Pach, R. Pollack, R. Seidel and M. Sharir, Arrangements of curves in the plane: Topology, combinatorics, and algorithms, *Proc. 15th Int. Colloq. on Automata, Languages and Programming*, 1988, pp. 214–229.

[6] H. Edelsbrunner, R. Seidel and M. Sharir, On the zone theorem for hyperplane arrangements, in preparation.

[7] B. Grünbaum, *Convex Polytopes*, Wiley, New York 1967.

[8] M.E. Houle and T. Tokuyama, On zones of flats in hyperplane arrangements, manuscript, 1991.

[9] M. Pellegrini, Combinatorial and algorithmic analysis of stabbing and visibility problems in 3-dimensional space, PhD Thesis, Courant Institute, New York University, February 1991.

[10] M. Pellegrini, Ray-shooting and isotopy classes of lines in 3-dimensional space, in *Proceedings of the 1991 Workshop on Algorithms and Data Structures*, Ottawa, 1991.

[11] M. Pellegrini, On the zone of a codimension p surface in a hyperplane arrangement, manuscript, 1991.

Ray-shooting and isotopy classes of lines in 3-dimensional space [1]

Marco Pellegrini[2]

Courant Institute, New York University

251 Mercer St. N.Y., N.Y. 10012

Abstract

A uniform approach to problems involving lines in 3-dimensional space is presented. This approach is based on mapping lines in R^3 into points and hyperplanes in 5-dimensional projective space (*Plücker space*). We improve previously known results on the following problems:

1. Preprocess n triangles so as to efficiently answer the query: "Given a ray, which is the first triangle hit?" (*Ray-shooting problem*). We discuss the ray-shooting problem for both disjoint and non-disjoint triangles and several space/time trade offs.

2. Efficiently detect the first face hit by any ray in a set of axis-oriented polyhedra.

3. Preprocess n lines (segments) so as to efficiently answer the query "Given 2 lines, is it possible to move one into the other without crossing any of the initial lines (segments)?" (*Isotopy problem*). If the movement is possible produce an explicit representation of it.

4. Construct the arrangement generated by n intersecting triangles in 3-space in an output-sensitive way, with a *subquadratic* overhead term.

5. Count the number of pairs of intersecting lines in a set of n lines in space.

1   Introduction

Several algorithms for variations of the ray-shooting problem are presented in this paper. Given n triangles in R^3, we want to build a data structure that efficiently answers queries on the interaction of rays (lines) with the set of triangle. Typically we are interested in knowing whether a query ray (line) hits at least one triangle (*decision problem*) and which is the first one (*report problem*). Another variation of this problem is the off-line case, when all queries are known in advance. Also, we might want to balance query time and storage, and give a spectrum of performances.

The first results on ray-shooting using the Plücker space approach are in [Pel90,Pel91]. The size of the data structure used in this paper to answer ray-shooting queries in logarithmic time depends on the total complexity $O(n^\beta)$ of all cells of a 5-dimensional arrangement of hyperplanes cut by a second degree algebraic surface II, called *Plücker hypersurface*. This set of cells is also called the *zone* of II in the arrangement of hyperplanes $\mathcal{A}(\mathcal{H})$ (denoted as $Z_\Pi(\mathcal{H})$). Since the zone of II can have $\Omega(n^4)$ cells in the worst case, this gives a lower bound for the method too. In [Pel90] the estimate for the complexity of $Z_\Pi(\mathcal{H})$ is $O(n^5)$, later improved in [Pel91] to $O(n^{4.669})$. Recently Aronov, Matoušek and Sharir [AMS91] have found the bound $O(n^{4.5})$. Finally, Aronov and Sharir [AS91b] have found an $O(n^4 \log n)$ upper bound. Since the logarithmic factor is going

---

[1] Research supported by NSF grant CCR-8901484.

[2] Current address: King's College, Comp. Sci. Dept., Strand, London WC2R 2LS, United Kingdom

to be dominated by other components of the method, this bound matches the lower bound for the algorithmic uses presented in this paper.

The algorithms in [Pel90,Pel91] to solve the *decision problem* for any set of triangles use $O(\log^2 n)$ query time and $O(n^{\beta+\epsilon})$ storage, for every $\epsilon > 0$, where the constants depend on ϵ. This paper incorporates the results of [AS91b] on the value of β into the methods in [Pel90]. Some of the results in [Pel90] on *report ray-shooting* were not stated correctly and some stronger hypothesis were needed for the algorithms to work within those stated bounds. The correct version of these results is in [Pel91]. This paper introduces new ideas and techniques that allow us to complete the proofs of the results in [Pel90] (Section 4), to extend those results to the case of intersecting triangles (Section 6) and to solve other problems on lines in 3-space.

When query lines are mapped into points and we locate these points in a special cell complex, we have the so called *primal approach*. Ray-shooting problems can be cast also in a *dual approach* where queries are mapped into hyperplanes, and we ask for the points in a point set that are above the query hyperplane (see [CSW90,AS91a]). Using the dual approach, Agarwal and Sharir [AS91a] obtain an algorithm to solve reporting line shooting queries that uses $O(m)$ storage and answers the queries in time $O(n^{16/15+\epsilon}/m^{4/15})$, for any $\epsilon > 0$. In this paper we modify the dual approach and we give a simpler solution for *disjoint* triangles within the same space/time trade off. In Section 2 an $O(n^{2+\epsilon})$ storage and $O(\log n)$ query time method is given for the ray shooting problem on axis-oriented polyhedra.

Independently, Mark de Berg, Dan Halperlin, Mark Overmars, J. Snoeyink and M. van Kreveld [dBHO+91] have obtained results similar to Theorem (1) for ray-shooting on axis-oriented boxes. Using more complicated techniques they obtained a result similar to Theorem (8) for ray-shooting on intersecting triangles. Independently and using similar techniques, Pankaj K. Agarwal and Micha Sharir [AS91a] have obtained results for on-line ray-shooting on intersecting triangles similar to those presented in this paper.

The methods used to solve the ray-shooting problems are applicable to a wide range of 3-dimensional problems. In Section 5 we discuss the *isotopy problem*: given n blue lines and two red lines decide whether the two red lines are in the same isotopy class (i.e. we can move continuously one into the other without crossing any blue line). McKenna and O'Rourke in [MO88] give an $O(n^4)$ storage and $O(n)$ query time method to solve this problem. Moreover, they use $O(n^4)$ elementary moves to find the actual movement of the 2 lines. At the best of my knowledge no better results were known for this problem. We present a method that uses $O(n^{4+\epsilon})$ storage and answers queries in $O(\log n)$ time. We use at most $O(n^{2+\epsilon})$ elementary moves to move one line into another, provided such movement is possible. These are sharp improvements over the bounds in [MO88]. We also extend the isotopy query data structure to deal with isotopy classes generated by sets of segments and polyhedra. In this case we have $O(n^{4+\epsilon})$ storage and $O(\log n)$ query time, but the movement consists of $O(n^{4+\epsilon})$ elementary moves in the worst case.

In Section 6.1 we give an output sensitive method for constructing arrangement of triangles in 3-space in $O(n^{8/5+\epsilon} + K \log K)$ time, where K is the output size. The interest of this result lies in the *subquadratic* overhead. At my knowledge no previous subquadratic overhead was know for this problem. An $O(n^2)$ overhead is trivial. Also, we sketch an algorithm that in $O(n^{8/5+\epsilon})$ expected time counts all pairs of intersecting lines.

2 Ray shooting on axis-oriented polyhedra

In this Section we give a solution to a simpler case of ray-shooting where the input objects are axis-oriented polyhedra (also called axis-oriented boxes).

Theorem 1 *Given n axis-oriented boxes in 3-space, there exists a data structure of size $O(n^{2+\epsilon})$ to answer ray shooting report queries in $O(\log n)$ time. The data structure can be built in $O(n^{2+\epsilon})$ time.*

Proof. First of all, we notice that the problem is easily decomposable. If $\mathcal{R} = \mathcal{R}_1 \cup \mathcal{R}_2$ is the set of rectangles bounding the boxes, the answer for $\mathcal{R}_1$ and the answer for $\mathcal{R}_2$ can be combined in $O(1)$ time to give the answer for $\mathcal{R}$. We partition the facets of the boxes into 3 sets of axis-parallel parallel rectangles, and we solve the problem on each set separately.

The second observation is that, given a rectangle R parallel to the xy-plane, a line l intersects R if and only if the projections of the line l and of the rectangle R on the xz-plane intersect and the same holds for projections on the yz-plane. The general strategy is to build a two-level data structure where each level is a data structure to answer queries about line stabbing of segments on the plane. Given n segments on the plane we dualize them into n double wedges, using standard duality [EMP+82]. We use Matoušek technique [Mat90] to partitions the plane into $O(r^2)$ triangles. Each triangle σ is covered by $O(n)$ of the original wedges and crossed by n/r of the wedges, as follows from [Mat90]. A query line on the plane dualizes into a point on the dual plane, which belongs to a region σ. The list of wedges covering the region σ corresponds to segments stabbed by the query line on the plane. We keep this list sorted in stabbing order. This order is well defined since the rectangles are parallel. We store these lists of wedges in space $S(n)$, which satisfies the recurrence: $S(n) \leq cr^2 S(n/r) + cr^2 n$. The solution is $S(n) = O(n^{2+\epsilon})$ for $r = n^{1/4}$, and the depth of the recursion is constant. We use the above data structure as a primary tree to store information about projections on the xz-plane. At each node we take the list of covering wedges and consider the associate rectangles and their projections on the yz-plane as an input to a similar secondary data structure. The total space used to build the secondary and the main data structures is: $S(n) \leq cr^2 S(n/r) + cr^2 n^{2+\epsilon}$. The solution is $S(n) = O(n^{2+\epsilon})$ for $r = n^{\epsilon'}$, with $\epsilon' < \epsilon$. The depth of the secondary tree is constant (depending on ϵ').

The query time on the secondary data structure satisfies this recurrence: $Q'(n) \leq Q'(n/r) + \log n$, which solves in $Q'(n) = O(\log n)$ because we chose $r = n^{\epsilon'}$. Similarly, the query time on the primary tree is: $Q(n) \leq Q(n/r) + Q'(n) \leq Q(n/r) + O(\log n)$, which again gives an $O(\log n)$ total query time.

Since we choose r non constant, we must build planar point location structures on the dual plane to locate the cell containing the query point. These additional structures do not asymptoticaly modify the space or the time needed to construct the 2-level tree. ∎

With techniques in [CSW90] and [AS91a] it is possible to trade off space and query time. Using $O(m)$ storage for $n^{1+\epsilon} \leq m \leq n^{2+\epsilon}$, we obtain query time $O(n^{1+\epsilon}/\sqrt{m})$.

3 Plücker Coordinates of lines

To solve ray-shooting on triangles with any orientation we use the *Plücker coordinates* of lines. This idea has been pioneered in [CEGS89a] and developed in [Pel90,Pel91,PS91]; a classical treatment of Plücker coordinates can be found in [Som51].

Two points $x = (x_0, x_1, x_2, x_3)$ $y = (y_0, y_1, y_2, y_3)$ in 3-dimensional homogeneous coordinates[3] define a line l in 3-space. The six quantities

$$\xi_{ij} = x_i y_j - x_j y_i \quad for \ ij = 01, 02, 03, 12, 23, 31$$

are called *Plücker coordinates* of the line l (oriented from x to y). They correspond to the two by two minors of the two by four matrix formed by the coordinates of the point x (on the first row) and y (on the second row).

The six parameters are not independent; they must satisfy the following equation (whose solution constitutes the Plücker hypersuperface or Klein quadric or Grassman manifold $\mathcal{F}_4^2$ [Sto89,Som51]):

$$\Pi \ : \ \xi_{01}\xi_{23} + \xi_{02}\xi_{31} + \xi_{03}\xi_{12} = 0 \tag{1}$$

The incidence relation between two lines l and l' can be expressed using the Plücker coordinates of l and l'. Let a_1, b_1 (resp. a_2, b_2) be two points on l (resp. l') oriented as l (resp. l'). The incidence between l and l' is expressed as the vanishing of the determinant of a four by four matrix whose rows are the coordinates of a_1, b_1, a_2, b_2 in this order from top to bottom.

$$\begin{vmatrix} a_{10} & a_{11} & a_{12} & a_{13} \\ b_{10} & b_{11} & b_{12} & b_{13} \\ a_{20} & a_{21} & a_{22} & a_{23} \\ b_{20} & b_{21} & b_{22} & b_{23} \end{vmatrix} = 0 \tag{2}$$

If we expand the determinant according to the two by two minors of the submatrix formed by the coordinates of the points a_1, b_1 and the minors of the submatrix formed by the points a_2, b_2, we obtain the following equation in which only Plücker coordinates are involved:

$$\xi_{01}\xi'_{23} + \xi_{02}\xi'_{31} + \xi_{03}\xi'_{12} + \xi'_{01}\xi_{23} + \xi'_{02}\xi_{31} + \xi'_{03}\xi_{12} = 0 \tag{3}$$

Let us introduce two mappings: $\pi : l \rightarrow \pi(l)$ maps a line in R^3 to an hyperplane in $\mathcal{P}^5$ (5-dimensional oriented projective space) whose plane coordinates are the Plücker coordinates of l appropriately reordered. $p : l \rightarrow p(l)$ maps a line in R^3 to a point in $\mathcal{P}^5$ whose coordinates are the Plücker coordinates of the line. The incidence relation between the two lines l, l' (expressed by Equation (3)) can be reformulated as an incidence relation between points and hyperplanes in $\mathcal{P}^5$. Equation (3) can be rewritten in the form $\pi_l(p_{l'}) = 0$, which is equivalent to to requiring point $p(l')$ to belong to hyperplane $\pi(l)$. Computations that are standard in Euclidean spaces can be done in oriented projective spaces using a method in [Sto89]. Form the analysis of [Pel90,Pel91], it is easy to prove the following lemma, which is the basis for our approach to ray-shooting.

Lemma 1 *Given a set of triangles T, and the arrangement $\mathcal{A}(\pi(\mathcal{L}_T))$ of the Plücker hyperplanes corresponding to lines spanning edges of T, for each cell c of $\mathcal{A}(\pi(\mathcal{L}_T))$ and any 2 lines l_1 and l_2, if $p(l_1) \in c$ and $p(l_2) \in c$ then l_1 and l_2 intersect the same subset of triangles in T.*

---

[3]A point in real 3-dimensional space has cartesian coordinates (x, y, z) and homogeneous coordinates (x_0, x_1, x_2, x_3). The relations between the two systems of coordinates is given by the following equations: $x = x_1/x_0$ $y = x_2/x_0$ $z = x_3/x_0$.

4 Ray-shooting on disjoint triangles

In [Pel91] the main difficulty in extending the decision ray-shooting algorithm into a reporting ray-shooting algorithm was the fact that a single cell in the zone of Π could contain Plücker points with different stabbing orders[4], therefore locating the cell containing the Plücker point was not enough to determine the first triangle hit. The refinement of the argument given in [Pel90,Pel91] is the following. If we divide each cell of $Z_\Pi(\pi(\mathcal{L}_T))$ into simplices, for each simplex s, $s \cap \Pi$ has a constant number (c) of components. In constant time we compute the components of $\Pi \cap s$ using the general approach for computing topological properties of real algebraic manifolds of Schwartz and Sharir [SS83].[5]

Since the Zone $Z_\Pi(\pi(\mathcal{L}_T))$ has complexity $O(n^4 \log n)$ we have at most $O(n^4 \log n)$ components of Π to deal with. Assuming that the triangles in T are pairwise disjoint, each component corresponds to a unique stabbing order. We simply associate a sorted list of triangles to each component.

We adopt the same random sampling approach ([Cla87]) as in [Pel90,Pel91]. Let r be the size of the sample of Plücker hyperplanes. We build the Zone of the Plücker surface in the sample, and we triangulate this zone. For each simplex we compute the components of $s \cap \Pi$ and the associated lists of triangles; then we recursively proceed in each simplex. The storage required satisfies this recurrence, where c and c_1 are constants:

$$S(n) \leq c_1 r^4 \log r S(\frac{n}{r} \log r) + ncc_1 r^4 \log r$$

For $r = n^{\epsilon'}$, $S(n) = O(n^{4+\epsilon})$, and the depth of the corresponding tree is constant. During the ray shooting query we locate the simplex s where the query Plücker point lies. For this task we use a point location data structure of size polynomial in r, which answers location queries in time $O(\log r) = O(\log n)$. This data structure is obtained expanding the facets of the simplices into full hyperplanes and using a standard method for point location in an arrangement of hyperplanes.

When we have the simplex s we perform a binary search in each of the at most c lists associated with components of $\Pi \cap s$. In $O(\log n)$ time we select one triangle in each list. In constant time we determine the closest triangle and compare it with the answer to the recursive data structure associated with the simplex. The first triangle can be determined in time $O(\log n)$. We summarize the above discussion with the following theorem:

Theorem 2 *Given a set T of n disjoint triangles in 3-dimensional space, there is a data structure $D(T)$ that uses $O(n^{4+\epsilon})$ storage and reports the first triangle hit by any ray ρ in $O(\log n)$ time. $D(T)$ can be built in $O(n^{4+\epsilon})$ expected time.*

4.1 Batched ray shooting

The main difficulty in extending batched ray shooting method of [Pel90] with the techniques used for Theorem (2) is that the batching method uses a dual space approach while we know how to deal with the Plücker surface only in the primal space.

Given m rays and n triangles, if $m \geq n^{4+\epsilon}$ we build the ray-shooting data structure of the previous section and we obtain an $O(m \log n)$ overall method. If $m \leq n^{4+\epsilon}$ we dualize the

---

[4]In [Pel90] it was implicitly assumed that this order was unique in a cell.
[5]Additional computational details are given in [CEGS89b] and [Cha85]. Furthermore, we can perform point location in the resulting data structure [SS83].

problem. Each triangle is mapped into three points and a query ray into an hyperplane. We select a random sample of r hyperplanes and we triangulate the zone of Π obtaining $O(r^4 \log r)$ simplices [AS91b]. Each simplex is cut by no more than $O(\frac{n}{r} \log r)$ hyperplanes [Cla87]. Now we consider every triple of simplices in turn (there are a constant number of them since r will be chosen constant). Let $\sigma = (\sigma_1, \sigma_2 \sigma_3)$ be a triple of simplices, N_σ be the set of triangles whose corresponding points are in σ, and n_σ be the cardinality of N_σ. We take all hyperplanes missing σ and we divide it into those above σ and those below σ. These two classes represent lines stabbing N_σ. We repeat the following argument for each stabbing class of hyperplanes separately. Let M_σ be such class of hyperplanes, and m_σ its cardinality (with M_σ^* we denote the corresponding set of dual points). We want to compare the m_σ rays against the n_σ triangles in time $O((m_\sigma + n_\sigma) \log n_\sigma)$.

A correct way to proceed, but too expensive, is the following: we dualize back again, returning to the primal space. We have n_σ Plücker hyperplanes and m_σ Plücker points. Since we know that all Plücker points in M_σ^* have the same relative position with respect to the hyperplanes, we can intersect the halfspaces supported by the hyperplanes containing the Plücker points. We obtain a polytope Q_σ of size n_σ^2 which we triangulate. We apply the query technique on lists of triangles used in the previous section. The problem with this approach is that Q_σ is too complex to get the desired time bound.

A better solution is the following. We dualize back to the primal space the vertices of the 3 simplices in σ (which are not necessarily Plücker points), obtaining a constant number of hyperplanes. These hyperplanes have a definite position with respect to the Plücker points M_σ^*, therefore we can intersect the corresponding halfspaces obtaining the polytope Q_σ'.

Lemma 2 $M_\sigma^* \subset Q_\sigma' \subset Q_\sigma$, and every component of $Q_\sigma' \cap \Pi$ maps to one component of $Q_\sigma \cap \Pi$.

Proof. $M_\sigma^* \subset Q_\sigma'$ is given by the fact that in dual space the hyperplanes M_σ were in the same relative position with the three simplices and, in defining Q_σ', we have consistently chosen the halfspace supported by any hyperplane dual to a vertex of σ.

Suppose now that exists a point $p \in Q_\sigma'$, $p \notin Q_\sigma$. In primal space this point corresponds to a hyperplane (not a Plücker hyperplane, though) that separates points in N_σ but it does not separates vertices of the simplices enclosing N_σ. This is obviously absurd.

Since $Q_\sigma' \subset Q_\sigma$ every point of $\Pi \cap Q_\sigma'$ is in $\Pi \cap Q_\sigma$. The only possible way we can violate the second part of the lemma is by merging in Q_σ' two components of $Q_\sigma \cap \Pi$. This is possible only if we have Plücker points in Q_σ' that are not in Q_σ, which is absurd. ∎

Since Q_σ' has constant complexity, the procedure for searching the source of the ray outlined in the previous section uses only logarithmic time. Lemma (2) ensures the correctness of the procedure. The total time used is $O((m_\sigma + n_\sigma) \log n_\sigma)$. Combining the above analysis with the analysis in [Pel91] (which, in turns, is similar to results in [GOS89]) we obtain the following theorem:

Theorem 3 *One can determine the first triangle hit by m rays in a set of n disjoint triangles using a randomized algorithm whose expected running time is bounded by*

$$[D m^{4/5-\delta} n^{4/5+4\delta} \log^2 n + A m \log^2 n + B n \log n \log m]$$

for any $\delta > 0$, where the coefficients A, B and D depend on δ. The storage is bounded by $O(m + n)$.

Also, from an observation in [Pel90] we obtain the following corollary

Corollary 1 *Given 2 non-convex polyhedra of total complexity $n = n_1 + n_2$ there is a randomized algorithm that in expected time $O(n^{8/5+3\delta})$ decides whether they intersect.*

Corollary 2 *Given a set T of n triangles we can test whether T is a set of pairwise disjoint triangles in time $O(n^{8/5+\epsilon})$.*

Proof. Partition T into two sets of roughly equal size and test them recursively. If they are both pairwise disjoint, apply the batched ray-shooting method, using in turns the edges one set as segment queries on the second set. ∎

4.2 Dual approach for reporting ray-shooting on disjoint triangles

In this section we simplify the dual method of [AS91a] to *report* the first triangle hit in a set of disjoint triangles. For sake of exposition we to outline the dual approach.

We map query lines into Plücker hyperplanes and triangles into three Plücker points. From [Pel90], we know that a line stabs a triangle if and only if the 3 points are all either above or below the hyperplane. We form three sets of Plücker points by drawing one edge from each triangle. For the first set we build the partition tree described in [CSW90] to answer efficiently halfspace range queries (form now on denoted as *CSW-tree*). The answer is formed by the union of certain prestored sets of Plücker points. To each of the prestored subsets we attach a secondary data structure which is a CSW-tree having as input the points in the second set drawn from the same triangles as the points of the first set prestored at that node. We repeat the same construction adding a third level that stores points from the third set. A line-shooting query consists in a descent in the data structure that checks if there is a triple of points, one from each of the three levels, drawn from the same triangle, which are on the same side of the query hyperplane. This data structure can be build with a continuous trade off between storage and query time.

At the leaves of the 3-layered partition tree we have many subsets of Plücker points. During the query we collect some of these sets and their union form the set of (representatives) of the triangles stabbed by the query line. We consider separately each set of Plücker points stored at the leaves as input to the data structure we are going to sketch. Let σ be a node of the third level and N_σ be the set of associated Plücker points (we also add the Plücker points coming from the same triangles collected from all 3 levels).

The points of the set N_σ are all contained in three simplices (resulting from the geometric partitioning done in the CSW-tree), therefore we can apply Lemma (2). Using the same procedure of Section 4.1, we build a data structure of linear size at each node to answer report ray shooting in $O(\log n)$ time.

Plugging this data structure into the space/time analysis of [CSW90] and using arguments similar to those in [AS91a] we obtain the following:

Theorem 4 *There is a family of data structures that, using $O(m)$ units of storage, answer report ray-shooting queries on disjoint triangles in time $O(n^{16/15+\epsilon}/m^{4/15})$. Each data structure can be built in $O(m^{1+\epsilon})$ randomized expected time.*

5 Querying isotopy classes

In this section we show how to adapt the ideas of the Section 4 in order to solve the following problem: given n blue lines in R^3, determine, for a pair of red lines if they belong to the same isotopy class.

For this problem we give an $O(n^2 \log n)$ time algorithm. If the problem is asked in repetitive mode, after $O(n^{4+\epsilon})$ preprocessing, we can answer in $O(\log n)$ time.

We check in $O(n)$ time that the Plücker points of the two red lines have the same sign with respect to any of the blue Plücker hyperplanes. If this is not the case we answer negatively. Otherwise, we construct the cell containing the two red Plücker points in the arrangement of the blue Plücker hyperplanes. We subdivide this cell into $O(n^2)$ simplices and compute the components of $s \cap \Pi$ for each simplex. If two components of Π in two adjacent simplices have a common boundary point, we put the two components in the same class. Using a union find data structure we can identify all the maximal components of Π into the cell (as opposed to the components of Π in the simplices of the triangulated cell), which represent isotopy classes. The second phase consists in locating the components of Π containing the 2 red points, and retrieving the associated isotopy class. Constructing the cell takes $O(n^2 \log n)$ time using a standard convex hull algorithm. The size of the cell is $O(n^2)$, and this is the time needed to triangulate the cell. The union find procedure takes $O(n^2 \alpha(n))$ time [Tar83] because we have $O(n^2)$ pairs of adjacent facets in the triangulation of the cell, and therefore $O(n^2)$ union operations.

Theorem 5 *Given a set $\mathcal{L}$ of n lines in R^3, we can determine if any 2 given lines, not necessarily in $\mathcal{L}$, are in the same isotopy class using $O(n^2 \log n)$ time and $O(n^2)$ storage.*

We define as an *elementary path* a path on Π connecting 2 points on one component of $\Pi \cap s$ completely contained in the component, where s is any simplex. It is easy to see that we can compute the movement that takes one red line into the other as the concatenation of at most $O(n^2)$ elementary paths.

Given two points on any component of $\Pi \cap s$ it is possible to compute an arc completely contained in the component and connecting the two points [SS83]. This implies that we can effectively compute the elementary steps required by Theorems (5), (6) and (7).

For the repetitive case, we adopt the random sample approach and we build a search tree structure of size $O(n^{4+\epsilon})$ in the first phase. The second phase consists in forming the equivalence classes, starting from the bottom of the search tree, and identifying components sharing boundary points.

The query algorithm locates the 2 query points in the Plücker arrangement, at the bottom of the decomposition tree, determines the components of $\Pi \cap s$ to which they belong. Then, using the auxiliary union-find data structure, determine the isotopy class.

Theorem 6 *Given a set $\mathcal{L}$ of n lines we can preprocess it into a data structure $D(\mathcal{L})$ of size $O(n^{4+\epsilon})$, so that, for any given pair of lines l_1 and l_2, in $O(\log n)$ time it is possible to decide whether l_1 and l_2 belong to the same isotopy class. $D(\mathcal{L})$ can be constructed in $O(n^{4+\epsilon})$ expected time. The path connecting the 2 lines, if exists, is composed by at most $O(n^{2+\epsilon})$ elementary paths.*

Let us consider a segment e in R^3. Let l_e be the line spanning e. Define $S_e = \{p(l) \mid l \cap e \neq \emptyset\}$. We observe that S_e is a connected semialgebraic set of Plücker points. From this fact it is easy to show that, given a simplex s on the Plücker hyperplane $\pi(l_e)$, $s \cap S_e$ has a constant number of connected components.

In order to decide whether 2 lines l_1 and l_2 are in the same isotopy class with respect to a set E of edges, we build the same data structure of the previous section for the set $\mathcal{L}(E)$ of lines spanning E. During the second phase of the algorithm, though, we modify the rule for identifying components of Π. Given 2 components $\mathcal{C}_1$ and $\mathcal{C}_2$, we identify them only if they have a common point on their boundary not in $\bigcup_{e \in E} S_e$. Clearly, if two components are separated by the hyperplane $\pi(l_e)$ we need to check only S_e. The complexity of each check is constant for each union find operation. We summarize with this theorem:

Theorem 7 *Given a set E of n segments in R^3, we can preprocess it into a data structure $D(E)$ of size $O(n^{4+\epsilon})$, so that, for any given pair of lines l_1 and l_2, in $O(\log n)$ time it is possible to decide whether l_1 and l_2 belong to the same isotopy class. $D(E)$ can be constructed in $O(n^{4+\epsilon})$ expected time. The path connecting two lines, if exists, is the concatenation of $O(n^{4+\epsilon})$ elementary paths in the worst case.*

6 Ray shooting on intersecting triangles

Let us consider again the data structure of Theorem (2). In every cell σ of the Plücker arrangement we have an associated set of triangles T_σ such that every Plücker point in σ meet every triangle of T_σ. From this follows that extending the triangles of T_σ into full planes does not introduce new intersections. We construct the full 3-dimensional arrangement of planes spanning T_σ, we process it for point location and each cell of the arrangement for fast polyhedral intersection [DK90]. The total cost is $O(n^{3+\epsilon})$ time and storage. It is easy now to locate the source of the ray in the 3-dimensional arrangement and find the first plane hit along the query ray. The space/time complexity $S(n)$ satisfies the following equation:

$$S(n) \leq c_1 r^4 \log r S(\frac{n}{r} \log r) + n^{3+\epsilon} cc_1 r^4 \log r$$

The solution is $S(n) = O(n^{4+\epsilon})$ setting $r = n^{\epsilon'}$, where $0 < \epsilon' < \epsilon$. Using the same tricks as in the proof of Theorem (2) we obtain an $O(\log n)$ query time.

Theorem 8 *Given any set of n (in general non-disjoint) triangles T in 3-dimensional space, there is a data structure $D(T)$ that uses $O(n^{4+\epsilon})$ storage and reports the first triangle hit by any ray ρ in $O(\log n)$ time. $D(T)$ can be built in $O(n^{4+\epsilon})$ expected time.*

6.1 Batched ray-shooting on intersecting triangles

The batching technique of [GOS89,EGS88] allows us to to divide recursively the problem into subproblems in which every line in the set of lines M_σ, of size m_σ, intersect all the triangles in the set N_σ of size n_σ. To solve this subproblem we adopt the following strategy:

1. If $m_\sigma \geq n_\sigma^{4+\epsilon}$, using the method of Theorem (8) we solve the problem in time $O(m_\sigma \log n_\sigma)$.

2. If $n_\sigma^{3+\epsilon} \leq m_\sigma \leq n_\sigma^{4+\epsilon}$, we build the 3-dimensional arrangement of the planes, and we process it for point location and ray-shooting in total time $O(n_\sigma^{3+\epsilon})$. The total time is therefore $O(m_\sigma \log n_\sigma)$.

3. If $m_\sigma \leq n_\sigma^{3+\epsilon}$, we solve the problem by dualizing in 3-space the rays into double spatial wedges and the planes into points. When we locate the points in the arrangement of the wedges, we are implicitly detecting the planes hit by the ray. For the planes intersecting a *set of rays* we generate the upper and lower cell of their arrangement and compute the first triangle hit by any ray in time $O(n_\sigma' \log n_\sigma' + m_\sigma' \log n_\sigma')$, where n_σ' is the number of planes intersected by m_σ' rays.

   Coming back to the original subproblem, the standard batching technique [GOS89,EGS88] gives us a solution $O(m_\sigma^{3/4-\delta} n_\sigma^{3/4+3\delta} \log n_\sigma + n_\sigma \log^2 n_\sigma + m_\sigma \log n_\sigma)$.

The partitioning technique of [GOS89,EGS88] at the external level of our construction gives us the following recursive equation:

$$T(m,n) \leq \begin{cases} 0 & for \ m = 0 \\ m \log n & for \ m \geq n^{4+\epsilon} \\ \Sigma_{i=1,M} T(m_i, n_i) + m^{3/4-\delta'} n^{3/4+3\delta'} \log n + m \log n + n \log^2 n & for \ m \leq n^{4+\epsilon} \end{cases}$$

Finally, solving the recurrence, we have the following theorem:

Theorem 9 *Given n triangles in 3-space, which may intersect, and m rays we can find the first triangle hit by any line in randomized expected time:*

$$[Dm^{4/5-\delta} n^{4/5+4\delta} \log^2 n + Am \log^2 n + Bn \log n \log m]$$

for any $\delta > 0$, where the coefficients A, B and D depend on δ. The storage is bounded by $O(m + n)$.

Theorem (9) can be easily modified to report all intersections of segments with the triangles. With this modification we can built an output sensitive algorithm to construct arrangements of triangles in 3-space. A trivial method to compute this arrangement would use $O(n^2 + K \log K)$ time. We improve on the overhead term.

Given n triangles in R^3, we use them as the triangles in our method an their edges as segments we shoot. In time $O(n^{8/5+\epsilon} + k)$ we find all the intersections of edges with triangles. Note that reported intersections are part of the final output. We use the segments cut on each triangle as input to a planar segment intersection reporting problem. If k_i is the number of segments on triangle t_i in $O(k_i \log k_i + K_i)$ time we can find all the intersections using the algorithm in [CE88], where K_i is the size of the contribution of t_i to the arrangement. Summing up over all triangles, we obtain the following theorem:

Theorem 10 *Given a set T of n triangles in R^3 we can build the arrangement $\mathcal{A}(T)$ in time $O(n^{8/5+\epsilon} + K \log K)$, where K is the size of the arrangement.*

Theorem 11 *Given n lines in 3-space it is possible to count the number of pairs of intersecting lines in expected time $O(n^{8/5+\epsilon})$, for any $\epsilon > 0$, where the constants depend on ϵ.*

Proof sketch. We partition the set of lines $\mathcal{L} = \mathcal{L}_1 \cup \mathcal{L}_2$ into two sets of roughly equal size and we solve recursively the problem in each set. Moreover, we locate the Plücker points of lines in $\mathcal{L}_1$ in the arrangement of the Plücker hyperplanes of lines in $\mathcal{L}_2$. We use the batching technique to perform this step in $O(n^{8/5+\epsilon})$ expected time. In order to handle the Plücker points which are on the sampled Plücker hyperplanes we also solve the problem recursively in the dimension of the space containing the query points. We repeat exchanging the roles of $\mathcal{L}_1$ and $\mathcal{L}_2$. ∎

7 Conclusions

This paper improves and generalizes previously known results on ray-shooting on triangles in 3-space. Also, we give the first fast isotopy test for lines among polyhedral obstacles. The unified approach of Plücker coordinates has a potential for applications to problems involving lines, segments and polyhedra in 3-dimensional space, which stretches beyond the algorithms presented in this paper.

8 Acknowledgments

I want to thank Richard Pollack, Janos Pach, Boris Aronov, Pankaj K. Agarwal and Peter Shor for useful discussions. I thank Mark de Berg and Pankaj K. Agarwal for giving me information about their recent results.

References

[AMS91]   B. Aronov, J. Matoušek, and M. Sharir. On the sum of squares of cell complexities in hyperplane arrangements. To appear in the Proceedings of the 7th ACM Symposium on Computational Geometry, 1991.

[AS91a]   P. K. Agarwal and M. Sharir. Applications of a new space partitioning technique. To appear in the Proceedings of the 1991 Workshop on Algorithms and Data Structures, 1991.

[AS91b]   B. Aronov and M. Sharir. On the zone of a surface in an hyperplane arrangement. To appear in the Proceedings of the 1991 Workshop on Algorithms and Data Structures, 1991.

[CE88]   B. Chazelle and H. Edelsbrunner. An optimal algorithm for intersecting line segments in the plane. In *Proc. of the 29th Ann. Symp. on Foundations of Computer Science*, 1988.

[CEGS89a]   B. Chazelle, H. Edelsbrunner, L. Guibas, and M. Sharir. Lines in space: combinatorics, algorithms and applications. In *Proc. of the 21st Symposium on Theory of Computing*, pages 382–393, 1989.

[CEGS89b]   B. Chazelle, H. Edelsbrunner, L. Guibas, and M. Sharir. A singly exponential stratification scheme for real semi-algebraic varieties and its applications. In *Proceedings of the 16th International Colloquium on Automata, languages and Programming*, pages 179–193, 1989. In Springer Verlag LNCS 372.

[Cha85]   B. Chazelle. *Fast searching in a real algebraic manifold with applications to geometric complexity*, volume 185 of *Lecture notes in computer science*, pages 145–156. Springer Verlag, 1985.

[Cla87]   K.L Clarkson. New applications of random sampling in computational geometry. *Discrete Computational Geometry*, (2):195–222, 1987.

[CSW90]   B. Chazelle, M. Sharir, and E. Welzl. Quasi-optimal upper bounds for simplex range searching and new zone theorems. In *Proceedings of the 6th ACM Symposium on Computational Geometry*, pages 23–33, 1990.

[dBHO+91]   M. de Berg, D. Halperlin, M. Overmars, J. Snoeyink, and M. van Kreveld. Efficient ray-shooting and hidden surface removal. To appear in Procedings of the 7th Annual Symposium on Computational Geometry, 1991.

[DK90]   D. Dobkin and D. Kirkpatrick. Determining the separation of preprocessed polyhedra: a unified approach. In *Proc. of the 17th International Colloqium on Automata, Languages and Programming*, pages 400–413, 1990.

[EGS88]   H. Edelsbrunner, L. Guibas, and M. Sharir. The complexity of many faces in arrangements of lines and segments. In *Proceedings of the 4th ACM Symposium on Computational Geometry*, pages 44–55, 1988.

[EMP+82]   H. Edelsbrunner, H. Mauer, F. Preparata, E. Welzl, and D. Wood. Stabbing line segments. *BIT*, (22):274–281, 1982.

[GOS89]   L. Guibas, M. Overmars, and M. Sharir. Ray shooting, implicit point location and related queries in arrangements of segments. Technical Report TR433, Courant Institute, 1989.

[Mat90]   J. Matoušek. Cutting hyperplane arrangements. In *Proceedings of the 6th ACM Symposium on Computational Geometry*, pages 1–9, 1990.

[MO88]   M. McKenna and J. O'Rourke. Arrangements of lines in 3-space: A data structure with applications. In *Proceedings of the 4th annual Symposium on Computational Geometry*, pages 371–380, 1988.

[Pel90]   M. Pellegrini. Stabbing and ray shooting in 3 dimensional space. In *Proceedings of the 6th ACM Symposium on Computational Geometry*, pages 177–186, 1990.

[Pel91]   M. Pellegrini. *Combinatorial and Algorithmic Analysis of Stabbing and Visibility Problems in 3-Dimensional Space*. PhD thesis, New York University–Courant Institute of Mathematical Sciences, February 1991.

[PS91]   M. Pellegrini and P. Shor. Finding stabbing lines in 3-dimensional space. In *Proceedings of the Second SIAM-ACM Symposium on Discrete Algorithms*, 1991.

[Som51]   D. M. H. Sommerville. *Analytical geometry of three dimensions*. Cambridge, 1951.

[SS83]   J.T. Schwartz and M. Sharir. On the piano mover's problem: II. General techniques for computing topological properties of real algebraic manifolds. *Adv. in Appl. Math*, 4:298–351, 1983.

[Sto89]   J. Stolfi. Primitives for computational geometry. Technical Report 36, Digital SRC, 1989.

[Tar83]   R.E. Tarjan. *Data Structures and Network Algorithms*, volume 44 of *CBMS-NSF Regional Conference Series in Applied Mathematics*. SIAM, 1983.

Finding Level-Ancestors in Dynamic Trees

Paul F. Dietz[1]
Department of Computer Science
University of Rochester
Rochester, NY 14627

dietz@cs.rochester.edu

1 Introduction

We examine the problem of answering *ancestor* queries on tree; that is, questions of the form "what is the ith ancestor of vertex v?" We solve this problem when the tree is static with linear preprocessing time and constant time per query (a previously known result), and when the tree is growing by the addition of leaves (the *addleaf* operation) and addition of new roots (*addroot* operations), where updates and queries take constant amortized time. In both cases, the machine model is the RAM with logarithmic word size.

The related problem of finding nearest common ancestors on growing trees has been addressed by Harel and Tarjan [6]. They present an algorithm for the static case in which, after $O(n)$ preprocessing, *nca* queries take constant time, and also gave a linear time algorithm for trees that are linked at the roots. Gabow [5] gave a linear time algorithm for the case of trees growing by addition of leaves, and an $O(m\alpha(m, n) + n)$ time algorithm (m the total number of operations, n the size of the forest) for arbitrary *link* operations (which cause the root of one tree to be a child of a vertex in another tree). The results in this paper make extensive use of the techniques presented by Gabow.

Previously, Berkman and Vishkin [2, 3] have solved this problem for the static case, and describe applications. The serial version of their algorithm achieves $O(1)$ query time after linear preprocessing. The algorithm presented here shares some features with their algorithm (for example, use is made of table lookup to accelerate the solution of small subproblems). However, it is not immediately obvious how to dynamize their algorithm.

2 Notation

A rooted, directed tree T is a digraph $(V(T), E(T))$ such that

---

[1]This work supported by NSF grant CCR-8909667.

1. Each vertex $v \in V(T)$ has out–degree one, except for a single vertex, the root, denoted $r(T)$.

2. For every $v \in V(T)$ there exists a directed path from v to $r(T)$.

A *forest* is a directed graph consisting of disjoint trees. For any $v \in V(T)$, the *ancestors* of v are the vertices on the directed path from v to $r(T)$. Let $Anc(v)$ be the set of ancestors of v. The set $Anc^+(v) = Anc(v) - \{v\}$ is the set of *proper ancestors* of v. A vertex is a *(proper) descendant* of its (proper) ancestors. The subtree of T rooted at v, denoted T_v, is the subtree of T induced by the descendants of v. The *depth* of v, $d(v)$, is the length of the directed path from v to $r(T)$, or, equivalently, $|Anc^+(v)|$. The ancestors of v are ordered along the path to $r(T)$, so the i-th ancestor of v $(0 \le i \le d(v))$, denoted $anc(v, i)$, is the unique $u \in Anc(v)$ such that $d(u) = d(v) - i$. The *parent* of v is $anc(v, 1)$, abbreviated $p(v)$. We say that v is a *child* of $p(v)$. The *height* of T is $\max\{d(v) | v \in V(T)\}$. For any function f on the nodes of a tree, f_T indicates the function refers to tree T. This subscript will be included when confusion is possible.

3 Finding Ancestors in Static Trees

We begin by considering the case of static trees. We show how to answer *ancestor* queries in $O(1)$ time after performing $O(n)$ preprocessing. In the following, $n = |V(T)| \ge 1$.

We use an auxiliary tree called a compressed tree. The following definitions are from [6, 5]. Let v be a vertex in a tree T. The *size* of v, $s(v)$, is the number of its descendants. A child v is *light* if $2s(v) \le s(p(v))$, and is *heavy* otherwise. Deleting all but the edges from heavy children to their parents decomposes T into a set of vertex disjoint paths, the *heavy paths* of T.

Let $\mathcal{P}$ be a partition of $V(T)$ into vertex disjoint paths. For each $v \in V(T)$, let $\mathcal{P}(v)$ be the path in $\mathcal{P}$ that contains v. For each path $z \in \mathcal{P}$, define the *apex* of z, $ap(z)$, to be the vertex in z of minimum depth. If u is a vertex in z then the predecessor of u in z is the vertex in z of depth $d(v) + 1$, if such a vertex exists. The *compressed tree for T and $\mathcal{P}$*, $C(T, \mathcal{P})$, is a tree with nodes $V(T)$ and root $r(T)$, such that for each $v \in V(T)$ other than the root, $p_C(v) = ap(\mathcal{P}(p_T(v)))$. If $\mathcal{P}$ is the set of heavy paths for T then we call this the *compressed tree for T*, denoted $C(T)$.

The height of a compressed tree $C(T)$ is at most $\lfloor \log n \rfloor$, since for any $v \ne r(T)$, $2s_C(v) \le s_C(p_C(v))$.

3.1 A One-Level Algorithm

We now show how to compute $p_T(v, i)$ in constant time, given $O(n \log n)$ preprocessing. We store at every vertex v its depth $d(v)$ and at every apex v the compressed path $\mathcal{P}(v)$. The path is stored as an array so that the i-th vertex in the path can be retrieved in constant time. At every vertex $v \in V(T)$, we store the set of pairs $S(v) = \{(u, d_T(v) - d_T(u)) | u \in Anc_C(v)\}$.

To find $anc_T(v, i)$, we do the following:

1. If $i = 0$ then return v.

2. Otherwise, find the deepest u in $Anc_C(v)$ such that $d_T(v) - d_T(u) \geq i$.

3. Let $j = d_T(v) - d_T(u) - i$. Return the jth predecessor of u in $\mathcal{P}(u)$.

Theorem 1 *This procedure correctly computes* $anc_T(v, i)$.

Proof: If $i = 0$ the algorithm is trivially correct. Otherwise, we can assume $0 < i \leq d_T(v)$. Since $d_T(r(T)) = 0$, and since $r(T) \in Anc_C(v)$, step (1) finds some $u \neq v$. Let w be the vertex in $Anc_C(v)$ such that $p_C(w) = u$. Then, $d_T(w) < i \leq d_T(u)$. Moreover, all the vertices that lie between w and u in T are in $\mathcal{P}(u)$; otherwise, $p_C(w)$ would not be u. Therefore, $anc_T(v, i)$ is on $\mathcal{P}(u)$. Step (3) finds the unique vertex of depth $d_T(v) - i$ on the path. ∎

We must now show how to implement this algorithm so that queries can be performed in constant time. The following result is a consequence of the results of Ajtai, Fredman and Komlos [1]:

Lemma 2 *Let n be a positive integer. There is a data structure on the RAM with logarithmic word size performing set insert, delete and predecessor (greatest lower bound) operations in $O(1)$ time on subsets of $\{0, \ldots, n - 1\}$, provided the subsets are of size $\log^{O(1)} n$ and $O(n)$ preprocessing can be performed.*

In fact, Ajtai *et. al.* showed how to do this problem in $O(1)$ time in the cell probe model of computation, but their algorithm can be implemented if their hash functions are precomputed, which can be done in $O(n^\epsilon)$ time for some $\epsilon < 1$. We call their data structure a *small priority queue*.

More recently, Fredman and Willard [4] have given a representation for small sets in which predecessor operations can be done in constant time on a RAM with certain arithmetic operations.

Since the sets $S(v)$ are of size at most $\lfloor \log n \rfloor$, small priority queues can be used to do queries in constant time. However, $O(n \log n)$ time and space is required to construct these sets.

3.2 A Two-Level Algorithm

The preprocessing time and space bounds of this algorithm can be improved by using a two-level scheme. Define the *rank* of a vertex v to be $rank(v) = \lfloor \log s_C(v) \rfloor$. One can show [7, 6]:

Lemma 3 *If v is an apex, then $s_C(v) = s_T(v)$; otherwise, $s_C(v) = 1$.*

Lemma 4 *For $v \neq r(T)$, $2s_C(v) \leq s_C(p_C(v))$ and $rank(v) < rank(p_C(v))$.*

Lemma 5 *For any i, there are at most $n/2^i$ vertices of rank i.*

Proof: See [6, lemma 8]. ∎

Let $c = \lfloor \frac{1}{2} \log n - \log \log n \rfloor$. We call a vertex v *major* if $s_C(v) > c$, otherwise, v is *minor*.

Lemma 6 *There are $O(n/\log n)$ major vertices.*

Proof: See [6, lemma 9]. ∎

Let F be the forest obtained by removing from T (not C) all major vertices. F consists of a collection of trees of size at most c containing a total of $n - O(n/\log n)$ vertices. Let $T_1, \ldots, T_k$ be the trees in F. Call these the *minor trees*. We find the T_i in $O(n)$ time by traversing T. For each minor v, we record the name of the T_i it is in.

For each T_i we do the following:

1. Order the children of each vertex.

2. Represent T_i by a string of $2|V(T_i)| \leq 2c \leq \log n$ bits. This is done by traversing the tree, emitting a 0 bit when a vertex is first visited and a 1 bit when it is last visited.

3. Represent $V(T_i)$ so that the vertices are stored in an array in preorder.

We precompute a function, which is stored as a table. The function takes as arguments the binary representation of a tree of size at most c, the preorder number i of some vertex in the tree, and some number j between 0 and the depth of vertex number i in U. It returns the preorder number of the j-th ancestor in U of vertex number i. This table has size at most $2^{2c}c^2 \leq n/4$, and it can be precomputed in $O(n)$ time.

For the subtree of C consisting of the major vertices, we precompute as in the one–level algorithm. Since there are $O(n/\log n)$ major vertices, this takes $O(n)$ time and space.

We find the i-th ancestor of vertex v as follows:

1. If $i = 0$, return v.

2. Otherwise, if v is a minor vertex,

   (a) Find the T_j containing v.

   (b) If $d_T(v) \geq i + d_T(r(T_j))$, then use the lookup table and the binary representation of T_j to find the i-th ancestor of v in T_j. Return it.

(c) Otherwise, set i to $i - d_T(v) + d_T(r(T_j))$, and set v to $r(T_j)$.

(d) Let $k = d_T(v) - d_T(p_C(v)) - i$. If $i \geq 0$ then return the k-th descendant of $p_C(v)$ in $\mathcal{P}(p_C(v))$. Otherwise, set i to $i - k$ and v to $p_C(v)$.

3. v is now a major vertex. Find the i-th ancestor of v using the algorithm of section 3.1.

Theorem 7 *The two–level algorithm is correct.*

Proof: We consider cases. Let v be a vertex, i a number in the range $\{0, \ldots, d_T(v)\}$, and let w be the deepest major ancestor of v in C.

1. If $i = 0$, then $anc(v, i) = v$.

2. Otherwise, if v is minor, there are several subcases. Let T_j be the minor tree containing v.

   (a) If $anc(v, i)$ is a descendant in C of $r(T_j)$, then $anc(v, i) \in T_j$, and statement 2.2 correctly computes it.

   (b) Otherwise, $anc(v, i) = anc(r(T_j), i - d_T(v) + d_T(r(T_j)))$. If $anc(v, i)$ is a descendant of w then it must be in $\mathcal{P}(w)$, so statement 3.4 correctly computes it. Otherwise, $anc(v, i) = anc(w, i - d_T(v) + d_T(w))$, so this reduces to the next case.

3. If v is a major vertex, then, by theorem 1, statement 3 computes $anc(v, i)$. ∎

It should be obvious that queries take constant time, given that the one–level algorithm takes constant time.

4 Finding Ancestors in Trees Growing by Leaf Addition

We now consider extending this algorithm to the case where the tree is growing by the operation $addleaf(x, y)$ (which causes a single vertex y to be made a child of x).

4.1 A One–Level Algorithm

Gabow [5] showed how to maintain an approximation of a compressed tree for a tree T being modified by $addleaf$ operations. He maintains a family of time-varying paths $\mathcal{P}$ and the compressed tree $D = C(T, \mathcal{P})$ induced by the paths. The paths have the property that the depth of D is $O(\log |V(T)|)$.

Gabow's algorithm maintains D by periodically *reconstructing* subtrees. When D is reconstructed, some subtree rooted a vertex v, D_v, is replaced by C_v (where C is the compressed tree defined from the heavy paths of T). If the cost of a reconstruction is $s_D(v)$, then Gabow shows that each *addleaf* operation has amortized cost $O(\log n)$.

If we store the sets $S(v)$ at the vertices of D as in the static one–level algorithm, then reconstructing a subtree D_v takes $O(s_D(v) \log n)$ time. Therefore, a sequence of m queries and n *addleaf* operations can be performed in $O(m + n \log^2 n)$ time.

When reconstructing D_v, we may change the lower part of $\mathcal{P}(v)$. This takes time at most $O(s_T(v))$, if we need not modify the part of the path above v. We can do this if each path in $\mathcal{P}$ is stored in the beginning of a large array, with sufficient space to store additional vertices if the path is lengthened. This increases the space to $O(n^2)$. A better technique is to store each path in an array with buffer space proportional to the length of the path. That is, when a path p gets too long (or too short) it is stored in an array of size $2|p|$. If the path shrinks to at most a half or grows to at least twice its length at the time of copying, the path is copied again. This increases the amortized running time by at most a constant factor, and uses at most $4n$ space.

4.2 An Improved One–Level Algorithm

We can modify the algorithm of the previous section to take only $O(\log n)$ amortized time per *addleaf* operation.

We change the algorithm so that the tree below D_v is reconstructed only when there is some child u of v that is an apex and for which $s_T(u) > \frac{2}{3} s_T(v)$. When this happens, the path $\mathcal{P}(v)$ is removed from $\mathcal{P}$ and split into two parts: a path P_1 consisting of ancestors of v, and a path P_2 consisting of proper descendants of v. The path P_2 is added to $\mathcal{P}$. The path P_1 is concatenated with $\mathcal{P}(u)$ and replaces $\mathcal{P}(u)$ in $\mathcal{P}$.

Consider the effect of this operation on D. If P_2 is nonempty, let $w = ap(P_2)$. For all vertices in $z \in P_2$ (except w), $p_D(z) = w$. Similarly, the vertices formerly in $\mathcal{P}(u)$ (except for u) now have a new parent in D, $p_D(u)$. These links can be updated in $O(s_D(v))$ time.

The update also causes changes in the sets $S(z)$ for all z in D_v. For vertices z in D_w (other than w), $S(z)$ gains a new element, the pair $(w, d_T(z) - d_T(w))$. For vertices formerly descendant from u in D, their sets lose the pair containing u. Because updates in the small sets can be done in constant time, all these updates can be performed in $O(s_D(v))$ time.

When an *addleaf*(x, y) operation is performed, this reconstruction is done for every ancestor v of x. When reconstruction is performed at vertex v, $\Omega(s_D(v))$ additional *addleaf* operations must be performed below v before v is again reconstructed. If the cost of a reconstruction is charged to the *addleaf* operations that made it necessary, each *addleaf* operation can be charged at most $O(\log n)$ work. Therefore,

Theorem 8 *A sequence of m queries and n addleaf operations can be performed in $O(m + n \log n)$ time.*

4.3 A Multilevel Incremental Algorithm

We can improve the amortized running time of this algorithm by exploiting a two–level scheme. We employ a scheme similar to Gabow's scheme for incremental nearest common ancestors.

The algorithm maintains a partition of T into subtrees $T_1, \ldots, T_k$, each of size at most $c(n) = \lfloor \frac{1}{2} \log n - \log \log n \rfloor$. A subtree T_i is a *minor* subtree if no vertex in any $T_j (i \neq j)$ is a descendant of $r(T_i)$ in T. Otherwise, the subtree is *major*. It will be the case that there are $O(n / \log n)$ major subtrees.

When an *addleaf*(x, y) operation is performed, the partition is maintained as follows:

1. $V(T) \leftarrow V(T) \cup \{y\}$, and $n \leftarrow n + 1$.

2. Let T_i be the subtree containing x. If $|V(T_i)| \leq c(n)$, add y to T_i. Otherwise, $k \leftarrow k + 1$, and create a new subtree T_k containing the single vertex y.

It is easily seen that there are $O(n / \log n)$ major subtrees. The function $c(i)$ can be computed incrementally. Whenever n is an integer power of 2, we compute $c(1), \ldots, c(2n)$ in $O(n)$ time. This adds $O(1)$ amortized time per update.

Ancestor queries inside subtrees can be handled in $O(1)$ time by the table lookup technique used in the static case. Again, we can compute the tables incrementally, rebuilding the table whenever n becomes an integer power of 2.

We must now show how to do the top level of the algorithm; that is, handle ancestor queries that are not confined to a single subtree. We define a tree $\hat{T}$ as follows. $\hat{T}$ has one vertex for every major subtree of T. $\hat{T}$ is obtained by deleting from T the minor subtrees, then collapsing each major subtree to a single vertex. For each vertex v in T in a major subtree, let $\hat{x}$ be the vertex in $\hat{T}$ into which it is collapsed.

We must solve a slightly generalized version of the problem on $\hat{T}$. This is because the edges $(\hat{x}, p_{\hat{T}}(\hat{x}))$ may correspond to more than one edge in T. Define the length of the edge from $\hat{x}$ to its parent $(\hat{x} \neq r(\hat{T}))$ to be the length of the path in T between the roots of the compressed trees corresponding to $\hat{x}$ and $p_{\hat{T}}(\hat{x})$. This is a number between 1 and $c(n)$. Then, we want to be able to answer generalized ancestor queries on tree $\hat{T}$:

gen-ancestor(x, k) Find the deepest ancestor y of x such that the sums of the lengths of the edges on the path from x to y is at least k.

Two straightforward modifications to the algorithm of section 4.2 will permit us to do this in constant time per query. First, we store, for each node in the compressed tree D, a small priority queue containing pairs in which the total lengths of paths replace differences in depths. Secondly, we must modify the representation of the compressed paths. Previously, the path could be stored in an array, so that the ith vertex in the path could be retrieved in $O(1)$ time. In the generalized problem, however, we would like to do ordered set predecessor operations on prefix sums of the lengths of edges in the path. Fortunately, we can do this in $O(1)$ time per operation.

Lemma 9 *Let $S = \{s_1, \ldots, s_q\}$ be a subset of $\{0, \ldots, n\}$ such that $0 = s_1 < \ldots < s_q$ and $s_{i+1} - s_i$ is $O(\log n)$, $i = 1, \ldots, q - 1$. Given $O(n)$ preprocessing, we can perform insertions, deletions and predecessor operations on S in $O(1)$ time.*

Proof: Divide the elements of S up into buckets. Element s_i is placed into bucket $\lfloor s_i / \log n \rfloor$. Each bucket contains $O(\log n)$ elements, and there are $O(1)$ empty buckets between nonempty buckets. If we represent each bucket with a small priority queue, we can search, insert into and delete from a bucket in $O(1)$ time. If we search for an element s that is smaller than anything currently in its bucket, we spend at most $O(1)$ time finding the previous nonempty bucket. Therefore, searches can also be done in constant time. ∎

So, by representing the prefix sums of distances along the paths with this data structure, we conclude

Theorem 10 *We can perform m generalized ancestor queries and n addleaf operations in $O(m + n \log n)$ time.*

Combining this upper level data structure with the bit string representation of the second level trees, we conclude

Theorem 11 *We can perform m ancestor and n addleaf operations in $O(m + n)$ time.*

4.4 Addroot Operations

We can easily extend the incremental data structure to also handle *addroot(y)* operations (which cause the new node y to become the parent of the previous root of the tree). We briefly outline how this is done. Without *addroot*, stored at each node is the depth of the node. With *addroot*, there is a number $\hat{d}(v)$ at each node such that (1) $-n \leq \hat{d}(v) \leq N$, and (2) $\hat{d}(v) = \hat{d}(p(v)) + 1$. Therefore, $d(v) = \hat{d}(v) - \hat{d}(r(T))$.

The array representing the path containing $r(T)$ must be modified so that new nodes can be inserted at its beginning. This can be done straightfowardly in constant amortized time in linear space.

5 Directions for Future Research

It should be possible to perform general links and answer level-ancestor queries in time $O(m\alpha(m,n)+n)$ time.

The original motivation for studying this problem was the study of fully persistent data structures. It would be interesting to explore the applications of level-ancestors in these data structures. It would also be interesting to eliminate amortization from the data structure.

6 Acknowledgements

I would like to thank Hal Gabow for bringing [5] to my attention, and Baruch Schrieber for bringing [2] to my attention.

References

[1] M. Ajtai, M. Fredman, and J. Komlós. Hash functions for priority queues. *Information and Control*, 63(3):217–225, December 1984.

[2] Omer Berkman and Uzi Vishkin. Recursive *-tree parallel data-structure. In *Proc. 30th Ann. IEEE Symp. on Foundations of Computer Science*, pages 196–202, October 1989.

[3] Omer Berkman and Uzi Vishkin. Finding level–ancestors in trees. Technical Report UMIACS-TR-91-9, Institute for Advanced Computer Studies, U. of Maryland, January 1991.

[4] Michael L. Fredman and Dan E. Willard. Blasting through the information theoretic barrier with fusion trees. In *Proc. 22nd ACM STOC*, pages 1–7, May 1990.

[5] Harold N. Gabow. Data structures for weighted matching and nearest common ancestor. Technical Report CU–CS–478–90, U. of Colorado at Boulder, Department of Computer Science, June 1990. An earlier version was presented at the 1990 Symp. on Discrete Algorithm.

[6] D. Harel and R. E. Tarjan. Fast algorithms for finding nearest common ancestors. *SIAM J. On Computing*, 13(2):338–355, 1984.

[7] Robert E. Tarjan. Applications of path compression on balanced trees. *Journal of the ACM*, 26(4):690–715, Oct. 1979.

TREEWIDTH OF CIRCULAR-ARC GRAPHS+
(ABSTRACT)

Ravi Sundaram, Karan Sher Singh,
and
C. Pandu Rangan

Department of Computer Science and Engineering
Indian Institute of Technology
Madras - 600 036, India

Abstract

The treewidth of a graph is one of the most important graph-theoretic parameters from the algorithm point of view. However, computing treewidth and constructing a corresponding tree-decomposition for a general graph is NP-complete. This paper presents an algorithm for computing the treewidth and constructing a corresponding tree-decomposition for cicular-arc graphs in $O(n^4)$ time.

+ Paper not available at time of printing.

FULLY DYNAMIC DELAUNAY TRIANGULATION IN LOGARITHMIC EXPECTED TIME PER OPERATION[1]

OLIVIER DEVILLERS
INRIA
2004 Route des Lucioles
F-06565 Valbonne cedex
Phone : +33 93 65 77 63
odevil@alcor.inria.fr

STEFAN MEISER
Max Planck Institut für
Informatik
D-6600 Saarbrücken
Phone : +49 681 302-5356
meiser@cs.uni-sb.de

MONIQUE TEILLAUD
INRIA
2004 Route des Lucioles
F-06565 Valbonne cedex
Phone : +33 93 65 77 62
teillaud@alcor.inria.fr

Abstract : The Delaunay Tree is a hierarchical data structure that has been introduced in [4] and analysed in [5,2]. For a given set of sites S in the plane and an order of insertion for these sites, the Delaunay Tree stores all the successive Delaunay triangulations. As proved before, the Delaunay Tree allows the insertion of a site in logarithmic expected time and linear expected space, when the insertion sequence is randomized.

In this paper, we describe an algorithm removing a site from the Delaunay Tree. This can be done in logarithmic expected time, where by expected we mean averaging over all already inserted sites for the choice of the deleted sites. The algorithm has been effectively coded and experimental results are given.

1 Introduction

The Delaunay triangulation and its dual, the Voronoï diagram, are subjects of major interest in Computational Geometry. A lot of algorithms compute it in optimal $\Omega(n \log n)$ time [16,12,11,8,15]. But these algorithms are rather complicated and difficult to implement effectively, so the sub-optimal algorithm [9] is often preferred. Furthermore, this algorithm is on-line and does not impose to compute again the whole triangulation at each insertion.

In the last few years some simpler algorithms have been proposed, non optimal in the worst case but with a good randomized complexity. Some of these algorithms [6,13] use a conflict graph and so are static. The others are on-line ; a first idea of on-line algorithms was presented in [4] and a randomized analysis can be found in [10,2].

Incremental randomized algorithms have also been used for constructing higher order Voronoï diagrams [14,3].

None of the above algorithms allows deletion. Using the algorithm of [1] a site can be removed from a Delaunay triangulation in time sensitive to the modification; their algorithm can however not be combined with an algorithm to insert sites with a good complexity.

---

[1] *This work has been supported in part by the ESPRIT Basic Research Action Nr. 3075 (ALCOM).*

In this paper we propose an extension of the Delaunay Tree [4,5] to allow insertion, deletion or location of a site in the Delaunay triangulation with expected complexities $O(\log n)$ where n is the number of sites actually present. The cost of a deletion can be improved (to $O(\log \log N)$ if the complexity is computed as a function of the total number N of sites. The bounds are randomized, i.e. all possible orders for already inserted sites are supposed to be equally likely, and when a site is deleted, it may be any site with the same probability.

Section 2 explains the principle of the Delaunay Tree, for insertion only, Section 3 defines the problem of deleting a site, Sections 3.2 and 3.3 describe the algorithm, and Section 4 gives the complexity analysis. Finally in Appendix A we present some practical results.

2 The Delaunay Tree

The Delaunay Tree was introduced in [4], studied in a randomized context [5] and also extended to higher order Voronoï diagrams [3] and to various problems (convex hulls, arrangements, Voronoï diagrams of line segments...) [2]. We first recall some basic ideas of this structure, before we take interest in the deletion algorithm. More details can be found in [5].

Let $\mathcal{E}$ be the euclidean plane, and $\mathcal{S}$ a set of n sites such that no four sites are cocircular. The Delaunay triangulation of $\mathcal{S}$ is the unique triangulation such that the circumscribing disk to each triangle does not contain any other site of $\mathcal{S}$. If a site lies inside the circumscribing disk to a triangle, we say that the site is *in conflict* with the triangle. In these terms, the Delaunay triangulation is the set of triangles without conflict. The algorithm described in [4,5] is an on-line algorithm to construct the Delaunay triangulation of $\mathcal{S}$ by adding sites one by one.

The Delaunay Tree is a hierarchical structure based on the incremental procedure of [9]. During the incremental algorithm, each site is introduced one after another and the triangulation is updated after each insertion. Let p be a site to be introduced in the triangulation. All the triangles in conflict with p can no longer be triangles of the triangulation (and are eliminated in the incremental algorithm). The union of these triangles is a star-shaped polygon $R(p)$ with respect to p. Let $F(p)$ denote the set of edges on the boundary of $R(p)$. The new triangles are obtained by linking p to the edges of $F(p)$.

The Delaunay Tree is constructed in a similar way. But, instead of eliminating triangles during the different steps of the construction, we store all the triangles which have been constructed as nodes of the Delaunay Tree, and at each step we define relationships between triangles of the successive Delaunay triangulations. The aim of this structure is to find $R(p)$ efficiently.

For the initialization step we take the first three sites. They generate one finite triangle and three half-planes (infinite triangles). These 4 triangles will be the sons of the root of the tree.

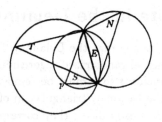

Figure 1: Inserting a new site

2.1 Structure of the Delaunay Tree

After the insertion of site p, the triangles in conflict with p are called *dead* and p is their *killer*. Observe that not every triangle must be incident to an edge belonging to $F(p)$ and thus gives rise to new triangles.

Let T be one of the triangles in conflict with p that has an edge E belonging to $F(p)$.

We construct the new triangle S as having vertex p and edge E. Let N be the triangle sharing edge E with T. Because the triangulation is a Delaunay one, the circumscribing disk of S is included in the union of the two disks circumscribing T and N (see Figure 1).

The newly created triangle S will be called : *son of T* and *stepson of N* through edge E. Notice that T is killed by p and is no longer a triangle of the Delaunay triangulation. We call p the *creator* of S.

If we now insert a new site p' in conflict with S but not with N, S will be killed by p' in turn, and its son S' having vertex p' and edge E will be another stepson of N. Thus a node has at most one son and one list of stepsons through each edge, that is at most three sons and at most three lists of stepsons.

We also maintain adjacency relationships between the triangles of the current triangulation.

This hierarchical structure is called a *Delaunay Tree* for short, but it is more exactly a rooted directed acyclic graph. This graph contains a tree : the tree whose links are the links between fathers and sons.

We will call a *leaf* of the Delaunay Tree a node associated with a triangle of the final triangulation. Such a triangle is not dead, and so a leaf has no son, but possibly stepsons. The other nodes will be called *internal nodes* (an internal node may have no son but the associated triangle is dead).

2.2 Inserting a new site p

Let p be a site to be introduced in the triangulation. If p is in conflict with a triangle T, we know that it is in conflict with the father of T or with its stepfather. So we will be able to find all the triangles which are killed by p by exploring the Delaunay Tree. For each leaf T in conflict with p, we create some sons if necessary : we look at each neighbor N of T ; if N is not in conflict with p, we create a triangle, son of T and stepson of N. Then we create the adjacency relationships between the triangles created by p.

See [4,5] for further details.

3   Deletion of a site in the Delaunay Tree

Let S be a set of n sites. We assume that the Delaunay Tree has been constructed for the set S, by using the incremental randomized algorithm. We now want to remove a site p of S. All the triangles incident to p must be removed from the Delaunay Tree : some of them are triangles of the Delaunay triangulation of S (so they are leaves of the Delaunay Tree), but other ones already died ; they correspond to internal nodes of the Delaunay Tree, and must be removed, too. Moreover, we must restore the Delaunay Tree in the same state it would be in if p had never been inserted, and if the other sites had been inserted in the same order. That way, we preserve the randomized hypothesis on the sequence of sites, and the conditions for further insertions or deletions are fulfilled.

We must thus reconstruct a past for the final triangulation in which p takes no part. The deletion of p creates a "hole" in each successive triangulation after the insertion of p, which the tree keeps a trace of. The idea of our algorithm is to fill each hole with the right Delaunay triangulation.

Let us describe the structure of a node of the Delaunay Tree (some fields of a node have not been used yet and will be defined in the following) :
• the triangle : creator vertex, two other vertices, circumscribed circle
• a mark **dead**
• pointers to the at most three sons and the list of stepsons
• pointers to the father and the stepfather
• the three current neighbors if the triangle is not dead, the three neighbors at the death otherwise
• the three neighbors at the time of the creation
• two special neighbors (defined in Section 3.3)
• a pointer **killer** to the site that killed the triangle
• a mark **to be removed**
• three pointers **star** to elements of structure **Star** (defined in Section 3.3)

3.1   A piece of zoology !

Let us describe all possible situations for a modified node (Figure 2).

Some nodes must be removed : they correspond to triangles having p as a vertex. Depending on its time of creation there are two cases for such a node : either it has been created by the insertion of p, or it has been created by the insertion of some site afterwards ; the latter occurs *iff* its father and stepfather both have p as a vertex and thus both the parents must be removed, too. During the construction of the Delaunay Tree, some sites did not create any triangle to be removed now, but if a site x created such a triangle, it created in fact two triangles to be removed : the two triangles created by x sharing edge px.

Some nodes are *unhooked* by the suppression of p : they are not removed, but they lose just one of their two parents. We must find a new parent in replacement of the removed one.

The sketch of the method is the following :
Search step : Find all nodes of the Delaunay Tree that have to be removed,

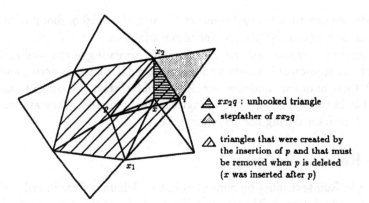

Figure 2: The two kinds of modified nodes : xx_1p and xx_2p where created by the insertion of x and must be removed when p is deleted, xx_2q is unhooked because its father px_2q is removed

and all unhooked nodes
Reinsertion step : Locally reinsert the sites that are creators of the triangles found during the Search step, and update the triangulation

3.2 The Search step

By the discussion above, the set of nodes to be removed can be found by searching the Delaunay Tree starting from the nodes that were created by p. At each node marked to be removed we visit all its sons and stepsons recursively. If one of them has p as a vertex, it will be marked to be removed as well. Otherwise it is an unhooked node. The creator of both these types of triangles must be reinserted, in order to replace the removed triangles by other triangles, and to hang up unhooked triangles again.

In order to be able to perform the Reinsert step, we must store the list of sites to be reinserted :

We need an auxiliary structure, **Reinsert**, which is a balanced binary tree consisting of the set of sites which created the nodes to be removed and the unhooked nodes ; the sites are sorted by order of insertion. This will allow us to reconstruct the triangles which will fill the holes in the successive triangulations, and to hang up again the unhooked nodes.

An element of **Reinsert** contains :
• the site x to be reinserted
• pointers to the two triangles xx_1p and xx_2p that were created by the insertion of x, if they exist (see Section 3.1). xx_1p is turning clockwise
• the list of unhooked triangles that were created by the insertion of x

The search is initialized by the set C of nodes created by p.

To this aim, we must maintain an auxiliary array, **Created**, containing, for each site s of S, a pointer to one of the nodes created by s.

From this node, we can then compute the set C using the neighborhood relations at the time of creation and examining the creator of the triangles.

We then simply recursively traverse the subgraph consisting of removed and unhooked nodes. Each son or stepson of a removed node is removed if it has p as vertex and unhooked otherwise. All these nodes are added in the element of **Reinsert** associated to their creator.

Observe that in the search step the triangles are only visited. They are removed from the Delaunay Tree later during the reinsertion step.

3.3 The Reinsertion step

The sites lying in **Reinsert** must be reinserted in the Delaunay Tree in order to construct the successive triangulations without site p. The scheme of the reinsertion of a site x is the same as the usual scheme of insertion, except that everything happens locally : the location of a site x to be reinserted in the whole Delaunay Tree is unnecessary and would be too expensive.

The location in a generally small set A (for active) of triangles is sufficient. At the beginning of the reinsertion set A is initialized with all triangles killed by p. They can be computed by looking at the fathers of the triangles in C and following their neighbor pointers at their death.

Then, during the Reinsertion step, A is maintained so that it is the set of triangles in conflict with p in the Delaunay triangulation at the time just preceding the reinsertion of x. In each step, A is modified as follows : all the triangles of A in conflict with x are killed by x and thus disappear from the Delaunay triangulation and from A, and the triangles created by the reinsertion of x appear in A. The triangles of A not in conflict with x still remain in A. The triangles outside A are not modified by a reinsertion (except for some neighborhood or stepson relations) since they are not in conflict with p.

We can notice that the triangles in A form a star-shaped polygon with respect to p, since they are in conflict with p, cf. the discussion in Section 2.

> It is sufficient to store only the edges of the boundary of A in a structure called **Star**, which is a circular list. **Star** is oriented counterclockwise.
>
> We can store, for each vertex of A, a pointer to the element of **Star** corresponding to the edge following it on the boundary of A.
>
> An element of **Star** consists of :
> - an edge of the boundary of A
> - a pointer to the triangle of A incident to this edge
> - two pointers to the two adjacent edges of the boundary of A
>
> Reciprocally, for a triangle of A having some edges on the boundary of A, we store in the node of the Delaunay Tree some pointers **star** linking the triangle to the corresponding elements of **Star**.

Some elements of A are not represented in **Star**, but the whole set A can nevertheless be traversed using **Star** and pointers to neighbors.

Star can be initialized by first choosing a vertex of a triangle in A. We then follow the boundary of the star-shaped polygon using the neighborhood relations, and the pointers **star**.

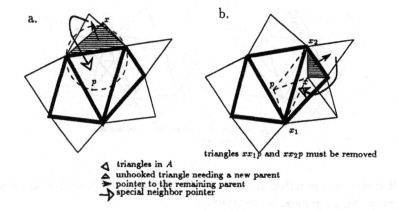

triangles xx_1p and xx_2p must be removed

◁ triangles in A
△ unhooked triangle needing a new parent
➤ pointer to the remaining parent
⇀ special neighbor pointer

Figure 3: An unhooked triangle (a) without or (b) with some removed triangles (x must be reinserted)

We know the current neighbors of each triangle of A. Each edge e of **Star** is an edge of a triangle U of A and of a triangle V that does not belong to A.

> The current neighbor of U through e is V, but the reciprocal relation does not always exist ; the neighbor pointer of V through e may reach another triangle W created a long time later.

> If W must not be removed, this pointer must remain after the deletion of p. So we do not want to systematically modify the current neighbors of triangles not belonging to A. We need to put a special neighbor pointer from V to U. Thus each triangle outside A, having an edge on the boundary of A, has a special neighbor pointer to the adjacent triangle in A. It is easy to see that two special neighbor pointers are enough : at most two edges of a given triangle lie on the boundary of a star-shaped polygon, if it is exterior to it. The special neighbor pointers store intermediate relations between triangles.

Everything is now set up to start the reinsertion. Each site in Structure **Reinsert** is reinserted in the right order (the order used for first insertion).

Processing the unhooked triangles
Each element of **Reinsert** contains a site x to be reinserted, and the list of corresponding unhooked triangles. To hang up such a triangle T again, we only have to go to the remaining parent of it, which must have an edge in **Star**, and then hang T up to the appropriate special neighbor of this parent. There may also exist some removed triangles created by x (Figure 3b). Notice that this is not always true (Figure 3a). If there is no removed triangle, the unhooked triangle necessarily needs a stepfather.

Replacing the triangles to be removed by new ones
For each element n of **Reinsert**, we check if triangle xx_1p (and xx_2p) exists. If xx_1p and xx_2p do not exist in the triangulation, then nothing has to be done. Otherwise we have to fill the gap of triangles incident at x between edges xx_1 and xx_2. We must look at **Star**

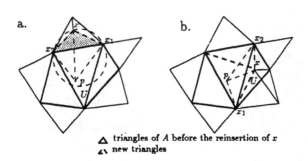

△ triangles of A before the reinsertion of x
↙ new triangles

Figure 4: Reinsertion - removed triangles - (a) First and (b) Second case (x must be reinserted, xx_1p and xx_2p must be removed)

in order to find the triangles that have to be created by the reinsertion of x. There are two cases : there may exist no triangle of A in conflict with x, or several such triangles. Let U be the triangle adjacent to x_1, and following x_1 in **Star** (remember that **Star** is oriented counterclockwise and xx_1p clockwise).

After the reinsertion of x, the edges xx_1 and xx_2 will be on the boundary of the new set A of triangles in conflict with p. So, if there are some vertices on the current boundary of A, between x_1 and x_2, the triangles adjacent to the edges of this chain of vertices must be in conflict with x. U is such a particular triangle. Thus, if U is not in conflict with x, x_1x_2 is an edge on the boundary of A, and consequently of U, and the first case occurs, otherwise the second case occurs. Pseudo-code procedures corresponding to the following algorithms can be found in [7].

First case, see Figure 4a :
In this case, the only way to fill the gap is to replace the removed triangles xx_1p and xx_2p around x by only one new triangle xx_1x_2. The new triangle xx_1x_2 has U as stepfather and the neighbor of U, which does not belong to A, as father.

Second case, see Figure 4b :
We know that U is in conflict with x. We must find all the triangles in A in conflict with x. Those triangles may be fathers for the nodes that will be created by x. They form a connected subset of A, so they will be found owing to neighbor pointers in the following way :

Starting with U we visit the triangles in A incident at x_1 in counterclockwise order until we reach a triangle not in conflict with x. Let V denote the last such triangle in conflict with x and let ϵ denote the edge of V at which the visit stops. Let $e = x_1x'$. We create triangle $W' = xx_1x'$ and start this process again at vertex x' with V as starting triangle. When vertex x_2 is reached, all the new triangles have been created. The iteration continues until vertex x_1 is reached in order to mark the triangles killed by x.

During the traversal, all relevant pointers (neighbor and star pointers) must be updated. Once this is achieved, it remains to compute all kinds of neighborhood relations involving edges xx_1 and xx_2. Particularly, as the current neighbor of the just created triangle having edge xx_1 we take the neighbor of the now removed triangle xx_1p at its creation. The same holds for edge xx_2.

4 Analysis

4.1 Randomized analysis of the insertion algorithm

This subsection aims at providing a randomized analysis of the space and time required to build the Delaunay Tree. Randomization here only concerns the order in which the inserted sites are introduced into the structure. Thus, if the current set of sites is a set S of cardinality n, our results are expected values that correspond to averaging over the $n!$ possible permutations of the inserted objects, each equally likely to occur.

The following lemmas are based on a result due to Clarkson and Shor [6]. They bound the number of triangles defined by three sites having at most j conflicts. The proof of this lemma uses the random sampling technique. Lemmas 4.1 and 4.2 are proved in [5,2] and in the [7].

Lemma 4.1 *If S has cardinality n, the expected size of the Delaunay Tree of S is $O(n)$.*

Lemma 4.2 *If S has cardinality n, the expected time for inserting the last site in the Delaunay Tree is $O(\log n)$.*

4.2 Randomized analysis of deleting a point

We assume that p is a random site in S, i.e. p is any of the precedingly inserted sites, with the same probability.

Lemma 4.3 *The expected number of removed nodes is constant.*

Proof : The expected number of removed nodes is

$$\sum_{T \text{ triangle}} Prob(p \text{ vertex of } T) Prob(T \text{ exists in the Delaunay Tree})$$
$$= \frac{3}{n}. \text{ expected size of the Delaunay Tree}$$
$$= O(1)$$

$\square$

Lemma 4.4 *The expected number of unhooked nodes is constant.*

Proof : The analogous computation using edges of the Delaunay Tree instead of nodes yields the result. $\square$

Lemma 4.5 *The expected number of nodes created by the deletion of p is constant.*

Proof : The number of created triangles during the deletion of p is

$$\sum_{T \text{ triangle}} Prob(T \text{ appears during the deletion of } p)$$

A triangle T of width j will appear *iff* p is one of the j sites in conflict with T, and p and the 3 vertices of T are introduced before the $j - 1$ other sites in conflict with T, and p is not inserted after the 3 vertices of T. So the probability that T appears is

$$\frac{j}{n} 3 \frac{3!(j-1)!}{(j+3)!} = \frac{3}{n} \frac{3!j!}{(j+3)!}$$

and it is, up to a factor $\frac{3}{n}$, the probability that T appears as a node of the Delaunay Tree during the usual insertion process. We can then deduce that the expected number of created triangles is constant. □

Lemma 4.6 *The expected cost of a deletion is $O(\log n)$.*

Proof : The expected number of triangles killed by p is constant using Lemma 4.3 (which also implies that the initialization of **Star** is achieved in constant time), and the traversal that is done during the Search step visits a constant number of nodes by Lemma 4.4. For each node, we must locate the creator of the node in **Reinsert**, which can be done in $O(\log n)$ worst case deterministic time, by using a balanced binary search tree.

The total cost of the work on unhooked triangles is constant, since we only have to reach the neighbor of the parent of each of them, and by Lemma 4.4.

For the triangles deleted by the reinsertion of x, the cost is linear in the number of triangles in conflict with both p and x, which is linear in the number of triangles created by the reinsertion of x. By Lemma 4.5, this expected cost is thus constant. There are a constant number of sites to reinsert (by Lemma 4.4).

The expected whole cost is then less than $O(\log n)$. □

> *Remark 4.7* The expensive part of this cost is the location in **Reinsert**. It is also possible to use for **Reinsert** a Van Emde Boas priority queue, in this case the complexity is $O(\log \log N)$ where N is the total number of insertions to be performed (even if the number of points present at the current step is smaller).

Lemmas 4.1, 4.2 and 4.6 yield the main theorem of this paper :

Theorem 4.8 *The Delaunay triangulation (or the Voronoï diagram) of a set S of n sites in the plane can be dynamically maintained in $O(\log n)$ expected time to insert, delete or locate a point. This result holds provided that, at any time, the order of insertion on the sites remaining in S may be each order with the same probability, and when a site is deleted, it may be any site with the same probability.*

5 Conclusion

We have shown that the Delaunay triangulation can be maintained in $O(\log n)$ expected time per insertion or deletion and $O(n)$ space (where n is the number of sites at the time of the operation).

The analysis is randomized, i.e. the result holds provided that at any time the order of insertion of the sites in the triangulation at that moment may be any possible order with the same probability. And when a site is deleted, it may be any site in the triangulation with the same probability. An important point is that our hypotheses are on the insertion order only ; there are no assumptions on the distribution of the sites.

This algorithm is practical and has been effectively coded (see Appendix A), the numerical computations involved are simple ; the only numerical calculation is to test if a site lies inside or outside a circle. The data structures involved by the algorithm are not too complicated ; besides the Delaunay Tree itself, we only need a balanced binary search tree.

This algorithm can be generalized to higher dimensions, provided that we maintain the adjacency relationships between faces of any dimension in the Delaunay triangulation, which is necessary to compute A at each step of the deletion algorithm.

Further investigations are to be done. In the same way as the insertion aspect of the Delaunay Tree [5] was generalized to various geometric problems [2], the deletion aspect can be generalized to other problems.

References

[1] A. Aggarwal, L. Guibas, J. Saxe, and P. Shor. A linear time algorithm for computing the Voronoï diagram of a convex polygon. *Discr. and Comp. Geom.*, 4:591–604, 1989.

[2] J. Boissonnat, O. Devillers, R. Schott, M. Teillaud, and M. Yvinec. Applications of random sampling to on-line algorithms in computational geometry. *Discr. and Comp. Geom.* To be published. Full paper available as Technical Report INRIA 1285. Abstract published in IMACS 91 in Dublin.

[3] J. Boissonnat, O. Devillers, and M. Teillaud. A semi-dynamic construction of higher order Voronoï diagrams and its randomized analysis. *Algorithmica.* To be published. Full paper available as Technical Report INRIA 1207. Abstract published in Second Canadian Conference on Computational Geometry 1990 in Ottawa.

[4] J. Boissonnat and M. Teillaud. A hierarchical representation of objects: the Delaunay Tree. In *ACM Symp. on Comp. Geom.*, Jun. 1986.

[5] J. Boissonnat and M. Teillaud. On the randomized construction of the Delaunay tree. *Theor. Comp. Sc.* To be published. Full paper available as Technical Report INRIA 1140.

[6] K. Clarkson and P. Shor. Applications of random sampling in computational geometry, II. *Discr. and Comp. Geom.*, 4(5), 1989.

[7] O. Devillers, S. Meiser, and M. Teillaud. *Fully dynamic Delaunay triangulation in logarithmic expected time per operation.* Technical Report 1349, INRIA, Dec. 1990.

[8] S. Fortune. A sweepline algorithm for Voronoï diagrams. *Algorithmica*, 2:153–174, 1987.

[9] P. Green and R. Sibson. Computing Dirichlet tesselations in the plane. *The Computer J.*, 21, 1978.

[10] L. Guibas, D. Knuth, and M. Sharir. Randomized incremental construction of Delaunay and Voronoï diagrams. In *ICALP 90*, pages 414–431, Springer-Verlag, Jul. 1990.

[11] L. Guibas and J. Stolfi. Primitives for the manipulation of general subdivisions and the computation of Voronoï diagrams. *ACM Trans. on Graphics*, 4(2):74–123, Apr. 1985.

[12] D. Lee and B. Schacter. Two algorithms for constructing a Delaunay triangulation. *International Journal of Computer and Information Sciences*, 9(3), 1980.

[13] K. Mehlhorn, S. Meiser, and C. Ó'Dúnlaing. On the construction of abstract Voronoï diagrams. *Discr. and Comp. Geom.*, 6:211–224, 1991.

[14] K. Mulmuley. On levels in arrangements and Voronoï diagrams. *Discr. and Comp. Geom.* To be published.

[15] F. Preparata and M. Shamos. *Computational Geometry : an Introduction.* Springer-Verlag, 1985.

[16] M. Shamos. *Computational Geometry.* PhD thesis, Department of Computer Science, Yale University, (USA), 1978.

A   Practical Results

The algorithm described in this paper has been effectively coded and was tested on various samples of data. We present here the result for a set of 15000 random sites in a square. The sites are first inserted in a random order, and afterwards they are all deleted in another random order. The graphic below shows the size of the Delaunay Tree in bold line, the size of the Delaunay triangulation in dotted line, and in thin line, a measure of the complexity of the operation. For insertions, it is the number of visited nodes, for deletions it is the number of unhooked triangles plus the numbers of triangles created during this deletion. The cost of deleting a site has a higher variance than the cost of inserting a site ; it may be important if the site had been inserted at the beginning of the construction, but this happens with a low probability.

The Delaunay triangulation of 15000 sites has been computed in 30s on a Sun 4/75 and the deletion phase has been computed in 60 s.

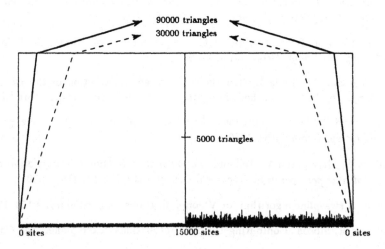

ON COMPUTING THE VORONOI DIAGRAM

FOR RESTRICTED PLANAR FIGURES

Hristo Djidjev
Center of Informatics & Comp. Technology
Bulgarian Academy of Sciences
Acad. G. Bonchev str. Bl 25-A
Sofia, Bulgaria

Andrzej Lingas
Department of Computer Science
Lund University
P.O.Box 118, 22100 Lund, Sweden

Abstract: We show that the Voronoi diagram of a finite sequence of points in the plane which gives sorted order of the points with respect to two perpendicular directions can be computed in linear time. In contrast, we observe that the problem of computing the Voronoi diagram of a finite sequence of points in the plane which gives the sorted order of the points with respect to a single direction requires $\Omega(n \log n)$ operations in the algebraic decision tree model. As a corollary from the first result, we show that the bounded Voronoi diagrams of simple n-vertex polygons which can be efficiently cut into the so called monotone histograms can be computed in $o(n \log n)$ time.

1. Introduction

Consider a set S of n points (*sites*), $S=\{s_1,...,s_n\}$, in the Euclidean plane. For $i=1,...,n$ let $R(s_i)$ be the region of all points p in the plane which are closer to s_i than to any other site in S. The region $R(s_i)$ can be seen as the common intersection of the $n-1$ half-planes induced by the perpendicular bisectors of the segment with endpoints s_i and s_j, $j=1,...,n$, $j \neq i$, that contain s_i. For this reason, the regions $R(s_i)$ are convex. They form a partition of the plane called the *Voronoi diagram* of S [PS] (*Vor(S)* for short). The maximal straight-line segments or half-lines on the boundaries of the regions in *Vor(S)* are called *edges* of *Vor(S)*. The endpoints of the edges are called *vertices* of *Vor(S)*. The straight-line dual of *Vor(S)* is called the *Delaunay triangulation* of S.

The problem of computing Voronoi diagrams and Delaunay triangulations and their diverse generalizations is central in computational geometry because of the variety of applications. There are few known algorithms for computing *Vor(S)* which run in time $O(n \log n)$ [F,PS]. They are time-optimal since for instance in the algebraic decision tree model the problem of sorting n real numbers easily reduces to the problem of computing *Vor(S)* [PS]. However, it might be possible to use special properties of a restricted class of point sets in order to obtain $o(n \log n)$-time algorithms. The problem of finding such classes class of point sets in order to obtain $o(n \log n)$-time algorithms. The problem of finding such classes seems to be very hard. The only previously known result is for a point set S, where the sites in S are given as vertices of a convex polygon in clockwise or counterclockwise order. Aggarwal, Guibas, Saxe and Shor have shown that $Vor(S)$ can be computed in linear time in this case [AGSS]. Actually, they proved a more general theorem about computing the convex hull of certain polygons in the three-dimensional Euclidean space in linear time (see Section 3 in [AGSS]). By the known relationship between the problem of computing $Vor(S)$ and the problem of computing the convex hull of projections of S on the paraboloid $U=\{(x,y,x^2+y^2) \mid (x,y) \in E^2\}$ (see [E,GS]), they could conclude that the problem of computing the Voronoi diagram of the vertices of a convex polygon is solvable in linear time. They also showed a bunch of other problems in the plane to be solvable in linear time, using their general theorem mentioned in the above (see Section 4 in [AGSS]).

In this paper, we derive a highly non-trivial consequence of the general theorem of Aggarwal *et al.* stating that if the sequence of sites on the plane is given in sorted order with respect to two perpendicular directions then their Voronoi diagram can be computed in linear time. Let a *histogram* (a *monotone histogram*) respectively denote a simple polygon $(h_1,...,h_n,h_{n+1})$ such that the subsequence $(h_1,...,h_n)$ is sorted with respect to a direction (or to two perpendicular directions, respectively). We can also rephrase our result as follows: the Voronoi diagram of the set of vertices of a monotone histogram given in clockwise or counterclockwise order along its perimeter can be computed in linear time. Note that a monotone histogram may have up to a linear number of reflex angles, and on the other hand any convex polygon can be trivially cut into four monotone histograms. Thus our result for monotone histograms is more general that the statement that the Voronoi diagram of convex polygon can be computed in linear time (as it implies the statement). In contrast, we observe that the problem of computing the Voronoi diagram of the set of vertices of a histogram requires $\Omega(n \log n)$ operations in the algebraic decision tree model (see [PS]).

In [AGSS], the authors also mention the open problem of computing the so called *generalized Delaunay triangulation* of a simple polygon in time $o(n \log n)$. The generalized Delaunay triangulation of a simple polygon can be defined as the straight-line dual of the so called *bounded Voronoi diagram* of a simple polygon [WS,L1]. In analogy to the concept of the standard Voronoi diagram, the latter is defined as follows. Let P be a simple polygon. The bounded Voronoi diagram of P, $Vorb(P)$ for short, is the partition of the interior of P into regions $BR(v)$, where v ranges over vertices of P, such that a point inside P belongs to $BR(v)$ if and only if (p,v) is the shortest straight-line segment that connects p with a vertex of P without crossing any edge of P. The edges of $Vorb(P)$ are maximal straight-line segments on the boundaries of its regions that do not overlap with the perimeter of P, and the vertices of $Vorb(P)$ are the endpoints of the edges. It is not difficult to see that if P is a convex polygon or a monotone histogram then $Vorb(P)$ is identical to the intersection of the standard Voronoi diagram of the vertices of P with the inside of P. This is not necessarily true for more complicated simple polygons like histograms, monotone polygons or star-shaped polygons (see [PS]). It is an open problem whether $Vorb(P)$ could be build in substantially less than $n \log n$ time for such polygons P. In this paper, we show that any simple polygon that can be efficiently cut into a small number of monotone histograms admits such a time-efficient construction of the bounded Voronoi diagram.

The paper is organized as follows. In Section 2 we formulate two previously known facts that we use to obtain our results. In Section 3 we develop a linear time algorithm for constructing Voronoi diagrams of monotone histograms. In Section 4 we provide an $\Omega(n \log n)$ lower bound for the problem of constructing Voronoi diagrams of general (non-monotone) histograms thus showing that our results can not be generalized to arbitrary histograms. Finally, in Section 5, we consider an application of our algorithm from Section 3.

2. Preliminaries

Guibas and Stolfi defined the so called *lifting mapping* μ of the Oxy plane into E^3 by $\mu(x,y) = (x,y,x^2+y^2)$ [GS]. The paraboloid which is the image of the whole plane under the mapping μ will be denoted by U.

Fact 2.1(see [E]): Let S be a finite set of points in the plane. There is a one-to-one correspondence between the edges of the lower part of the convex hull of $\mu(S)$ and the edges of the Delaunay triangulation of S. Given the convex hull of $\mu(S)$, we can compute the Delaunay triangulation of S in linear time.

Aggarwal *et al.* used the above fact to derive their linear upper bound on the time needed to construct the Voronoi diagram of a convex polygon by proving the following statement.

Fact 2.2 [AGSS]: Let P be a polygon $(p_1,...,p_n)$ in E^3. Suppose that for each edge of any subpolygon P' of P given by a subsequence of $(p_1,...,p_n)$ there exists a plane that contains the edge and leaves all other vertices of P' in the same half-space. Then the convex hull of the vertices of P can be constructed in time $O(n)$.

3. The main result

Recall from Introduction that a *histogram* is a simple polygon $H=(h_1,...,h_n,h_{n+1})$ $(n>1)$ such that there is a direction in which the subsequence $(h_1,...,h_n)$ is sorted. If there are two perpendicular directions in which the subsequence $(h_1,...,h_n)$ is sorted then H is a monotone histogram. See Fig. 3.1. (Note that our definition of a histogram and a monotone histogram is slightly more general than the usual one where one would require the distinguished directions to be perpendicular to the edge (h_1,h_{n+1}) or to the edges (h_1,h_{n+1}) and (h_n,h_{n+1}) respectively).

To simplify the exposition we assume throughout the paper that no three points from H are colinear. We also assume that the sequence of vertices of monotone histograms form *increasing* sequences with respect to the two perpendicular directions.

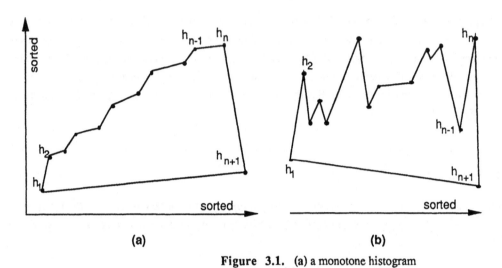

Figure 3.1. (a) a monotone histogram
(b) a non-monotone histogram

Theorem 3.1: Let H be a monotone histogram. The Voronoi diagram of any k-vertex subsequence H' of H can be computed in time $O(k)$.

Proof: Let $H=(h_1,...,h_n,h_{n+1})$ and $H'=(h'_1,...,h'_k)$. We may assume without loss of generality that h_{n+1} is not in H' since we can always update in linear time the Voronoi diagram after adding a single site [PS]. Consider the projection $\mu(H')=(\mu(h'_1),...,\mu(h'_k))$ of H on the paraboloid U. By Facts 2.1, 2.2, it is sufficient to prove that for each edge of $\mu(H')$ there is a plane that includes the edge and leaves all other vertices of H' in the same half-space.

The above property is easily seen to hold for the edges $(\mu(h'_i),\mu(h'_{i+1}))$, where $i=1,...,k-1$, by the relationship between the convex hull of $\mu(H')$ and the Delaunay triangulation of the vertices of H' following from Fact 2.1. To prove this consider a point p on (h'_i,h'_{i+1}). Since H' is a subsequence of a monotone histogram then all vertices of H' below h'_i have a larger distance to p than h'_i has. Similarly, all vertices of H' above h'_{i+1} have a larger distance to p than h'_{i+1} has. It follows that the edge (h'_i,h'_{i+1}) has to intersect an edge of the Voronoi diagram of the vertices of H' separating the region of h'_i from the region of h'_{i+1}. Consequently, (h'_i,h'_{i+1}) is an edge of the Delaunay triangulation of the vertices of H' which implies that $(\mu(h'_i),\mu(h'_{i+1}))$ is an edge of the convex hull of $\mu(H')$ by Fact 2.1. This implies the existence of a plane satisfying the required condition.

The hard part of the proof is to show that there exists a plane that includes $(\mu(h'_1),\mu(h'_k))$ and leaves all other vertices of H' in the same half-plane. This follows from the lemma below. Q.E.D.

Lemma 3.1: Let $h'_1,h'_2,...,h'_k$ be a sequence of points in the plane which is sorted in increasing order with respect to two perpendicular directions. Then there exists a plane Π containing the segment with endpoints $\mu(h'_1)$ and $\mu(h'_k)$ such that all points $\mu(h'_i)$, $i=2,...,k-1$, belong to the same halfspace determined by Π.

Proof: We will prove the lemma by explicitly constructing Π. Choose a coordinate system in E^3 such that the points of H' are in the Oxy plane and the two perpendicular directions are axes Ox and Oy. For any $J:E^3 \rightarrow \{true, false\}$ by $Q\{J(x,y,z)\}$ we shall denote the set of points of E^3 for which $J(x,y,z)=true$. For instance, $Q\{x>x^*\}$ and $Q\{x<x^*\}$ will denote the two halfspaces determined by the plane $x=x^*$. Let $h'_i=(x_i,y_i)$, $i=1,...,k$, and let D denote the part of the paraboloid $z=x^2+y^2$ that is contained in the region $Q\{(x_1<x<x_k) \wedge (y_1<y<y_k)\} = Q\{x>x_1\} \cap Q\{x<x_k\} \cap Q\{y>y_1\} \cap Q\{y<y_k\}$. It is obvious that $\mu(h'_i) \in D$, $i=2,...,k-1$, and therefore it will be enough to define Π so that all points of D belong to the same half space determined by Π.

The intersection of D with any plane Π_σ perpendicular to Oxy and incident to the point $\mu(h'_1)=(x_1,y_1,x_1^2+y_1^2)$, where $\sigma \in [0,\pi/2]$ is the angle between Ox and Π_σ, is a portion of a parabola in Π_σ with endpoints $(x_1,y_1,x_1^2+y_1^2)$ and some point on $Q\{x=x_k\}$ or $Q\{y=y_k\}$ depending on σ. The endpoints of these parabolas are $\mu(h'_k)$ and $a=(x_k,y_1,x_k^2+y_1^2)$, in $Q\{x=x_k\}$, and $\mu(h'_k)$ and $b=(x_1,y_k,x_1^2+y_k^2)$, in $Q\{y=y_k\}$. From the convexity of function $f(x)=x^2$ in E^2 all points of $D\cap Q\{x=x_k\}$ are below the segment joining a and $\mu(h'_k)$ and all points of $D\cap Q\{y=y_k\}$ are below the segment joining b and $\mu(h'_k)$. Then also in Π_σ any point of $D \cap \Pi_\sigma$ lies below the segment s_σ joining $\mu(h'_1)$ with the intersection point between Π_σ and $(a,\mu(h'_k))$ for $\sigma \in [0,\text{arctg}(y_k/x_k)]$, or $(b,\mu(h'_k))$ for $\sigma \in [\text{arctg}(y_k/x_k),\pi/2]$. We will complete the proof by showing that points $\mu(h'_1)$, $\mu(h'_k)$, a, and b are coplanar and all segments s_σ belong to the plane, Π, defined by $\mu(h'_1)$, $\mu(h'_k)$, a, and b (notice that $(\mu(h'_1),\mu(h'_k)) = s_\sigma$ if $\sigma=\text{arctg}(y_k/x_k)$). As

$$\mu(h'_1)=(x_1,y_1,x_1^2+y_1^2),$$

$$\mu(h'_k)=(x_k,y_k,x_k^2+y_k^2),$$

$$a=(x_k,y_1,x_k^2+y_1^2),$$

$$b=(x_1,y_k,x_1^2+y_k^2),$$

then the segments $(\mu(h'_1),\mu(h'_k))$ and (a,b) intersect at point $(\frac{x_1+x_k}{2},\frac{y_1+y_k}{2},\frac{x_1^2+y_1^2+x_k^2+y_k^2}{2})$,

whence the claim follows. Q.E.D.

Remark 3.1: Lemma 3.1 can also be proved by purely algebraic means by showing that the determinant of the matrix

$$C = \left\{ \begin{matrix} x_1 & y_1 & x_1^2+y_1^2 & 1 \\ x_1 & y_k & x_1^2+y_k & 1 \\ x_k & y_1 & x_k^2+y_1^2 & 1 \\ x & y & x^2+y^2 & 1 \end{matrix} \right\}$$

is negative for $x_1<x<x_k$ and $y_1<y<y_k$. By elementary linear algebra computations we find

$$\det(C)=(x_k-x_1)(y_k-y_1)[(x-x_1)(x-x_k)+(y-y_1)(y-y_k)],$$

whence $\det(C)<0$ for $x_1<x<x_k$ and $y_1<y<y_k$. Actually, an analogous determinant is examined in the proof of a known fact implying that the image of any circle under the lifting mapping μ is contained in a plane such that the two resulting halfspaces respectively contain the images of the interior and the exterior of the circle under μ (see Lemma 8.1 in [GS]). As it was pointed by an unknown referee, we can also reduce the proof of Lemma 3.1 to the above fact by constructing the circle that passes through (x_1,y_1) and (x_k,y_k) and encircles all the points (x,y) where $x_1<x<x_k$ and $y_1<y<y_k$.

Remark 3.2: If a histogram $H=(h_1,...,h_n,h_{n+1})$ is monotone, the corresponding two perpendicular directions can be easily found in $O(n)$ time (if not given) by considering the set of the angles between some fixed half-line and the half-lines determined by (h_i,h_{i+1}), $i=1,...,n-1$.

4. The lower bound

To derive an $\Omega(n \log n)$ lower bound on the problem of computing the Voronoi diagram of vertices of a histogram in the algebraic decision tree model, we consider the following auxiliary decision problem:

Given a sequence $x_1,..,x_n$ of n reals, decide whether the smallest difference between any two elements in the sequence is smaller than ε, where ε is any positive constant.

The above problem is called ε-*closeness problem* in [PS], where also the following claim is stated.

Fact 4.1: For any $\varepsilon>0$ the ε-closeness problem requires $\Omega(n \log n)$ tests in the algebraic decision tree model.

On the other hand, we have the following lemma.

Lemma 4.1: The problem of 1-closeness can be reduced to the problem of computing the Voronoi diagram of vertices of a histogram using a linear number of operations.

Proof: Let $x_1, x_2,...,x_n$ be a sequence of n reals. We may assume without loss of generality that the reals are greater than $1/2$. Form the histogram $H = ((0,0),(1/(3n),x_1),(2/(3n),x_2),..., ((n/(3n),x_n),(1/2,0))$ (see Figure 4.1). Compute $Vor(H)$ and its intersection with the Oy axis. Test whether all intervals $R(x_i) \cap Oy$, $i=1,...,n$, are not empty and appear in a sorted order with respect to Oy (it can easily be done using a linear number of tests). If so, we can trivially find the minimum difference between any two x_i's using a linear number of tests, and compare it with 1. Otherwise, there exists j, $1 \leq j \leq n$, such that the projection of the site $(j/(3n),x_j)$ on the Oy axis, i.e. $(0,x_j)$, belongs to the region of another site. The other site can be neither $(0,0)$ nor $(1/2,0)$ since $x_j>1/2$ and the site $(j/(3n),x_j)$ is in the distance at most $1/3$ from $(0,x_j)$. It follows from the construction of H that there is a site $(k/(3n),x_k),1 \leq k \leq n$, $k \neq j$, which is at distance at most $1/3$ from $(0,x_j)$. This immediately implies that $(0,x_k)$ is at distance smaller than $1/3$ from $(0,x_j)$. Thus, the minimum difference is definitely smaller than 1 in this case. Q.E.D.

Note that the problem of 1-closeness is actually reduced to the problem of computing the Voronoi diagram of vertices of the histogram H outside H in the proof of Lemma 4.1. By moving H slightly to the left such that the topmost vertex is placed on the Oy axis, we can obtain an analogous reduction of the problem of 1-closeness to the problem of computing the Voronoi diagram of vertices of H inside H using a linear number of operations.

Combining Fact 4.1 with Lemma 4.1, we obtain our lower bound.

Theorem 4.1: The problem of computing the Voronoi diagram or Delaunay triangulation of the vertices of an n-vertex histogram requires $\Omega(n \log n)$ operations in the algebraic decision tree model.

Remember that the above lower bound does not give any evidence that the bounded Voronoi diagrams of histograms or even of more general simple polygons cannot be computed in time $o(n \log n)$.

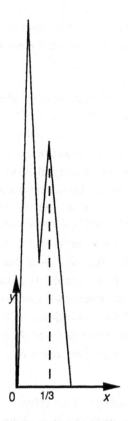

Figure 4.1. The histogram H from the proof of Lemma 4.1

By similar arguments we can also derive the following theorem.

Theorem 4.2: The problem of computing the Voronoi diagram or Delaunay triangulation of the vertices of a star-shaped n-vertex polygon requires $\Omega(n \log n)$ operations in the algebraic decision tree model.

5. Computing bounded Voronoi diagrams for certain polygons

By a *chord* of a simple polygon P, we shall mean a straight-line segment s such that the endpoints of s lie on the perimeter of P and the inside of s lies inside P. Note that the endpoints of a chord of P are not necessarily vertices of P. In order to compute the bounded Voronoi diagram of a simple polygon (*Vorb(P)*, see Introduction), we shall recursively split it into monotone histograms along chords. Then, we shall compute the bounded Voronoi diagrams of the monotone histograms and recursively merge them employing the following lemma.

Lemma 5.1: Let P be a simple polygon, and let s be a chord of P. Let P_1 and P_2 be the two polygons resulting from splitting P along s. Given the bounded Voronoi diagrams of P_1 and P_2, the bounded Voronoi diagram of P can be computed in linear time.

Proof: Recall that the generalized Delaunay triangulation of a simple polygon is the straight-line dual of the bounded Voronoi diagram of the polygon (see [WS,L1]). In [LL], Lee and Lin showed how to merge the generalized Delaunay triangulations of two polygons, resulting from splitting a larger polygon along a diagonal in linear time (this enabled them to derive an $O(n \log n)$-time divide-and-conquer algorithm for the generalized Delaunay triangulation of a simple polygon). As we can compute the generalized Delaunay triangulation from the bounded Voronoi diagram and *vice versa* in linear time, Lee-Lin's result on merging implies the lemma in case the endpoints of s are vertices of P. Otherwise, after merging the bounded Voronoi diagrams of P_1 and P_2 into the bounded Voronoi diagram M of P with an endpoint (or both endpoints, respectively) of s added to the set of vertices of P, we have to divide the region of each added vertex between the adjacent regions of vertices of P. Note that if we consider the standard Voronoi diagram of the vertices defining the adjacent regions and the added vertex (added vertices, respectively) then the boundaries of the region of any added vertex will overlap with those from M. For this reason, via Fact 2.1, the projections of the vertices defining the adjacent regions on the paraboloid U lie on a convex cone (whose apex is the projection of the added vertex on U). Hence, by Fact 2.2, we can update the diagram M after deleting the added vertex (vertices, respectively) in linear time (see also Section 4 in [AGSS]). Q.E.D.

To formalize the notion of recursively cutting a simple polygon into smaller polygons along chords we need the following definition. By a *full binary tree* we mean a binary tree with any non-leaf node of degree two.

Definition 5.1: Let P be a simple polygon. A full binary tree T with each node v labelled with a simple polygon $P(v)$ and each non-leaf node additionally labelled with a chord $s(v)$ of $P(v)$ is said to *cut* P into the set of polygons assigned to its leaves if it satisfies the following conditions:

(1) if v is the root of T then $P(v)=P$;

(2) if a node v has sons w, u in T then the simple polygons $P(w)$ and $P(u)$ can be obtained by splitting $P(v)$ along $s(v)$.

Theorem 5.1: Let T be a full binary tree of height d which cuts a simple n-vertex polygon P into monotone histograms each containing at least one vertex of P. The bounded Voronoi diagram of P can be computed in time $O(nd)$.

Proof: First, we compute the Voronoi diagrams of the monotone histograms and convex polygons labelling the leaves of T. Their total size is $O(n)$. Therefore, it takes $O(n)$ time by Theorem 3.1. Then, for $i=1,...,d$, we compute the bounded Voronoi diagrams of the polygons that label the i-th level of T assuming that the bounded Voronoi diagrams of the polygons that label the $(i-1)$-th level of T are already computed and using Lemma 5.1. As the total size of all polygons and diagrams involved in the above computation corresponding to the i-th level of T is again $O(n)$, the computation takes also $O(n)$ time by Lemma 5.1. Since there are d levels in T, the total time is $O(nd)$. Q.E.D.

Since any convex polygon can be trivially cut into four histograms using three chords, we obtain the following generalization of the above theorem as an immediate corollary from it.

Corollary 5.1: Let T be a binary tree of height d which cuts a simple polygon P into monotone histograms and convex polygons each containing at least one vertex of P. The bounded Voronoi diagram of P can be computed in time $O(nd)$.

Corollary 5.2: Let P be a simple polygon on n vertices. If P can be cut by r parallel chords into $r+1$ monotone histograms or convex polygons then its bounded Voronoi diagram can be computed in time $O(n \log r)$.

Proof: We may assume without loss of generality that $r \le n$. Sort the r parallel lines in the direction perpendicular to them. Form a binary tree T that cuts P into the $r+1$ monotone histograms assigning the middle chord to the root of T, and then, for each intermediate node v of T, recursively set $s(v)$ to the middle chord of the chords cutting the polygon $P(v)$ (see Definition 5.1). Now, by Theorem 5.1, the thesis follows from the fact that T has $O(\log r)$ height. Q.E.D.

Corollary 5.3: Let P be a simple polygon with r reflex angles. The bounded Voronoi diagram of P can be computed in time $O(n \log r)$.

Proof: Assign weight 1 to each vertex of P at a reflex angle and weight 0 to the remaining vertices of P. Recall Chazelle's theorem on polygon cutting in its general weighted form [C]. Given a triangulation of P, we can find a diagonal that splits P into two sub-polygons, each of total weight not greater than two thirds of the total weight of P plus one, and also the whole family of such diagonals for sub-polygons recursively created, everything in linear time. The above family of diagonals immediately induces a binary tree T of height $O(\log r)$ that cuts P into sub-polygons with $O(1)$ reflex angles. As a triangulation of P can be constructed in linear time [C1], the tree T can also be constructed in linear time. Each of the final sub-polygons with $O(1)$ reflex angles can be easily split into $O(1)$ convex parts, for instance by using at most two diagonals from the preprocessing triangulation per each reflex angle for the splitting. By augmenting T with $O(1)$ levels that corresponds to the cuts into the convex parts, we obtain a binary tree T' that cuts P into $O(r)$ convex polygons and has height $O(\log r)$. Now, the thesis follows from Corollary 5.1 . Q.E.D.

6. Final remarks

Similarly as the Delaunay triangulation or the Voronoi diagram of a planar point set S corresponds to the lower part of the convex hull of $\mu(S)$ (see Fact 2.1), the so called furthest-site Delaunay triangulation or furthest-site Voronoi diagram of S correspond to the upper part of the convex hull of $\mu(S)$ (see [E]). This combined with the proof of Theorem 3.1 via Fact 2.2 implies that the furthest-site Voronoi diagram of a monotone histogram can be computed in linear time.

We believe that our main result about computing the Voronoi diagram of a monotone histogram in linear time will lead to an $o(n \log n)$ time algorithm for computing the bounded Voronoi diagram of a histogram. In contrast to the case of the standard Voronoi diagram of a histogram, we do not know any non-trivial lower bound for the latter problem.

Acknowledgements

We thank un unknown referee and Rolf Klein for valuable comments.

References

[AGSS] A. Aggarwal, L.J. Guibas, J. Saxe and P.W. Shor, *A Linear-Time Algorithm for Computing the Voronoi Diagram of a Convex Polygon*. Discrete and Computational Geometry 2 (1987), Springer Verlag.

[C] B. Chazelle, *A Theorem on Polygon Cutting with Applications*. Proc. 23rd IEEE FOCS Symposium, 1982.

[C1] B. Chazelle, *Triangulating a Simple Polygon in Linear Time*. Proc. 31st IEEE FOCS Symposium, 1990.

[E] H. Edelsbrunner, *Algorithms in Combinatorial Geometry*. EATCS Monographs on Theoretical Computer Science 10, Springer Verlag, Berlin (1987).

[F]   S. Fortune, *A Sweepline Algorithm for Voronoi Diagrams*, Algorithmica 2(1987), pp. 153-174.

[GS]   L.J. Guibas and J. Stolfi, *Primitives for the Manipulation of General Subdivisions and the Computation of Voronoi Diagrams*. In ACM Trans. Graphics 4 (1985), 74-123.

[LL]   D.T. Lee and A. Lin, *Generalized Delaunay Triangulations for Planar Graphs*. In Discrete and Computational Geometry 1 (1986), Springer Verlag, pp. 201-217.

[L]   A. Lingas, *On Partitioning Polygons*. Proc. 1st ACM Symposium on Computational Geometry, Baltimore, 1985.

[L1]   A. Lingas, *Voronoi Diagrams with Barriers and the Shortest Diagonal Problem*. In Information Processing Letters 32(1989), pp. 191-198.

[PS]   F.P. Preparata and M.I. Shamos, *Computational Geometry, An Introduction*. Texts and Monographs in Theoretical Computer Science , Springer Verlag, New York (1985).

[WS]   C. Wang and L. Schubert, *An Optimal Algorithm for Constructing the Delaunay Triangulation of a Set of Line Segments*. In Proc. 3rd ACM Symposium on Computational Geometry, Waterloo, pp. 223-232, 1987.

The MINSUMCUT problem[*]

J. Díaz[†]   A.M. Gibbons[‡]   M.S. Paterson[‡]   J. Torán[†]

May 22, 1991

Abstract

In this paper we first present a sequential linear algorithm for a linear arrangement problem on trees, MINSUMCUT, and then an $O(\log n)$-time parallel algorithm for MINSUMCUT on trees, which uses $n2/(\log n)$ processors.

1 Introduction

Linear arrangement problems have played an important role in computer science ([NK72]) and ([AH73]). Let us recall that a *linear arrangement* (or *layout*) of a graph $G(V, E)$ is an enumeration φ from V onto $\{1, 2, \ldots, |V|\}$. We describe some of the well-known arrangement problems. The minimum linear arrangement problem (also known as the EDGESUM problem ([Har77])) is: given a graph $G(V, E)$, find the enumeration φ which minimizes $\sum_{\{u,v\} \in E} |\varphi(u) - \varphi(v)|$. The decision version of this problem is NP-complete for general graphs ([GJ76]), and has an $O(n2.2)$ solution for unrooted trees ([Shi79]). Another problem is the MINCUT problem: given a graph $G(V, E)$ find the enumeration φ that minimizes the maximum over all i, $1 \le i \le |V|$, of the number of edges in the set $cut_\varphi(i) = \{(u, v) \in E \mid \varphi(u) \le i \le \varphi(v)\}$. This problem is also NP-Complete for general graphs ([GJ79]), and has a polynomial time solution for trees ([Yan83]). The BANDWIDTH problem; given a graph $G(V, E)$ find an enumeration φ which minimizes the maximum of $|\varphi(u) - \varphi(v)|$ over all the edges $\{u, v\} \in E$. This problem is NP-Complete for general graphs ([GJ79]) and also for trees ([GGJK78]).

In this paper, we consider another layout problem, the minimal sumcut problem, or MINSUMCUT for short.

Given a graph $G(V, E)$ with a layout φ of V, for any integer i, $1 \le i \le |V|$, let S_i be the set defined by $S_i = \{v \in V \mid \varphi(v) \le i\}$. Then we define

$$\delta(S_i) = |\{w \in V - S_i \mid \exists v \in S_i, (v, w) \in E\}|$$

Notice that given a graph with n nodes, any layout of it determines in a unique way, a nested sequence $S_0, S_1, \cdots, S_n$. Notice that S_0 is the empty set, S_n is the set of all nodes, and for any i, $1 \le i \le n$, $|S_i| = i$.

---

[*]This research was supported by the ESPRIT BRA Program of the EC under contract no. 3075, Project ALCOM.

[†]Departament de Llenguatges i Sistemes, Universitat Politècnica Catalunya, Pau Gargallo 5, 08028-Barcelona, Spain

[‡]Department of Computer Science, University of Warwick, Coventry, CV4 7AL, UK

Definition 1.1. Given a graph $G(V, E)$, $|V| = n$, the MINSUMCUT problem consists in finding a layout which minimizes

$$\sum_{i=1}^{n} \delta(S_i)$$

where $\{S_i\}_{i=1}n$ is the nested sequence determined by the layout.

A related problem, the Delta Operator, was studied in ([Dia79]).

In Section 2 we present a linear algorithm to solve the problem for an unrooted tree. In the following section we sketch the proof of the optimality of the algorithm and in the remaining section we present a parallel algorithm to solve the MINSUMCUT problem for trees.

2   The MINSUMCUT problems for trees

The enumeration of a tree with n nodes can be considered as a process of n steps. At step k we assign label k to a node that has not yet been enumerated.

Definition 2.1. At a certain stage of the enumeration a node is said to be *taken* if a label has already been assigned to it. A node is said to be *marked* if one or more of its immediate neighbours has already been taken.

For a tree T we will denote by $w(T)$ the number of nodes of the tree.

We introduce three basic enumerations which will be used later by the algorithm. These enumerations are defined for rooted trees. The enumeration algorithm works however for trees in which no orientation is given. For the enumeration of T, the algorithm will first give an orientation to T and then recursively enumerate the different subtrees defined by the orientation. Assume that T' is a subtree of T with root r, and from r hang the subtrees $A_1, \ldots, A_m$. Denote by $f(r)$ the father of r in T. (See Figure 1 in the annex.) Depending on whether $f(r)$ has already been taken or marked, we will consider three types of enumeration of the subtrees: β-mode, μ-mode and τ-mode.

The numbering in β-mode will apply when $f(r)$ is neither marked nor taken. It consists of choosing in a certain way which optimizes the numbering a subtree A_i, enumerating recursively A_i in β-mode, enumerating the other descendants of r in τ-mode and finally enumerating r. This way of numbering can be considered as the way in which the algorithm starts enumerating a tree, when none of its nodes has been taken.

The numbering in τ-mode will apply when $f(r)$ is marked. One of the subtrees of T', A_{i_j}, is enumerated in β-mode, then the rest of the subtrees except one are recursively enumerated in τ-mode, then the root r and finally the last subtree A_j is enumerated in μ-mode. The way to choose in which order the subtrees are enumerated will be explained later. This way of numbering corresponds to the middle of the enumeration when some nodes have already been taken.

The numbering in μ-mode applies when $f(r)$ has been taken (and therefore r is marked). In the enumeration all the subtrees except one are enumerated in τ-mode, then r is taken, and finally the last subtree of r is enumerated recursively in μ-mode. This way of numbering corresponds to the end of the enumeration when the algorithm is finishing the enumeration of a subtree in which the root is already taken.

The cost of such enumerations can be computed in the following way:

Let T' be a subtree of T with root r. If T' is a leaf then $cost(\beta(T')) = 1$, $cost(\tau(T')) = 1$, and $cost(\mu(T')) = 0$. Observe that in the first two cases we have to count node $f(r)$, which by hypothesis is not taken, while in the μ case, this node is already taken.

If T' is not a leaf, and it consists of a root r and subtrees $A_1, \ldots, A_m$ hanging from r, the costs of the enumerations are:

- $cost(\beta(T')) = min_i\{cost(\beta(A_i)) + \sum_{j \neq i} cost(\tau(A_j)) + 1\}$.

- $cost(\tau(T')) = min_{i,j(i \neq j)}\{cost(\beta(A_i)) + cost(\beta(A_j)) + \sum_{k \neq i,j} cost(\tau(A_k)) + w(T')\}$ if $m \geq 2$.

- $cost(\tau(T')) = cost(\beta(A_1)) + w(T')$ if $m = 1$.

- $cost(\mu(T')) = min_i\{cost(\mu(A_i)) + \sum_{j \neq i} cost(\tau(A_j)) + 1\}$.

The following technical lemmas, which will be used later, follow from the definition of the enumerations.

Lemma 2.2. For every tree T, $cost(\beta(T)) = cost(\mu(T)) + 1$.

Lemma 2.3. For every tree T, $cost(\beta(T)) + w(T) - 1 \geq cost(\tau(T))$.

Lemma 2.4. For every tree T, if $w(T) > 1$ then $cost(\tau(T)) > cost(\beta(T))$.

We give now a description of the algorithm, which finds the optimal enumeration in linear time. An example of the algorithm can be followed in Figures 2a, 2b and 2c, in the annex.

Algorithm Part 1: Input a tree T. Output $Cost(T)$.

1. For every undirected edge $e = (u, v)$ of T, replace e by two directed edges $e_1 = (u, v)$ and $e_2 = (v, u)$.

   We will associate with every directed edge $e = (u, v)$ the subtree with root u defined when node v is deleted from the tree. We will denote this subtree by T_e.

   For every directed edge e denote by $w(e)$ the number of nodes in the subtree T_e.

2. To every directed edge e we will assign a 5-tuple $[B(e), C(e), F(e), S(e), W(e)]$. The first two parameters are values which denote the cost of a β- and a τ-enumeration of T_e, respectively. The third and fourth parameters are pointers to the two direct descendants of the root of T_e, T_{e_1} and T_{e_2} which have the smallest and second smallest values of $(C(e_i) - B(e_i))$. These two pointers will be needed later since T_{e_1} will be the first subtree taken in the case of a β or a τ-enumeration and the last taken in the case of a μ-enumeration, while T_{e_2} will be the last subtree enumerated in the τ case. $W(e)$ will denote the number of nodes of the subtree T_e.

   The values B, C, F, S are recursively computed in the following way:

(a) If $e = (u, v)$ and u is a leaf of T then $B(e) = 1$ $C(e) = 1$, and $F(e)$ and $S(e)$ are the empty string, and $W(e) = 1$.

(b) If $e = (u, v)$ and u is not a leaf then let $e_1, e_2, \ldots, e_m$ be the edges pointing to u, not considering the inverse edge of e. Assign to $F(e)$, $S(e)$, $W(e)$, $B(e)$ and $C(e)$ values as follows:

$F(e)$ is a pointer to the direct descendant of u, T_{e_i}, such that for every $j \neq i$ $(C(e_i) - B(e_i)) \geq (C(e_j) - B(e_j))$,

$S(e)$ is a pointer to the direct descendant of u, T_{e_j} $(j \neq i)$ such that for every $k \neq i, j$ $(C(e_j) - B(e_j)) \geq (C(e_k) - B(e_k))$,

$W(e) = 1 + \sum_i W(e_i)$ with $1 \leq i \leq m$,

$B(e) = B(e_i) + \sum_{j \neq i} C(e_j) + 1$ with $e_i = F(e)$,

$C(e) = B(e_i) + B(e_j) + \sum_{k \neq i, j} C(e_k) + W(e)$ with $e_i = F(e)$ and $e_j = S(e)$, If $m = 1$ then $C(e) = B(e_i) + w(e)$.

3. $\text{Cost}(T) = min_{(u,v) \in E}\{B(u, v) + B(v, u)\} - 1$.

Algorithm Part 2: Input a tree T, and the output of Part 1.

Let $e_1 = (u, v)$ be the edge selected by the previous part to minimize cost, and let $e_2 = (v, u)$.

1. Recursively do the following:

   (a) enumerate T_{e_1} in β-mode,
   (b) enumerate T_{e_2} in μ-mode.

2. For a subtree T_e in a β-enumeration do:

   (a) enumerate $F(e)$ in β-mode,
   (b) enumerate the rest of the subtrees of T_e in τ-mode,
   (c) enumerate the root of T_e.

3. For a subtree T_e in a τ-enumeration do:

   (a) enumerate $F(e)$ in β-mode,
   (b) enumerate the rest of the subtrees of T_e (except $S(e)$) in τ-mode,
   (c) enumerate the root of T_e,
   (d) enumerate $S(e)$ is μ-mode.

4. For a subtree T_e in a μ-enumeration do:

   (a) enumerate all the subtrees of T_e (except $F(e)$) in τ-mode,
   (b) enumerate the root of T_e,
   (c) enumerate $F(e)$ is μ-mode.

3 Optimality of the enumeration

In this section we shall prove that the algorithm given in the previous section produces an enumeration which has optimal cost. For this we need the following definitions.

Definition 3.1. Let T be a tree and $\{a_1, \ldots, a_i\}$ and $\{b_1, \ldots, b_j\}$ be two disjoint sets of nodes of T. Let $neigh(a_1, \ldots, a_i)$ denote the set of nodes which are neighbours in $T - \{a_1, \ldots, a_i\}$ of $\{a_1, \ldots, a_i\}$. Let $cost(a_1, \ldots, a_i|_{b_1, \ldots, b_j})$ denote the cost of enumerating $\{a_1, \ldots, a_i\}$ in this order, supposing that $\{b_1, \ldots, b_j\}$ have been already enumerated. If $\{a_1, \ldots, a_i\} = T$ then $cost(a_1, \ldots, a_i)$ denotes the cost of enumerating T in this order.

Theorem 3.2. Let T be an undirected tree. Let s be an interior node of T, and $T_1, \ldots, T_i$ be the set of subtrees defined by deleting s. Let φ be an enumeration that does not start and finish enumerating T by the same subtree T_i. There is an enumeration φ' of cost smaller than or equal to φ, and such that for every T_j, φ' enumerates all the nodes from T_j consecutively, before enumerating any node from a different subtree.

Proof. Let φ be an enumeration of T. We will show that for any subtree T_s defined by deleting node s in T, if $\varphi(1) \notin T_s$ or $\varphi(|T|) \notin T_s$, then there is an enumeration φ' satisfying the following properties:

- φ' enumerates all the nodes of T_s consecutively,

- $cost(\varphi') \leq cost(\varphi)$, and

- for every pair of nodes u, v which belong to the same subtree of T (defined by deleting node s), if φ enumerates u and v consecutively, then φ' also enumerates u and v consecutively.

Theorem 3.2 follows from this fact.

We proceed by induction on the size of T_s. The induction basis (T_s has just one node) is trivial. For the induction step, suppose that T_s is composed of a root r (hanging from s) and subtrees $A_1, \ldots, A_m$ hanging from r. Let φ be an enumeration of T, and suppose that $\varphi(1) \notin T_s$ or $\varphi(|T|) \notin T_s$. Since all the subtrees A_i have size smaller than T_s, by the induction hypothesis there is an enumeration φ' which for every i enumerates all the nodes of A_i consecutively, and has cost smaller than or equal to the cost of φ.

We show now that there is another enumeration which takes all the nodes of T_s consecutively. The proof is divided into different claims which study all the possible ways in which φ can enumerate T_s. In this version of the paper the proofs of the claims are omitted. They are not hard to prove using a case analysis of the costs of the different enumerations considered.

Claim 1. The enumeration φ defined by

$$a_1, \ldots, a_i, A_1, \ldots, A_j, b_1, \ldots, b_k, A_{j+1}, c_1, \ldots, c_l$$

with $\{a_1, \ldots, a_i\} \cap T_n = \emptyset$ and $\{b_1, \ldots, b_k\} \cap T_n = \emptyset$ has a cost greater than or equal to an enumeration φ' in which $A_1, \ldots, A_j, A_{j+1}$ are enumerated consecutively. Also, for every pair of nodes u, v which belong to the same subtree of T (defined by deleting node s), if the first enumeration takes u and v consecutively, then so does φ'.

Claim 2. The enumeration φ defined by

$$a_1, \ldots, a_i, r, b_1, \ldots, b_k, A_j, c_1, \ldots, c_l$$

with $\{b_1, \ldots, b_k\} \cap T_n = \emptyset$ has a cost greater than an enumeration φ' in which A_j is enumerated directly after r. Also, for every pair of nodes u, v which belong to the same subtree of T (defined by deleting node s), if the first enumeration takes u and v consecutively, then so does the φ'.

Claim 3. The enumeration $a_1, \ldots, a_i, r, A_j, b_1, \ldots, b_k$ where A_j is not the last tree enumerated from T_s has a cost greater than the enumeration $a_1, \ldots, a_i, A_j, r, b_1, \ldots, b_k$. Also, for every pair of nodes u, v ($u, v \neq r$) which belong to the same subtree of T (defined by deleting node s), if the first enumeration takes u and v consecutively, then so does the second.

Claim 4. The enumeration given by

$$a_1, \ldots, a_i, A_1, \ldots, A_{m-1}, b_1, \ldots, b_k, r, A_m, c_1, \ldots, c_l$$

where $A_1, \ldots, A_m$ are all the subtrees in T_s hanging from the root, and $\{a_1, \ldots, a_i, c_1, \ldots, c_l\} \neq \emptyset$ has a cost greater than or equal to an enumeration in which $A_1, \ldots, A_{m-1}, r, A_m$ are enumerated consecutively in this order. Also, for every pair of nodes u, v which belong to the same subtree of T (defined by deleting node s), if the first enumeration takes u and v consecutively, then so does the second.

Claim 5. The enumeration $a_1, \ldots, a_i, A_1, \ldots, A_m, b_1, \ldots, b_k, r, c_1, \ldots, c_l$ where $A_1, \ldots, A_m$ are all the subtrees in T_s hanging from the root, has a cost greater than or equal to an enumeration φ' in which $A_1, \ldots, A_m, r$ are enumerated consecutively in this order. Also, for every pair of nodes u, v which belong to the same subtree of T (defined by deleting node s), if the first enumeration takes u and v consecutively, then so does φ'.

Theorem 2.3 follows from the previous claims, since they consider all possible ways of enumerating T. After any one of the claims is applied we obtain a new enumeration of cost smaller than or equal to the previous one, and in which some new set of nodes of T_s is enumerated consecutively, maintaining a consecutive enumeration of nodes of the same subtree which were consecutively taken by the first enumeration. The fact that the first and last nodes taken by the first enumeration do not belong to the same subtree is used in Claim 4 (the fact implies $\{a_1, \ldots, a_i, c_1, \ldots, c_l\} \neq \emptyset$). $\square$

Lemma 3.3. Let φ be an enumeration of T and let s be an interior node of the tree such that φ does not start and finish enumerating T in the same subtree defined by deleting s. There is an enumeration φ' of T, of cost less than or equal to φ such that for every subtree T_j (defined by deleting s), all the nodes of T_j are consecutively enumerated by φ' in the following way:

1. If in the stage in which the first node of T_j is taken, s has neither been marked nor taken then T_j is enumerated in β-mode.

2. If in the stage in which the first node of T_j is taken, s has been marked but has not been taken then T_j is enumerated in τ-mode.

3. If in the stage in which the first node of T_j is taken, s has already been taken then T_j is enumerated in μ-mode.

Proof. Let s be a node of T. By Theorem 3.2, we know that there is an enumeration in which for every subtree T_j defined by deleting s, all the nodes of T_j are consecutively enumerated by it.

Let T_s be one of these subtrees. We argue by induction on the size of the subtree T_s. The induction basis ($|T_s| = 1$) is a trivial case, since there is just one possible enumeration of T_s. For the induction step, suppose that T_s is composed of a root r (hanging from s) and subtrees $A_1, \ldots, A_m$ hanging from r. Since the nodes of T_s are consecutively enumerated, all the possible enumerations of T have the form $A_1, A_2, \ldots, A_i, r, A_{i+1}, \ldots, A_m$.

1. If s is neither marked nor taken, by induction hypothesis the cost of such an enumeration is

$$cost(\beta(A_1)) + \sum_{k=2} icost(\tau(A_k)) + 1 + (m - i) +$$
$$+ \sum_{k=i+1} m[cost(\mu(A_k)) + (m + 1 - k) \times w(A_k)]$$

Using Lemmas 2.2 to 2.4, it follows that the minimum value for this expression is for the case $i = m$, i.e, an enumeration in which the root is the last node of T which is taken. This is the case of a β-enumeration.

2. If s is marked but not taken, the cost of such an enumeration is

$$cost(\beta(A_1)) + w(A_1) + \sum_{k=2} i[cost(\tau(A_k)) + w(A_k)] + 1 + (m - i) +$$
$$+ \sum_{k=i+1} m[cost(\mu(A_k)) + (m + 1 - k) \times w(A_k)]$$

The minimum value for this expression is for the case $i = m - 1$, i.e., when the root is the node taken before the last one (τ-enumeration).

3. If s is taken, the cost of the enumeration is

$$\sum_{k=1} icost(\tau(A_k)) + (m - i) + \sum_{k=i+1} m[cost(\mu(A_k)) + (m - k) \times w(A_k)]$$

Again the minimum value for this expression is for the case $i = m$, i.e, an enumeration in which the root is taken just before the last subtree This is the case of a μ-enumeration.

$\square$

Definition 3.4. An edge e of T is called a bridge for an enumeration φ, if φ starts at T_{e_1} and finishes at T_{e_2} where T_{e_1} and T_{e_2} are the two subtrees generated when e is deleted from T.

Lemma 3.5. Let T be an undirected tree and let φ be an optimal enumeration of T. Let e be a bridge for φ. Let s be the root of T_{e_1}. There is an optimal enumeration φ' in which s is enumerated by φ' after all its subtrees except T_{e_2} have been enumerated.

Proof. Let $T_1, T_2, \ldots, T_m$ be the set of subtrees of T defined by deleting s. (Observe that T_{e_2} is one of the subtrees of this list). Since e is a bridge for φ, and s is one of the nodes of e, by Lemma 3.3 there is an optimal enumeration φ' of T enumerating every subtree T_i either in β, τ or μ-mode. We can suppose that φ starts the enumeration by T_{e_1} and finishes in T_{e_2}, and therefore the last subtree enumerated by φ' is T_{e_2}. φ' can enumerate T in the order $T_1, T_2, \ldots, T_i, n, T_{i+1}, \ldots, T_{m-1}, T_{e_2}$, and the cost of the enumeration is

$$cost(\beta(A_1)) + w(A_1) + \sum_{k=2}^{} i[cost(\tau(A_k)) + w(A_k)] + (m-i) +$$

$$+ \sum_{k=i+1}^{} m - 1[cost(\mu(A_k)) + (m-k) \times w(A_k)] + \mu(T_{e_2})$$

Using Lemmas 2.2 to 2.4, it follows that the minimum value for this expression is for the case $i = m - 1$, i.e, an enumeration in which s is taken after all the subtrees except T_{e_2} have been enumerated. Observe that in this case the cost of the enumeration is exactly $cost(\beta(T_{e_1})) + cost(\mu(T_{e_2}))$. □

Lemma 3.6. Let T be an undirected tree. An enumeration φ of T is optimal if and only if

$$cost(\varphi) = \min_{(u,v) \in E} \{cost(\beta(T_{(u,v)})) + cost(\beta(T_{(v,u)}))\} - 1$$

Proof. If φ is optimal and e is a bridge for φ, then by Lemma 2.2, $cost(\varphi) = cost(\beta(T_{e_1})) + cost(\mu(T_{e_2})) = cost(\beta(T_{e_1})) + cost(\beta(T_{e_2})) - 1$.

If there is an edge $a \in E$ with $cost(\beta(T_{a_1})) + cost(\beta(T_{a_2})) < cost(\varphi) + 1 = cost(\beta(T_{e_1})) + cost(\beta(T_{e_2}))$, enumerating T_{a_1} in β-mode and T_{a_2} in μ-mode, we would obtain an enumeration φ' starting at some leaf of T_{a_1} and finishing at T_{a_2} with cost $cost(\beta(T_{a_1})) + cost(\beta(T_{a_2})) - 1 < cost(\varphi)$, contradicting the fact that φ is an optimal enumeration. □

4 A parallel algorithm for the MINSUMCUT problem on trees

Throughout this section we assume the CREW PRAM model of parallel computation (see [GR88] for example). Theorem 4.3 places the MINSUMCUT problem on trees in the complexity class $NC1$ using $n2/(\log n)$ processors. Throughout this section we use the terminology employed in the description of the sequential algorithm.

Theorem 4.1. For any tree, given the edge e_1 and values $F(e)$, $S(e)$ and $W(e)$ for all edges e of the tree as determined by the preprocessing step (step 1) of the sequential algorithm, we can find an optimal enumeration of the tree in $O(\log n)$ parallel time using n processors.

Proof. As for Part 2 of the sequential algorithm, the optimal enumeration requires us to enumerate T_{e_1} in β-mode followed by an enumeration of T_{e_2} in μ-mode. It is sufficient to show that given any rooted tree T (with k nodes), then we can find a φ-enumeration of T (where φ is β, τ or μ) in $O(\log k)$ time using k processors.

The algorithm starts on the basis of the following observation. Given any subtree $T_{(u,v)}$ (rooted at u and with $v = f(u)$), suppose that the sons of u are $(s-1, s_2, \ldots, s_m)$, $F(u,v) = s_1$ and $S(u,v) = s_m$. In a "stand-alone" evaluation of $T_{(u,v)}$:

(a1) if the evaluation is in β-mode, then u will receive the value $\left(\sum_{i=1}^m W(s_i)\right) + 1$

(b1) if the evaluation is in τ-mode, then u will receive the value $\left(\sum_{i=1}^m - 1 W(s_i)\right) + 1$

(c1) if the evaluation is in μ-mode, then u will receive the value $\left(\sum_{i=2}^m W(s_i)\right) + 1$

In an enumeration of T, suppose that $T_{u,v}$ is enumerated in β-mode. Then u will receive the value of $\left(\sum_{i=1}^m W(s_i)\right) + 1 + NB(u)$ where $NB(u)$ denotes the number of nodes of T enumerated before $T_{(u,v)}$ is enumerated. At the start of the algorithm we do not know in which mode $T_{(u,v)}$ is to be enumerated for an enumeration of T and, moreover, we do not know the value of $NB(u)$. By the end of the algorithm, for a given mode of enumeration of T, both these unknowns will have been determined. The algorithm starts with three copies of each subtree of T of height 1 (one copy for each possible mode of enumeration) and proceeds to join trees together to form larger trees, these larger trees will generally be twice the height of those joined together.

After at most $\lceil \log^{(} \text{depth of } T) \rceil = O(\log n)$ parallel joining phases, T will be reconstructed. At each stage of the process, for each of the three copies of any constructed tree, the following three invariants hold:

(a2) For every non-leaf node v, $E(v)$ is the order in which v is visited in a "stand alone" enumeration of the constructed tree.

(b2) Each leaf l of the constructed tree "knows" $N(l)$, the number of nodes of this tree which would have been enumerated before the subtree rooted at l in T is enumerated.

(c2) Each leaf l "knows" what mode of enumeration would be required (along with the two values F and S) for the subtree rooted at l in T.

The construction process starts with the subtrees of height 1. Initialising the $E(v)$ with the appropriate values given by (a1), (b1) and (c1) ensures that invariant (a2) holds at the outset. In order that invariant (b2) holds for the trees of height 1, we need to assign to the $N(l)$ values obtained from an appropriate prefix sum of the W's of the sons in these trees. Which prefix sum is appropriate will depend (in an obvious way) on the prevailing mode of enumeration and the local F and S. Invariant (c2) obviously holds initially.

Before describing the overall pattern of rejoining trees to ensure that only $O(\log k)$ parallel joining steps are required, we will describe how just two (copies of) trees are joined to ensure that the invariants (a2), (b2) and (c2) will hold. Suppose that a leaf of T_1 is identified with the root of T_2 to produce T_3. For a copy of T_1 corresponding to one enumeration, the leaf of T_1 identified with the root of T_2 will determine which copy of T_2 is required in order that invariant (c2) holds in T_3. in order that invariant (a2) continues to hold, we add to each $E(v)$ in T_3 that used to belong to T_2 the value of $N(z)$ where z is the leaf of T_1 identified with the root of T_2. Similarly, adding $N(z)$ to each $N(u)$, where u are those leaves of T_3 which used to be leaves of T_2, will ensure that invariant (b2) continues to hold. Notice that these additions require concurrent reads of $N(z)$.

We have described what operations are necessary when joining two trees. Consider now what trees are joined together in one parallel step of tree joining. First use the Euler tour technique to determine at which level of T each internal vertex resides. We assume that the root is at level 1. Let $S(i)$ denote the set of all roots of constructed trees that are to be identified with leaves of the other trees in the ith step of parallel tree-joining. We specify that $S(i)$ contains all those roots at depth d in T such that $(d - 2i - 1)$ is exactly divisible by $2i$. Thus $S(1)$ contains roots at depth $(2, 4, 6, 8, \ldots)$, $S(2)$ those at depths $(3, 7, 11, 15, \ldots)$, $S(3)$ those at depths $(5, 13, 21, 29, \ldots)$ and so on. This construction of one copy of T is shown schematically in Figure 3. Because of invariant (a2), when each copy of T has been constructed, we have each of the three possible enumerations of T. A similar logarithmic reconstruction of a tree was employed in ([GR90]).

The construction of the initial trees of height 1 may be achieved in constant time with $O(k)$ processors and this is easily simulated in $O(\log k)$ time using $k/(\log k)$ processors. Their initialisation is only complicated by the need to perform some prefix sums, but there are well known optimal logarithmic-time algorithms (see [GR88], for example) for this type of computation. Each parallel tree-joining phase takes constant time with $O(k)$ processors ($O(k)$ additions are all that is essentially required) and since there are $O(\log k)$ phases, the overall reconstruction can be achieved in $O(\log k)$ time using k processors. □

Theorem 4.2. For a given directed edge e, the values of $B(e)$, $C(e)$, $F(e)$, $S(e)$ and $W(e)$ required for a φ-enumeration of T_e can be found in $O(\log k)$ time using $k/(\log k)$ processors, where T_e has k nodes.

We give a sketch of the proof.
Proof. The value of $W(e)$ is the number of descendants of u, where $e = (u, v)$. It is well known (for example using an Euler tour technique, see [GR88]) that the number of descendants of all nodes of a rooted tree with k nodes may be found in $O(\log k)$-time using $k/(\log k)$ processors. If T_e is a binary tree then we note that $B(e)$, $C(e)$, $F(e)$ and $S(e)$ can be found by employing a straightforward variation of the logarithmic-time, optimal expression evaluation algorithm of ([GR89]). In general T_e is not a binary tree. this complication may be handled by local reconstruction of the tree (to produce a binary version) and by employing a more complicated version of ([GR89]). The local reconstruction is indicated in Figure 4. in the algorithm of ([GR89]), one functional form (which we call $f_v(x)$) is associated with each internal node v of the tree which is modified during so-called "leaves-cutting" phases. The $f_v(x)$, when evaluated, is the current value that v must pass to its father if the value of the subtree rooted

at v is x. In the more complicated version of the algorithm required here, when v needs to be re-formulated in a leaves-cutting phase, there may be more than one $f_v(x)$ associated with v (although only one in each phase) and more rules for re-formulation. The complications arise because $f_v(x)$ may, for example, need reformulation when v is associated with a leaf whose father is coalescent with v in the original tree or whose father is actually the father of v in the original tree. The complications, although straightforward, are technically intricate and so we omit the details. With carefully arranged book-keeping, each leaves-cutting operation still takes constant time and so, just as in ([GR89]), the algorithm takes $O(\log k)$ time using $k/(\log k)$ processors. □

Theorem 4.3. The MINSUMCUT problem for any tree T, with n nodes, can be solved in $O(\log n)$ time using $n2/(\log n)$ processors.

Proof. In order to solve the MINSUMCUT problems for trees, we need first to identify an edge (u, v) which provides a minimum value of the expression $(B(u, v) + B(v, u))$, as stated in Part 1 of the sequential algorithm. there are $O(n)$ edges in T so that these expressions may be evaluated after $O(\log n)$ time using $n2/(\log n)$ processors by applying the algorithm of Theorem 4.2 to all the edges in parallel. Finding the minimum is easily achieved in $O(\log n)$ using $n/(\log n)$ processors by standard algorithms. Having found an appropriate (u, v) and using the F, S and W found by the algorithm of Theorem 4.2, we employ the algorithm of Theorem 4.1 to find the enumeration. □

Remark. Although we have demonstrated that there is a logarithmic-time algorithm for the MINSUMCUT problem on trees, we would like to reduce the number of processors used. the bottle-neck in this regard is the evaluation of $(B(u, v) + B(v, u))$ for all edges of T. In fact, in $O(\log n)$ time using $n/(\log n)$ processors it is possible to find the B's for half the directed edges of T. this task is accomplished by a further modification of the algorithm in Theorem 4.2. The modification is equivalent to that required to the algorithm used for expression evaluation ([GR89]) when there is a need to evaluate all the subexpressions represented by the internal nodes of the expression tree. After the root has been evaluated, it is possible to reconstruct the tree by the process which is the reverse of leaves-cutting. As internal nodes re-appear in the tree, prior book-keeping can ensure that the associated expressions are evaluated and overall this requires the same time and processor requirements as leaves-cutting. The implication is that, if T is rooted arbitrarily, the Bs may be found for all edges directed towards the root by one application of the modified algorithm. The problem, however, is that an (arbitrary) tree would still have to be rooted $O(n)$ times to capture all the edges.

References

[AH73]   D. Adolphson and T.C. Hu. Optimal linear ordering. *SIAM J. on Applied Mathematics*, 25(3):403–423, Nov 1973.

[Dia79]   J. Diaz. The δ-operator. In L. Budach, editor, *Fundamentals of Computation Theory*, pages 105–111. Akademie-Verlag, 1979.

[GGJK78] M.R. Garey, R.L. Graham, D.S. Johnson, and D. Knuth. Complexity results for bandwidth minimization. *SIAM J on Applied Mathematics*, 34:477–495, Sept. 1978.

[GJ76]   M.R. Garey and D.S. Johnson. Some simplified NP-complete graph problems. *Theoretical Computer Science*, 1:237–267, 1976.

[GJ79]   M.R. Garey and D.S. Johnson. *Computers and Intractability: A Guide to the Theory of NP-Completeness*. Freeman, San Francisco, 1979.

[GR88]   A. Gibbons and W. Rytter. *Efficient Parallel Algorithms*. Cambridge University Press, Cambridge, 1988.

[GR89]   A. Gibbons and W. Rytter. Optimal parallel algorithms for dynamic expression evaluation and context free recognition. *Information and Computation*, 81(1):32–45, April 1989.

[GR90]   A. Gibbons and W. Rytter. Optimal edge-colouring outerplanar graphs is in NC. *Theoretical Computer Science*, 71:401–411, 1990.

[Har77]   L.H. Harper. Stabilization and the edgesum problem. *Ars Combinatoria*, 4:225–270, Dec. 1977.

[NK72]   M. Nanan and M. Kurtzberg. A review of the placement and quadratic assignment problems. *SIAM Review*, 14(2):324–341, April 1972.

[Shi79]   Yossi Shiloach. A minimum linear arrangement algorithm for undirected trees. *SIAM J. on Computing*, 8(1):15–31, February 1979.

[Yan83]   Mihalis Yannakakis. A polynomial algorithm for the min cut linear arrangement of trees. In *IEEE Symp. on Found. of Comp. Sci.*, volume 24, pages 274–281, Providence RI, Nov. 1983.

ANNEX

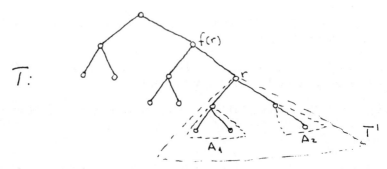

Figure 1. Example of rooted tree T with subtree T'. T' has itself two subtrees A_1 and A_2 hanging from its root r.

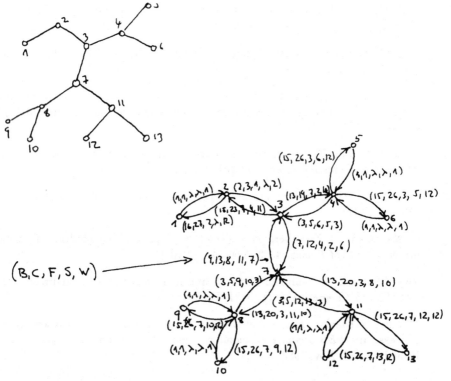

Figure 2a. An unrooted tree T, and the result of applying part 1 of the sequential algorithm.

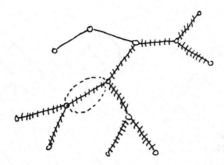

Figure 2b. In shadow the edges which minimize $\{B(u,v) + B(v,u)\}$. The algorithm selects one of them, for example the one surounded with doted line.

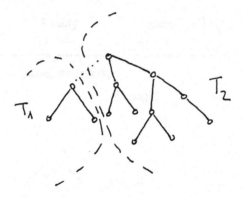

Figure 2c. Enumerate T_1 in β mood and T_2 in μ mood. The resulting enumeration is: 9,10,8,12,11,13,7,1,2,3,5, 4,6, and the cost is 16.

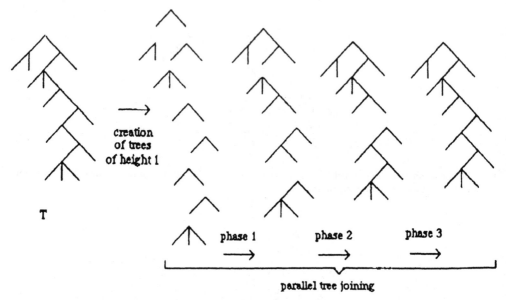

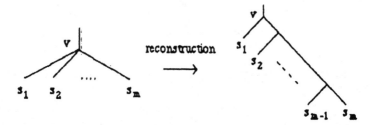

T

creation
of trees
of height 1

phase 1 phase 2 phase 3

parallel tree joining

Figure 3

reconstruction

Figure 4

Efficient Algorithms for the Minimum Range Cut Problems

(Extended Abstract)

Naoki Katoh*, Kazuo Iwano†

* Department of Management Science, Kobe University of Commerce
Gakuen-Nishimachi 8-2-1, Nishi-ku, Kobe 651-21, Japan

†Tokyo Research Laboratory, IBM Japan
5-11 Sanbancho, Chiyoda-ku, Tokyo 102, Japan

1 Introduction

Let $G = (V, E)$ be a connected undirected graph with n vertices and m edges, in which a real-valued *weight*, denoted by $w(e)$, is associated with every edge e. This paper studies the problem that asks to find an elementary cut in G such that the range of all edge weights in the cut is minimum. Here an *elementary cut* C in G is a subset of E induced by a bipartition $(X, V - X)$ for $X \subset V$ with $X \neq \emptyset, V$ such that C is the set of all edges connecting X and $V - X$. In the subsequent discussion, when we speak of a cut, it means an elementary cut unless confusion occurs. The *range* of a cut C is defined to be the maximum difference among weights of edges in the cut, i.e., $\max_{e \in C} w(e) - \min_{e \in C} w(e)$. We call this problem the *minimum range cut problem*.

Recently several researchers studied a number of minimum range problems, e.g., minimum range spanning tree problems by Camerini et al. [1] and Galil and Schieber [5], minimum range assignment problems by Martello et al. [6], and so on. Martello [6] presented a formal description of a general class of minimum range problems and gave an efficient algorithm for it. To the authors' knowledge, no one has ever considered the minimum range cut problem. Simply applying the algorithm of [6], our problem can be solved in $O(m^2)$ time. We shall propose in this paper an $O(MST(m, n) + n \log n)$ algorithm for the minimum range cut problem, where $MST(m, n)$ denotes the time required to compute minimum and maximum spanning trees of G. The best known bound for $MST(m, n)$ is $O(m \log \beta(m, n))$ by Gabow at al. [4], where $\beta(m, n)$ is defined to be $\min\{i \mid \log^{(i)} n \leq m/n\}$, and $\log^{(i)} x$ denotes the log function iterated i times.

The first important observation is that only the edges of minimum and maximum spanning trees are sufficient to find a minimum range cut. Thus, after preprocessing the original graph by computing minimum and maximum spanning trees, the edge set E can be reduced to the set of at most $2(n - 1)$ edges. This result is rather surprising because a closely related problem, the minimum range spanning tree problem, does not have such a property.

For a closed interval $[\alpha, \beta]$ with $\alpha \leq \beta$, suppose that there is a cut C such that all edge weights of C lie in the interval. If there is no closed interval $[\alpha', \beta']$ properly included in $[\alpha, \beta]$ that satisfies the above property, $[\alpha, \beta]$ is called *critical*. Such a cut C is called a *critical cut*. It is clear that the minimum range cut problem can be reduced to the problem of finding a critical interval $[\alpha, \beta]$ such that $\beta - \alpha$ is minimum. Thus, a naive approach is to enumerate all critical intervals, and then identify a minimum range cut as a critical cut in the critical interval with the minimum range. Enumerating all critical intervals is done in our paper by following the systematic approach proposed by Martello et al. [6]. However, the algorithm in this paper accelerates their algorithm by making use of the special structure of our problem. A critical interval can be characterized in our problem by a maximum spanning tree of a graph G in which edge weights are appropriately modified. With this characterization, given a critical interval, a next critical interval can be efficiently computed by updating the current maximum spanning tree by using dynamic trees proposed by Sleator and Tarjan [9]. This approach leads to an $O(MST(m,n) + n \log n)$ time algorithm.

If G is planar, a minimum range cut can be obtained in $O(n \log n)$ time because $MST(m,n) = O(n)$ for planar graphs [2]. If G is Euclidean, i.e., V is the set of points in the plane and the weight of an edge is equal to the distance between the corresponding vertices, both minimum and maximum spanning trees can be obtained in $O(n \log n)$ time [7], [8]. A minimum range cut is thus obtained in $O(n \log n)$ time. Notice that, in case G is Euclidean, the output of the algorithm is not the set of edges in the cut, but a partition of vertices induced by the cut, because reporting the set of edges in the cut may require $O(n^2)$ time.

This paper then studies the problem of finding a minimum range cut with a target, which asks to find a minimum range cut such that the interval of edge weights therein (i.e., $[\min_{e \in C} w(e), \max_{e \in C} w(e)]$) contains a prespecified value (called *target*). We call this problem the *minimum range target cut problem*. The algorithm for the previous problem cannot be applied to this problem because no critical interval computed by the algorithm for the previous problem may contain a target value. However, we can show that this problem can also be solved in a manner similar to the above problem with the same time complexity.

This paper is organized as follows. Section 2 presents an $O(MST(m,n) + n \log n)$ algorithm for the minimum range cut problem. Section 3 studies the minimum range target cut problem and develops an $O(MST(m,n) + n \log n)$ algorithm.

2 Minimum Range Cut Problem

Since it takes $O(m)$ time to test the connectivity of a given graph, our problem can be solved in $O(m^2)$ time by simply applying the algorithm in [6]. In this section, we improve this trivial time bound to $O(MST(m,n) + n \log n)$.

Let (a, b) denote the interval $\{x \mid a < x < b\}$. When we include a boundary of the interval, we use "[" (resp. "]") instead of "(" (resp. ")"), for example, $[a, b) = \{x \mid a \le x < b\}$. Given an interval $[a, b)$, we define $E[a, b)$ as $\{e \in E \mid a \le$

$w(e) < b\}$. $E(a, b]$ is similarly defined. For the simplicity we use $E(a)$ instead of $E[a, a]$.

A cut C is $[a, b]$-*critical* when the minimum (resp. maximum) weight of edges in C is a (resp. b), and there is no cut in $E(a, b]$ or $E[a, b)$. A value a is said to be *lower-critical* if there exists a $[a, b]$-critical cut C for some b with $a \le b$. Analogously, we define an *upper-critical* value. A closed interval $[a, b]$ is *critical* if there exists a $[a, b]$-critical cut. For a cut C, let $\max(C) \equiv \max_{e \in C} w(e)$, $\min(C) \equiv \min_{e \in C} w(e)$, and *max-edge(C)* (resp. *min-edge(C)*) denote the edge $e \in C$ with $w(e) = max(C)$ (resp. $w(e) = min(C)$). We use the following facts in our algorithm.

Fact 1. If there exists a $[a, b]$-critical cut, then there exists a spanning tree in $E - E[a, b)$ and $E - E(a, b]$, but not in $E - E[a, b]$. □

Fact 2. A minimum range cut is critical to some interval. □

Since a critical interval cannot be contained in any other critical interval, we can order critical intervals by their lower boundaries. We say that a critical interval $[a_1, b_1]$ is *lower* (resp. *higher*) than a critical interval $[a_2, b_2]$ when $a_1 < a_2$ (resp. $a_1 > a_2$).

Let T_{max} (resp. T_{min}) denote a maximum (resp. minimum) spanning tree of G. We use the convention throughout the paper that T_{max} (resp. T_{min}) represents the set of edges in T_{max} (resp. T_{min}). The following Lemma shows that we only have to process the edges of $T_{max} \cup T_{min}$ in order to compute all critical intervals. That is, after obtaining maximum and minimum spanning trees of G, we can reduce the edge set E to $T_{max} \cup T_{min}$.

Lemma 1. For any cut C, there exists an edge $e \in T_{max} \cap C$ (resp. $e \in T_{min} \cap C$) with $w(e) = \max(C)$ (resp. $w(e) = \min(C)$).

Proof. Let $w(e') = \max\{w(e) \mid e \in T_{max} \cap C\}$, and let *max-edge(C)* $= (u, v)$. Suppose $w(e') < max(C)$. Since *max-edge(C)* $\notin T_{max}$, there exist a path $p(u, v)$ from u to v on T_{max} and an edge $\hat{e} \in p(u, v) \cap C$. Since T_{max} is a maximum spanning tree, $w(\hat{e}) \ge \max(C)$ follows. This contradicts $w(e') < \max(C)$. Thus, there exists an edge $e \in T_{max} \cap C$ with $w(e) = \max(C)$. Similarly, we can show that there exists an edge $e \in T_{min} \cap C$ with $w(e) = \min(C)$. □

From the above lemma, our algorithm first computes a minimum and maximum spanning tree T_{min} and T_{max}, and then reduces the edge set E to $T_{min} \cup T_{max}$. For the simplicity, we assume throughout the succeeding discussion that E is already reduced to $T_{min} \cup T_{max}$.

Our algorithm enumerates critical intervals $[a_1, b_1], [a_2, b_2], \ldots$ in the increasing order of their lower boundaries. The general scheme of our algorithm follows the algorithm by Martello et al. [6]. However, we elaborate on that algorithm in order to speed up the computation of the next higher critical interval $[a_{i+1}, b_{i+1}]$ when the current critical interval $[a_i, b_i]$ is given. The following Lemma shows how to compute b_1.

Lemma 2. The minimum edge-weight β of a maximum spanning tree is the lowest upper-critical value.

Proof. For any $x \leq \beta$, there exists a spanning tree in $E[x, v_p]$. $\qquad \square$

Since $E[v_1, b_1]$ has a cut but $E(a_1, b_1]$ does not have a cut from the definition of a_1, the lower boundary a_1 is characterized as the minimum value x such that $G' = (V, E[v_1, x] \cup E(b_1, v_p])$ is connected. It is computed by starting with graph $G' = (V, E(b_1, v_p])$, and then adding the edges of $E(v_1)$, $E(v_2)$, ... to G' in this order until the augmented graph first becomes connected. During this process, we maintain connected components, which are represented as a spanning forest, of the corresponding graph. When a_1 is obtained, we have a spanning tree in $G = (V, E[v_1, a_1] \cup E(b_1, v_p])$.

After obtaining the lowest critical interval $[a_1, b_1]$, the next higher critical interval $[a_2, b_2]$ is computed. This is done in general as follows. The algorithm maintains the current critical interval $[\alpha, \beta]$, the graph $G_{\alpha, \beta} \equiv (V, E[v_1, \alpha] \cup E(\beta, v_p])$, and a spanning tree T in $G_{\alpha, \beta}$. The computation of the next higher critical interval $[\alpha', \beta']$ consists of the following two phases; that is, (1) the first phase computes the next higher upper-critical value β', and (2) the second phase computes the next higher lower-critical value α' when given an associated upper-critical value β'. We use the following properties to compute α' or β' in each phase. Since β' is an upper-critical value next higher than β, there is a cut in $E(\alpha, \beta']$ but there does not exist a cut in $E(\alpha, \gamma]$ for $\gamma < \beta'$. Therefore, β' is the minimum value y such that $E[v_1, \alpha] \cup E(y, v_p]$ is disconnected. Since α' is the lower-critical value next higher than α, there exists a cut in $E[\alpha', \beta']$ but not in $E[\gamma, \beta']$ for $\gamma > \alpha'$. Thus, α' is the minimum value x such that $E[v_1, x] \cup E(\beta', v_p]$ is connected. Let M be a constant such that $M > v_p$.

In the first phase, we build a spanning tree T' of maximum weight in $G_{\alpha, \beta}$ in which weights of edges in $E[v_1, \alpha]$ are modified as $w'(e) = M$ and those in $E(\beta, v_p]$ are not changed. (We call a spanning tree T' defined above *the modified maximum spanning tree in $G_{\alpha, \beta}$*.) Then β' is the minimum of modified edge weights in T', i.e., the minimum of original edge weights in $T' \cap E(\beta, v_p]$. This is justified by the following Lemma.

Lemma 3. Let $[\alpha, \beta]$ be the current critical interval and T' be a modified maximum spanning tree in $G_{\alpha, \beta}$. If $[\alpha, \beta]$ is not the highest critical interval, then the minimum of modified edge-weights in T' is the upper-critical value next higher than β.

Proof. Let y be the minimum of modified edge weights in T'. Let β' be the upper-critical value next higher than β with the associated critical interval $[\alpha', \beta']$. Since β' is upper critical, there exists an edge $e = (u, v)$ of weight β'. Suppose $\beta' < y$. Since $E[v_1, \alpha] \cup E(\beta, v_p]$ is connected and β' is an upper-critical value, β' is the minimum value such that $\beta' > \beta$ and $E[v_1, \alpha] \cup E(\beta', v_p]$ is disconnected. From the definition of T', there exists a path from u to v on T' using edges of modified weights larger than β'. Therefore, u and v are connected by a path in $E[v_1, \alpha] \cup E[y, v_p]$. This contradicts to that there exists a critical cut in $(\alpha, \beta']$. $\qquad \square$

The description of the entire algorithm is given in Figure 1. The procedure

$Refine(x, y, T)$ in Figure 2 executes the second phase that finds a next higher lower-critical value when a modified maximum spanning tree T in $G_{x,y}$ is given. In this phase, we first obtain a spanning forest F' of maximum modified weight in $E[v_1, \alpha] \cup E(\beta', v_p]$, (which is simply obtained from T' by deleting edge(s) of weight β'). Let $v_i = \alpha$. We then add edges of $E(v_{i+1})$, $E(v_{i+2})$, ... with modified weight M until F' with augmented edges contains a spanning tree. Note that since augmented edges are of the maximum weight M, a spanning forest in F' with augmented edges is of maximum modified weight, and moreover, a spanning tree finally obtained is also of maximum modified weight in $E[v_1, \alpha'] \cup E(\beta', v_p]$. The role of modified weights is to ensure that edges to be added in the algorithm have a higher priority than those in $E(\beta', v_p]$ to be included in a spanning forest maintained by the algorithm.

Procedure *Minimum-Range-Cut*
begin

(1)   Compute minimum and maximum spanning trees T_{min} and T_{max}, and let $E = T_{min} \cup T_{max}$.
       $\alpha = 0$; $z = v_p - v_1$;

(2)   Let $v_1, v_2, \cdots, v_p$ be the increasing sequence of distinct weights of edges in E.
       $\alpha = 0$; $z = v_p - v_1$;

(3) $T \leftarrow T_{max}$;

(4)   do until (all edges in T have weights M)

(4.1) $\beta = w(e)$ such that $e \in T$ and $w'(e)$ is minimum among modified edge weights in T;

(4.2) $(\alpha, \beta, T) \leftarrow Refine(\alpha, \beta, T)$;

(4.3)   if $z > \beta - \alpha$ then $\{z = \beta - \alpha$; $\alpha^* = \alpha$; $\beta^* = \beta$;$\}$
         end

(5)   Compute the connected components of $E[\alpha^*, \beta^*]$ and report a set of edges which connect a connected component and the others.

end

Figure 1. Algorithm *Minimum-Range-Cut*.

When a critical interval $[\alpha^*, \beta^*]$ with minimum range is obtained, a minimum range cut is computed as follows. If a subgraph $G' = (V, E[\alpha^*, \beta^*])$ consists of two components, a minimum range cut is the set of edges connecting two components. Otherwise (i.e., G' consists q connected components with $q \geq 3$), let V_1, V_2, ..., V_q be the sets of vertices of the connected components C_1, C_2, ..., C_q in G'. Then for any subset I of $\{1, 2, \ldots, q\}$, consider a bipartition $(\cup_{j \in I} V_j, V - \cup_{j \in I} V_j)$ and a cut induced by this partition. It can be shown that this cut is the desired minimum range cut. Then all edges in the cut belong to $E[\alpha^*, \beta^*]$ and hence the weights of those edges lie in the interval $[\alpha^*, \beta^*]$. In addition, the range of the cut cannot be smaller than $\beta^* - \alpha^*$ because the interval $[\alpha^*, \beta^*]$ is critical. Therefore, the algorithm computes one connected component of G' and reports the set of edges connecting this component and the other components as the minimum range cut.

Theorem 1. We can compute a minimum range cut in $O(MST(m,n) + n \log n)$.

Proof. It is clear from Lemmas 1, 2 and 3 and from the explanation given above that the algorithm correctly computes an minimum range cut. We now analyze the time complexity. Finding T_{min} and T_{max} requires $O(MST(m,n))$ time. Sorting weights of at most $2(n-1)$ edges in T_{min} and T_{max} requires $O(n \log n)$ time. Finding the next higher upper-critical value β' can done in $O(n \log n)$ time in total by using a heap [10]. The other time-consuming part is to update a modified maximum spanning tree to find the next higher lower-critical value. This can be implemented by using dynamic trees [9]. We associate each modified edge-weight with an edge of dynamic trees. Initially, we build a dynamic tree associated with a maximum spanning tree. When we add an edge $e = (u,v)$ to T', we first check if the addition of this edge creates a cycle or not. If so, we delete a minimum weight edge on the cycle by $evert(u)$ and $cut(findmin(v))$ operations. Then we link two trees by $link(evert(u), v)$ operation (see [9] for the details of operations $link$, $evert$, and $findmin$). Since performing a sequence of n operations to the dynamic trees takes $O(n \log n)$ amortized time, it takes $O(n \log n)$ in total. □

In particular, the time complexity shown in the above theorem becomes $O(n \log n)$ for planar or Euclidean graphs, since $MST(m,n) = O(n)$ for planar graphs [2] and $MST(m,n) = O(n \log n)$ for Euclidean graphs [7], [8].

---

procedure $Refine(x,y,T)$
begin
(1)　　　　$T' \leftarrow T - E(y)$;
(2)　　　　Suppose $x = v_i$.
(3)　　　　do $k = i + 1$ to p until (T' becomes a spanning tree)
(3.1)　　　　　add $E(v_k)$ to T' and maintain T' as a modified maximum
　　　　　　　spanning tree;
　　　　end
(4)　　　　$\alpha = v_k$;
(5)　　　　$return(\alpha, y, T')$;
end

Figure 2. Procedure $Refine$.

3　Minimum Range Target Cut Problem

In this section, we consider the following variant of the minimum range cut problem. We shall show that this problem can be solved in the same time complexity as the algorithm in Section 3.

The minimum range target cut problem: Given a connected undirected graph $G = (V, E)$ with a real valued edge-weight function w and a target value γ, we find a cut with the minimum range containing γ. □

We may consider that we can solve this problem by picking up an appropriate critical interval from intervals obtained by the algorithm in the previous section. However, this approach may not work. This is because that there may be no feasible solution among critical intervals generated by the algorithm in the previous section, since γ may not be contained in any critical interval even if γ satisfies $v_1 \leq \gamma \leq v_p$. To overcome this difficulty, we define new concepts upper-critical interval (or cut) with respect to e for each $e \in T_{min}$ in the next paragraph.

First notice that as in the minimum range cut problem, the edge set E can be reduced to $T_{min} \cup T_{max}$ by Lemma 1. To avoid an complicated argument, we assume that no two edges in E have the same weight (the other case can be similarly treated). From Lemma 1, we can assume without loss of generality that max-edge(C) $\in T_{max}$ and min-edge(C) $\in T_{min}$. The interval $[w(e), \beta]$ is called upper-critical with respect to e if there exists a cut C such that min-edge(C) = e (i.e., $\min(C) = w(e)$) and $\max(C) = \beta$, but there does not exist a cut C' such that min-edge(C') = e and $max(C') < \beta$. Such C is called a upper-critical cut with respect to e and is denoted by $C(e)$. Let upper(e) denote max-edge(C). Since all edge weights are assumed to be distinct, such C is uniquely determined. For each $f \in T_{max}$, lower-critical interval (or cut) with respect to f is similarly defined.

The algorithm first computes upper-critical intervals with respect to e and upper(e) for each $e \in T_{min}$. We will then show that, if $e \neq e'$ for $e, e' \in T_{min}$, upper(e) $\neq$ upper(e') holds. This fact is a key to finding a minimum range target cut.

Let C denote the set of cuts $C(e)$ for all $e \in T_{min}$. We now define the following three subsets of C:

$$C_1 = \{C(e) \in C \mid \min(C(e)) \leq \gamma \leq \max(C(e))\}, \tag{1}$$
$$C_2 = \{C(e) \in C \mid \min(C(e)) > \gamma\}, \tag{2}$$
$$C_3 = \{C(e) \in C \mid \max(C(e)) < \gamma\}. \tag{3}$$

For the sake of simplicity, we assume that none of C_1, C_2, or C_3 is an empty set. Let also

$$r_1 = \min\{\max(C(e)) - \min(C(e)) \mid C(e) \in C_1\}, \tag{4}$$
$$r_2 = \min\{\max(C(e)) \mid C(e) \in C_2\}, \tag{5}$$
$$r_3 = \max\{\min(C(e)) \mid C(e) \in C_3\}. \tag{6}$$

Let e^* (resp. e^{**}, e^{***}) be an edge which realizes the above value r_1 (resp. r_2, r_3). Then a cut $C(e^*)$ is a candidate for a minimum range target cut, but cannot be concluded to be a minimum range target cut. The reason is that, for any $C(e') \in C_2$ and $C(e'') \in C_3$, a cut in $C(e') \cup C(e'')$ clearly contains a target γ, and such a cut may have a shorter range. Among those cuts, the best candidate for a minimum range target cut is a cut C' in $C(e^{**}) \cup C(e^{***})$. Between these two cuts $C(e^*)$ and C', the one with smaller range is output as the desired minimum range target cut (the rigorous proof will be given later in Lemma 7).

We first give the outline of the algorithm for finding all upper-critical intervals with respect to all $e \in T_{min}$. The algorithm is similar to Algorithm *Minimum-Range-Cut* given in the previous section. Let $e_1, e_2, \ldots, e_{n-1}$ be the list of edges in T_{min} with $w(e_1) < w(e_2) < \cdots < w(e_{n-1})$ (recall that we assume all edge weights are distinct). Define T_i for i with $1 \leq i \leq n-1$ to be a spanning tree that has the maximum weight among all spanning trees that contain $\{e_1, e_2, \ldots, e_i\}$. We use the convention that T_0 stands for T_{max}. Starting with T_0, the algorithm computes $T_1, T_2, \ldots, T_{n-1}$ in this order by performing an edge-exchange one at a time. When T_i is obtained from T_{i-1} by performing an edge-exchange (e_i, f_i) (i.e., $T_i = (T_{i-1} - f_i) \cup e_i$), $upper(e_i)$ is output as f_i. Since outputting the edge set of all T_i requires $O(n^2)$ time, we simply report (e_i, f_i) for each i. The algorithm is described as follows.

Since we assume that all edge weights are distinct, for each $f_i \in T_{max}$, there exists the unique e_i such that $upper(e_i) = f_i$. We also denote such e_i as $lower(f_i)$. Here M used in Step 2 is a big constant defined in the previous section. A tree T newly obtained in Step 2 is equal to T_i as will be proved in Lemma 4. We shall show in Lemma 5 that $f_i = upper(e_i)$ holds for all i.

Procedure *Upper-Critical-Cuts*
begin
(1)   Let $T = T_{max}$ and $i = 1$.
(2)   For $e_i = (u_i, v_i)$, let f_i be the edge such that $w(f_i)$ is minimum among edges on the unique path between u_i and v_i in T. Let $T = (T - f_i) \cup e_i$ and $w(e_i) = M$. Output f_i as $upper(e_i)$. Let $i = i + 1$.
(3)   If $i \leq n - 1$ return to Step 2. Else stop.
end

Lemma 4. The algorithm explained above correctly computes T_i for all i with $1 \leq i \leq n - 1$.

Proof. Note that f_i computed by the algorithm belongs to T_{max} since $w(f_i)$ is of minimum weight among all edges on the path of T_{i-1} between u_i and v_i and all edges $e_1, e_2, \ldots, e_{i-1}$ have the weight M which is larger than $w(f_i)$ from the definition of M. Then the proof is done in a straightforward manner by induction on i. Thus the details are omitted. $\quad\square$

Let $f_i = (x_i, y_i)$ and let $T(x_i)$ and $T(y_i)$ be the two subtrees obtained by deleting f_i from T_{i-1} such that $x_i \in T(x_i)$ and $y_i \in T(y_i)$. Let $V(x_i)$ (resp. $V(y_i)$) be the set of vertices of $T(x_i)$ (resp. $T(y_i)$). Assume without loss of generality that $u_i \in T(x_i)$ and $v_i \in T(y_i)$.

Lemma 5. For each e_i with $1 \leq i \leq n - 1$, let the cut C be the set of edges connecting $V(x_i)$ and $V(y_i)$. Then (1) C is the upper-critical cut with respect to e_i, and (2) C is the lower-critical cut with respect to f_i.

Proof. We first prove (1). Note that $e_i, f_i \in C$ holds. Suppose that $max(C(e_i)) < w(f_i)$ holds for the upper-critical cut $C(e_i)$ with respect to e_i. Thus $C(e_i)$ does not contain f_i. Since $C(e_i)$ must contain at least one edge on the path of T_{i-1} between u_i and v_i and all edges on the path have larger weights than $w(f_i)$ or smaller weights than $w(e_i)$, this is a contradiction. Therefore, $max(C(e_i)) = w(f_i)$ holds. If $max(C) > w(f_i)$, the edge $max\text{-}edge(C) \in T_{max}$ connects $V(x_i)$ and $V(y_i)$. Exchanging f_i and $max\text{-}edge(C)$ creates a spanning tree that has a weight larger than T_{i-1} and contains $\{e_1, e_2, \ldots, e_{i-1}\}$, which contradicts Lemma 4. This proves (1). We can also prove (2) in a similar way as above. □

Note that this lemma shows that Step 2 in the algorithm correctly computes a critical interval with respect to the current edge e_i. From this lemma we assume in the succeeding discussion that $C(e_i)$ is the one defined in the lemma statement.

Procedure *Minimum-Range-Target-Cut*
begin
(1) Compute T_{min} and T_{max} and let $E = T_{min} \cup T_{max}$.
(2) Compute an upper-critical interval with respect to e for each $e \in T_{min}$ and three sets C_1, C_2, and C_3. Compute r_1, r_2 and r_3 as well as the corresponding three cuts $C(e^*)$, $C(e^{**})$ and $C(e^{***})$.
(3) If $r_1 \leq r_2 - r_3$, output $C(e^*)$ as a minimum range target cut. Else output $C(e^{**}) \cup C(e^{***})$.
end

Lemma 6. The above algorithm runs in $O(n \log n)$ time by using dynamic trees of [9].
Proof. Similarly done as in the proof of Theorem 1. □

After obtaining all pairs of $(e, upper(e))$ for each $e \in T_{min}$, the desired cut is computed as explained at the beginning of this section. The following lemma justifies this. We need a definition of non-crossing cuts in the lemma. Suppose a cut C_1 (resp. C_2) is induced by a bipartition $(X_1, V - X_1)$ (resp. $(X_2, V - X_2)$). When $X_1 \subset X_2$ or $X_1 \subset V - X_2$, we say that two cuts C_1 and C_2 *do not cross*. Otherwise, two cuts *cross* each other.

Lemma 7. Either $C(e^*)$ or $C(e^{**}) \cup C(e^{***})$ is a minimum range target cut.
Proof. Suppose C' is a desired minimum range target cut, and let $f_* = min\text{-}edge(C')$ and $f^* = max\text{-}edge(C')$.

If $w(upper(f_*)) \geq \gamma$ or $w(lower(f^*)) \leq \gamma$, then $upper(f_*) = f^*$ since otherwise $C(f_*)$ or $C(lower(f^*))$ belongs to C_1 and has a range shorter than that of C'. Therefore, in this case $C(e^*)$ is a minimum range target cut.

Suppose $w(upper(f_*)) < \gamma$ and $w(lower(f^*)) > \gamma$. Notice that $C(f_*) \in C_3$ and $C(lower(f^*)) \in C_2$. We now prove that two cuts $C(f_*)$ and $C(lower(f^*))$ do not cross. Suppose they cross. Since we assume that every edge weight is distinct, there is a cut C'' in $C(f_*) \cup C(lower(f^*))$ whose range contains γ and is less than $w(f^*) - w(f_*)$. This contradicts to that C' is a minimum range target cut. Therefore, $C(f_*) \in C_3$ and $C(lower(f^*)) \in C_2$ do not cross, and thus $C(f_*) \cup C(lower(f^*))$ is a minimum range target cut. From the definition of $C(e^{**})$ and $C(e^{***})$, a cut $C(f_*) \cup C(lower(f^*))$ must be a cut $C(e^{**}) \cup C(e^{***})$. □

The entire algorithm is constructed as follows.

Theorem 2. The above algorithm correctly computes a minimum range target cut in $O(MST(m, n) + n \log n)$ time. $\qquad\qquad\qquad\qquad\square$

As in the minimum range cut problem, the time complexity shown in the above theorem becomes $O(n \log n)$ for planar or Euclidean graphs.

4 Conclusion

In this paper, we devised $O(MST(m, n) + n \log n)$ time algorithms for the minimum range cut problem and its variant. Our algorithms improved the previous $O(m^2)$ time bound, which can be obtained by Martello et al.'s general approach [6]. The following two ingredients lead to this improvement: they are, (1) we only have to process edges in a minimum or maximum spanning tree to solve range cut problems; (2) for any cut, its minimum (*resp.* maximum) weight edge belongs to a minimum (*resp.* maximum) spanning tree. We hope that these are of self interest, and are now investigating on the further applications of these observations.

Acknowledgements

This work was partially supported by Grant in Aid for Scientific Research of the Ministry of Education, Science and Culture of Japan under Grant-in-Aid for Co-operative Research (A) 02302047 (1990).

References

[1] P.M. Camerini, F. Mafioli, S. Martello, and P. Toth, Most and Least Uniform Spanning Trees. *Discrete Applied Mathematics*, Vol. 15, 181-197. 1986.

[2] D. Cheriton and R.E. Tarjan, Finding Minimum Spanning Trees. *SIAM Journal on Computing*, Vol. 5, 724-742. 1976.

[3] M.L. Fredman and D.E. Willard, Trans-Dichotomous Algorithms for Minimum Spanning Trees and Shortest Paths. *Proceedings of the IEEE 31st Annual Symposium on Foundations of Computer Science*, 719-725. 1990.

[4] H.N. Gabow, Z. Galil, T. Spencer, and R.E. Tarjan, Efficient Algorithms for Finding Minimum Spanning Trees in Undirected and Directed Graphs. *Combinatorica*, Vol. 6, No. 2, 109-122. 1986.

[5] Z. Galil and B. Schieber, On Finding Most Uniform Spanning Trees. *Discrete Applied Mathematics*, Vol. 20, 173-175. 1987.

[6] S. Martello, W.R. Pulleyblank, P. Toth, and D. de Werra, Balanced Optimization Problems. *Operations Research Letters*, Vol. 3, No. 5, 275-278. 1984.

[7] C. Monma, M. Paterson, S. Suri, and F. Yao, Computing Euclidean Maximum Spanning Trees. *Proceedings of the Fourth Annual ACM Symposium on Computational Geometry*, 241-251. 1988.

[8] F.P. Preparata and M.I. Shamos, *Computational Geometry*, Springer Verlag, New York, NY. 1985.

[9] D.D. Sleator and R.E. Tarjan, A Data Structure for dynamic Trees. *Journal of Computer and System Sciences*, Vol. 26, 362-391. 1983.

[10] R.E. Tarjan, *Data Structures and Network Algorithms*, Society for Industrial and Applied Mathematics, Philadelphia, PA. 1983.

Appendix A. Minimum Range Cut Algorithm

Figure A illustrates the behavior of the algorithm in Section 3 for an input graph G. Notice that T_{min} (resp. T_{max}) is a minimum (resp. maximum) spanning tree. Since the minimum edge weight in T_{max} is 8, we have $b_1 = 8$. During the execution of the algorithm, a modified maximum spanning tree is indicated by thick lines. After the deletion of edge 8 and additions of edges 2, 3, 4, and 5, a subgraph with thick lines becomes a spanning tree. Therefore, we obtain the lowest critical interval [5, 8]. In the same way, we obtain critical intervals [12, 26] and [16, 30]. Then we report the interval [5, 8] as a minimum range interval.

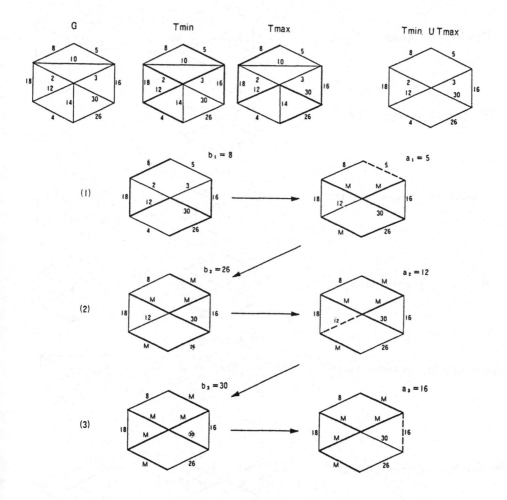

Figure A.

Appendix B. Minimum Range Target cut

Figure B shows the behavior of the algorithm in Section 4 with a target value 11. Figures in the left-hand side illustrate the current modified maximum spanning trees T_i, while figures on the other side illustrate the associated cuts. For example, in Figure (B1) indicates that the minimum edge weight 2 in T_{min} is associated with an edge of weight 12 in T_1. Then the corresponding cut is obtained by deleting the edge of weight 12 from T_1. We now add the edge of weight 2 with a new weight M into T_1, create T_2, and then proceed to the next step. Finally, we obtain the desired minimum range $[2, 12]$.

Modified max. spanning trees   Corresponding cuts

(1) $w(e_1) = 2$, $w(upper(e_1)) = 12$

(2) $w(e_2) = 3$, $w(upper(e_2)) = 16$

(3) $w(e_3) = 4$, $w(upper(e_3)) = 18$

(4) $w(e_4) = 5$, $w(upper(e_4)) = 8$

(5) $w(e_5) = 12$, $w(upper(e_5)) = 26$

(6) $w(e_6) = 16$, $w(upper(e_6)) = 30$

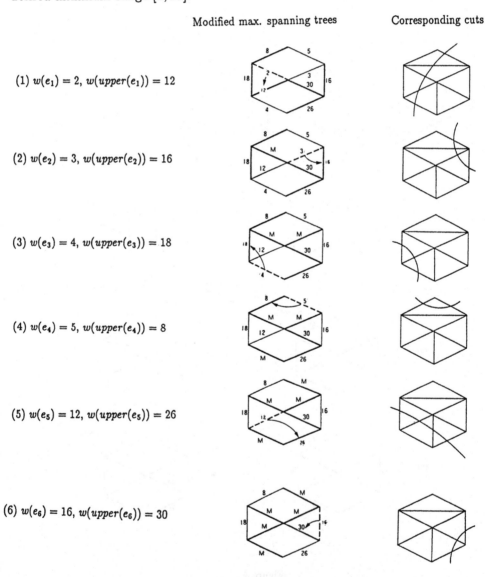

Figure B.

Memory Access in Models of Parallel Computation: From Folklore to Synergy and Beyond

Selim G. Akl
Department of Computing and Information Science
Queen's University
Kingston, Ontario Canada K7L 3N6

Abstract

We investigate various aspects of the role played by memory in the parallel random access machine (PRAM) model. Three results are obtained, each exhibiting a different consequence of the power of concurrent memory access in the PRAM: (i) two folk theorems are revisited and found not to hold for a class of inherently parallel problems; (ii) for a wide family of computations, the amount of work done by a parallel solution is shown to be smaller than that done by the best possible sequential one; and (iii) a model of computation is introduced which is an extension of the strongest variant of the PRAM, yet it requires no more resources than its weakest variant.

1. Introduction

Over the last few years it has become increasingly obvious that parallel computers provide an opportunity for the efficient use of a memory system that is larger, faster, and more cost-effective than its uniprocessor counterpart. This, coupled with the realization that memory is indeed the costly part of a computer, has been one of the main motivations behind the growing number of applications of parallelism to the solution of real problems.

In this paper we attempt to justify the above claim from a theoretical point of view. Specifically, we investigate various aspects of the role played by memory in a theoretical model of parallel computation, the parallel random access machine (PRAM). The paper offers three results, each exhibiting a different consequence of the power of concurrent memory access in the PRAM.

In section 2, we define the two models of computation used, namely the (sequential) random access machine (RAM), and the PRAM. Two folk theorems in parallel computation (the *speedup* and *slowdown* theorems) are revisited in section 3 and shown not to hold, as usually formulated in the literature, for a class of inherently parallel problems. In section 4, we demonstrate that for a wide family of computations the amount of work done by a parallel solution is smaller than that done by the best possible sequential one.

This research was supported by the Natural Sciences and Engineering Research Council of Canada under grant A3336.

This contradicts the currently established belief and does not appear to be amenable to explanation using existing approaches. We use the term *parallel synergy* to refer to this phenomenon. A model of computation called *Broadcasting with Selective Reduction* (BSR) is introduced in section 5. This model is an extension of the strongest variant of the PRAM, yet it requires no more resources than its weakest variant. BSR derives its power from an effective exploitation of the network necessary in the PRAM for linking processors to memory locations. Efficient parallel algorithms for several problems are described in order to illustrate the use of the model. Section 6 offers a number of conclusions and open problems.

2. Models of Computation

The results derived in the next three sections use the two models that have established themselves as the most widely accepted for sequential and parallel computation, namely the RAM [Hopcroft] and the PRAM [Akl 1], respectively.

2.1 The Random Access Machine

As shown in Fig. 1, the main components of interest in a RAM are:

1) a *memory* consisting of an arbitrary number of *memory locations*;
2) an *interconnection unit* (IU) which allows access to any selected memory location;
3) a *processor* capable of loading and storing data from and into the memory and executing a number of arithmetic and logical operations.

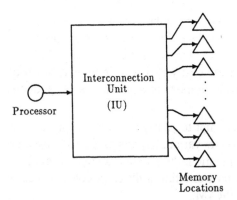

Fig. 1 The RAM

The processor possesses a constant number of local registers to perform computations on data, and operates under the control of a sequential algorithm. Each step of such an algorithm consists of (up to) three phases:

(i) a READ phase, where the processor reads a datum from memory and stores it in one of its local registers;

(ii) a COMPUTE phase, where the processor performs an elementary operation (such as comparison, addition, etc...) on data in its local registers;

(iii) a WRITE phase, where the processor writes into the random access memory the result of some computation.

Each of the phases is assumed to take constant time, leading to a constant execution time per step. Note that two registers are required to perform a computation of the form a := f(a,b), where f is an elementary operation. It should also be clear that all data movements in memory must be done through the processor. Thus in order to copy the contents of one memory location into another, one step is needed which uses one internal register.

2.2 The Parallel Random Access Machine

As shown in Fig. 2, the PRAM consists of a number of identical processors (of the type used in the RAM), sharing a common memory (also of the type used in the RAM). An interconnection unit allows the processors access to memory. The latter stores data and serves as the communication medium for the processors. These operate under the control of a parallel algorithm. Each step of such an algorithm consists of (up to) three phases:

(i) a READ phase where processors read data from the shared memory and store them in their local registers;

(ii) a COMPUTE phase where processors perform elementary operations on local data;

(iii) a WRITE phase where processors write results to the shared memory.

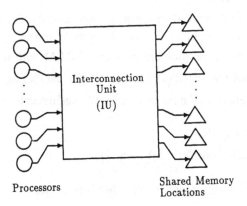

Fig. 2 The PRAM

Each of these phases requires constant time, again leading to a constant time per step of the algorithm. Note that some processors may not execute one or two phases of a given step. Also, the model is further specified by defining the mode of memory access. Three variants are most commonly used. In the Exclusive-Read Exclusive-Write (EREW) PRAM, no two processors can gain access to the same memory location simultaneously whether for reading or for writing. The Concurrent-Read Exclusive-Write (CREW)

PRAM allows more than one processor to read from, but not write into, the same memory location at the same time. Finally, in the Concurrent-Read Concurrent-Write (CRCW) PRAM it is possible for several processors to gain access to the same memory location either for the purpose of reading or for the purpose of writing. (Note that the fourth logical variant, the Exclusive-Read Concurrent-Write PRAM, has a very restricted range of applications [Akl 1].) When two or more processors attempt to write into the same memory location, the CRCW PRAM uses one of several conflict resolution rules in order to determine what actually ends up in the memory location. Examples of such rules include the COMMON rule (whereby the processors are allowed to write provided that they are attempting to write the same value), the ARBITRARY rule (whereby a random processor is selected for writing), the PRIORITY rule (whereby the processor with the highest priority wins), and the SUM rule (whereby the sum of the values being written is stored).

3. Data Movement Intensive Problems

A computational problem is said to be *data-movement-intensive* if the number of operations required to move data from one location to another in memory dominates all other computations. In this section, we reconsider two folk theorems in parallel computation in the context of data-movement-intensive problems. Our analysis uncovers some peculiarities typically hidden by asymptotic interpretations of these two folk theorems. Specifically, we exhibit problems for which the two folk theorems do not hold on the EREW PRAM.

3.1 A Folk Speedup Theorem

Given a certain computational problem, assume that the fastest possible sequential algorithm for solving this problem on a RAM runs in time t_s. Assume further that a parallel algorithm for the same problem runs on a PRAM with n processors in time t_p. The ratio t_s/t_p is called the *speedup* provided by the parallel algorithm. The first folk theorem we consider relates to the maximum speedup achievable theoretically [Akl 1].

FOLK SPEEDUP THEOREM:

For any computational problem, $t_s/t_p \leq n$. []

We now exhibit a problem for which the speedup is larger than n.

PROBLEM 1:

An array $X[1]$, $X[2]$, ..., $X[n]$ is given which contains n distinct integers I_1, I_2, ... I_n in the range ($-\infty$, n], such that $X[i] = I_i$ for all $1 \leq i \leq n$. It is required to modify array X so that for all i, $1 \leq i \leq n$, $X[I_i] = I_i$ if and only if $1 \leq I_i \leq n$, otherwise $X[i] = I_i$.

It is shown in [Akl 3] that Problem 1 cannot be solved in less than 2n - 1 steps on the RAM, and that this bound is tight since an algorithm requiring this many steps can be easily derived. On the other hand, a PRAM with n processors can solve the problem in the obvious way in *one step*. Thus, while the maximum possible speedup by the Folk Speedup Theorem is n, we have managed to achieve a speedup of 2n - 1. Although this is greater than n only by a multiplicative factor, the result is remarkable in that it uncovers the existence of problems for which the amount of computational effort (also referred to as the *cost* or *work*, and meaning the total number of steps) spent by a parallel algorithm can be *smaller* in absolute terms than that of any sequential algorithm for the same problem (in our example n versus 2n - 1). We return to this theme in section 4.

3.2 A Folk Slowdown Theorem

The second theorem appears in various forms in the literature [Gibbons, Karp] and is usually attributed to Brent [Brent].

FOLK SLOWDOWN THEOREM:

If n processors can perform a computation in one step, then p processors can perform the same computation in $\lceil n/p \rceil$ steps, for $1 \leq p \leq n$. []

Again, we exhibit a problem for which this theorem does not hold.

PROBLEM 2:

Given an array X[1], X[2], ..., X[n] containing arbitrary data, and an integer q that divides n, the contents of every sequence of q consecutive elements of X are to be cyclically shifted.

Clearly an EREW PRAM with n processors can solve this problem in one step, while, as shown in [Akl 3], $\lceil n/p + n/pq \rceil$ steps are necessary and sufficient on an EREW PRAM with $2 \leq p < q$ processors. This means that when q = 3 and p = 2, for example, the number of steps is $\lceil 2n/3 \rceil$ instead of the $\lceil n/2 \rceil$ predicted by the Folk Slowdown Theorem. This result has two noteworthy consequences:

1. It implies that one can no longer simply design an algorithm for a given number of processors and assume that the algorithm can be used with an arbitrary number of processors by applying the Folk Slowdown Theorem, since the latter may, in fact, not apply.
2. It gives renewed importance to so-called *adaptive* algorithms, i.e. algorithms specifically designed to run on an arbitrary number of processors.

4. Synergy

In our definitions of the RAM and PRAM an important part of the two models was left unspecified, namely the IU connecting processors to memory locations. In order to make this notion more precise, we undertake to implement the IU in the form of a *combinational circuit* [Parberry]. Such a circuit is a device which takes a number of inputs at one end and produces a number of outputs at the other end. It is made up of a number of interconnected *components* (comparators, adders, switches, etc...) arranged in columns called *stages*. Each component has a constant number of input lines and a constant number of output lines: in *one time unit* it computes a function of its input and produces the result as output. The circuit has no *feedback;* this means that each component is used once. The *depth* of a circuit is the maximum number of components on a path from input to output. Its *width* is the maximum number of components in a stage. The *size* of a combinational circuit is defined as the number of components it uses, and is bounded from above by the product of the circuit's depth and width. Examples of combinational circuits are the circuits of [Batcher] and [Ajtai] for sorting n data values: both have width $O(n)$, but their depths are $O(\log^2 n)$ and $O(\log n)$, respectively.

An IU for a RAM with m memory locations can be implemented as a binary tree of switches: the tree's root is connected to the processor, and each of its m leaves to a distinct memory location [Kuck]. The circuit decodes a log m bit address in $O(\log m)$ time. It has width $O(m)$, depth $O(\log m)$, and size $O(m)$, which is clearly optimal. An IU for a PRAM with m memory locations and $n = O(m)$ processors is given in [Vishkin]. Using the sorting circuit of [Ajtai] as a main building block, this IU has $O(m)$ width, $O(\log m)$ depth and $O(m \log m)$ size, and implements all the variants of the PRAM. Furthermore, its size is optimal (even for the EREW PRAM) in view of a well-known lower bound on the number of switches needed to connect m processors to m memory locations [Shannon].

Let us define the *cost* of an algorithm as the product of the number of processors it uses and its running time. It is usually said that if c is the cost of running an algorithm A on a PRAM, then the cost of simulating A on a RAM is (asymptotically) equal to c. We argue in [Akl 2] that this statement is no longer true once the cost of the IU is taken into consideration. More precisely, we show that for a wide family of computational problems, the cost of a PRAM solution is smaller (asymptotically) than that of the best possible sequential solution. Furthermore, the cost of the PRAM solution grows when simulated on a RAM.

4.1 True Efficiency

Assume that a problem P of size n is given, for which A_s is a sequential algorithm running in sequential time $t_s(n)$ on a RAM. The cost of A_s is $c_s(n) = 1 \times t_s(n) = t_s(n)$. Note that, because there is only one processor, the cost of a sequential algorithm is exactly its running time. Further, let A_p be a parallel algorithm for P running in parallel time $t_p(n) = t_s(n)/n$ on an n-processor PRAM. The cost of A_p is $c_p(n) = n \times t_p(n) = t_s(n)$.

As $c_s(n)$ and $c_p(n)$ are derived for the RAM and PRAM, respectively, they do not take into consideration the IU. Indeed, both the RAM and PRAM ignore the IU despite the fact that its cost (i.e. the product of the number of processors it uses and the time required to traverse it) dominates that of many computations. We now propose to examine what happens to $c_s(n)$ and $c_p(n)$ once the cost of the IU is taken into account.

The revised cost of A_s is:

$$c'_s(n) = (1 + O(n)) \times (t_s(n) \times O(\log n)) = t_s(n) \times O(n \log n),$$

while the revised cost of A_p is:

$$c'_p(n) = (n + O(n \log n)) \times (t_s(n)/n \times O(\log n)) = t_s(n) \times O(\log^2 n).$$

In other words, the cost of the parallel algorithm is asymptotically smaller than that of the sequential algorithm! It is important to point out that, since $t_p(n)$ is defined as $t_s(n)/n$, the exact value of $t_s(n)$ is irrelevant when computing the ratio $c'_s(n)/c'_p(n)$. If so needed, one may of course assume that $t_s(n)$ is the running time of the fastest possible sequential algorithm for P.

As noted above, the traditional approach to analyzing parallel algorithms does not take into consideration the cost of the IU. In that approach, the *efficiency* of a parallel algorithm for a given problem is defined as the ratio of the running time of the fastest sequential algorithm for that problem to the cost of the parallel algorithm. Because of the assumption that a RAM can simulate a PRAM algorithm in no more time than the cost of the latter, the efficiency of a parallel algorithm is at most 1 (or some constant larger than 1, in view of our results in section 3). By contrast, we refer in this paper to the ratio $c'_s(n)/c'_p$ as the *true efficiency* of a parallel algorithm. This ratio in the case of A_s and A_p is $O(n/\log n)$. For example, consider Problem 1. As discussed in section 3.1, $t_s(n) = O(n)$ and $t_p(n) = O(1)$. Consequently, $c'_s(n) = O(n^2 \log n)$ and $c'_p(n) = O(n \log^2 n)$. It follows that $c'_s(n)/c'_p(n) = O(n/\log n)$. Note that simulating the PRAM algorithm on the RAM leads to an algorithm whose cost is:

$$O(n) \times (c'_p(n) \times O(\log n)) = O(n^2 \log^3 n).$$

This cost is larger than the (optimal) costs of both the RAM and PRAM solutions.

In general, for any computational problem of size n, $c'_s(n) = t_s(n) \times O(n \log n)$ is asymptotically larger than $c'_p(n) = t_p(n) \times O(n \log^2 n)$, provided that $t_s(n)$ is asymptotically larger than $t_p(n) \times O(\log n)$, and the number of steps where a READ and/or a WRITE phase is executed dominates the computation. This condition is satisfied by a wide family of problems. These problems include selection, merging, sorting, and a variety of computations in numerical analysis, graph theory and computational geometry [Akl 1].

4.2 Discussion

Traditional analyses of cost either ignore the existence of the IU, or (implicitly) assume that its cost is $O(1)$. Both approaches are clearly unrealistic: any reasonable model of computation must include as an integral part a means of linking processors to memory locations, whose cost is a function of the number of these processors and memory locations.

On the other hand, in deriving $c'_s(n)$ and $c'_p(n)$ we have taken into account both the number of processors required for the IU and the time elapsed during memory access. It may be argued that since the processors used to build the IU are simpler than those actually doing the arithmetic and logical computations, the two ought not be treated equally in the cost analysis. The fallacy in this argument is that the complexity (i.e. size and number of internal components such as registers) of a (RAM or PRAM) processor used for computing is only a constant multiple of that of an IU processor. It is therefore quite reasonable in an asymptotic cost analysis to lump the two kinds of processors together and view them (as we did) as *active agents* of a computational model (memory locations being the *passive agents*). As a result, we arrived at a conclusion contradicting the established belief whereby the cost of a parallel algorithm for a given problem cannot be smaller than that of the best possible sequential algorithm for that problem.

From the above discussion we conclude that the PRAM allows for a synergistic phenomenon to occur. This phenomenon manifests itself by a reduction in computational cost (as defined in this paper) for a wide variety of problems. These problems are characterized by an intensive movement of data from and into memory during the course of a computation. On the RAM, each access to *one* of n memory locations requires $O(\log n)$ time, and uses an interconnection network of size $O(n)$. The PRAM, on the other hand, allows access to *all* n memory locations simultaneously (also in $O(\log n)$ time) via an interconnection network of size $O(n \log n)$, *and not $O(n^2)$*. Through what we call *parallel synergy*, a PRAM with n processors is therefore more efficient than n RAMs.

5. Broadcasting with Selective Reduction

In sections 3 and 4 we saw how memory access is the key to the power of the PRAM whether in terms of speed or efficiency. We also established that the IU linking processors to memory locations is an integral part of the model. In this section we show that once the IU has been adopted in the model definition it can be exploited to obtain an even more powerful model than the PRAM, namely the Broadcasting with Selective Reduction (BSR) model. BSR is of the shared memory family. It possesses all the features of the CRCW PRAM. One additional instruction, BROADCAST, makes it more powerful: The IU is used to allow *all* processors to *broadcast* data (one datum per processor) to all memory locations. Each memory location *selects* a subset of the received data and *reduces* it to one value eventually stored in that location [Akl 4]. This instruction is implemented in time T, where T is the time for memory access in the PRAM. In what follows we describe the model in detail and present algorithms for it that run in time $O(T)$. If $T = O(1)$, as usually assumed in the PRAM (see section 2.2), then the BSR algorithms run in constant time. It is important to stress here that while BSR is more powerful than the CRCW PRAM, it requires no more resources than the EREW PRAM.

5.1 Model, Notation and Algorithms

As mentioned above, the BROADCAST instruction allows *all* processors to simultaneously write to *all* shared memory locations. This is accomplished as follows. Each processor i produces a datum d_i (to be broadcast) accompanied by a tag t_i. Each memory location j is associated with a selection rule σ, a limit value l_j, and a reduction operator R. All data d_i for which $t_i \sigma l_j$ is true are combined by R and stored in memory location j. In what follows, let x_1, x_2, ..., x_m denote a sequence of variables stored in shared memory. We typically take $\sigma \in \{<, \leq, =, \geq, >, \neq\}$, and $R \in \{\Sigma, \Pi, \cap, \cup, \text{AND}, \text{OR}, \text{XOR}\}$, denoting the binary associative operations: sum, product, maximum, minimum, and the logical and, or, and exclusive-or. For a BSR model with n processors, the BROADCAST instruction can therefore be written as follows:

$$x_j := \underset{t_i \sigma l_j}{R} \ d_i \qquad (1 \leq j \leq m, 1 \leq i \leq n).$$

We illustrate the use of this notation through the following examples. All algorithms given below use n processors and require a constant number of memory accesses, i.e. O(T) time. This performance is superior to that of the best PRAM solutions.

5.1.1 Prefix Sums

Given a sequence of n numbers x_1, x_2, ..., x_n, it is required to compute $s_j = \sum_{i=1}^{j} x_i$, for all j, $1 \leq j \leq n$.

BSR Algorithm: $\quad s_j := \underset{i \leq j}{\Sigma} \ x_i \quad (1 \leq j \leq n, 1 \leq i \leq n).$ []

5.1.2 Sorting

Given a sequence of n distinct numbers x_1, x_2, ..., x_n, it is required to permute them so that $x_i < x_{i+1}$ for all $1 \leq i \leq n - 1$.

BSR Algorithm: 1. $c_j := \underset{x_i < x_j}{\Sigma} \ 1 \qquad (1 \leq j \leq n, 1 \leq i \leq n)$

              2. $x_{c_j} := x_j \qquad\qquad (1 \leq j \leq n).$ []

5.1.3 Maximal Vectors

Given the Cartesian coordinates of n points in the plane as two sequences x_1, x_2, ..., x_n and y_1, y_2, ..., y_n, it is required to determine which points are maximal.

BSR Algorithm: 1. $z_j := y_j \qquad\qquad (1 \leq j \leq n)$

              2. $z_j := \underset{x_i > x_j}{\cap} \ y_i \qquad (1 \leq j \leq n, 1 \leq i \leq n)$

              3. if $z_j \leq y_j$ then (x_j, y_j) is maximal $(1 \leq j \leq n).$ []

5.1.4 Maximal Sum Subsegment

Given a sequence of n numbers d_1, d_2, ..., d_n, it is required to find two indices x and y such that $\sum\limits_{k=x}^{y} d_k$ is maximal.

BSR Algorithm: 1. {Compute the prefix sums}

$$s_j \quad := \quad \sum\limits_{i \leq j} \quad d_i \quad (1 \leq j \leq n, 1 \leq i \leq n)$$

2. {For each number find the largest sum to its right}

$$2.1) \, m_j \quad := \quad \bigcap\limits_{i \geq j} \quad s_i \quad (1 \leq j \leq n, 1 \leq i \leq n)$$

$$2.2) \, e_j \quad := \quad \bigcap\limits_{s_i = m_j} \quad i \quad (1 \leq j \leq n, 1 \leq i \leq n)$$

3. {Starting at each element compute the sum of a maximal sum subsegment}

$$m_j := m_j - s_j + d_j \quad (1 \leq j \leq n)$$

4. {Find an overall maximal sum subsegment}

$$4.1) \, t := \bigcap\limits_{i > 0} \quad m_i \quad (1 \leq i \leq n)$$

$$4.2) \, x := \bigcap\limits_{m_i = t} \quad i \quad (1 \leq i \leq n)$$

$$4.3) \, y := \quad e_x . \, []$$

5.2 An IU for BSR

We now show how the BSR model can be implemented at a cost matching Shannon's lower bound mentioned above. Our design for the IU uses two circuits as building blocks: a sorting circuit [Ajtai], and a circuit for computing prefix sums [Akl 1].

Assume for simplicity of exposition that it is desired to build an IU to implement the following BROADCAST instruction on a BSR model with 4 processors and 4 memory locations:

$$v_j := \sum\limits_{t_i < l_j} \quad d_i \quad (1 \leq j \leq 4, 1 \leq i \leq 4).$$

Each processor i, $1 \leq i \leq 4$, produces a record of the form (i, t_i, d_i), and each memory location j, $1 \leq j \leq 4$, is associated with a record of the form $(j + 4, l_j, v_j)$. For our example, let these be (1,4,6), (2,2,-4), (3,6,-2), (4,9,15), and $(5,20,v_1)$, $(6,8,v_2)$, $(7,11,v_3)$, $(8,5,v_4)$, respectively. When the broadcast is completed the values of v_1, v_2, v_3, and v_4 should be 15, 0, 15, and 2, respectively.

Fig. 3 shows a circuit for implementing this instruction. It consists of six *boxes*. In box A, the (i, t_i, d_i) records are sorted on the t_i, while in box C the $(j+4, l_j, v_j)$ records are sorted on the l_j. In box B a prefix sums computation is performed on the d_i. Box D receives two lists of records, one sorted on the t_i, the other on the l_j, which it merges into one list. Box E is essentially a prefix sums circuit augmented with additional links to copy a datum from each *leader*, i.e. datum holder, in a list of records to all the records separating it from the next (alternatively, previous) leader in the list. It is in box E that the memory records receive their v_j values from the processor records (the leaders). Finally, in box F sorting is performed on the first field in order to place the processor and memory records once again in their original positions. The BROADCAST instruction is complete.

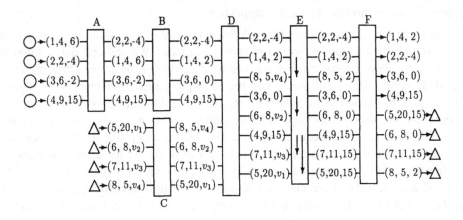

Fig. 3 Implementing the BROADCAST Instruction

The same circuit can be used to handle other BROADCAST instructions, for various combinations of σ and R by simply modifying the mode of operation of boxes B and E. In addition, the same circuit can accommodate the case where shared memory is divided into more than one array, and only one of these arrays at a time is to be affected by a BROADCAST instruction. Finally, it is not difficult to see that this circuit is also capable of implementing all variants of the PRAM. Further details pertaining to the functioning of the various boxes and the way in which each of the BROADCAST cases is implemented can be found in [Fava Lindon].

For a BSR model with m memory locations and n = O(m) processors, the IU described above has a width of O(m), a depth of O(log m), and a size of O(m log m). This is easily seen since these bounds apply to the four sorting circuits (boxes A, C, D and F) and the two prefix sums circuits (boxes B and E). Note that, as with Vishkin's design for the PRAM, the above analysis suggests a memory access time of T = O(log m), a quantity that we have assumed all along to be a constant in conformity with the definition of the PRAM.

6. Conclusions

It has been known for some time that memory access is the key to the power of the PRAM. In this paper we have exploited this power and pushed it further in three directions. Our first result was to show that two of the most fundamental beliefs in parallel computation, namely the speedup and slowdown theorems, are not universally true. Furthermore, we have exhibited the class of data-movement-intensive problems, which may be characterized as being *inherently parallel*, i.e. inefficient to solve sequentially or with fewer than the maximum possible number of processors. This suggests two directions for further research. The study of data-movement-intensive problems is to be continued with the purpose of exhibiting other members, and discovering additional properties, of the class. It should also be interesting to investigate the validity of the two folk theorems for other models of parallel computation.

Our second result was obtained by redefining the cost of a computation to include the cost of the IU linking processors to memory locations. We showed that - through a phenomenon called parallel synergy - a parallel computer can be more efficient than the sum of its parts. This result has a number of implications. Our work was originally motivated by the observation that both the RAM and the PRAM, *as theoretical models of computation*, are severely lacking. Indeed, neither model takes into account the cost of such a fundamental operation as memory access. By defining the abstract model more precisely to include all important operations, not only do we get a more realistic and meaningful analysis, but we also uncover hitherto unknown phenomena. However, both the RAM and the PRAM are idealized computers. From the practical point of view, it may be useful to conduct analyses of true efficiency, as defined in this paper, for real computers.

Our third and final result was to show that once the IU is included explicitly as part of the definition of the PRAM, a significantly more powerful model, namely BSR, can be obtained. There are several open questions in connection with BSR. One of the most important is to characterize the class of problems that can take advantage of the powerful BROADCAST instruction. It would also be useful to know whether there are computations for which fast BSR (and hence PRAM) algorithms do not exist. So far, problems that have resisted constant time solution with a polynomial number of processors include matrix problems (e.g. iterated matrix multiplication), graph theoretic problems (e.g. minimum spanning tree), list problems (e.g. list ranking), and sequence problems (e.g. longest common subsequence).

Acknowledgements

I wish to thank my colleagues and friends Michel Cosnard, Lorrie Fava Lindon, Afonso Ferreira, and Grant Guenther, in whose delightful company many of the ideas presented in this paper were developed.

References

[Ajtai] Ajtai, M., Komlós, J., and Szemerédi, E., An O(n log n) sorting network, Proceedings of the 15th Annual Symposium on Theory of Computing, Boston, Massachusetts, May 1983, pp. 1 - 9.

[Akl 1] Akl, S.G., The Design and Analysis of Parallel Algorithms, Prentice-Hall, Englewood Cliffs, New Jersey, 1989.

[Akl 2] Akl, S.G., Parallel synergy: Can a parallel computer be more efficient than the sum of its parts?, Technical Report No. 90-285, Department of Computing and Information Science, Queen's University, Kingston, Ontario, Canada, September 1990.

[Akl 3] Akl, S.G., Cosnard, M., and Ferreira, A.G., Data-movement-intensive problems: Two folk theorems in parallel computation revisited, Technical Report No. 90-18, Laboratoire de l'Informatique du Parallélisme, Ecole Normale Supérieure de Lyon, Lyon, France, June 1990.

[Akl 4] Akl, S.G., and Guenther, G.R., Broadcasting with selective reduction, Proceedings of the 11th IFIP Congress, San Francisco, California, August 1989, pp. 515 - 520.

[Batcher] Batcher, K.E., Sorting networks and their applications, Proceedings of the AFIPS 1968 Spring Joint Computer Conference, Atlantic City, New Jersey, April 1968, pp. 307 - 314.

[Brent] Brent, R.P., The parallel evaluation of general arithmetic expressions, Journal of the ACM, Vol. 21, No. 2, April 1974, pp. 201 - 206.

[Fava Lindon] Fava Lindon, L., and Akl, S.G., An optimal implementation of broadcasting with selective reduction, Technical Report No. 91-298, Department of Computing and Information Science, Queen's University, Kingston, Ontario, Canada, March 1991.

[Gibbons] Gibbons, A., and Rytter, W., Efficient Parallel Algorithms, Cambridge University Press, Cambridge, England, 1988.

[Hopcroft] Hopcroft, J.E., and Ullman, J.D., Introduction to Automata, Languages, and Computation, Addison-Wesley, Reading, Massachusetts, 1979.

[Karp] Karp, R.M., and Ramachandran, V., A survey of parallel algorithms for shared memory machines, in: Handbook of Theoretical Computer Science, North-Holland, Amsterdam, 1990, pp. 869 - 941.

[Kuck] Kuck, D.J., The Structure of Computers and Computations, Vol. 1, John Wiley & Sons, New York, 1978.

[Parberry] Parberry, I., Parallel Complexity Theory, John Wiley & Sons, New York, 1987.

[Shannon] Shannon, C.E., Memory requirements in a telephone exchange, Bell Systems Technical Journal, Vol. 29, 1950, pp. 343 - 349.

[Vishkin] Vishkin, U., A parallel-design distributed-implementation (PDDI) general-purpose computer, Theoretical Computer Science, Vol. 32, 1984, pp. 157 - 172.

Farthest Neighbors, Maximum Spanning Trees and Related Problems in Higher Dimensions

Pankaj K. Agarwal
Duke University

Jiří Matoušek
Charles University and Georgia Tech

Subhash Suri
Bellcore

Abstract

We present a randomized algorithm of expected time complexity

$$O(m^{2/3}n^{2/3}\log^{4/3}m + m\log^2 m + n\log^2 n)$$

for computing bi-chromatic farthest neighbors between n red points and m blue points in $\mathcal{E}^3$. The algorithm can also be used to compute all farthest neighbors or external farthest neighbors of n points in $\mathcal{E}^3$ in $O(n^{4/3}\log^{4/3}n)$ expected time. Using these procedures as building blocks, we can compute a Euclidean maximum spanning tree or a minimum-diameter two-partition of n points in $\mathcal{E}^3$ in $O(n^{4/3}\log^{7/3}n)$ expected time. The previous best bound for these problems was $O(n^{3/2}\log^{1/2}n)$. Our algorithms can be extended to higher dimensions.

We also propose fast and simple approximation algorithms for these problems. These approximation algorithms produce solutions that approximate the true value with a relative accuracy ε and run in time $O(n\varepsilon^{(1-k)/2}\log n)$ or $O(n\varepsilon^{(1-k)/2}\log^2 n)$ in k-dimensional space.

1 Introduction

In this paper, we consider problems in the Euclidean space $\mathcal{E}^k$, for some fixed dimension k. We present efficient algorithms for the following problems:

1. **All farthest neighbors:** Given a set S of n points in $\mathcal{E}^k$, for each point $p \in S$, compute a point $q \in S$ such that $d(p,q) \geq d(p,q')$, for all $q' \in S$; q is called a *farthest neighbor* of p.

2. **Bi-chromatic farthest neighbors:** Given a set R of n "red" points and another set B of m "blue" points in $\mathcal{E}^k$, for each point $r \in R$, compute a point $b \in B$ such that $d(r,b) \geq d(r,b')$ for all $b' \in B$; b is called a *bi-chromatic farthest neighbor* of r.

3. **External farthest neighbors:** Given a set S of n points in $\mathcal{E}^k$ and its partition S_1, S_2, ..., S_m into m subsets, for each $p \in S$, if $p \in S_i$ then compute a point $q \in S - S_i$ such that $d(p,q) \geq d(p,q')$, for all $q' \in S - S_i$; q is called an *external farthest neighbor* of p.

4. **Euclidean maximum spanning tree (EMXST):** Given a set S of n points in $\mathcal{E}^k$, compute a spanning tree of S whose edges have the maximum total length among all spanning trees, where the length of an edge is the Euclidean distance between its endpoints.

5. **Minimum diameter two-partition:** Given a set S of n points in $\mathcal{E}^k$, partition S into two sets such that the larger of the two diameters is minimized.

Computing neighbors (nearest, farthest, or some intermediate ones) is a fundamental problem in Computational Geometry, with many applications. While the all nearest neighbors problem in $\mathcal{E}^k$ can be solved in (optimal) $O(n \log n)$ time [21], no algorithm of similar efficiency is known for computing the all farthest neighbors, for $k \geq 3$. Many applications, however, require a more general formulation of this problem, such as the *bi-chromatic* and the *external* neighbors problems, where for points in one set we want to find the neighbors in some other set. The bi-chromatic and external neighbors problems also have resisted optimal algorithms for dimensions $k \geq 3$. A recent result of Agarwal et al. [1] computes a "diametral pair" in $O(n^{2-\frac{2}{\lceil k/2 \rceil + 1 + \gamma}})$ time[1] for n points in k dimensions, but their algorithms does not seem to generalize to the all farthest (or, nearest) neighbors problem. The best algorithm currently known for computing all farthest neighbors (bi-chromatic or otherwise) in three dimensions runs in time $O(n^{3/2} \log^{1/2} n)$. In this paper, we present improved, albeit randomized, algorithms for the all farthest neighbors and some related problems in three and higher dimensions.

Our main result is a randomized algorithm for computing bi-chromatic all farthest (or nearest) neighbors in expected time $O(m^{2/3} n^{2/3} \log^{4/3} m + m \log^2 m + n \log^2 n)$ in $\mathcal{E}^3$. The algorithm can be used to compute in $O(n^{4/3} \log^{4/3} n)$ expected time the farthest neighbors and the external farthest neighbors for n points in $\mathcal{E}^3$. Our algorithms can be extended to k dimensions, $k \geq 4$, for solving the bi-chromatic farthest neighbors problem in expected time $O((mn)^{1-\frac{1}{\lceil k/2 \rceil + 1 + \gamma}} + (m+n)^{1+\gamma})$, and for solving all farthest neighbors and external farthest neighbors problems in expected time $O(n^{2-\frac{2}{\lceil k/2 \rceil + 1 + \gamma}})$.

An Euclidean maximum spanning tree of n points can be computed by repeatedly invoking the (external) farthest neighbors algorithm. Thus, our result for computing the farthest neighbors leads to an improved algorithm for maximum spanning tree as well. The algorithm has expected time complexity $O(n^{4/3} \log^{7/3} n)$ in $\mathcal{E}^3$ and $O(n^{2-\frac{2}{\lceil k/2 \rceil + 1 + \gamma}})$ in $\mathcal{E}^k$, for $k \geq 4$. A variant of our algorithm can also compute an Euclidean minimum spanning tree of a set of points within the same time bound. The new algorithm is simpler than the one in Agarwal et al. [1], although unlike their algorithm, we do not know how to make ours deterministic without significantly increasing the running time. Euclidean maximum spanning tree are useful for determining a minimum diameter two-partitioning of a point set, and hence we obtain a similar result for the latter problem as well.

We also propose a simple approximation scheme that, given a set S of n points and a real parameter $0 < \epsilon < 1$, computes, for each point $p \in S$, a point $q' \in S$ that is a farthest neighbor of p with a relative error $\leq \epsilon$. In particular, if the farthest neighbor of p is $q \in S$, then q' satisfies the following inequality

$$\frac{d(p, q')}{d(p, q)} \geq 1 - \epsilon.$$

The algorithm runs in time $O(n\epsilon^{(1-k)/2})$ in k-dimensional space, and does not use any data structure beyond a linked list. The method also works for computing an ϵ-approximation of the diameter and the maximum spanning tree.

Egecioglu and Kalantari [11] recently proposed an iterative algorithm for approximating the diameter of a k-dimensional set of points. Each iteration of their algorithm runs in time $O(n)$, and the first iteration produces a distance Δ such that the diameter is between Δ and $\sqrt{3}\Delta$. The interesting aspect of this approximation is that the bound is independent of the dimension. However, in the worst case, their algorithm cannot guarantee an approximation factor better than $\sqrt{3}$, no matter how many iterations are allowed. By contrast, our algorithm achieves ϵ-approximation for arbitrarily small $\epsilon > 0$.

Our paper is organized as follows. In Section 2 we describe algorithms to compute farthest neighbors. For simplicity, we describe our algorithms in three dimensions, and sketch the details for extension to higher dimensions. In Section 3, we describe our algorithm for computing an EMXST of a set of points. Section 4 presents our approximation algorithms.

---

[1] Throughout this paper, γ will denote an arbitrarily small but positive fixed constant.

2 Computing Farthest Neighbors

In this section, we develop a randomized algorithm for computing bi-chromatic farthest neighbors, and then apply it to compute all farthest neighbors and external farthest neighbors. The structure of our algorithm follows a pattern that is fairly standard for random-sampling based geometric algorithms (e.g., see [1]). First we give an efficient algorithm for the "unbalanced" case, where the blue points greatly outnumber the red points. Then we use a randomized divide-and-conquer to convert a "balanced" problem into several instances of the unbalanced problem.

2.1 An algorithm for many blue and few red points

Recall that we are given a set R of n red points and another set B of m blue points, and for each $r \in R$, we want to compute its farthest neighbor in B. We randomly choose a subset $B' \subset B$ of $\lfloor m/4 \rfloor$ blue points. We partition the set R of red points into subsets R_1, R_2 of sizes $\lfloor n/2 \rfloor$ and $\lceil n/2 \rceil$, respectively, and then recursively solve the problems for R_1, B' and for R_2, B'. This gives us, for each $r_i \in R$, its farthest neighbor $b'_i \in B'$. Let S_i denote the (closed) ball of radius $d(r_i, b'_i)$ centered on r_i, and let $\bar{S}_i$ denote the (open) exterior of S_i. Since b'_i is a farthest neighbor of r_i in B', $B' \subset S_i$, for all i, and the farthest neighbor of r_i in B is either b'_i or lies in $\bar{S}_i$. We are therefore interested in the points of B that fall outside the common intersection $I = \bigcap_{i=1}^{n} S_i$.

Imagine that the points of B are sorted in the order of increasing distance from a point $r_i \in R$. Then the points of $B \cap \bar{S}_i$ are those lying beyond all points of the sample B', and it is easy to see that the expected number of such points is $O(1)$, for every r_i. By the additivity of expectation, we get that the expected size of $B'' = B \setminus I$ is at most $O(n)$.

The above discussion implies that we only need to compute a farthest neighbor of each $r \in R$ in the set B''. We do this in a straight-forward manner by computing $d(r, b)$, for all $b \in B''$.

The correctness of the algorithm is obvious. As for the running time of our algorithm, I can be computed in $O(n^2)$ time by mapping the balls to half-spaces in $\mathcal{E}^4$ and computing their common intersection [8, 19]. In $O(n^2)$ time and space we can preprocess I into a data structure that supports $O(\log^2 n)$ point-in-common-intersection queries, see Aurenhammer [2]. Thus, the set B'' can be computed in total time $O(n^2 + m \log^2 n)$. Finally, the farthest neighbor of each $r \in R$ in B'' can be computed in expected $O(|R| \cdot |B''|) = O(n^2)$ time. If $T(n, m)$ denotes the expected time complexity of the algorithm for the input sets of cardinalities n and m, then we have the following recurrence relation:

$$T(n, m) = O(n^2 + m \log^2 n) + 2T(\lceil n/2 \rceil, \lfloor m/4 \rfloor),,$$

which solves to $T(n, m) = O(n^2 + m \log^2 n)$.

Now we make a technical modification of our algorithm, which will give the running time in a more convenient form. Namely, we partition R into $\lceil n/s \rceil$ subsets, each of size at most $s = \sqrt{m} \log m$ and, for each i, compute the farthest neighbors of R_i in B separately using the algorithm described above. The expected running time of the whole procedure is now

$$O\left(\lceil n/s \rceil (s^2 + m \log^2 s)\right) = O(n\sqrt{m} \log m + m \log^2 m).$$

Theorem 2.1 *Let R and B be two sets of points in $\mathcal{E}^3$ with n and m points, respectively. There is an $O(n\sqrt{m} \log m + m \log^2 m)$ expected time randomized algorithm that computes, for each $r \in R$, its farthest neighbor in B.*

The same approach can be extended to higher dimensions. The only change that we need to make is to use a different point-location data structure for computing the set B''. In particular, in $\mathcal{E}^k$ we map the spheres S_i, for $1 \leq i \leq n$, to the half-spaces

$$U_i : z \leq 2\alpha_1 x_1 + \cdots + 2\alpha_k x_k + (r^2 - \alpha_1^2 - \cdots - \alpha_k^2)$$

in $\mathcal{E}^{k+1}$, where $(\alpha_1, \ldots, \alpha_k)$ and r are the center and the radius of S_i, respectively. A point $p = (p_1, \ldots, p_k)$ lies outside S_i if and only if the point $(p_1, \ldots, p_k, p_1^2 + \cdots + p_k^2)$ does not lie in U_i. Thus, $p \in I$ if and only if $p \notin \bigcap_{1 \leq i \leq n} U_i$. We preprocess the polytope formed by the intersection of half-spaces U_i $(1 \leq i \leq n)$ for point location queries using a variant of Clarkson's algorithm for point location in hyperplane arrangements [4]. This structure answers a point location query in $O(\log n)$, and requires $O(n^{\lceil k/2 \rceil + \gamma})$ time and space for preprocessing. Therefore B'' can be computed in time $O(n^{\lceil k/2 \rceil + \gamma} + m \log n)$. Analyzing in the same way as above, one can show the total running time to be $O(nm^{1 - \frac{1}{\lceil k/2 \rceil + \gamma}} + m \log n)$.

Theorem 2.2 *Let R and B be two sets of points in $\mathcal{E}^k$ with n and m points, respectively. Then, for each point $r \in R$, its farthest neighbor in B can be computed by a randomized algorithm in expected time $O(nm^{1 - \frac{1}{\lceil k/2 \rceil + \gamma}} + m \log n)$.*

Remark: The above approach can be used to solve *bi-chromatic all nearest neighbor* problem in the same expected time. The only change that we need to make in the above algorithm is replace $I = \bigcap_{i=1}^n S_i$ with $I = \bigcap_{i=1}^n \bar{S}_i$. We leave it to the reader to fill in the details.

2.2 An algorithm for the balanced case

We now describe our final algorithm, which is significantly faster than the previous algorithm when n and m are of the same order. We map each blue point $b = (b_1, b_2, b_3)$ to the hyperplane $\mathcal{D}(b)$ in $\mathcal{E}^4$, where

$$\mathcal{D}(p) : x_4 = 2b_1 x_1 + 2b_2 x_2 + 2b_3 x_3 - (b_1^2 + b_2^2 + b_3^2).$$

We identify $\mathcal{E}^3$ with the plane $x_4 = -\infty$, so the red points lie on this plane. A crucial property of the above transform is that if we treat the hyperplane $\mathcal{D}(b)$ as a 3-variate linear function

$$h_b(x_1, x_2, x_3) = 2b_1 x_1 + 2b_2 x_2 + 2p_3 x_3 - (b_1^2 + b_2^2 + b_3^2),$$

then b is a farthest neighbor of a red point $r = (r_1, r_2, r_3)$ if and only if

$$h_b(r_1, r_2, r_3) = \min_{b' \in B} h_{b'}(r_1, r_2, r_3).$$

The problem of computing the farthest neighbor of r thus reduces to finding the blue hyperplane that lies immediately above r (in x_4 direction). This property is well-known and the reader is referred to [8, 9] for details. Our second algorithm for the bi-chromatic all-farthest neighbors problem can now be outlined as follows.

1. Transform B to the set of hyperplanes $\mathcal{D}(B)$ in $\mathcal{E}^4$ and identify $\mathcal{E}^3$ with the hyperplane $x_4 = -\infty$ in $\mathcal{E}^4$.

2. Randomly choose a subset $H \subset \mathcal{D}(B)$ of size t, for some parameter t; all subsets of H of size t are chosen with equal probability.

3. Let h^- be the half space lying below the hyperplane h. Compute the common intersection $P = \bigcap_{h \in H} h^-$; P is a convex polyhedron unbounded from below.

4. Triangulate P as follows. A ridge (2-face) of P is triangulated by connecting its lowest vertex (in x_4 direction) to all other vertices. Each triangle is thus bounded by two edges incident to the lowest vertex of the 2-face and a third edge which is an original edge of the face. Similarly, a facet of P is triangulated by connecting its lowest vertex to all vertices, edges and triangles in the boundary of the triangulated facet. Each tetrahedron is now incident to its lowest vertex and is bounded by a triangle not incident to it. Finally the interior of P is decomposed by extending each of its 2-faces τ vertically downwards, that is, erecting a vertical prism whose top face is τ and which is unbounded from below. Let $\{\Delta_1, \ldots, \Delta_s\}$ be the set of cells in P. Each cell Δ_i is bounded by vertical facets and by a top facet, which is a portion of a hyperplane of H.

5. Given a cell Δ_i, let $R_i \subseteq R$ denote the set of red points contained in Δ_i.

6. Let $B_i \subseteq B$ denote the set of points b such that $\mathcal{D}(B)$ either contains the top facet of Δ_i or intersects the interior of Δ_i.

7. Solve each subproblem (R_i, B_i) separately using the previous algorithm. That is, for points in R_i, find their farthest neighbors in B_i.

The correctness of the algorithm follows from the observation that if b is a farthest neighbor of $r \in R_i$, then among all hyperplanes of B, $\mathcal{D}(b)$ lies immediately above r, and therefore $\mathcal{D}(b)$ either contains the top facet of Δ_i or intersects the interior of Δ_i. In either case $b \in B_i$ and therefore Step 7 correctly computes a farthest neighbor of r.

Lemma 2.3 *For a suitable choice of the parameter t, the expected running time of the above algorithm is $O(m^{2/3}n^{2/3}\log^{4/3} m + m \log^2 m + n \log^2 n)$.*

Proof: If $T(n, m)$ denotes the maximum running time of the algorithm for the input sets R and B of cardinalities n and m, respectively, then we have the following recurrence:

$$T(n, m) = \sum_{i=1}^{s} T(n_i, m_i) + M(n, m),$$

where $M(n, m)$ is the time needed to break up the problem of size n and m into s subproblems of sizes n_i and m_i.

The polyhedron P can be computed in time $O(t^2)$, using Seidel's convex hull algorithm [19] in the dual space. It is known that the triangulation described in Step 3 partitions P into $s = O(t^2)$ cells and such a triangulation can be obtained within the same time bound [4, 1].

To compute the sets R_i, for $i = 1, 2, \ldots, s$, we orthogonally project the triangulated polytope onto $x_4 = -\infty$ and preprocess the resulting cell-complex for point location. The projection of P is acyclic, by a result of Edelsbrunner [7], and therefore we can use the data structure of Preparata and Tamassia [18] for point location. This data structure requires $O(t^2 \log t)$ preprocessing and supports $O(\log^2 t)$ time point location queries. Thus, Step 5 can be accomplished (deterministically) in time $O(t^2 \log t + n \log^2 t)$.

Finally, in Step 6 we trace each hyperplane of $B - H$ through the triangulation to determine the sets B_i. Since the triangulation is a cell complex, the total time spent in tracing these hyperplanes is $O(\sum_{i=1}^{O(t^2)} m_i)$ (see [1]). The running time of the above algorithm is therefore bounded by

$$T(n,m) = \sum_{i=1}^{O(t^2)} T(n_i, m_i) + O\left(t^2 \log t + n \log^2 t + \sum_{i=1}^{O(t^2)} m_i\right).$$

Since we solve each subproblem directly by invoking the algorithm of the previous section, the expected running time is

$$E[T(n,m)] = E\left[\sum_{i=1}^{O(t^2)} O(n_i\sqrt{m_i}\log m_i + m_i\log^2 m_i)\right] + O(t^2\log t + n\log^2 t)$$

$$= O\left(E\left[\sum_{i=1}^{O(t^2)}(n_i\sqrt{m_i}\log m + m_i\log^2 m)\right]\right) + O(t^2\log t + n\log^2 t).$$

The theory of random sampling [6, 10, 1] implies that

$$E[\sum_{i=1}^{O(t^2)} m_i] = O(mt), \quad E[\sum_{i=1}^{O(t^2)} n_i\sqrt{m_i}] = O\left(n\sqrt{\frac{m}{t}}\right).$$

Therefore,

$$E[T(n,m)] = O\left(n\sqrt{\frac{m}{t}}\log m\right) + O\left(t^2\log t + n\log^2 t + mt\log^2 m\right)$$

By setting the parameter value $t = \left\lceil \frac{n^{2/3}}{m^{1/3}\log^{2/3} m}\right\rceil$, we obtain

$$E[T(n,m)] = O(m^{2/3}n^{2/3}\log^{4/3} m + m\log^2 m + n\log^2 n).$$

We can state the main result of this section as follows.

Theorem 2.4 *Let R and B two sets of points in $\mathcal{E}^3$ with n and m points, respectively. There is an $O(m^{2/3}n^{2/3}\log^{4/3} m + m\log^2 m + n\log^2 n)$ expected time randomized algorithm that computes, for each $r \in R$, its farthest neighbor in B.*

As in the previous section, the same approach works also in higher dimensions, with slight modifications of Steps 4 and 7. We now choose t to be some suitable constant. In Step 7, we solve the subproblem using Theorem 2.1 only if $n_i < m_i^{\frac{1}{\lceil k/2\rceil+7}}\log m_i$. Otherwise, we solve the subproblem recursively.

In Step 4, we first triangulate the boundary of P using a recursive procedure, and then "pull" each simplex downwards in the negative x_d-direction to form a prism. We triangulate the boundary of P, by triangulating its faces in increasing order of their dimensions. Suppose that we have already triangulated all the j-faces, for some $j < h$, and we now want to triangulate the h-faces. For every h-face f, we choose its lowest vertex v and form a h-simplex by connecting it to every $(h-1)$-simplex of f's $(h-1)$-faces. Note that the $(h-1)$-faces of f have already been triangulated. It is well known that the above procedure generates $O(t^{\lceil k/2\rceil})$ cells and requires $O(t^{\lceil k/2\rceil})$ time; see [4, 1] for details.

Since t is now a constant, the sets R_i, B_i can be easily computed in $O(m+n)$ time. We get the following recurrence:

$$T(n,m) = \begin{cases} \displaystyle\sum_{i=1}^{O(t^{\lceil k/2 \rceil})} T(m_i, n_i) + O(m+n) & \text{if } n \geq m^{\frac{1}{\lceil k/2 \rceil + \gamma}} \log m \\ O(nm^{1-\frac{1}{\lceil k/2 \rceil + \gamma}} + m \log m) & \text{if } n < m^{\frac{1}{\lceil k/2 \rceil + \gamma}} \log m \end{cases}$$

where $\sum_i n_i = n$ and, by the theory of random sampling [14], $m_i \leq \frac{cm}{r} \log r$. The solution of this recurrence is $O((mn)^{1-\frac{1}{\lceil k/2 \rceil + 1 + \gamma'}} + (m+n) \log m)$, for another but arbitrarily small positive constant γ'. We have established the following theorem.

Theorem 2.5 *Let R and B two sets of points in $\mathcal{E}^k$ with n and m points, respectively. Then there is an $O((mn)^{1-\frac{1}{\lceil k/2 \rceil + 1 + \gamma'}} + (m+n) \log m)$ expected time randomized algorithm that computes, for each $r \in R$, its farthest neighbor in B.*

Remark: Our second algorithm can also be modified for computing bi-chromatic nearest neighbors problem within the same time bound. In this case, we map $\mathcal{E}^3$ to the hyperplane $x_4 = +\infty$ and, for every red point r, we determine the blue hyperplane lying immediately below r.

By setting $B = R$, we can compute, for each point $r \in R$, its farthest neighbor within R using Theorems 2.4 and 2.5. We have the following result.

Corollary 2.6 *Given a set S of n points in $\mathcal{E}^k$, one can compute, for each point $p \in S$, its farthest neighbor in S in randomized expected time $O(n^{4/3} \log^{4/3} n)$ for $k = 3$, and in time $O(n^{2 - \frac{2}{\lceil k/2 \rceil + 1 + \gamma}})$ for $k > 3$.*

2.3 Computing external farthest neighbors

We are given a set S of n points and its partition $\Pi = \{S_1, S_2, \ldots, S_m\}$ into m subsets. For each $p \in S$, we want to compute its external farthest neighbor.

We will use the standard divide and conquer technique. If $m = 1$, there is nothing to do and we stop, otherwise we divide Π into two subsets Π', Π'' as follows. Let $S' = \bigcup_{S_i \in \Pi'} S_i$ and $S'' = \bigcup_{S_i \in \Pi''} S_i$. If there is an S_i such that $|S_i| \geq 2n/3$, we set $\Pi' = \{S_i\}$ and $\Pi'' = \Pi - \Pi'$. Otherwise, we divide Π into two subsets Π', Π'', such that $|S'|, |S''| \leq 2n/3$. Such a partition can be computed in $O(n)$ time. We recursively solve the subproblems (S', Π'), (S'', Π''), that is, for each $p \in S'$ (resp. $p \in S''$) compute its external farthest neighbor q' with respect to Π' (resp. Π'').

Finally, for each $p \in S'$ ($p \in S''$), we compute its bi-chromatic farthest neighbor q in S'' (resp. S') using Theorem 2.4. We return q, if $d(p,q) \geq d(p,q')$, and q' otherwise.

The correctness of the algorithm is evident, and the running time $T(m,n)$ obeys the following recurrence, assuming that $M(n)$ denotes the time complexity of the bi-chromatic farthest neighbors problem:

$$T(m,n) \leq T(m_1, n_1) + T(m - m_1, n - n_1) + M(n),$$

where $T(1,n) = O(1)$, $|S'| = n_1$, $|\Pi'| = m_1$, and either $m_1 = 1$ or $n_1, n - n_1 \leq 2n/3$. This recurrence solves to $M(n)$, because $M(n) \geq n^{1+\delta}$, for some fixed $\delta > 0$. We thus obtain the following result.

Theorem 2.7 *Given a set S of n points in $\mathcal{E}^k$ and its partition $S_1, \ldots, S_m$ into m subsets, we can compute, for each $p \in S$, its external farthest neighbor in randomized expected time $O(n^{4/3} \log^{4/3} n)$ for $k = 3$, and in time $O(n^{2 - \frac{2}{\lceil k/2 \rceil + 1 + \gamma}})$ for $k > 3$.*

3   Euclidean Maximum Spanning Tree and Clustering

In this section, we give a randomized algorithm for computing a Euclidean maximum spanning tree of n points in $\mathcal{E}^k$. EMXSTs find applications in clustering. For instance, given an EMXST of n points, we can find a *minimum diameter* 2-clustering of the points in $O(n)$ additional time, see Monma and Suri [16]. In particular, they show that any two coloring of an EMXST of S gives a minimum diameter 2-clustering of S.

A maximum spanning tree can be computed by repeatedly invoking the external farthest neighbors algorithm. The algorithm is based on the following well-known property.

Let $\{S_1, S_2\}$ be an arbitrary partition of the point set S, and suppose that the pair p_1^*, p_2^* maximizes the distance over all pairs of points $p_1 \in S_1$ and $p_2 \in S_2$. Then there exists a maximum spanning tree of S containing the edge (p_1^*, p_2^*). It is used in the following well-known algorithm for computing a maximum spanning tree, see [13].

Algorithm EMXST

Initialization: Make each point $p \in S$ a component consisting of a singleton vertex p.

1. If the number of components in the spanning forest is more than one, then execute Steps (2) through (5). Otherwise, output the component and stop.

2. Let $V_1, V_2, \ldots, V_m$ be the components of the spanning forest. Solve the external farthest neighbors problem for S with respect to $V_1, \ldots, V_m$.

3. For each component V_i, let p_i^*, q_i^* be a pair of points that maximizes the distance over all pairs $p_i \in V_i$ and $q_i \in S - V_i$.

4. Pick the pairs (p_i^*, q_i^*) one by one in the non-increasing order of their distances, and add the edge $p_i^* q_i^*$ to the spanning forest if it does not create a cycle.

5. Go to Step (1).

end Algorithm EMXST.

Algorithm EMXST clearly computes a Euclidean maximum spanning tree of S: each edge that is added in Step 4 is a maximum weight edge for some partition of S and it does not create a cycle. Since each time Step 1-4 are executed, the number of components reduces by at least half, and since Step 1-4 requires $O(n^{4/3} \log^{4/3} n)$ time in $\mathcal{E}^3$ and $O(n^{2 - \frac{2}{\lceil k/2 \rceil + 1 + \tau}})$ in $\mathcal{E}^k$. We obtain

Theorem 3.1 *Given a set S of n points in $\mathcal{E}^k$, its Euclidean maximum spanning tree and a minimum-diameter two-partition can be computed by a randomized algorithm in expected time $O(n^{4/3} \log^{7/3} n)$ for $k = 3$, and in time $O(n^{2 - \frac{2}{\lceil k/2 \rceil + 1 + \tau}})$ for $k > 3$.*

4   Approximation Algorithms

We give a simple approximation scheme for computing bi-chromatic all-farthest neighbors. By the results of the previous sections, this leads to similar approximation algorithms also for the all-farthest neighbors, the external farthest neighbors, the maximum spanning tree, and the minimum-diameter two-partitioning.

Let R and B be two sets of points in $\mathcal{E}^k$, where $|R| = n$ and $|B| = m$, and let ε be a real parameter. We want to approximate the farthest neighbors of R in B. Specifically, let $b \in B$ be a farthest neighbor of $r \in R$, and let $b' \in B$ be another point. We say that b' is an ε-approximate farthest neighbor of r if

$$\frac{d(r, b')}{d(r, b)} \geq 1 - \varepsilon.$$

To illustrate the method, we first consider the problem in two dimensions.

We form a net Φ of uniformly spaced s directions by partitioning the *unit sphere* (circle in two dimensions) into s equal parts. In particular, the ith member of Φ, say, ϕ_i is directed from the origin to $(\cos \frac{2\pi i}{s}, \sin \frac{2\pi i}{s})$.

For direction ϕ_i, let $\ell(\phi_i)$ denote the line whose normal vector is ϕ_i, and let $B(\phi_i)$ denote a point of B that maximizes the linear function $\ell(\phi_i)$: $x \cos \frac{2\pi i}{s} + y \sin \frac{2\pi i}{s} = 1$. (That is, $B(\phi_i)$ is an extreme point of B in direction ϕ.) At a high level, our approximation algorithm consists of the following steps.

Algorithm APPROXIMATION

1. Compute $B(\Phi) = \{B(\phi) \mid \phi \in \Phi\}$.

2. For each $r \in R$, find its farthest neighbor in $B(\Phi)$.

end Algorithm

The approximation potential of the above algorithm is established by the following lemma.

Lemma 4.1 *For $s \geq \pi/\sqrt{2\varepsilon}$, the algorithm* APPROXIMATION *computes ε-approximate farthest neighbors of R in B.*

Proof: Consider a point $r \in R$, and assume that $b \in B$ is a (true) farthest neighbor of r. Since $b \in B$ is a farthest neighbor of r, all points of B lie in the disk of radius $d(r, b)$ and center r. There is a direction $\phi \in \Phi$ that makes an angle of no more than π/s with the direction determined by the pair (r, b). Consider the line $\ell(\phi)$ and the support point $B(\phi)$. We claim that $B(\phi)$ is an ε-approximate farthest neighbor of r. Indeed, let δ be the distance from r to the line $\ell(\phi)$. Clearly, $d(r, b) \geq d(r, B(\phi)) \geq \delta$. Hence,

$$\frac{d(r, B(\phi))}{d(r, b)} \geq \frac{\delta}{d(r, b)} \geq \cos \frac{\pi}{s} = 1 - 2\sin^2 \frac{\pi}{2s}$$

$$\geq 1 - \frac{\pi^2}{2s^2} \geq 1 - \varepsilon.$$

Since Step 2 of algorithm APPROXIMATION returns a farthest neighbor among $B(\Phi)$, $d(r, B(\phi))$ is a lower bound on our approximation, which proves the lemma. ∎

Theorem 4.2 *There is an $O((n + m)\varepsilon^{-1/2})$ time algorithm for computing ε-approximation of bi-chromatic farthest neighbors in $\mathcal{E}^2$.*

Proof: A straightforward implementation of the algorithm APPROXIMATION leads to $O((n + m)\varepsilon^{1/2})$ bound: the set $B(\Phi)$ can be computed at the cost of $O(m)$ per direction, and a farthest neighbor of $r \in R$ can be found by checking each of the $O(\varepsilon^{-1/2})$ candidates in the set $B(\Phi)$. ∎

Remark: Let us remark that at a slight increase in the conceptual complexity, we can improve the time complexity of the algorithm to $O((n+m)\log\frac{1}{\varepsilon})$, as follows. Instead of spending linear time per direction, we do a divide-and-conquer on the set of directions. This approach computes the set $B(\Phi)$ in $O(m\log\frac{1}{\varepsilon})$ time. Then, to find farthest neighbors, we compute the farthest-point Voronoi diagram of $B(\Phi)$ and locate points of R in it by point-location techniques. The latter step takes $O(n\log\frac{1}{\varepsilon} + \frac{1}{\sqrt{\varepsilon}}\log\frac{1}{\varepsilon})$.

The algorithm for higher dimensions is essentially the same as in two dimensions: we form a dense net Φ of direction vectors, by dividing the direction sphere into a grid, compute the support point $B(\phi)$ for each direction $\phi \in \Phi$, and report, for each point $r \in R$, its farthest neighbor neighbor in $B(\Phi)$. (The support point $B(\phi)$ is the one that maximizes the linear function $\mathbf{x}\cdot\phi = 1$.) The following lemma is straightforward.

Lemma 4.3 *There exists a set Φ of unit vectors in $\mathcal{E}^k$, where $|\Phi| = O(\varepsilon^{-\frac{k-1}{2}})$ such that for an arbitrary unit vector α, we can find a vector $\phi \in \Phi$ satisfying $\arccos(\alpha,\phi) \leq \varepsilon^{1/2}$. The set Φ can be found in $O(|\Phi|)$ time for any fixed dimension k.*

To show that the farthest neighbor reported by the algorithm APPROXIMATION, indeed, is an ε-approximation even for higher dimensions, we proceed as follows.

Consider a point $r \in R$ and its (true) farthest neighbor $b \in B$. Choose a direction $\phi \in \Phi$ such that the angle between ϕ and the vector $\vec{rb}$ is at most $\sqrt{\varepsilon}$, as guaranteed by Lemma 4.3. Let H be the two-dimensional linear space spanned by the vectors ϕ and $\vec{rb}$. Observe now that we are in the two-dimensional case, and the analysis of Lemma 4.1 can be applied. We thus have the following theorem.

Theorem 4.4 *Given two sets of points R and B in $\mathcal{E}^k$, with $|R| = n$ and $|B| = m$, and an $\varepsilon > 0$, there is an $O((n+m)\varepsilon^{-\frac{k-1}{2}})$ time algorithm for computing an ε-approximation of bi-chromatic all-farthest neighbors between R and B.*

Of course, the diameter and all-farthest neighbors of a set of n points can also be found within the same bounds. We can also compute an ε-approximation of external farthest neighbors by repeatedly finding the ε-approximate bi-chromatic farthest neighbors, instead of the true neighbors, in the algorithm of Section 2.3.

Theorem 4.5 *Given a set S of n points in $\mathcal{E}^k$ and an $\varepsilon > 0$, we can compute ε-approximate external farthest neighbors of S in time $O(n\varepsilon^{(1-k)/2}\log n)$. An ε-approximation of a maximum spanning tree or a minimum diameter two-partitioning of S can be found in time $O(n\varepsilon^{(1-k)/2}\log^2 n)$.*

5 Conclusion

We have considered the problem of computing bi-chromatic farthest or nearest neighbors in k-dimensional space. These problems arise quite naturally in computational geometry applications. Despite their fundamental nature, optimal algorithms are still not known for solving them in dimensions greater than two. A traditional method for solving these problem is the so-called *locus* approach: build a Voronoi diagram for the set of points B, and then find a neighbor for each point $r \in R$ through point-location. When combined with a batching technique, this scheme yields a (slightly) subquadratic algorithm in any fixed dimension k and for any reasonable metric. Our contribution in this paper has been to improve the exponent significantly by incorporating random-sampling, which allows us to break the problem into several smaller subproblems. A key difference between the batching-based locus approach and the random-sampling approach is that while, in the former method, only one of the sets is partitioned among the subproblems, in the latter, both sets are simultaneously subdivided.

We suspect that the correct time complexity for these problems is close to linear. Achieving that goal, however, seems quite difficult at this point, at least for high dimensions. At the same time, no lower bound better than $\Omega(n \log n)$ is known.

The topic of approximation algorithms is still mostly unexplored. Our result shows that linear or near-linear algorithms are indeed possible if one is willing to settle for an ε-approximation. The "constant factor" of our algorithm grows exponentially with dimension in $\frac{1}{\varepsilon}$. It may be possible to reduce this constant to something like $(\log \frac{1}{\varepsilon})^k$, which would be interesting.

References

[1] P. Agarwal, H. Edelsbrunner, O. Schwarzkopf, and E. Welzl. Euclidean minimum spanning trees and bichromatic closest pairs. *Proc. of 6th ACM Symp. on Computational Geometry*, 1990, pp. 203–210.

[2] F. Aurenhammer. Improved algorithms for discs and balls using power diagrams. *Journal of Algorithms*, 9 (1988), 151-161.

[3] B. Chazelle. How to search in history. *Information and Control*, 64 (1985), 77-99.

[4] K. Clarkson. Fast expected-time and approximate algorithms for geometric minimum spanning tree. *Proc. 16th Annual Symposium on Theory of Computing*, 1984, pp. 342–348.

[5] K. Clarkson. A randomized algorithm for closest-point queries. *SIAM J. Computing*, 17 (1988), 830–847.

[6] K. Clarkson and P. Shor. Applications of random sampling in computational geometry, II. *Discrete and Computational Geometry*. 4 (1989), pp. 387-421.

[7] H. Edelsbrunner. An acyclicity theorem for cell complexes in d dimensions. *Proc. of 5th ACM Symp. on Computational Geometry*, 1989, pp. 145–151.

[8] H. Edelsbrunner. *Algorithms in Combinatorial Geometry*, Springer-Verlag, 1987.

[9] H. Edelsbrunner and R. Seidel. Voronoi diagrams and arrangements. *Discrete and Computational Geometry*, 1 (1985), 25–44.

[10] H. Edelsbrunner and M. Sharir, A hyperplane incidence problem with applications to counting distances. *Proc. of International Symposium on Algorithms*, LNCS 450, Springer-Verlag, 1990.

[11] O. Egecioglu and B. Kalantari. Approximating the diameter of a set of points in the Euclidean space. Technical report, Rutgers University, 1989.

[12] H. N. Gabow and R. E. Tarjan. A linear-time algorithm for a special case of disjoint set union. *Journal of Computer and System Sciences*, 30 (1985), 209–221.

[13] R. L. Graham and P. Hell. On the history of minimum spanning tree problem. *Annals of History of Computing*, 7 (1985), 43-57.

[14] D. Haussler and E. Welzl. ε-nets and simplex range queries, *Discrete and Computational Geometry*. 2 (1987), 127–151.

[15] C. Monma, M. Paterson, S. Suri and F. F. Yao. Computing Euclidean maximum spanning trees. *Algorithmica*, 5 (1990), 407-419.

[16] C. Monma and S. Suri. Partitioning points and graphs to minimize the maximum or the sum of diameters. *Proceedings of 6th International Conference on the Theory and Applications of Graphs,* John Wiley and Sons, 1989,

[17] F. P. Preparata and M. I. Shamos. *Computational Geometry.* Springer Verlag, New York, 1985.

[18] F. Preparata and R. Tamassia. Efficient spatial point location. *Workshop on Algorithms and Data Structures,* LNCS 382, Springer-Verlag, (1989), 3–11.

[19] R. Seidel. A convex hull algorithm optimal for point sets in even dimensions. University of British Columbia, Vancouver, 1981.

[20] P. Vaidya. Minimum spanning trees in k-dimensional space. *SIAM J. of Computing*, 17 (1988), 572-582.

[21] P. Vaidya. An $O(n \log n)$ algorithm for the all-nearest-neighbor problem. *Discrete and Computational Geometry*, 4 (1989), 101–115.

[22] A. Yao. On constructing minimum spanning trees in k-dimensional spaces and related problems, *SIAM J. Computing*, 11 (1982), 721–736.

SHALLOW INTERDISTANCE SELECTION
AND INTERDISTANCE ENUMERATION

Jeffrey S. Salowe

Department of Computer Science
University of Virginia
Charlottesville, Virginia 22903

ABSTRACT

Shallow interdistance selection refers to the problem of selecting the kth smallest interdistance, $k \leq n$, from among the $\binom{n}{2}$ interdistances determined by a set of n points in $\Re^d$. Shallow interdistance selection has a concrete application — it is a crucial component in the design of a linear-sized data structure that dynamically maintains the minimum interdistance in sublinear time per operation (Smid [9]). In addition, the study of shallow interdistance selection may provide insight into developing more efficient algorithms for the problem of selecting Euclidean interdistances (Agarwal *et al.* [1]). We give a shallow interdistance selection algorithm which takes optimal $O(n \log n)$ time and works in any L_p metric. To do this, we prove two interesting related results. The first is a combinatorial result relating the rank of x to the rank of $2x$. The second is an algorithm which enumerates all pairs of points within interdistance x in time proportional to the rank of x (plus $O(n \log n)$). A corollary to our work is an algorithm which, given a set of n points and an integer k, outputs all interdistances having rank at most k in $O(n \log n + k)$ time.

1. Introduction

We consider the *shallow selection* problem of selecting the kth smallest interdistance, $k \leq n$, from among the $\binom{n}{2}$ determined by a set P of n points in $\Re^d$. This problem is in part motivated by recent progress by Smid [9] in the design of an efficient, linear-sized data structure supporting the following three operations:

insert (x,P): update point set P to $P \cup \{x\}, x \in \Re^d$;

delete (x,P): update point set P to $P - \{x\}$;

minimum (P): return a pair of points having minimum interdistance with respect to a given metric.

Smid's data structure supports these three operations in $O(n^{2/3}\log n)$ time if the interpoint distance is given by an L_p metric. Smid's algorithm depends on an algorithm for "shallow" interdistance selection.

Specifically, Smid shows how to find an ordered sequence of the $O(n^{2/3})$ smallest distances determined by P in $O(n \log n)$ time, and he notes that the update times for his data structure can be improved to $O(\sqrt{n} \log n)$ time if the ordered sequence of n smallest interdistances can be found in $O(n \log n)$ time. (We note that Smid has recently developed a second dynamic algorithm which has polylogarithmic update times but uses $O(n \log^{O(1)} n)$ storage [8], so the actual impact of shallow interdistance selection on closest point problems is primarily theoretical.)

A second motivation to study this problem is to determine the true complexity of interdistance selection. Salowe [7] presented an $O(n \log^d n)$ time algorithm to select L_∞ interdistances in d-dimensional space. L_1 interdistances in the plane may be selected in $O(n \log^2 n)$ time. Recently, Agarwal, Aronov, Sharir and Suri [1] gave an $O(n^{3/2} \log^{5/2} n)$ time algorithm for L_2 interdistances in the plane. The best result for points in d-dimensional space (with respect to the Euclidean metric) is due to Chazelle [3], who devised a nearly-quadratic algorithm based on Yao's technique [11] for constructing minimum spanning trees in d-dimensional space. What are the true bounds? What is the relationship between the apparently easy selection in the L_∞ metric to the seemingly harder selection in the L_2 metric?

In this paper, we devise an $O(n \log n)$ time algorithm to select the kth smallest interdistance determined by P, where $k \leq n$. This algorithm works for any L_p metric and its running time is optimal; it relies on an interesting combinatorial result and an algorithm which enumerates interdistances less than or equal to x in time proportional to the rank of x (plus $O(n \log n)$). Work on the latter problem, sometimes referred to as "fixed-radius near neighbor searching" has appeared before; Bentley, Stanat, and Williams proposed three related algorithms [2], and their "array-based" algorithm has the same worst-case time bound as ours but possibly uses much more space (their space bound depends on x and the actual coordinate values). Dickerson and Drysdale [4] presented a two-dimensional algorithm using Delaunay triangulation techniques.

Dickerson and Drysdale [5] recently studied a slightly different problem which they dub "enumerating k distances." In the enumerating k distances problem, the input consists of a point set of size n and an integer k, and the output is all interdistances with rank k or less. Dickerson and Drysdale proposed an algorithm which enumerates the k smallest distances in the plane in sorted order in $O(n \log n + k \log k)$ time. A corollary to our work is a more general algorithm which enumerates the k smallest distances in $O(n \log n + k)$ time, not necessarily in sorted order. Our approach has two advantages over the Dickerson and Drysdale algorithm. First, it is faster by a $\log k$ factor, and second, it is more general in that it can be applied in any dimension. Our algorithm may not, however, be as practical as the Dickerson and Drysdale algorithm for planar problems.

The organization is the following. In Section 2, we state and prove the combinatorial result. Section 3 contains the interdistance selection algorithm. The first subsection describes an algorithm for the fixed-radius near neighbors problem, the second subsection presents an optimal shallow interdistance selection algorithm for the L_∞ metric, the third subsection presents an optimal algorithm for L_p metrics. Section 4 sketches our algorithm to enumerate k distances in d-space. Some remarks are made in Section 5.

2. Combinatorial Results

We begin with a question involving interdistance ranks. Given point set P with distances measured in an L_p metric, the rank $r_p(x)$ of distance x is $\#\{(u,v) : d_p(u,v) \leq x, u,v \in P, u \neq v\}$. Here, (u,v) is considered to be an unordered pair so that (u,v) and (v,u) are not double-counted. Note that the set of points within distance x of point u in the L_∞ metric are contained in a *hypercube* centered at u of *side length* $2x$. Suppose that distance x has rank k. What is the rank of distance $2x$? It is easy to describe a point set in the plane where distance x is the smallest interdistance, but distance $2x$ has rank $\Omega(n)$ (see Figure 1). In general, however, the ranks of x and $2x$ do not differ significantly.

Theorem 1: Let P be a set of n points in $\Re^d$; then if $r_\infty(x) = k$, $k \leq r_\infty(2x) \leq c(d)(k+n)$, where $c(d) = 5^{2d}$.

Proof: The lower bound is obvious. For the upper bound, divide up $\Re^d$ into hypercubes of side x. Note that any two points inside a given hypercube are within distance x of each other. Label the non-empty hypercubes arbitrarily with integers $1, 2, \ldots$ Let n_i be the number of points of P inside the ith hypercube. Then

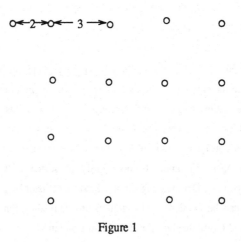

Figure 1

$$k \geq \sum_i \binom{n_i}{2}.$$

Now consider distance $2x$. With respect to hypercube i, the points within distance $2x$ of hypercube i are contained in a larger hypercube C_i of side $5x$ centered at the same point as the center of hypercube i. There are at most 5^d original hypercubes intersecting C_i, and an upper bound on the number of interdistances less than or equal to $2x$ involving points in hypercube i is:

$$B_i = \sum_{j \in C_i} n_i n_j$$

Among these hypercubes having non-empty intersection with C_i, charge $5^d n_k^2$ to the hypercube k containing the most points. Note that $B_i \leq 5^d n_k^2$.

Summing up all charges, no hypercube can be charged more than 5^d times, so

$$\sum_i B_i \leq 5^{2d} \sum_i n_i^2.$$

However,

$$\sum_i n_i^2 = n + 2 \sum_i \binom{n_i}{2}.$$

Recalling that distances are double-counted, we have

$$r_\infty(2x) \leq \frac{1}{2} \sum_i B_i \leq \frac{1}{2} 5^{2d} \sum_i n_i^2$$

$$= \frac{1}{2} 5^{2d} \left(n + 2 \sum_i \binom{n_i}{2} \right)$$

$$\leq 5^{2d} (n + k).$$

□

Recall that the L_p metrics are related by constant factors; that is if u and v are points in $\Re^d$, $d_\infty(u,v) \leq d_p(u,v) \leq d \cdot d_\infty(u,v)$.

Corollary 1:

$$r_\infty(tx) \leq c(d)^{\lceil \lg t \rceil} r_\infty(x) + n \frac{c(d)^{\lceil \lg t \rceil + 1} - c(d)}{c(d) - 1},$$

where $t \geq 1$ and $c(d) = 5^{2d}$.

Proof: We first note that $r_\infty(tx) \leq r_\infty(2^{\lceil \lg t \rceil} x)$ since ranks are monotonically nondecreasing with distance. We now show by induction on i that

$$r_\infty(2^i x) \le c(d)^i \, r_\infty(x) + n \, \frac{c(d)^{i+1} - c(d)}{c(d)-1}$$

for $i \ge 1$. Theorem 1 proves the basis. In general,

$$r_\infty(2^i x) \le c(d)(r_\infty(2^{i-1}x) + n).$$

Applying the inductive hypothesis,

$$r_\infty(2^i x) \le c(d)((c(d)^{i-1} \, r_\infty(x) + n \, \frac{c(d)^i - c(d)}{c(d)-1}) + n).$$

Therefore,

$$r_\infty(2^i x) \le c(d)^i \, r_\infty(x) + n \, \frac{c(d)^{i+1} - c(d)}{c(d)-1},$$

as asserted. $\square$

Corollary 2: Let $y = x/d$. Then

$$r_\infty(y) \le r_p(x) \le c(d)^{\lceil \lg d \rceil} r_\infty(y) + n \, \frac{c(d)^{\lceil \lg d \rceil+1} - c(d)}{c(d)-1}.$$

Proof: Recall that $r_p(x) = \#\{(u,v) : d_p(u,v) \le x\}$. Since $d_\infty(u,v) \le d_p(u,v)$, $r_p(x) \le \#\{(u,v) : d_\infty(u,v) \le x\}$, so $r_p(x) \le r_\infty(x)$. Since $d_p(u,v) \le d \cdot d_\infty(u,v)$ and $r_\infty(x/d) = \#\{(u,v) : d_\infty(u,v) \le x/d\}$. We therefore have $r_\infty(x/d) = \#\{(u,v) : d \cdot d_\infty(u,v) \le x\}$ and $r_\infty(x/d) \le \#\{(u,v) : d_p(u,v) \le x\}$. Consequently, $r_\infty(x/d) \le r_p(x)$.

Now let $y = x/d$. We now have $r_\infty(y) \le r_p(x) \le r_\infty(d \cdot y)$, and we get the stated result by applying Corollary 1. $\square$

Corollary 2 states that if y is the nth smallest interdistance with respect to the L_∞ metric, $d \cdot y$ has rank $\Theta(n)$ with respect to any L_p metric.

3. Shallow Selection Algorithms

Using Corollary 2, we give an efficient shallow selection algorithm based on Salowe's L_∞ interdistance selection algorithm [7].

$SELECT_p(P,k)$
{ P is the input set, $k \le n$ is the rank, p indicates the metric. }

1. Preprocess P for orthogonal range queries using the layered range tree [6]
2. $y = SELECT_\infty(P,n)$
3. $L = \emptyset$
   For each $u \in P$
   $L = L \cup Range-query-report(P,u,d\cdot y)$
   { $Range-query-report$ returns those points inside the orthogonal
   hypercube of dimension $2d\cdot y$ centered at u. }
4. Select $2k^{th}$ smallest element in L
   { Note that interdistances are double counted. }

Step 1 takes $O(n\log^{d-1}n)$ time [6], and step 2 takes $O(n\log^d n)$ time [7]. Step 3 takes $O(n\log^{d-1}n + n)$ time, since Corollary 2 states that $\Theta(n)$ points are reported. Step 4 takes $O(n)$ time by the linear-time selection algorithm. Therefore, the algorithm takes $O(n\log^d n)$ time. Note that the actual k smallest interdistances all appear on L.

We now show how to speed up this algorithm to obtain an optimal $O(n\log n)$ time algorithm for shallow selection. Optimality is justified by observing the time required to find the smallest interdistance is $\Omega(n\log n)$ in the algebraic decision tree model [6].

The improved algorithm is based on an efficient algorithm which, given x, reports all pairs of points within L_∞ distance x of each other. After $O(n\log n)$ preprocessing, the algorithm returns all pairs whose interdistance is at most x in $O(n + r_\infty(x))$ time. In fact, we can modify the algorithm to do a bit more — after preprocessing, the algorithm can report in $O(n)$ time that $r_\infty(x) > c\cdot n$ for some constant c. This algorithm, the *fixed-radius near neighbors algorithm*, is used to speed up steps 1, 2, and 3 above.

3.1. The Fixed-Radius Near Neighbors Algorithm

Bentley, Stanat, and Williams' array-based algorithm for the fixed-radius near neighbors problem requires too much space for use in Smid's linear-sized data structure. We therefore devise a new algorithm inspired by Vaidya's optimal-time all-nearest-neighbors algorithm; we assume the reader is familiar with Vaidya's paper [10], though we briefly sketch the relevant results to be self-contained. (Terminology is adopted from Vaidya's paper.)

Given a set P of n points in $\Re^d$, Vaidya's algorithm finds a nearest neighbor to each point in P. The algorithm makes use of hypercubes called "boxes." During the course of the algorithm, P is partitioned by a set of disjoint boxes B called a *box list*. Initially, B consists of a single box b_0 which is a smallest box containing P. During a *stage*, a largest box in B, say b, is subdivided by d mutually orthogonal hyperplanes passing through its center. The resulting 2^d boxes $\overline{b}_1, \ldots, \overline{b}_{2^d}$ make up the set *immediate–successors*(b). Discarding the empty immediate successors and shrinking the remaining boxes as much as possible without changing the contents, one arrives at a set of boxes $b_1, \ldots, b_j$ called *successors*(b). B then becomes $B \cup successors(b) - \{b\}$. This process stops when each box in B contains exactly one point from P. Vaidya associates two sets with each box; these sets are irrelevant for the application considered here.

The fixed-radius near neighbor algorithm consists of three steps. Step 1, the box-subdivision step described above, takes $O(n \log n)$ time [10]. The result is a "tree-of-boxes" containing at most $2n - 1$ boxes rooted at b_0 with the property that the children of box b are precisely *successors*(b). Constructing this tree-of-boxes is the preprocessing step for the fixed-radius near neighbors algorithm. Along with each box b in the tree, we associate the quantities $d_{max}(b) = \max \{ d_\infty(u,v) : u,v \in b \}$ and $n(b) = \# \, b \cap P$. The boxes are also labeled in the order they were subdivided. The points $b \cap P$ can be found in time proportional to $n(b)$ by traversing the subtree rooted at b (points are contained in 0-volume boxes which are leaves in the tree-of-boxes).

Step 2 finds a specific partition of the boxes and determines pairs of these boxes within interdistance x of each other, where x is the input distance. Specifically, the tree-of-boxes obtained from step 1 is traversed in the order the boxes were subdivided, and the sequence of box lists is re-created. Let $d_{min}(b,b') = \min \{d_\infty(u,v) : u \in b, v \in b'\}$. Let B be the box list just before $\bar{b}$, a largest box in B, is subdivided. With each box b on the box list B, a set of boxes $B'(b) \subseteq B$ for which $B'(b) = \{b' \mid d_{min}(b,b') \leq x \text{ and } b \neq b'\}$ is maintained. After $\bar{b}$ is subdivided to create *successors*$(\bar{b})$, the set $B'(\bar{b})$ is examined to update the remaining sets $B'(b)$ to reflect $B = B \cup successors(\bar{b}) - \{\bar{b}\}$. Specifically, a new list $B(b_i)$ is created for each $b_i \in successors(\bar{b})$, and $\bar{b}$ is replaced by an appropriate subset of *successors*$(\bar{b})$ in each $B'(b)$ such that $\bar{b} \in B'(b)$. This process stops the first time a box to be subdivided has side length less than or equal to x. (Implementation details are left to the reader.) We show below that at most a constant number of changes are made to B and all the $B'(b)$ during a stage, and since there are $O(n)$ stages, step 2 takes $O(n)$ time.

Step 3 uses the output from step 2 to actually compute the pairs of points having interdistance x or less. At the end of step 2, there is a box list B consisting solely of boxes of side length less than or equal to x, and, for each box $b \in B$, there is a list $B'(b)$ containing all of the boxes in B within distance x of a point in b. The L_∞ rank of x is

$$\sum_{b \in B} \binom{n(b)}{2} + \frac{1}{2} \sum_{b \in B} \sum_{b' \in B'(b)} \#\{u,v : u \in b \cap P, v \in b' \cap P, d_\infty(u,v) \leq x\}.$$

It is clear that the rank and the pairs of points within x of each other can be computed from B and the $B'(b)$ in time

$$|B| + \sum_{b \in B} \binom{n(b)}{2} + \frac{1}{2} \sum_{b \in B} \sum_{b' \in B'(b)} n(b)n(b').$$

We show that this term is $O(n + r_\infty(x))$; this is the cost of step 3.

We first prove that step 2 takes $O(n)$ time. The proof of the following lemma appears in Vaidya [10] where it is asserted for any L_p metric, so the statement is a bit weaker than one specific to the L_∞ metric.

> **Lemma 1:** [Paraphrased from Vaidya, Packing Lemma 1] Let r be a positive integer. Let $b \in B$, and let b_L be a largest box in B which is about to be subdivided in step 2. Then the number of boxes b' in B such that $d_{min}(b,b') \le rd_{max}(b_L)$ is at most $2^d(2rd+3)^d$.

A corollary to Lemma 1 is the following:

> **Corollary 3:** Let $b \in B$. During any stage in step 2, the number of boxes b' in B such that $d_{min}(b,b') \le x$ is at most $2^d(2d+3)^d$.

Proof: Let $r = 1$ and recall that $d_{max}(b_L) > x$. □

In step 2, since $O(n)$ boxes are subdivided and the number of changes made to the $B'(b)$ sets is bounded by Corollary 3, the $O(n)$ time bound follows.

The time bound for step 3 is a consequence of Corollary 1. All boxes in B have side length at most x, so if $b' \in B'(b)$, then any point in b' must be within distance $3x$ of any point in b. This is because $d_{max}(b) \le x$, $d_{max}(b') \le x$, and $d_{min}(b,b') \le x$. Therefore,

$$\sum_{b \in B} \binom{n(b)}{2} + \frac{1}{2} \sum_{b \in B} \sum_{b' \in B'(b)} n(b)n(b').$$

is bounded by $r_\infty(3x)$, and Corollary 1 implies $r_\infty(3x)$ is $O(n + r_\infty(x))$.

> **Theorem 2:** Let P be a set of n points in $\Re^d$. The the L_∞ interdistances of length at most x can be found in $O(n \log n + r_\infty(x))$ time.

Using Corollary 2, we can easily generalize Theorem 2 to L_p metrics.

How do we use the fixed-radius near neighbors algorithm for shallow selection? Suppose the preprocessing step, step 1, has already been done. If $r_\infty(x)$ is $O(n)$, then the rest of the fixed-radius near neighbors algorithm takes $O(n)$ time. However, if $r_\infty(x)$ is large, say greater than $c \cdot n$ for some c, we can force the fixed-radius near neighbors algorithm to stop in $O(n)$ time and report "too large" once the bound on $r_\infty(3x)$ is exceeded in step 3.

3.2. Shallow L-Infinity Interdistances

Salowe's L_∞ interdistance selection algorithm uses a technique called parametric search [7]. The reader is referred to that paper for details. In Salowe's algorithm, a sequentialized parallel sorting algorithm determines $O(\log n)$ distances to be ranked, each in $O(n \log^{d-1} n)$ time. In fact, the actual rank of a distance is not needed. Instead, one needs the relationship between distance x and the unknown distance x^* with rank k. (Is $x < x^*$, is $x = x^*$, or is $x > x^*$?) For shallow selection, we can therefore make use of the "too large" feature explained above.

Specifically, after $O(n \log n)$ preprocessing, we can select a shallow interdistance in $O(n \log n)$ time using parametric search and the fixed-radius near neighbors algorithm. Suppose we seek an interdistance with rank $k \le n$. $O(\log n)$ distances x_i are ranked. If $r_\infty(x_i) > n$, then the fixed-radius near neighbors algorithm stops in $O(n)$ time and reports "too large." If $r_\infty(x_i) \le n$, then the fixed-radius near neighbors algorithm computes and counts the pairs of points within interdistance x, giving the exact rank. The total time expended is therefore $O(n \log n)$, and we have:

> **Lemma 2:** Given a set P of n points in $\Re^d$, the kth smallest L_∞ interdistance can be selected in $O(n \log n)$ time, $k \le n$.

3.3. Shallow L-p Interdistances

An efficient algorithm to select shallow L_p interdistances is the following.

$SELECT_p(P,k)$
{ P is the input set, n is the rank, p indicates the metric. }

1. Perform Step 1 of the Fixed-Radius Near Neighbors Algorithm
2. $y = SELECT_\infty(P,n)$
3. Perform Steps 2 and 3 of the Fixed-Radius Near Neighbors Algorithm for distance $d \cdot y$. Call the reported pairs L.
4. Select k^{th} smallest interdistance in L

Vaidya's analysis and Lemma 2 imply that steps 1 and 2 take $O(n)$ time. Corollary 2 and Theorem 2 imply that step 3 takes $O(n)$ time and the size of L is $O(n)$. The linear-time selection algorithm proves the rest, giving:

Theorem 3: Given a set P of n points in $\Re^d$, the kth smallest L_p interdistance can be selected in $O(n \log n)$ time, $k \leq n$, and $O(n)$ space.

As an immediate corollary, we improve on Smid's result:

Corollary 4: There is a linear-sized data structure for point sets which supports *insert*, *delete*, and *minimum* in $O(\sqrt{n} \log n)$ time per operation.

4. Enumerating k Distances in Space

We now sketch our algorithm to enumerate k distances in d-space. Recall that the input is a set P of n points in $\Re^d$ and an integer k, and the output is the set of interdistances with rank at most k. Consider the following two quantities needed during step 2 of the fixed-radius near neighbors algorithm:

$$r_\infty(x) = \sum_{b \in B} \binom{n(b)}{2}$$

$$\overline{r_\infty}(x) = \sum_{b \in B} \binom{n(b)}{2} + \frac{1}{2} \sum_{b \in B} \sum_{b' \in B'(b)} n(b) \, n(b').$$

It is clear that $r_\infty(x) \leq r_\infty(x)$, and it was shown in Section 3.1 that $r_\infty(x) \leq \overline{r_\infty}(x) \leq r_\infty(3x)$. Note that $r_\infty(x)$ and $\overline{r_\infty}(x)$ can be found in $O(n)$ time after $O(n \log n)$ preprocessing by simply stopping at step 2 of the fixed-radius near neighbors algorithm.

Using Salowe's interdistance selection algorithm, we can in $O(n \log n)$ time compute a value of x such that

$$r_\infty(x) \leq k \leq \overline{r_\infty}(x).$$

Given $O(\log n)$ interdistances x to rank, either $k > \overline{r_\infty}(x)$, in which case $k > r_\infty(x)$ and the corresponding comparison can be resolved, $k < r_\infty(x)$, in which case $k < r_\infty(x)$ and the corresponding comparison can be resolved, or the above condition holds.

To enumerate k distances, we run the fixed-radius near neighbors algorithm for distance $3x$, followed by the linear-time selection algorithm. Since $k \leq \overline{r_\infty}(x) \leq r_\infty(3x)$, the k smallest interdistances will be output. We must also ensure that $r_\infty(3x)$ is $O(n+k)$. To prove this, note that $|r_\infty(x) - k| \leq \overline{r_\infty}(x) - r_\infty(x)$. We claim that $\overline{r_\infty}(x) \leq C(d)(r_\infty(x) + n)$ for some constant $C(d)$ depending on the dimension d. Corollary 3 asserts that prior to the last box-subdivision step, $|B'(b)| \leq 2^d(2d+3)^d$ for each box $b \in B$. The final subdivision step increases the number of boxes by at most 2^d-1, so $|B'(b)| \leq 2^d((2d+3)^d + 1)$ for each box b present at the end of step 2. To complete the argument, we require Vaidya's Packing Lemma 3:

Lemma 3: [Paraphrased from Vaidya, Packing Lemma 3] Let $b \in B$. At the beginning of a stage, the number of boxes b' containing b in $B'(b')$ is bounded by $12^d + 2^d (8d + 3)^d$.

Using a method similar to the proof of Theorem 1 and Vaidya's Packing Lemma 3, we bound $\frac{1}{2} \sum_{b \in B} \sum_{b' \in B'(b)} n(b)\,n(b')$. Charge $c_1(d) = 2^d((2d+3)^d + 1)\,n(b')^2$ to the box b' in $B'(b) \cup \{b\}$ with the largest value of $n(b')$. By Lemma 3, no box is charged more than $c_2(d) = 12^d + 2^d((8d+3)^d + 1)$ times. Therefore,

$$\frac{1}{2} \sum_{b \in B} \sum_{b' \in B'(b)} n(b)\,n(b') \le \frac{1}{2}\, c_1(d) c_2(d) \sum_{b \in B} n(b)^2,$$

and

$$\overline{r_\infty(x)} \le \underline{r_\infty(x)} + \frac{1}{2}\, c_1(d) c_2(d) n + c_1(d) c_2(d) r_\infty(x),$$

yielding

$$\left| r_\infty(x) - k \right| \le \overline{r_\infty(x)} - \underline{r_\infty(x)} \le c_1(d) c_2(d)(\frac{n}{2} + \underline{r_\infty(x)}).$$

Since $r_\infty(x) \le k$, $r_\infty(x)$ is $O(n + k)$, and Corollary 1 implies that $r_\infty(3x)$ is also $O(n + k)$.

To summarize, we have the following theorem.

Theorem 4: Given a set P of n points in $\Re^d$ and an integer k, the k smallest L_p interdistances can be enumerated in $O(n \log n + k)$ time.

5. Remarks

We have presented an algorithm which, given n points in $\Re^d$, selects the kth smallest interdistance, $k \le n$, in optimal $O(n \log n)$ time. The algorithm works in any L_p metric. To obtain the algorithm, we proved a combinatorial result interrelating distances and ranks. We also described a nontrivial algorithm for the problem of enumerating all interdistances less than or equal to x in time proportional to the rank of x (including an $O(n \log n)$ term).

There are several interesting questions. First, what is the complexity of selecting the median? Second, can a result analogous to the one above be obtained when selecting the kth largest interdistance?

Acknowledgement: Scot Drysdale kindly pointed out or provided several references, and his help is greatly appreciated.

6. References

1. P. K. Agarwal, B. Aronov, M. Sharir and S. Suri, Selecting Distances in the Plane, *Sixth ACM Symposium on Computational Geometry*, 1990, pp. 321-331.

2. J. Bentley, D. Stanat and E. Williams, The Complexity of Finding Fixed-Radius Near Neighbors, *Inf. Proc. Letters*, **6**, 1977, pp. 209-213.

3. B. Chazelle, Some Techniques for Geometric Searching with Implicit Set Representations, *Acta Informatica*, **24**, 1987, pp. 565-582.

4. M. T. Dickerson and R. L. S. Drysdale, Fixed-Radius Near Neighbors Search Algorithms for Points and Segments, *Inf. Proc. Letters*, **35**, 1990, pp. 269-273.

5. M. T. Dickerson and R. L. S. Drysdale, Enumerating k Distances for n Points in the Plane, *Seventh ACM Symposium on Computational Geometry*, 1991.

6. F. P. Preparata and M. I. Shamos, *Computational Geometry: An Introduction*, Springer Verlag, New York, NY, 1985.

7. J. S. Salowe, L-Infinity Interdistance Selection by Parametric Search, *Inf. Proc. Letters*, **30**, 1989, pp. 9-14.

8. M. Smid, Maintaining the Minimal Distance of a Point Set in Polylogarithmic Time, *Universitat des Saarlandes 13/90*, 1990.

9. M. Smid, Maintaining the Minimal Distance of a Point Set in Less Than Linear Time, *Universitat des Saarlandes 06/90*, 1990.

10. P. M. Vaidya, An $O(n \log n)$ Algorithm for the All-Nearest-Neighbors Problem, *Discrete Comput. Geom.*, **4**, 1989, pp. 101-115.

11. A. C. Yao, On Constructing Minimum Spanning Trees in k-Dimensional Spaces and Related Problems, *Siam J. on Computing*, **11**, 1982, pp. 721-736.

Sharing Memory in Asynchronous Message Passing Systems

Oscar R. Aguilar* University of Nevada, Las Vegas, NV 89154

Ajoy Kumar Datta* University of Nevada, Las Vegas, NV 89154

Sukumar Ghosh University of Iowa, Iowa City, IA 52242

Abstract

We present an algorithm to simulate Read-Modify-Write registers in a message passing system with unreliable asynchronous processors and asynchronous communication. The algorithm works correctly in the presence of a strong adversary that can stop up to T processors, or stop the delivery of their messages where $T = \lceil N/2 \rceil$-1 and N is the number of processors in the system. This is the best resilience that can be achieved in the message passing systems. The high resilience of the algorithm is obtained by using randomized consensus algorithms and a robust communication primitive. The use of this primitive allows a processor to exchange local information with a majority of processors in a consistent way and therefore, make decisions safely. The simulator makes it possible to translate algorithms for the shared memory model to that for the message passing model. With some minor modifications the algorithm can be used to robustly simulate shared queues, shared stacks, etc.

Key words: Atomicity, asynchronous systems, consensus, message passing model, mutual exclusion, randomized algorithms, resiliense, shared memory model, synchronization primitives, Read-Modify-Write registers.

1 Introduction

In the shared memory model for intercommunication in multiprocessor systems, several synchronization primitives have been proposed to coordinate actions among the processors. The most common use of these primitives is to resolve conflicts when there are concurrent requests of access to a shared resource. These primitives are also used when the processors need to select some unique values, e.g., id's, time stamps, etc. On the other hand, the agreement problems in which all processors select the same value can also be solved using these synchronization primitives. In all cases the purpose of a primitive is to let a processor execute some specific operation without being interrupted. The type of operations we are referring to always involves access to a shared data object and the uninterrupted (atomic) access to it is absolutely necessary to guarantee the correct execution of the operation.

The most basic synchronization primitives are the atomic read and atomic write instructions. These instructions guarantee that concurrent reads and writes to the object will be executed in a serial manner according to some definite order to maintain consistency of the shared data. A thorough study of these primitives can be found in [3], [4], [5], [11], [14],

*This research was supported in part by the U. S Army Research Office under grant DAAL03-87-G-0004 and by the Information Science Research Institute, University of Nevada, Las Vegas.

[15], [17], [18]. A very useful primitive is the read-modify-write instruction which atomically reads an object and writes a new value that is a function of the current value. Instructions that belong to this category are Test-and-Set, Fetch-and-Add, etc. A good description of these primitives can be found in [9], [10], [11]. Another category of synchronization primitives includes shared queues, shared stacks, shared lists etc. [11].

Can we use similar concepts to achieve synchronization in the message passing systems? The use of synchronization primitives greatly simplifies the solutions of problems (e.g., mutual exclusion) in shared memory environments. It would be desirable to be able to simulate these synchronization primitives in a message passing system. If this is possible, then the solution to a problem in a shared memory environment can be adapted to work in a message passing environment in a straightforward manner. The motivation behind this paper resides in the search for a general way to simulate any synchronization primitive in a message passing system. By simulation it is meant that a processor can invoke an operation on an abstract shared data object, the synchronization primitive in a message passing system, and obtain the same result as if the object had been implemented in a shared memory system.

Bar-Noy and Dolev [2] gave the first step in the direction of developing a shared memory—message passing translation scheme. They proposed the use of building blocks — one for the shared memory model and one for the message passing model. If an algorithm for the shared memory model is written using the building block, it is possible to translate the algorithm to work in the message passing model by using the equivalent block, and vice versa. In a more recent work [1] the authors presented an algorithm to simulate atomic single writer multireader registers in a message passing system. Besides translation, another important implication of the development of shared memory—message passing translation schemes is to prove that if there is a solution to a given problem in one of the models then a solution to the same problem can be developed in the other model also. This becomes especially interesting in the context of fault tolerance, where it is not evident that the existence of a resilient solution in one model implies the existence of a resilient solution in the other model.

In this paper, an algorithm to simulate shared memory synchronization primitives is presented. It can be used to translate solutions to problems in the shared memory model to work in the message passing model. It should be noted however that since the algorithm uses randomized consensus routines, a deterministic algorithm in the shared memory model will translate into a randomized algorithm in the message passing model. The rest of the paper is organized as follows: in the next section, the message passing model is described. In Section 3, a robust communication primitive called *Exchange* is discussed. An algorithm for the simulation of Read-Modify-Write registers using this primitive *Exchange* in an asynchronous message passing system is presented in Section 4. Finally, Section 5 provides some concluding remarks.

2 The model

A message passing multiprocessor system can be characterized by a set of parameters that describes the behavior of its components. These parameters model the environment in which an algorithm will work. The type of parameters of special importance in this paper is the one that describes the degree of synchronism among the processors and in the communication mechanism in the system. The algorithm presented in Section 4 handles asynchronous processors and asynchronous communication. The message delivery order can also be asynchronous. In addition, we assume that the processors are fail-stop processors.

Processors in a system are said to be asynchronous if any processor can wait an arbitrary amount of time between two of its own steps. If processors are synchronous then there is a constant ϕ such that in any time interval in which some processor makes $\phi + 1$ steps all nonfaulty processors make at least one step.

The communication in the system is asynchronous if messages can take an arbitrary amount of time to be delivered. Synchronous communication implies that there is a constant t_d such that every message sent has to be delivered within time t_d to its destination. Another type of asynchrony in the communication is the message delivery order. If messages can be delivered out of order then the message delivery order is said to be asynchronous. If messages are delivered in the same order as they were sent then the delivery is synchronous. Notice that the synchronous message delivery order guarantees that if two messages are sent by two different processors to the same destination, the one that was sent at the earlier time arrives first.

Processor and communication asynchrony are potential sources of nondeterminism that an algorithm must be able to handle when solving a specific problem. Another source of nondeterminism is the type of processor failures that the system can have. Two types of processor behavior under failures are usually considered: Fail-stop and Byzantine. A processor has fail-stop behavior if any failure makes the processor stop doing any computation or sending messages. A processor has Byzantine behavior if a failure makes it behave erratically, making wrong computations or sending contradicting or corrupted messages.

All these sources of nondeterminism in the system can be represented by a game between an adversary and a given algorithm. The goal of the adversary is to make the algorithm fail. For this purpose the adversary is assumed to have some or all the following privileges:

- Instruct the processors when to operate or even to stop up to T processors and not to let them restart at all (processor asynchrony). The factor T is a resilience parameter. If an algorithm can stand up to T processor failures and continue producing correct results then it is T-resilient.

- If a processor is waiting for N-1 messages from other processors, the adversary can suspend up to T messages (communication asynchrony), $T = \lceil N/2 \rceil$-1, where N is the total number of processors.

- Dictate the delivery time for messages sent and thus make them arrive out of order (message delivery order asynchrony).

- Arrive at the decisions by observing the messages to be sent and the internal state of all the processors. If however, an algorithm takes random steps like flipping a coin, the adversary will not be able to predict the outcome of future random steps.

3   The Communication Primitive Exchange

The processors use the primitive *Exchange* to exchange their knowledge. *Exchange* will ensure that the global information collected by a processor is consistent with that collected by a majority of processors. *Exchange* requires every processor to have an array to store information about all processors in the system including itself. By invoking *Exchange*, a processor sends its array and receives the arrays from other processors. This allows the processor to update its local information. In order to know which information is the most recent one, the timestamp must be used. The timestamps are generated locally by the processors. With all the elements in the array timestamped, it is possible to compare them and keep the most recent one.

Notation

From this point onward, $Ts(A[i])$ will denote the timestamp of the element i in the array A. The procedure *Exchange* for any processor is given below:

Procedure *Exchange* (*Myarray*)
```
    begin
        Counter:=0;
        Done:= false;
        while not Done do begin
            Broadcast(Myarray);
            Exit:=false;
            while not Exit do begin
                Getarray(Otherarray);
                            { Otherarray is sent by a processor P_j, j∈[1..N]}
                if Less(Otherarray, Myarray) then
                    Discard(Otherarray);
                else
                    if Equal(Otherarray, Myarray) then begin
                        Counter := Counter + 1;
                        if Counter >= N − T then Exit := true;
                            { N is the number of processors, T ≤ ⌈N/2⌉ − 1 }
                        end
                    else
                            Update(Myarray, Otherarray);  Exit := true;
                end-while;
            if Counter >= N − T then Done := true;
        Return();
    end-Exchange;
```

Procedure *Update* (*Array1, Array2*)
```
    begin
        For j:=1 to N do
            if Ts(Array1[j]) < Ts(Array2[j]) then Array1[j] := Array2[j];
    end-Update;
```

Procedure *Less* (*Array1, Array2*)
```
    begin
    Isless := true;
        while Isless and j ≤ N do
            if Ts(Array1[j]) >= Ts(Array2[j]) then Isless := false;
    end-Less;
```

Procedure *Equal* (*Array1, Array2*)
```
    begin
    Isequal := true;
        while Isequal and j ≤ N do
            if Ts(Array1[j]) ≠ Ts(Array2[j]) then Isequal := false
    end-Equal;
```

Procedure *Getarray* deals with the messages arriving out of order. It has to be implemented such that it screens messages based on their time stamps. In the algorithm described in Section 4, the processes execute in rounds and the timestamps used are the round numbers. At any given time, *Getarray* is looking for messages containing information about a particular round. If we assume that messages are queued as they arrive to a process, then *Getarray* will discard messages from previous rounds and will leave messages from later rounds (untouched) in the queue. It will only remove the messages containing information about the current round and pass them to procedure *Exchange*. Details about the implementation of *Getarray* are omitted.

Procedure *Exchange* assumes that a majority of the processors will always remain connected with each other. A disconnected processor's messages cannot reach other processors directly or indirectly. When the processors invoke *Exchange*, they all have an empty slot in their arrays corresponding to the disconnected processor. They agree in that there is no information available for that processor and will continue with their jobs without taking the disconnected processor in consideration.

The following lemmas show that a processor belonging to a majority will successfully execute *Exchange* in a finite time. It is assumed that the processors pass a correctly initialized array to *Exchange*. Specifically, the array passed to *Exchange* is initialized such that the only element in the array containing information corresponds to the local processor. The other elements are initialized to default values and their timestamps are initialized to a value which is less than any timestamp used in the system. Also, it is assumed that a majority of processors are executing *Exchange*.

Lemma 3.1
Every invocation of the procedure *Exchange* terminates.

The following lemma shows that the arrays a processor obtains by using *Exchange* are linearly ordered in time. This implies that the decisions taken based on the information in the array are safe because they are done according to the most recent information about the system.

Lemma 3.2
Let X and Y be two arrays collected by processor P_i at times t_0 and t_1, $t_0 < t_1$. Then for all $j \epsilon [1..N]$, $Ts(X[j]) \leq Ts(Y[j])$.

4   The Simulator

The algorithm in this section simulates Read-Modify-Write registers. This type of registers allows the atomic execution of the following procedure: (*Reg* is a Read-Modify-Write register)

```
Procedure RMW
begin
        Temp := Reg;
        Reg := f(Temp);
        return(Temp);
end-RMW;
```

In the above procedure RMW, f is a function from register values to register values. If f is the increment function then the register is a fetch-and-add register. If f sets the content of the *Reg* to 1 only when *Temp* is equal to 0 then the register is a test-and-set register. To simulate a Read-Modify-Write register in a message passing system every processor keeps a copy of the register. Since several processors may attempt to execute a RMW operation at the same time, it must be ensured that all processors execute the same operation on their local copies. So, at any time, all the copies have the same value and the consistency is maintained. In our algorithm, when a processor wants to execute an operation, it broadcasts a request message including its *ID*. The message also contains a timestamp to distinguish old request messages from new ones. Each processor maintains two set of flags $Opreq1_j$ and $Opreq2_j$, $j \epsilon [1..N]$ to keep track of the requesting processors. The request messages are collected by a procedure *Receive*. The procedure *Receive* after receiving a message from a processor P_k reads $Opreq1_k$ and writes $Opreq2_k$ with the opposite value. *Receive* also stores the timestamp included in the message in a variable $Tstamp_k$. When *Receive* gets a request for the first time, it retransmits it to ensure that all processors receive it. *Receive* knows if the request is old by comparing the timestamps. If an incoming message carries a timestamp older than or equal to the timestamp of the last message received, it is a duplicate request and must be discarded. Every processor has a variable called *Opnum*. This variable is incremented every time a processor executes an operation in its local copy of the register. *Opnum* is used as the timestamp for the request messages. The procedure *Receive* is described below ($Tstamp_j$ is the variable where *Receive* stores the timestamp of the last request message received from the processor P_j):

Procedure *Receive*;
   begin
         upon receiving operation request message $< Req, Opnum, ID >$
            if $(Opreq1_{ID} = Opreq2_{ID})$ and $(Tstamp_{ID} < Opnum)$ then begin
               $Opreq2_{ID} := \neg Opreq1_{ID}$;
               $Tstamp_{ID}:=$Opnum;
               Broadcast($< Req, Opnum, ID >$);
            end;
      end-receive;

To resolve conflicts when two or more processors want to execute a RMW operation at the same time, the processors are forced to reach an agreement. In the algorithm, every time a process wants to execute an operation it goes into a leader election process. If a processor is elected as the leader, it knows that all other processors agree that it is its turn to execute an operation. The leader election procedure works based on the processor $ID's$. The processors reach an agreement on a single ID by going through several rounds of consensus. In order to defeat an adversary that can stop processors and hide messages, a randomized consensus algorithm must be used. In the *Leader_election* procedure described in this section, a highly resilient consensus protocol is used. A similar prorocol was proposed in [2]. This protocol was proven to have resilience $T = \lceil N/2 \rceil - 1$ in an environment of asynchronous processors and asynchronous communication.

In the procedure *Leader_election* a processor starts with a suggested ID. To decide on which processor will be elected, it goes into $\lceil logN \rceil$ rounds of consensus. In every round it reaches an agreement with the rest of the processors on one bit of the leader's ID. If a processor loses a round of consensus, it selects at random another ID among the processors that have requested operations. The ID it selects must still have the chance to be selected as the leader's ID. The $ID's$ of the processors that have requested operations are kept in a set called *Reqset*.

The procedure *Bitconsensus* used inside *Leader_election* works in the following way: A processor calls this procedure with an initial value. This value is one of the bits of the ID it is suggesting as the leader's ID. The processor maintains an array Bcr with N elements. Every element of Bcr has three fields. The first one is *Sugval* which contains the value the associated processor is suggesting. The second variable *Round* keeps the number of attempts a processor has made to reach an agreement. *Round* is initialized to zero. The third field is *Timestamp* where a processor writes *Opnum* and *Bitnum*, the index of the leader's ID bit on which the processors are trying to reach an agreement. These information will be used by the procedure *Exchange* to generate labels that identify the messages. Processors exchange the values and check if all are suggesting the same value. If this is true, *Bitconsensus* terminates. If there is a disagreement, they flip a coin and adopt a new value according to the result. They exchange the values and flip the coin again if the disagreement persists. The processors keep repeating these steps until all flip the coin with identical values.

Ideally if all processors flip the same coin, they all get the same result. This corresponds to flipping a globally shared coin. With such a coin *Bitconsensus* terminates at most in two rounds. It has been proven however [11] that a perfectly unbiased global coin cannot be built. The procedure *Flip_Global_Coin* implements a biased globally shared coin. For $\gamma > 1$, a biased global coin has the property that with probability greater than $\frac{\gamma+1}{2\gamma}$ all the processors flip the same value. In the *Flip_Global_Coin* procedure every processor flips a local coin and depending on the result it increments or decrements a counter. The processor then gets the values of the counters of other processors and adds them up. If the result is greater than γN then it decides one. If the result is less than $-\gamma N$ then it decides zero. Otherwise, it repeats the steps above. To keep information about other processors, every processor maintains an array $Coin$ of size N. Every element in $Coin$ has two fields. The first field *Contribution* keeps the value of a processor's counter. The second field is *Timestamp* where a processor stores the value of *Opnum*, *Bitnum*, *Coinnum*, and *Iteration*. The variable *Coinnum* contains the number of times a processor has flipped a coin in a particular consensus round. *Iteration* keeps the number of times a processor has flipped a local coin while trying to decide a value in *Flip_Global_Coin*. These values are used by *Exchange* as labels to distinguish old messages from new ones. The expected number of rounds the processors will go through before flipping the same coin is $(\gamma + 1)^2 N^2$. The proofs of these results can be found in [2]. *IDpool* is a set containing the $ID's$ of all processors. *MYID* is the ID of the executing processor.

```
Procedure Leader_election(Opnum, ID);
    begin
        for i:=1 to ⌈logN⌉ do begin
        { IDᵢ and Leaderᵢ are the iₜₕ bit of ID and Leader, respectively}
            Leaderᵢ := Bitconsensus(Opnum, i, IDᵢ);
            if Leaderᵢ ≠ IDᵢ then
                ID := Randomsel{NewID∈Reqset | NewIDⱼ = Leaderⱼ ∀j∈(1..i) }
            if ID =NULL then begin
                Found := false;
                repeat
                    for all NewID∈{IDpool − Reqset}, NewID ≠ MYID do begin
                        Requestarrived:=( Opreq₂ ≠ Opreq₁);
                        if Requestarrived and (NewIDⱼ = Leaderⱼ∀j∈(1..i) ) then begin
```

```
                ID := NewID;
                Found := true;
                Exit for_ all loop;
            end-if;
        end-for_ all;
    until found;
    end-if ;
  end-for;
end-Leader_ election;
```

Procedure *Bitconsensus(Opnum, Bitnum, Val)*;
```
    begin
        for j:=1 to N do begin
            Bcr[j].Round := 0;
            Bcr[j].Sugval := 0;
        end-for;
        while true do begin
            Bcr[MYID].Round := Bcr[MYID].Round + 1;
            Bcr[MYID].Sugval := Val;
            Bcr[MYID].Timestamp :=(Opnum, Bitnum);
            Exchange(Bcr);
            Maxround := max₁≤k≤N (Bcr[k].Round);
```
if ($(Bcr[MYID].round = Maxround)$ and $(\forall k, k\epsilon(1..N)$ and
$k \neq MYID, (Bcr[k].Sugval \neq Bcr[MYID].Sugval) \rightarrow$
$(Bcr[k].Round < Bcr[MYID].Round - 1))$ then
```
                return( Bcr[MYID].Sugval);
            else
```
if $(\exists v \mid \forall k, (Bcr[k].Round = Maxround) \rightarrow (Bcr[k].Sugval = v))$ then
```
                    Val := v;
                else
                    Val := Flip_Global_Coin(Opnum, Bitnum, Bcr[MYID].Round);
            end-while;
    end-bitconsensus;
```

procedure *Flip_Global_Coin(Opnum, Bitnum, Coinnum)*;
```
    begin
        for j:=1 to N do
            Coin[j].Contribution := 0;
        Iteration := 0;
        while true do begin
            Iteration := Iteration + 1;
            Coin[MYID].Contribution := Coin[MYID].Contribution + Localcoinflip
                { Localcoinflip returns +1 or -1 with equal probability}
            Coin[MYID].Timestamp := (Opnum, Bitnum, Coinnum, Iteration);
            Exchange(Coin);
```
$Globalvalue := \sum_{j=1}^{N} Coin[j].Contribution;$
if $Globalvalue > \gamma N$ then return(1);
if $Globalvalue < -\gamma N$ then return(0);

end-while;
end-FlipGlobalCoin;

The procedure used to execute operations on a processor's local copy is *Execop*. *Execop* continuously checks the flags $Opreq1_j$ and $Opreq2_j$ to find out if any processor P_j wants to execute an operation. If these flags have different values, the procedure *Receive* received a request message from P_j. All the requesting processors $ID's$ are written in the set *Reqset* by *Execop*. If *Execop* receives a request from its local processor, it broadcasts a request message. Then it calls *Leader_election*. Next it executes the operation requested by the winner of the election (leader) in the local copy. *Execop* also resets the request flags of the leader by reading $Opreq2$ and writing $Opreq1$ with the same value. If the executing processor P_i is not the leader, *Execop* calls *Leader_election* again, and repeats the above steps until P_i is elected the leader. After P_i executes its local operation, *Execop* looks into the processors in *Reqset* and tries to elect them as leaders. By making *Execop* help other processors before executing another operation requested by its own processor, it is guaranteed that any processor succeeds in executing its operation in finite time. The local processor P_i running *Execop* requests to execute an operation by calling the procedure *Myrequest*. *Myrequest* signals *Execop* by reading $Opreq1_i$ and writing $Opreq2_i$ with the opposite value. The function *Randomsel* selects at random an ID from the set *Reqset*. The processors use the function f to change the value of the register.

Procedure *Execop*;
var *Opnum*: integer;
    Reqset: Set;
    Object, Retval: Register;
    ID:integer;

    begin
        Opnum := 0;
        Reqset := $\emptyset$
        while *true* **do begin**
            for *j*:=1 **to** *N* **do**
                if (($Opreq1_j \neq Opreq2_j$) **and** ($j \neq MYID$)) **then** *Reqset* := *Reqset* + [*j*];
            Myop := *false*;
            Otherop := *false*;
            if $Opreq1_{MYID} \neq Opreq2_{MYID}$ **then begin**
                Myop := *true*;
                ID := *MYID*;
                Broadcast(< *Req, Opnum, ID* >);
            end
            else
                if *Reqset* $\neq \emptyset$ **then begin**
                    Randomsel{$ID \epsilon Reqset$};
                    Otherop := *true*;
                  end;
            if *Myop* **or** *Otherop* **then begin**
                Leader := *Leader_election*(*Opnum, ID*);
                if *Leader* $\neq$ *ID* **then wait until** $Opreq1_{Leader} \neq Opreq2_{Leader}$;

```
        Retval := Object;
        Object := f(Retval);
        Opnum := Opnum + 1;
        Opreq1_Leader := Opreq2_Leader;
        if Leader = MYID then
            repeat
                Randomsel{ID ε Reqset}
                Leader := Leader_election(Opnum, ID);
                if Leader ≠ ID then wait until Opreq1_Leader ≠ Opreq2_Leader;
                Retval := Object;
                Object := f(Retval);
                Opnum := Opnum + 1;
                Opreq1_Leader := Opreq2_Leader;
                if (ID = Leader) then Reqset := Reqset − [ID];
            until Reqset = ∅;
        end-if;
    end-while;
end-execop;
```

Procedure *Myrequest*;
```
    begin
        Opreq2_MYID := ¬Opreq1_MYID;
        wait until Opreq2_MYID = Opreq1_MYID;
        return(Retval);
    end-myrequest;
```

4.1   Correctness

To prove the correctness of the simulation algorithm of the Read-Modify-Write registers we need to prove that it satisfies the following conditions:

C1) Any processor will execute a requested operation in finite time.

C2) A requested operation is not executed more than once.

C3) If a processor executes an operation that it requested in its local copy, the other processors execute the same operation in their local copies. This guarantees that the processors will see the same value for the object at any time and the consistency is maintained.

The proofs of the following lemmas and theorems are omitted due to the page limit.

Lemma 4.1

Every invocation of the procedure *Leader_election* terminates in $O((\gamma + 1)^2 N^2 log(n))$ time.

Lemma 4.2

Any processor will succeed in becoming a leader within a finite number of rounds of leader election.

Lemma 4.3

Every invocation of the procedure *Myrequest* terminates.

Theorem 4.1

Condition C1 is satisfied.

Theorem 4.2

Condition C2 is satisfied.

Lemma 4.4
For every invocation of the procedure *Leader_election* at most one leader is elected.
Theorem 4.3
Condition C3 is satisfied.
Theorem 4.4
The algorithm simulates Read-Modify-Write registers in a message passing system.
Theorem 4.5
The algorithm works correctly in the presence of up to $\lceil N/2 \rceil$-1 processor failures.

5 Conclusion

We have presented an algorithm that simulates Read-Modify-Write registers in message passing systems. Algorithms for simulating other types of synchronization primitives like atomic multiwriter registers, shared queues, etc. can be obtained by making some minor modifications to the algorithm presented here. For these primitives, it is necessary that a processor that wants to execute an operation pass to others the value it wants to enqueue or write. This value could be included in the message the processor broadcasts requesting to execute an operation. The receiving processors buffer this value and use it later when the sender succeeds in being elected the leader. The algorithm proposed in this paper is very resilient. It guarantees that any processor will be able to execute an operation even in the presence of up to $\lceil N/2 \rceil$-1 processor failures. This resilience is achieved through the use of a robust communication primitive *Exchange* and a randomized consensus algorithm. The algorithm is based on the fact that if a majority of processors can always communicate and exchange information, then they would be able to decide which processors have failed. The price paid for achieving high resilience is the large number of messages exchanged and the large size of the messages. The existence of algorithms for simulation of synchronization primitives in message passing systems implies that it is possible to construct emulators that translate algorithms for the shared memory model to that for the message passing model. It also implies that the existence of the solution for a problem in one model guarantees the existence of a solution in the other model. After the translation is done the optimization should be done in order to reduce the amount of message exchanges. One point to emphasize here is that when translating algorithms from the shared memory model to the message passing model, the degree of resilience cannot be maintained. Chor and Moscovici [6] have proven that the shared memory model is more powerful in this respect. One potential application of this algorithm is in the implementation of resilient mutual exclusion algorithms.

6 References

[1] H. Attiya, A. Bar-Noy, and D. Dolev, Sharing Memory Robustly in Message passing systems. *Proc. Ninth Annual ACM Symp. on Principles of Distributed Computing*, Quebec City, Montreal, August 1990, 363-375.

[2] A. Bar-Noy and D. Dolev, Shared Memory vs. Message-Passing in an Asynchronous Distributed environment. *Proc. Eighth Annual ACM Symp. on Principles of Distributed Computing*, Edmonton, Alberta, August 1989, 307-318.

[3] B. Bloom, Constructing Two Writer Atomic Register. *Proc. Sixth Annual ACM Symp. on Principles of Distributed Computing*, Vancouver, B.C., August 1987, 249-259.

[4] J. Burns and G. Peterson, Constructing Multireader Atomic Values from Non Atomic Values. *Proc. Sixth Annual ACM Symp. on Principles of Distributed Computing*, Vancouver, B.C., August 1987, 222-231.

[5] J. Burns and G. Peterson, Concurrent Reading while Writing II: The Multiwriter Case. *TR GIT-ICS-86/26*, School of Information and Computer Science, Georgia Tech., December 1986.

[6] B. Chor and L. Moscovici, Solvability in Asynchronous Environments, 30th *Annual Symp. on Foundations of Computer Science*, Research Triangle Park, North Carolina, Oct/Nov 1989, 422-427.

[7] D. Dolev, C. Dwork, and L. Stockmeyer, On the Minimal Synchronism Needed for Distributed Consensus. *Journal of the ACM*, Vol 34, No 1, January 1987, 77-97.

[8] M. Fisher, N. Lynch, and M. Paterson. Impossibility of distributed consensus with one faulty process. *Journal of the ACM*, Vol 32, No. 2, April 1985, 374-382.

[9] A. Gottlieb, R. Grishman, C. Krustal, K McAuliffe, L. Rudolph, and M. Snir. The NYU Ultracomputer- Designing an MIMD parallel computer. *IEEE Transactions on Computers*, Vol 32, No. 2, Feb. 1984, 175-189.

[10] C.P. Kruskal, L. Rudolph, and M. Snir, Efficient Synchronization on Multiprocessors with Shared Memory. *Proc. Fifth Annual ACM Symp. on Principles of Distributed Computing*, Vancouver, B.C. , August 1986, 218-227.

[11] M. Herlihy, Impossibility and Universality Results for Wait-Free Synchronization. *Proc. Seventh Annual ACM Symp. on Principles of Distributed Computing*, Vancouver, B.C. , August 1988, 276-290.

[12] M. Herlihy and J. Wing. Axioms for Concurrent Systems. *Proc. 14th ACM Symp. on Principles of Programming Languages*, Jan 1987, pp 13-26.

[13] L. Lamport, On Interprocess Communication Part I and II. *Distributed Computing*, Vol 1, 77-101, 1986.

[14] R. Newmann-Wolfe, A Protocol for Wait-Free, Atomic Multireader Shared Variables. *Proc. Sixth Annual ACM Symp. on Principles of Distributed Computing*, Vancouver, B.C., August 1987, 232-247.

[15] G. Peterson, Concurrent Reading while Writing. *ACM Transactions on Programming Languages and Systems*, Vol 5, No 1, January 1983, 46-55.

[16] S. Plotkin, Sticky Bits and the Universality of Consensus. *Proc. Eighth Annual ACM Symp. on Principles of Distributed Computing*, Edmonton, Alberta, August 1989, 159-175.

[17] A. Singh, J. Anderson, and M. Gouda, The Elusive Atomic Register. *Technical Report TR 86.29*, Department of Computer Science, University of Texas at Austin, December 1986.

[18] A. Singh, J. Anderson, and M. Gouda, The Elusive Atomic Register Revisited. *Technical Report TR 86.30*, Department of Computer Science, University of Texas at Austin, December 1986.

A Linear-Time Scheme for Version Reconstruction [†]
(Extended Abstract)

Lin Yu,   Daniel J. Rosenkrantz

Department of Computer Science
State University of New York at Albany
Albany, NY 12222
{linyu, djr}@cs.albany.edu

SUMMARY

An efficient scheme to store and reconstruct versions of sequential files is presented. The reconstruction scheme involves building a data structure representing a complete version, and then successively modifying this data structure by applying a sequence of specially formatted differential files to it. Each application of a differential file produces a representation of an intermediate version, with the final data structure representing the requested version.

The scheme uses a linked list to represent an intermediate version, instead of a sequential array, as is used traditionally. A new format for differential files specifying changes to this linked list data structure is presented. Algorithms are presented for using such a new format differential file to transform the representation of a version, and for reconstructing a requested version. Algorithms are also presented for generating the new format differential files, both for the case of a forward differential specifying how to transform the representation of an old version to the representation of a new version, and for the case of a reverse differential specifying how to transform the representation of a new version to the representation of an old version.

This new version reconstruction scheme takes time linear in the sum of the size of the initial complete version and the sizes of the file differences involved in reconstructing the requested version. In contrast, the classical scheme for reconstructing versions takes quadratic time. The time cost of the new differential file generation scheme is comparable to the time cost of the classical differential file generation scheme.

I. Introduction

In software systems and database systems, especially in computer aided design (CAD) database systems, many different versions of a data object often coexist. Version control systems are employed to manage these coexisting versions, e.g., for software systems, SCCS [Ro], RCS [Ti1] in Unix [U], and CMS [DEC] in VMS, and for database systems, those described in [BK, DL, KB, KL]. The efficiency of the underlying version control system directly affects the performance of a system, so the capability and efficiency of version control systems are of research interest ([BK, DL, HG, Ka1, Ka2, KB, KL, Ro, Ti1, YR1]). For a recent survey on version control issues, see [Ka2].

The data objects stored in software systems or in databases are often *textual* in nature. Each such textual object appears to the users as a sequential file consisting of a sequence of lines. Among different versions of such a data object, the difference is

---

† Research supported in part by the National Science Foundation under Grants CCR88-03278 and CCR90-06396.

often relatively small. To efficiently utilize storage space, many existing version control systems use *differential files* to physically represent versions of sequential files ([DEC, He, HM, KL, Ro, Ti1]).

In differential files for version control systems, the differences between two versions are usually described on a line basis [DEC, Ro, Ti1]. (In contrast, in some database systems, a file is regarded as a vector of pages, and a differential file contains those pages which have changed.)

The differences between two given sequential files can be computed by algorithms such as those in [He, HM, Ti2] and the utility program *diff* in Unix system [U]. In this paper, we also use line differences. The *difference* from text file F_1 to text file F_2, denoted Δ_{F_1,F_2}, is a list containing lines to be deleted from and all lines to be inserted into F_1 in order to transform F_1 into F_2. We assume that this list is presented in the format used by utility *diff* [U]. Our approach is independent of which algorithm is actually used to compute this list of differences. Define the *size* of a text file to be the *number of lines* contained in that text file, and the *size of* Δ_{F_1,F_2}, denoted S_{F_1,F_2}, to be the number of lines to be deleted from F_1 by the deletion commands contained in Δ_{F_1,F_2} plus the number of lines to be inserted into F_1 by the insertion commands contained in Δ_{F_1,F_2}.

In version control systems that use Δ-files, the configuration of versions is usually a tree (e.g., SCCS [Ro], RCS [Ti1]), where each tree vertex corresponds to a version and each tree edge corresponds to a Δ-file between two versions. The format for representing the text files of versions and version differences can vary considerably, and may not involve separate Δ-files. For instance, to store the difference between two versions, SCCS [Ro] uses *interleaved* Δs and RCS [Ti1] uses separate differential files (Δs). For a given textual object, RCS maintains a set of Δs and the text file of a version that is stored intact; this intact version is called the *complete version*.

The central data structure used by the scheme in [Ti1] to reconstruct text files of versions from RCS files is called a *piece table*. A piece table for version r is a one dimensional array, denoted as PT_r. Each element of PT_r is a pointer to the beginning of a line of text in r. Given version r of a file, let $r[i]$ denote the i'th line of version r. The i'th element of PT_r points to $r[i]$ and is denoted as $PT_r[i]$.

When a version is requested, an initial piece table is constructed from the text file of the complete version, and then a sequence of Δs are successively applied to the piece table, transforming it into successive piece tables, each of which corresponds to the text file of a version in the sequence. The resulting final piece table corresponds to the text file of the requested version.

In RCS, a Δ contains a sequence of text edit commands and lines of text to be inserted against a given text file. The commands in a Δ are of the form

$$command \quad i \quad j$$

where *command* is either *a*(dd) or *d*(elete), *i* is a nonnegative integer indicating a line number in the given file to which the Δ is being applied, and *j* is an integer greater than 0 representing the number of lines affected. Given version *r* and a Δ against *r*, an add command says to insert into the new version being constructed, right after *r*[*i*], the *j* lines of the Δ immediately following that command. A delete command says to delete *j* consecutive lines of *r* starting at *r*[*i*]. The line number in each command refers to a position in the original file (rather than the file that would be produced by executing previous commands in the Δ).

An advantage of using a piece table, rather than directly modifying text files, is that modifying the piece table only involves insertions and deletions of line pointers in PT_r, instead of text lines. Note that such insertions and deletions require shifting piece table elements in the array. Hence, using the method of applying a Δ as described in [Ti1], application of a Δ file command for line *i* involves shifting the entire portion of the piece table after the entry for *i*. Thus the time to apply a Δ-file can be quadratic. However, the algorithm as described by [Ti1] can be modified so as to essentially perform a merge of the piece table and the Δ-file. The merge can be implemented so that each entry of the original piece table not deleted will be shifted at most once in the entire process. Such a modified algorithm would take time linear to the size of the given piece table and given Δ-file. However, even if the algorithm to apply a single Δ-file to a piece table, as described in [Ti1], is modified to run in linear time, applying a sequence of Δ-files can still require quadratic time, since each application of a Δ-file may require shifting the entire piece table.

II. Version Reconstruction

In this section we describe our scheme to store and reconstruct versions of sequential objects. Due to space limitations, we omit some details, which can be found in [YR2]. As in RCS, we assume that the set of versions is configured as a tree, with one node corresponding to a complete version, and the edges corresponding to differential files. We propose a data structure called a *P-array* (for *Piece-array*), and a new format for differential files. To differentiate our files from RCS files, we call ours *VCS files* (*Version Control System* files). We also describe the new file format for the complete version which is used to construct the initial P-array. We describe our scheme for constructing an initial P-array and for applying a sequence of Δs to a P-array. We assume that a requested version is reconstructed by applying a sequence of Δs to the complete version, as in RCS.

A *P-array L* of size *m* is an array containing *m*+2 cells: $L[-1]$, $L[0]$, $L[1]$, . . . , $L[m]$. Each cell has two fields: *next* and *item*. The field *item* can hold a physical pointer pointing to a line in some file, and $L[0]$.*item* always points to an empty string λ. The field *next* can hold an integer in the range $[0 .. m]$ (this integer is generally interpreted as the index of a cell in the P-array). A P-array is assumed to have embedded in it a list, called the *P-list*, consisting of a linked list of cells, beginning with cell $L[0]$. The *next* field of each P-list cell points to the next P-list cell. The *next* field of the last P-list cell has the special value 0, indicating end-of-list. Note that the cells in the P-list need not be contiguous. A P-array represents a text file via its P-list. Cell $L[0]$ begins the P-list, and can be thought of as representing the text file spot just before the first text line. To recreate the text file a P-array represents, one need only traverse its P-list and collect the text string pointed to by the *item* field of each traversed cell in that P-list. The *P-area* in a P-array is the set of cells whose indices are in the range $[1 .. m]$. We define the size of a P-array to be the number of cells in the P-area, and the size of a P-list to be the number of P-area cells on the P-list. Cell $L[-1]$ is used as the head of a linked list containing *free cells* in the P-array, i.e., cells in the P-area that are not in the P-list. This free cell linked list is used only by the process of generating VCS files (see Section III).

As an example, consider the two text files F_1 of version V_1 and F_2 of version V_2, shown in Figure 2.1. Figure 2.2 shows two P-arrays, L_1 representing F_1 and L_2 representing F_2.

| apple tree |
| black forest |
| cold spring |
| dog fight |
| engineering school |
| flight simulator |
| ground rule |
| house salad |

| jingle bell |
| cold spring |
| dog fight |
| moonlight sonata |
| x-ray machine |
| year book |
| flight simulator |
| ground rule |
| zebra code |

(a) F_1 of version V_1 (b) F_2 of version V_2

Figure 2.1. Two files F_1 and F_2

A *VCS file* is either a file for a complete version or a file representing a Δ. A VCS file for a complete version, a *C-file* for short, describes a P-list in a P-array. It consists of a sequence of records, beginning with record zero, where the k'th record in the sequence describes the k'th P-list cell. A C-file record consists of two fields,

i
string

where the value of i identifies the index of an array cell, and the value of *string* is a text line that is to be pointed to by the *item* field of that cell. The value of the *next* field of the k'th P-list cell is the P-array index of the $(k+1)$'th P-list cell, i.e., it is the i-value of the $(k+1)$'th record in the C-file. The last P-list cell described by the last C-file record has value 0 in its *next* field. For example, for the C-file shown in Figure 2.3(a), the corresponding P-array is shown in Figure 2.2(a), and the corresponding text file is F_1 as shown in Figure 2.1(a).

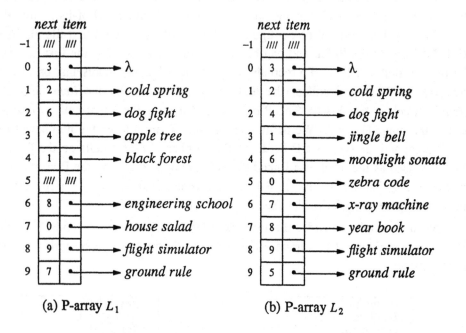

Figure 2.2. Two P-arrays L_1 and L_2

A VCS differential file, a *D-file* for short, contains a section of deletion commands followed by a section of insertion commands. A deletion section begins with a line of one word

deletion

followed by a (possibly null) sequence of deletion commands, each of the form

i   j

where i and j are positive integers. The meaning of a deletion command (i, j) is that, given a target P-array L, a sequence of j cells are to be deleted from the P-list embedded in L, where the first cell to be deleted is the cell pointed to by $L[i].next$. After executing the deletion command, $L[i].next$ will point to the cell previously pointed to by the *next* field of the last cell in the deleted sequence of cells. An insertion section begins with a line of one word

insertion

followed by a (possibly null) sequence of insertion commands, each of the form

i   j
string

where i and j are integers, $i > 0$, $j \geq 0$, and *string* is a line of text. The meaning of an insertion command is to insert a new cell into the P-list. The new cell is cell $L[i]$, and is to be inserted into the P-list immediately after $L[j]$. Hence, after the execution of an insertion command, $L[i].item$ will point to *string*, $L[i].next$ will point to the cell

146

previously pointed to by $L[j].next$, and $L[j].next$ will equal i, i.e., $L[j].next$ will point to cell $L[i]$.

The purpose of placing all of the deletion commands before all of the insertion commands in a D-file is that the free cells resulted from the execution of deletion commands can be reused by the insertion commands.

To apply a D-file to its target P-array, we sequentially apply the deletion and insertion commands contained in that D-file to the P-array.

For example, given the P-array L_1 shown in Figure 2.2(a), and the D-file D shown in Figure 2.3(b), after applying the deletion and insertion commands in D to L_1, we obtain P-array L_2 shown in Figure 2.2(b).

0 λ	deletion 0 2 2 1 9 1
3 apple tree	insertion 3 0
4 black forest	jingle bell 4 2 moonlight sonata
1 cold spring	6 4 x-ray machine
2 dog fight	7 6 year book
6 engineering school	5 9 zebra code
8 flight simulator	
9 ground rule	
7 house salad	

(a) C-file C for F_1 (b) D-file D for Δ_{F_1,F_2}

Figure 2.3. Two VCS files

Now consider version reconstruction. Given the *name* of a requested version, the C-file for the complete version is used to construct an initial P-array L. We then successively apply a sequence of D-files to L to obtain a P-array for the requested version. The size of the P-array created from the C-file must be at least the maximum cell index referred to in any of the files to be processed.

Suppose the size (number of records) of a C-file is m. Then a corresponding P-array can be constructed in $\Theta(m)$ time. We observe that given a P-array L representing text file F_1 and a D-file D_{F_1,F_2} for Δ_{F_1,F_2}, our scheme takes $\Theta(S_{F_1,F_2})$ time to transform L into a P-array representing text file F_2, independent of the sizes of F_1 and F_2. Suppose the sequence of files involved in recreating a requested file F_k is F_0, F_1, ..., F_k. Then our scheme takes $\Theta(S_{F_0,F_1}+S_{F_1,F_2}+ \ldots +S_{F_{k-1},F_k})$ time to reconstruct a P-list representing F_k from the initial P-array for F_0, i.e., the time is proportional to the sum of the differences between the text files of adjacent versions in the sequence. Thus the VCS linear time version reconstruction scheme is faster than the RCS scheme, even if the RCS scheme, as described in [Ti1], is modified to run in linear time for processing a single Δ over a piece table.

III. VCS File Generation

In this section we present our scheme for generating VCS files. Algorithms are presented in this section in a simplified form; for details, see [YR2].

The scheme requires the use of a file difference computing algorithm. There are well known fast algorithms to compute the minimum difference between two text files, e.g., the algorithm described in [Ti2]. In this section, we assume the use of the Unix utility *diff* [U], but any such algorithm can be used. We make no assumption about the minimality of the differences computed by the chosen algorithm, and leave which algorithm to choose as an orthogonal design issue.

First we discuss the generation of a VCS C-file for the very first version in the system. This task is relatively easy. Suppose we are to generate a C-file C for text file F. Record zero of C is given the pair of values 0 and λ for the two fields i and *string*. Suppose the j'th line of F, where $j \geq 1$, is *string-j*; then record j of C is given the pair of values j and *string-j*.

The generation of a C-file C from a given P-array L is also easy: we traverse the P-list in L. For each cell c traversed, we place in C a record where the values of the two fields are the index of c in L and the string pointed to by the *item* field of c.

Next, we discuss the generation of VCS D-files. Suppose the version tree is nonempty, i.e., there are a C-file and perhaps some D-files already stored in the system. We want to add a new version V_2 into the version control system, adjacent to version V_1 in the tree. This task involves the creation of a D-file.

We consider two cases of this problem: one is generating a *forward D-file*, and the other is generating a *reverse D-file*. Suppose we are given two text files, F_1 of version V_1 and F_2 of version V_2, and a P-array L_1 representing F_1. V_1 is an old version (a version already residing in the VCS system) and V_2 is a new version to be added to the system. The goal for generating a forward D-file is to

G1:   create a D-file that transforms L_1 into a P-array L_2 for F_2;

and the goal for generating a reverse D-file is to

G2:   create C-file specifying a P-array L_2 for F_2, and create a D-file D, such that D transforms L_2 into L_1.

In the latter case, we assume that V_1 is the old complete version. The reason for reconstructing L_1 from L_2 by applying D is to avoid recomputing the D-files already in the system. Since these D-files are based on applying Δs to L_1, their validity depends on the system's ability to reconstruct L_1 (which was reconstructed directly from the old C-file C_1 for F_1). Thus, in order to avoid the recomputation of existing D-files in the system after "demoting" F_1 to a Δ version, it is necessary that after applying the generated D-file to L_2, the P-array obtained for F_1 contain the same P-list as would be constructed from the old C-file C_1 for F_1, that is, each element in the P-list for F_1 must occupy exactly the same array cell as would be occupied should the P-list be constructed directly from C_1. This requirement is formalized as follows.

We say that two P-lists A and B are *equivalent* if they consist of the same sequence of cells having exactly the same contents in their *next* fields, and having exactly the same values of the strings pointed to by their *item* fields. Note that the contents of the *next* field of a P-list cell is the index of the next cell in the P-list. Thus, equivalent P-lists consist of cells occupying exactly the same position in their P-arrays: each pair of corresponding cells in the lists have the same index in the P-array, the same *next* value, and the same string value pointed to by *item*. We say that two P-arrays are *equivalent* if their corresponding P-lists are equivalent. Note that equivalent P-arrays can differ in the number of empty cells whose index is greater than the highest index of any P-list cell, and can differ in the contents of free cells. We now refine goal $G2$ as follows:

G2': create a C-file specifying a P-array L_2 for F_2, and create a D-file D such that D transforms L_2 into a P-array equivalent to that which would be obtained by processing C_1.

In processing insertion commands in a *diff* file, careful handling of free cells in the P-area is required. There are two types of free cells. Given a P-array L, we say that a P-area cell is an *original free cell* if it is not included in the original P-list of L. We call P-list cells that are deleted from the P-list because of *diff* deletion commands *released free cells*.

The original free cells are kept track of using a classical technique for implementing sparse arrays [AHU]. This technique uses a data structure involving two arrays S and T. The two arrays are used to distinguish cells which are originally on the P-list from original free cells. The size of S is the number of P-area cells, and the size of T is the number of P-list cells. The arrays are initialized so that P-array cell $L[i]$ is on the original P-list iff $S[i]$ contains an index into array T and $T[S[i]] = i$. A scan of the S array, in ascending index order, can then be used to obtain the original free cells, if needed, in ascending index order. Since the number of original free cells needed is no greater than the size of F_2, and the number of original P-list cells that need be processed is no greater than the size of F_1, this technique for handling original free cells ensures a linear time algorithm, even if the size of the P-array is much larger than the size of the P-list. The released free cells are organized in a *queue* called *FL* (for *Free cell List*), whose initial cell is $L[-1]$. The *next* field of each cell in *FL* points to the next cell in *FL*, and the *next* field of the last cell in *FL* is assigned the special value 0.

When an original P-list cell is released during the processing of a deletion command, it is inserted into *FL* at the rear of *FL*. When a free cell is needed during the processing of an insertion command, if *FL* is non-empty, the front cell from *FL* is returned; if *FL* is empty, i.e., the insertion commands require more free list cells than there are released free cells, the next available original free cell is returned.

We now present our VCS D-file generation algorithm, which produces either a forward D-file or a reverse D-file on demand. The algorithm is presented in the Appendix in a simplified form; its details are described in [YR2]. The inputs of the algorithm are the name of a version V_1, a text file F_2, and a flag, where V_1 is a version already in the system whose text file is F_1, F_2 is the text of a new version to be added into the system, and the value of the flag indicates whether to generate a forward D-file or a reverse D-file. In the case of forward D-file generation, the algorithm produces a forward D-file D that is a VCS differential file from F_1 to F_2. In the case of reverse D-file generation, where F_1 is assumed to be the old complete version whose P-array L_1 can be constructed directly from a C-file C_1 already stored in the system, the algorithm produces a C-file C_2 for the new complete version F_2, produces a reverse D-file D that is a VCS differential file from F_2 to F_1, and replaces C_1 by D. If the D-file D is a forward D-file, then when it is applied to the P-array L_1 constructed by the algorithm, the deletion and insertion commands in D transform L_1 into a P-array L_2 representing F_2. If D is a reverse D-file, then when it is applied to the P-array L_2 for F_2 constructed by the algorithm directly from C_2, the deletion and insertion commands in D transform L_2 into a P-array equivalent to the P-array L_1 that was previously constructed directly from C_1.

During Step 1 of the algorithm, a P-array L_1 representing F_1 is constructed, and F_1 is recreated from L_1. The P-array L_1 is created with size at least the maximum of the size of F_1 and the size of F_2. *Diff* is utilized to generate a preliminary differential file from F_1 to F_2, Δ-*file*. If the size of F_2 is larger than the size of F_1, so that original free cells are needed for the extra lines in F_2, then data structures S and T are constructed for original free cells.

The algorithm scans Δ-*file* twice, once during Step 2 and once during Step 3. For forward D-file generation, during Step 2, forward D-file deletion commands are generated and put into D, and during Step 3, forward D-file insertion commands are generated and put into D. For reverse D-file generation, during Step 2, reverse D-file insertion commands are generated and saved in file *R-insert*, during Step 3, reverse D-file deletion commands are generated and put into D, and during Step 4, the reverse D-file insertion commands from *R-insert* are put into D. In Steps 2 and 3, the *logical line positions* in a deletion or insertion command in Δ-*file* are *translated* into *physical cell indices*, and a forward VCS D-file command is obtained. During Step 2, the forward D-file deletion commands are applied to the P-array, and during Step 3, the forward D-file insertion commands are applied to the P-array. When Step 3 has been completed, L_1 has been transformed into P-array L_2.

In the case of generating a forward D-file, during Steps 2 and 3 each forward D-file command is directly *recorded* into D-file D. In the case of generating a reverse D-file, during Step 2 each forward D-file deletion command is converted into a reverse D-file insertion command, and this converted D-file command is saved in file *R-insert*; and during Step 3 each forward D-file insertion command is converted into a reverse D-file deletion command and this converted D-file command is *recorded* in D-file D. During Step 2, an additional task performed is to modify each *diff* insertion command in Δ-*file*, so that the insertion command refers to the cell index of the P-list cell representing the line after which the insert is to take place. Originally the *diff* insert command in Δ-*file* contains the line number i in F_1 after which the insert is to take place, but Step 2 replaces this line number by the cell index corresponding to the *undeleted* line after which the insertion is to take place. During Step 3, P-array cells for the lines inserted by *diff* commands are obtained. Finally, in Step 4, in the case of forward D-file generation, the algorithm stores the forward D-file D into the system; and in case of reverse D-file generation, the algorithm creates a C-file C_2 according to L_2, stores C_2 into the system for F_2, appends to file D the insertion commands that had been saved in file *R-insert*, and replaces C-file C_1 by the reverse D-file D.

For example, suppose version V_1 has text F_1 as shown in Figure 2.1(a) and has P-array L_1 (as constructed by the version regeneration algorithm) shown in Figure 2.2(a). Suppose text file F_2 is shown in Figure 2.1(b). For forward D-file generation, the algorithm produces the forward D-file D from F_1 to F_2 shown in Figure 2.3(b).

We next consider the cost of generating VCS files. First, to generate a C-file our scheme takes time linear in the size of the file. Now consider the generation of D-files. Suppose for given files F_1 and F_2 with size n_1 and n_2, respectively, the cost of producing a *diff* file over F_1 and F_2 is $O(f(n_1,n_2))$, for some function f. Let $n = \max(n_1,n_2)$. Since *diff* has to completely examine both F_1 and F_2, $f(n_1,n_2) \geq n$.

Step 1 of the algorithm involves reconstruction of a P-array and the execution of *diff*. In case the construction of the data structure involves processing some differential files, we know from Section II that the P-array can be constructed in time linear in the size of the C-file and the total size of the D-files involved. In practice, when a user constructs a new version, a data structure representing the old version would usually be built in order to produce the old version as a text file, which the user would edit. Thus, usually the major cost in Step 1 would be the cost of *diff*. The remaining steps of *D_File_Generator* can be done in $O(n)$ time, since the total number of bypassed, deleted, and inserted cells is linearly bounded by n.

Note that during Step 1, the time spent by the algorithm in constructing a P-array L_1 is no more than the time spent by RCS in constructing a corresponding piece table. Also note that an RCS Δ file *is* a *diff* file, and so also is generated in $O(f(n_1,n_2))$ time. Thus RCS and VCS are comparable in the time to generate the stored differential files; the time saving in VCS over RCS is in version reconstruction.

Appendix: D-File Generation Algorithm

Input V_1 : version name, F_1 : file name, *flag* : { F, R }

Step 1.

> construct P-array L_1 for V_1 by version reconstruction algorithm;
> recreate F_1 from L_1;
> Δ-*file* ← *diff*(F_1, F_2);
> if size of F_2 is greater than size of F_1
>> construct arrays S and T;
>
> open file D;
> if (*flag* = R)
>> open file *R-insert*; { use *R-insert* to temporarily store reverse insertion commands }

Step 2.

> sequentially scan Δ-*file*, during which
>> convert logical line positions into physical cell indices;
>>
>> for each insertion command,
>>> replace its logical insertion position by its physical insertion position, and write back into Δ-*file*;
>>
>> for each deletion command,
>>> perform forward deletion on L_1, releasing freed P-list cells;
>>> if (*flag* = F)
>>>> record the forward deletion command into D;
>>> if (*flag* = R)
>>>> convert the forward deletion command into a reverse insertion command, and record the latter into *R-insert*;

Step 3.

> sequentially scan Δ-*file*, during which
>> for each Δ-*file* insertion command (as modified in Step 2),
>>> find free cells and perform forward insertion on L_1;
>>> if (*flag* = F)
>>>> record the forward insertion command into D;
>>> if (*flag* = R)
>>>> convert the forward insertion command into a reverse deletion command and record it into D;
>
> .   { L_1 has been transformed into L_2 for F_2 }

Step 4.

    if ($flag$ = F)

        store D as the forward D-file from F_1 to F_2;

    if ($flag$ = R)

        remove C_1 from the system;

        use L_2 to create a C-file C_2 and store C_2 for F_2;

        append *R-insert* to D, and store D as the reverse D-file from F_2 to F_1;

end_of_algorithm.

REFERENCES

[AHU]   Aho, A.V., Hopcroft, J.E., and Ullman, J.D., *The Design and Analysis of Computer Algorithms*, Addison-Wesley, Reading, MA, 1974.

[BK]   Batory, D.S., and Kim, W., "Modeling Concepts for VLSI CAD Objects," *ACM Trans. on Database Systems, Vol.10(3), pp.322-346*, Sept. 1985.

[DEC]   *Code Management System*, Digital Equipment Corporation, document number EA-22134-82, 1982.

[DL]   Dittrich, K.R., and Lorie, R.A., "Version Support for Engineering Database Systems," *IEEE Trans. on Software Engineering, Vol.14(4), pp.429-437*, April 1988.

[HG]   Haynie, M., and Gohl, K., "Revision Relations Maintaining Revision History Information," *IEEE Quarterly Bulletin on Database Engineering, 7(2), pp.26-34*, June, 1984.

[He]   Heckel, P., "A Technique for Isolating Differences Between Files," *Comm. ACM Vol.21(4), pp.264-268*, April 1978.

[HM]   Hunt, J.W., and McIlroy, M.D., "An Algorithm for Differential File Comparison," *Computer Science Technical Report 41*, AT&T Bell Laboratories, Murray Hill, June 1976.

[Ka1]   Katz, R.H., "A Database Approach for Managing VLSI Design Data," *ACM IEEE 19th Design Automation Conf., pp.274-282*, Las Vegas, Nevada, June 1982.

[Ka2]   Katz, R.H., "Toward a Unified Framework for Version Modeling in Engineering Databases," *ACM Computing Surveys, Vol.22(4), pp.375-408*, Dec. 1990.

[KB]   Katz, R.H., Bhateja, R., Chang, E.E., Gedge, D., and Trijanto, V., "Design Version Management," *IEEE Design & Test, Vol.4(1), pp.12-22*, Feb. 1987.

[KL]   Katz, R.H., and Lehman, T.J., "Database Support for Versions and Alternatives of Large Design Files," *IEEE Trans. on Software Engineering, Vol.SE-10(2), pp.191-200*, March 1984.

[Ro]   Rochkind, M.J., "The Source Code Control System," *IEEE Trans. on Software Engineering, Vol.SE-1(4), pp.364-370*, Dec. 1975.

[Ti1]   Tichy, W.F., "RCS - A System for Version Control," *Software - Practice and Experience, Vol.15(7), pp.637-684*, July 1985.

[Ti2]   Tichy, W.F., "The String-to-String Correction Problem with Block Moves," *ACM Trans. on Computer Systems, Vol.2(4), pp.309-321*, Nov. 1984.

[U]   *Unix User's Reference Manual*, Dept. E.E. & Comp. Sci., UC Berkeley, CA, April 1986.

[YR1]   Yu, L., and Rosenkrantz, D. J., "Minimizing Time-Space Cost for Database Version Control," *Acta Informatica, Vol.27(7), pp.627-663*, July, 1990.

[YR2]   Yu, L., and Rosenkrantz, D. J., "A Linear-Time Scheme for Version Reconstruction," *TR90-22*, Dept. Comp. Sci., State University of New York at Albany, NY, 1990.

The Interval Skip List: A Data Structure for Finding All Intervals That Overlap a Point *

Eric N. Hanson[1,2]
[1] USAF Wright Laboratory
WL/AAA-1
Dayton, OH 45433

[2] Wright State University
Dept. of Computer Science
Dayton, OH 45435

Abstract

A problem that arises in computational geometry, pattern matching, and other applications is the need to quickly determine which of a collection of intervals overlap a query point. The *interval binary search tree* (IBS-tree) has been proposed as a solution to this problem. A recently discovered randomized data structure called the *skip list* provides functionality and performance similar to balanced binary trees (e.g., AVL-trees), but is much simpler to implement than balanced trees. This paper introduces an extension of the skip list called the *interval skip list*, or IS-list, to support interval indexing. IS-lists remain dynamicly balanced, and show similar performance to IBS-trees, but can be implemented in about one-fourth as much high-level language code. Searching an IS-list containing n intervals to find intervals overlapping a point takes expected time $O(\log n + L)$ where L is the number of matching intervals. Inserting or deleting an interval takes expected time $O(\log^2 n)$.

1 Introduction

An important problem that arises in a number of computer applications is the need to find all members of a set of intervals that overlap a particular point. Queries of this kind are also called *stabbing queries* [Sam90]. This paper introduces a data structure called the *interval skip list* (IS-list) designed to handle stabbing queries efficiently. The need for a dynamic, balanced data structure for indexing intervals arose in designing a pattern matcher for a database rule system [Han89, HCKW90]. The IS-list is an extension of the randomized list structure known as the *skip list* recently discovered by Pugh [Pug90]. In

*This work was supported by the Air Force Office of Scientific Research under grant number AFOSR-89-0286.

Section 2, other methods for solving stabbing queries are discussed. Section 3 describes the interval skip list data structure and methods for searching and updating it. Section 4 gives an analysis of the complexity of algorithms for manipulating IS-lists. Finally, Section 5 presents conclusions.

2 Review of Stabbing Query Solution Methods

Formally, we describe the stabbing query problem as the need to find all intervals in the set $Q = \{i_1, i_2, ... i_n\}$ which overlap a query point X.[1] Several different approaches to solving the stabbing query problem have been developed. The most trivial solution is to place all the intervals in a list, and traverse the list sequentially, checking each interval to see if it overlaps the query point. This algorithm has search complexity of $O(n)$.

A more sophisticated approach is based on the *segment tree* [Sam90]. To form a segment tree, the set of all end points of intervals in Q is formed, and an ordered complete binary tree is built which has the end points as its leaves. To index an interval, the identifier of the interval is placed on the uppermost nodes in the tree such that all values in the subtrees rooted at those nodes lie completely within the interval. In this way, an interval of any length can be covered using $O(\log n)$ identifiers. In order to solve a stabbing query using a segment tree, the tree is traversed starting from the root, examining the path in the tree from the root to the location the query value X would occupy. The interval identifiers on all nodes traversed are returned as the answer to the query. A query takes $O(\log n + L)$ time where n is the total number of intervals, and L is the number of intervals retrieved. The segment tree works well in a static environment, but is not adequate when it is necessary to dynamically add and delete intervals in the tree while processing queries.

Another data structure that can be used to process stabbing queries is the *interval tree* [Ede83a, Ede83b]. Unfortunately, as with the segment tree, all the intervals must be known in advance to construct an interval tree.

A data structure that can index intervals dynamically is the R-tree [Gut84]. R-trees are a multi-dimensional extension of B-trees in which each tree node contains a set of

---

[1]The notation used here for intervals shows a pair of values with inclusive boundaries indicated by square brackets, and non-inclusive boundaries indicated by parentheses. Open intervals have one boundary at positive or negative infinity, and points have both boundaries equal. Examples of intervals are [17,19), [12,12], [-inf,22].

possibly overlapping n-dimensional rectangles. Subtrees of each index node contain only data that lies within a containing rectangle in the index node. Since rectangles in each node may overlap, on searching or updating the tree, it may be necessary to examine more than one subtree of any node. An important part of the R-tree algorithm involves use of heuristics to decide how to partition the rectangles in a subtree to determine the best set of index rectangles for an index node. Due to its generality, and the indexing heuristics required, the R-tree is challenging to implement. A useful property of R-trees is that they require only $O(n)$ space. Their performance should be good for rectangles (or intervals in the 1-dimensional case) with low overlap, but when there is heavy overlap, search time can degenerate rapidly.

Another data structure which solves the stabbing query problem efficiently (among others), and does allow dynamic insertion and deletion of intervals is the *priority search tree* [McC85]. An advantage of the priority search tree is that it requires only $O(n)$ space to index n intervals. However, the priority search tree in its balanced form is very complex to implement. In addition, To handle intervals with the same lower bound, priority search trees must use a special transformation from pairs with non-unique lower bounds to pairs with unique lower bounds. This transformation is not trivial, and it must be created for each different data type to be indexed.

The *interval binary search tree* (IBS-tree) can handle stabbing queries, and is somewhat more flexible and easier to implement than the priority search tree, although it requires $O(n \log n)$ storage [HC90]. A data structure closely related to the IBS-tree called the *stabbing tree* has been developed to find the *stabbing number* for a point given a collection of intervals [GMW83]. The stabbing number is the number of intervals that overlap a point. In contrast, the IBS-tree and the IS-list return a *stabbing set* containing all the intervals overlapping the query point, not just the number.

The interval skip list is quite similar in principle to the IBS-tree, but it inherits the simplicity of skip lists, making it much easier to implement than balanced IBS-trees. In the next section, we present the details of the IS-list.

3   Interval Skip Lists

In this section we introduce a method for augmenting a skip list with additional information to make it possible to rapidly find all intervals that overlap a query point. The

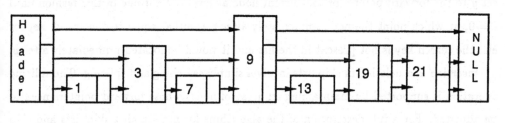

Figure 1: Example skip list.

IS-list can accommodate points as well as open and closed intervals with inclusive and exclusive boundaries. First, we will review the skip list data structure [Pug90] and then discuss the extensions needed to index intervals.

3.1 Review of Skip Lists

The skip list is similar to a linked list, except that each node on the list can have one or more forward pointers instead of just one forward pointer. The number of forward pointers the node has is called the *level* of the node. When a new node is allocated during a list insertion, its level is chosen at random. The probability distribution of the random function that returns a new node level x is defined as follows (p is a parameter of the function with an appropriate value, e.g., $1/2$):

$$P(x = k) = \begin{cases} 0 & \text{for } k < 1 \\ (1 - p) \cdot p^{k-1} & \text{for } k \geq 1 \end{cases}$$

With $p = 1/2$, the distribution of node levels will allocate approximately $1/2$ the nodes with one forward pointer, $1/4$ with two forward pointers, $1/8$ with three forward pointers, and so on.

A skip list is normally organized with values in increasing order. A node's pointer at level l points to the next node with l or more forward pointers. An example skip list is shown in Figure 1. Searching in a skip list involves "stair-stepping" down from the beginning of the list to the location of the search key. To search in a skip list, start at the list header at the level i equal to the maximum level of a node in the list. Call the current node being visited y (y initially points to the header). If the value of the search key at the node pointed to by the level i pointer of y is $\geq$ the search key, decrement i. Otherwise, set y to the forward pointer of the current node at level i. Continue in this fashion until $i = 0$, at which point the node immediately after y is either equal to the search key, or else the search key is not present in the list and it would be located immediately after y.

Insertion and deletion in skip lists involves simply searching and splicing. The splicing operation is supported by maintaining an array of nodes whose forward pointers need to be adjusted. For a full description of the algorithms for maintaining skip lists and skip lists extended to support additional capabilities such as search fingers, efficient merging, finding the k'th item in a list etc. the reader is referred to [Pug90, Pug89].

The performance of skip lists is quite similar to that of AVL trees [AVL62]. The expected value of times for search, insertion and deletion in a skip list with n elements are all $O(\log n)$. The variance of search time is also quite low, making the probability that a search will take significantly longer than $\log n$ time vanishingly small. Comparing actual implementations of skip lists and AVL trees, skip lists perform as well or better than highly-tuned non-recursive implementations of AVL trees, yet programmers tend to agree that skip lists are significantly easier to implement than AVL trees [Pug90]. A discussion of extensions to skip lists to support interval indexing is given below.

3.2   Extending Skip Lists to Support Intervals

The basic idea behind the interval skip list is to build a skip list containing all the end points of a collection of intervals, and in addition place markers on nodes and forward edges in the skip list to "cover" each interval. The placement of markers on edges and nodes in an interval skip list can be stated in terms of the following invariant:

Interval skip list marker invariant: Consider an interval $I = (A,B)$ to be indexed. End points A and B are already inserted into the list. Consider some forward edge in the skip list from a node with value X to another node with value Y. The interval represented by this edge is (X,Y). A marker containing the identifier of I will be placed on edge (X,Y) if and only the following conditions hold:

1. **containment:** I contains the interval (X,Y).

2. **maximality:** There is no forward pointer in the list corresponding to an interval (X', Y') that lies within I and contains (X,Y).

In addition, if a marker for I is placed on an edge, then any node that is an endpoint of that edge and has a value contained in I will also have a mark placed on it for I.

Example intervals:

   a. [2,17]
   b. (17,20]
   c. [8,12]
   d. [7,7]
   e. [-inf,17)

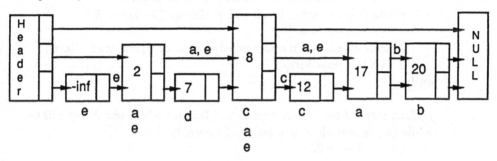

Figure 2: Example interval skip list for intervals shown.

An example set of intervals and the IS-list for those intervals is shown in Figure 2. Searching an IS-list to find all intervals that overlap a search key can be done efficiently given a skip list with markers on it satisfying this invariant. The challenge in inserting and deleting intervals into an interval skip list is to perform the operations efficiently while maintaining the interval skip list invariant. The remainder of this section describes the procedures for searching, insertion and deletion in IS-lists.

3.3 Searching

In order to search an IS-list L to find all intervals that overlap a search key K, and return those intervals in a set S, search along the same path that would be visited by the standard skip list search procedure, and add markers to S as the search proceeds. Whenever dropping from level i to $i-1$ during the search, add to S the markers on the forward pointer at level i of the current node. This is valid since markers on the forward pointer at level i must belong to an interval that contains K. At the final destination, if K is present in the list, add the markers on node K to S. Otherwise, (K is not present) add the markers on the lowest pointer of the current node to S. When the search terminates, exactly one marker for every interval that overlaps K will be in S. No duplicates will be

```
procedure findIntervals(K,L,S)
    x = L.header; S = φ;
    // Step down to bottom level.
    for i=maxLevel down to 1 do
        // Search forward on current level as far as possible.
        while (x→forward[i] ≠ null and x→forward[i]→key < K)
            x = x→forward[i];
        // Pick up interval markers on edge when dropping down a level.
        S := S ∪ x→markers[i];
    end;

    // Scan forward on bottom level to find location where search key will lie.
    while (x→forward[0] ≠ null and x→forward[0]→key < K)
        x = x→forward[0];

    // If K is not in list, pick up interval markers on edge,
    // otherwise pick up markers on node with value = K.
    if (x→forward[0] = null or x→forward[0]→key ≠ K)
        S := S ∪ x→markers[0];
    else
        S := S ∪ x→forward[0]→eqMarkers;
end findIntervals
```

Figure 3: Procedure to find intervals that overlap a point using an interval skip list.

found. An outline of an implementation of this search algorithm is shown as the procedure findIntervals(K,L,S) in Figure 3. Each node of level i in an interval skip list contains the following:

key a key value,

forward an array of forward pointers, indexed from 0 to $i-1$, as in a regular skip list,

markers an array of sets of markers, indexed from 0 to $i-1$,

owners a bag (multi-set) of identifiers of the intervals that have an endpoint equal to the key value of this node (note that one interval identifier can appear twice here if the interval is a point).

eqMarkers a set of markers for intervals that have a marker on an edge that ends on this node, and which contain the key value of this node.

We now turn to a discussion of the algorithm for inserting an interval into an IS-list.

3.4 Insertion

To insert an interval (A,B) into an IS-list, insert A and B separately if they are not already in the list, adjusting existing markers as necessary. Then, start at A, search for B, and place markers for (A,B). To place an interval end-point E into the list if it is not already there: 1. Search to find the location where E *will* be located. Place a copy of all the markers on pointers that will be adjusted to point to E in a set S. Let L be the nearest node in front of the position for E which is at least as tall as E. Let R be the nearest node after the position for E that is at least as tall as E. For an interval I, let left(I) and right(I) denote the left and right end points of I, respectively. For each interval I in S, start at max$(L,$left$(I))$, search to min$(R,$right$(I))$, and delete the markers for I found on the search path. Insert E into the list in the normal fashion. Finally, For each interval I in S, start at max$(L,$left$(I))$, search to min$(R,$right$(I))$, and place markers for I on the search path. For brevity, a detailed description of the insertion and deletion algorithms is omitted.

3.5 Deletion

To delete an interval (A,B) do the following: 1. Delete its markers. To do so, search for A, then search forward for B, deleting markers owned by (A,B) from pointers as you go. 2. Delete node A. Markers on the pointers to A and out of A may need to be adjusted. Since A is not the end point of any interval with a marker in the list, any interval with a marker on a pointer coming into A must also have a marker on a pointer leaving A. Consider a marker for an interval I on a pointer coming into A. Call this marked pointer P-IN. Similarly, call the pointer marked by interval I that is leaving A P-OUT. Let $X =$ source(P-IN) and $Y =$ destination(P-OUT). Place the triple (I,X,Y) in a set S. Do this for every pair of pointers into and out of A that both have a mark for interval I. Now, delete all the markers on edges into A, and remove A from the list, adjusting the pointers appropriately. Finally, for each triple in S, starting at X, search forward for Y, placing markers for interval I as you go. 3. Delete node B in the same manner as node A.

4 Performance Analysis

The expected cost to search an IS-list to find all intervals that overlap a key K is $O(\log n + L)$, including $O(\log n)$ to search the skip list, plus $O(L)$ where L is the number of intervals that overlap K. It is quite a bit more difficult to determine the expected cost of insertion and deletion operations on IS-lists. The components of the insertion cost are as follows:

1. the cost to insert the left end point and adjust markers,

2. the cost to insert the right end point and adjust markers,

3. the cost to place markers for the inserted interval.

Inserting the left and right end point, and placing markers for the inserted interval all take expected time $O(\log n)$. We need to determine the cost to adjust the markers. Let $P(\text{level} = i)$ and $\text{AdjustCost}(i)$ denote the the probability that an inserted node has i levels, and the expected cost to adjust markers in the list when a node with i levels is inserted, respectively. The expected cost to adjust the markers disturbed by the insertion of a new node with is:

$$\sum_{i=1}^{\infty} P(\text{level} = i) \cdot \text{AdjustCost}(i)$$

Let $\text{DisturbedIntervals}(i)$ and $\text{RemarkingCost}(i)$ denote the expected number of interval markers disturbed and the cost to remove and replace the markers for each interval that had a marker disturbed, respectively, when a node of level i is inserted. We make the simplifying assumption that $\text{RemarkingCost}(i)$ and $\text{DisturbedIntervals}(i)$ are expected values for independent random variables. Thus, using these terms we can give the following formula for $\text{AdjustCost}(i)$:

$$\text{AdjustCost}(i) = \text{DisturbedIntervals}(i) \cdot \text{RemarkingCost}(i)$$

Replacing the markers for a disturbed interval requires $O(1)$ time for each of the i levels of the inserted node, so

$$\text{RemarkingCost}(i) = O(i)$$

To find $\text{DisturbedIntervals}(i)$, we make the simplifying assumption that all end points in the list are unique. As a further simplification, let us assume that $P(\text{level} = i)$ is $1/2^i$. Given these assumptions, the expected number of markers on an edge at level j is $\approx 2^j$.

This can be derived from the fact that there are approximately $2n$ markers placed on each level in the list, and there are approximately $2^{\log n - (j-1)}$ forward edges at level j. The expected number of markers on a level j edge is $2n/2^{\log n - (j-1)}$, which simplifies to 2^j. The expected total number of markers disturbed during the insertion of a level i node is

$$\text{DisturbedIntervals}(i) \approx \sum_{j=1}^{i} 2^j = 2^{i+1} - 2$$

Moreover, DisturbedIntervals(i) is always no greater than n, the number of total intervals in the list. Hence, DisturbedIntervals(i) $\approx \min(n, 2^{i+1} - 2)$. Now, we can restate the expected cost to adjust markers when a node is inserted as follows:

$$\sum_{i=1}^{\infty} P(\text{level} = i) \cdot \text{DisturbedIntervals}(i) \cdot \text{RemarkingCost}(i)$$

$$= O\left(\sum_{i=1}^{\infty} \frac{1}{2^i} \cdot \min(n, 2^{i+1} - 2) \cdot i \right)$$

$$\approx O\left(\sum_{i=1}^{\lceil \log n \rceil} \frac{1}{2^i} \cdot (2^{i+1} - 2) \cdot i + \sum_{i=\lceil \log n \rceil + 1}^{\infty} \frac{1}{2^i} \cdot n \cdot i \right)$$

$$\approx O\left(\sum_{i=1}^{\lceil \log n \rceil} 2i + \sum_{i=\lceil \log n \rceil + 1}^{\infty} \frac{1}{2^i} \cdot n \cdot i \right)$$

The first of the two sums in the expression above is clearly $O(\log^2 n)$. Substituting variables, the second sum can be written as follows:

$$O\left(\sum_{j=1}^{\infty} \frac{1}{2^{j + \lceil \log n \rceil}} \cdot n \cdot (\lceil \log n \rceil + j) \right)$$

$$\approx O\left(\sum_{j=1}^{\infty} \frac{1}{2^j} \cdot \frac{1}{n} \cdot n \cdot (\lceil \log n \rceil + j) \right)$$

$$\approx O\left(\sum_{j=1}^{\infty} \frac{1}{2^j} \cdot (\lceil \log n \rceil + j) \right)$$

Splitting this into two sums, we have

$$\approx O\left(\sum_{j=1}^{\infty} \frac{1}{2^j} \cdot \lceil \log n \rceil + \sum_{j=1}^{\infty} \frac{j}{2^j} \right)$$

The first of the two sums in the expression above converges to $O(\log n)$. Using the ratio test, it can be shown that the second sum converges to a constant. The complete expression above thus simplifies to $O(\log n)$. Hence, the expected cost to adjust markers after inserting a node with an arbitrary number of levels is:

$$O(O(\log^2 n) + O(\log n))$$

$$= O(\log^2 n)$$

The cost to adjust markers dominates the other costs associated with inserting a new interval into an IS-list. Hence, the expected cost of an insertion in an IS-list is $O(\log^2 n)$. A similar analysis shows that the expected cost of a deletion in an IS-list is also $O(\log^2 n)$. The expected storage utilization in an IS-list is $O(n \log n)$ since there are n total intervals, and each of them can place $O(\log n)$ markers in the list.

5 Conclusion

The interval skip list is an efficient and relatively simple algorithm for indexing intervals to handle stabbing queries efficiently. We have implemented IS-lists in about 700 lines of C++ code, which is about one-fourth the amount of C++ code required in our implementation of interval binary search trees. The main drawback of IS-lists is their potentially large storage utilization of $O(n \log n)$. Fortunately, this worst-case storage utilization does not occur when interval overlap is low. In fact, if intervals do not overlap, only $O(n)$ storage is needed. It is likely that many real applications will only need to index intervals with limited overlap.

References

[AVL62]   G. M. Adel'son-Vel'skii and E. M. Landis. An algorithm for the organization of information. *Soviet Math. Dokl.*, 3, 1962.

[Ede83a]   H. Edelsbrunner. A new approach to rectangle intersections: Part I. *International Journal of Computer Mathematics*, 13(3-4):209–219, 1983.

[Ede83b]   H. Edelsbrunner. A new approach to rectangle intersections: Part II. *International Journal of Computer Mathematics*, 13(3-4):221–229, 1983.

[GMW83] Gaston H. Gonnet, J. Ian Munro, and Derick Wood. Direct dynamic stru-
cutures for some line segment problems. *Computer Vision, Graphics, and
Image Processing*, 23, 1983.

[Gut84] A. Guttman. R-trees: A dynamic index structure for spatial searching. In
*Proceedings of the 1984 ACM SIGMOD International Conference on Man-
agement of Data*, June 1984.

[Han89] Eric N. Hanson. An initial report on the design of Ariel: a DBMS with an
integrated production rule system. *SIGMOD Record*, 18(3), September 1989.

[HC90] Eric N. Hanson and Moez Chaabouni. The IBS-tree: A data structure for
finding all intervals that overlap a point. Technical Report WSU-CS-90-
11, Dept. of Computer Science and Eng., Wright State Univ., April 1990.
Submitted for publication.

[HCKW90] Eric N. Hanson, Moez Chaabouni, Chang-ho Kim, and Yu-wang Wang. A
predicate matching algorithm for database rule systems. In *Proceedings of
the 1990 ACM SIGMOD International Conference on Management of Data*,
May 1990.

[McC85] Edward M. McCreight. Priority search trees. *SIAM Journal of Computing*,
14(2):257–278, 1985.

[Pug89] William Pugh. A skip list cookbook. Technical Report CS-TR-2286, Dept.
of Computer Science, Univ. of Maryland, July 1989.

[Pug90] William Pugh. Skip lists: A probabilistic alternative to balanced trees. *Communications of the ACM*, 33(6), June 1990.

[Sam90] Hanan Samet. *The Design and Analysis of Spatial Data Structures*. Addision
Wesley, 1990.

Geometric Knapsack Problems

Esther M. Arkin [*]
School of Operations Research
& Industrial Engineering
Cornell University
Ithaca, NY 14853

Samir Khuller [†]
Institute for Advanced Computer Studies
University of Maryland
College Park, MD 20742

Joseph S.B. Mitchell [‡]
School of Operations Research
& Industrial Engineering
Cornell University
Ithaca, NY 14853

Abstract

We study a variety of geometric versions of the classical knapsack problem. In particular, we consider the following "fence enclosure" problem: Given a set S of n points in the plane with values $v_i \geq 0$, we wish to enclose a subset of the points with a fence (a simple closed curve) in order to maximize the "value" of the enclosure. The value of the enclosure is defined to be the sum of the values of the enclosed points minus the cost of the fence. We consider various versions of the problem, such as allowing S to consist of points and/or simple polygons. Other versions of the problems are obtained by restricting the total amount of fence available and also allowing the enclosure to consist of up to K connected components. When there is an upper bound on the length of fence available, we show that the problem is NP-complete. Additionally we provide polynomial-time algorithms for many versions of the fence problem when an unrestricted amount of fence is available.

1 Introduction

The *knapsack problem* asks one to pack a "knapsack" of a given capacity with a collection of items in such a way that the total value of the items packed is maximized. Knapsack problems are of fundamental importance and have been studied for many years in the fields of operations research and computer science ([Chv 83, Da 63, GN 72, PS 82]). The knapsack problem, and its

---

[*]Partially supported by NSF Grants DMC–8451984, ECSE–8857642, and DMS–8903304. Email: estie@orie.cornell.edu

[†]Partially supported by NSF grant CCR-8906949. Part of this research was done while this author was at Cornell University, and supported by NSF Grant DCR 85–52938, an IBM Graduate Fellowship, and PYI matching funds from AT&T Bell Labs and Sun Microsystems. Email: samir@umiacs.umd.edu

[‡]Partially supported by NSF Grants IRI-8710858, ECSE-8857642 and a grant from Hughes Research Labs. Email: jsbm@cs.cornell.edu

generalization to the "cutting-stock" problem ([Chv 83, Da 63]), have a broad range of important applications and have motivated several methodologies in the field of combinatorial optimization.

In this paper, we introduce a new class of *geometric* knapsack problems. The general statement of these problems asks us to find a maximum-value subset of a given set of objects (points, polygons, etc.) that are to be placed within a "knapsack" while obeying a given "capacity" constraint. For our purposes, a knapsack is a simple closed curve, and its *capacity* is usually defined to be its perimeter or area. In some cases, we have a finite upper bound on the capacity of the knapsack, while in other cases, we allow arbitrarily large capacity knapsacks, but we incorporate a cost per unit capacity into the objective function of our optimization problem.

An application motivating this line of research was the following "oil lease" problem. Given that we know the locations and values of oil wells and of oil reserves, and given that land costs a certain amount per acre to lease, what is our optimal strategy in purchasing an oil lease? We can consider the known oil wells to be point sites with known values (their estimated production) and oil reserve pools to be polygonal objects with known value (based on estimated total reserves). Our goal is to determine an optimal connected oil field to lease, such that the net value of our field is maximized. In some situations, we are allowed to draw the boundary of our leased field through the middle of a reserve pool, while in other cases, the royalty laws forbid it.   If we allow several disconnected pieces of land to be leased, then we arrive at the K-enclosure version of our problem.

Problem Formulation/Taxonomy: Given a set $S = \{s_1, \ldots, s_n\}$ of n objects in the plane, each with a given value, v_i, we wish to construct one or more "fences" that enclose some or all of the objects in such a way so as to maximize "net profit", subject to some form of capacity constraints and/or costs associated with the fences. The net profit is defined to be the sum of the values of the objects enclosed *minus* the cost (if any) of the fence used. By adjusting the value of c, it is easy to give a varying amount of importance to the cost of the fence relative to the value of the points. There are many different problems in this general class, depending on the following parameters:

- The objects S may be points, line segments, simple polygons, etc. For objects other than points, we may consider the objects to be obstacles, or we may permit the fence to pass through the objects.

- The values v_i may be constrained to be nonnegative, or they may be unrestricted in sign.

- There may be an upper bound $L \leq +\infty$ on the total length of fence available.

- There may be an upper bound $A \leq +\infty$ on the total area allowed to be enclosed.

- There may be a cost c (≥ 0) per unit length of fence we use.

- There may be a cost α (≥ 0) per unit area that is enclosed by the fence.

- We may be permitted to construct more than one enclosure. For example, we may be allowed to construct up to K (disjoint) enclosures. In addition, there may be a fixed charge per enclosure.

Many of our results also hold when we use the area of the enclosure in the objective function rather than the perimeter length. Note that area objective functions are interesting only in the case that we impose the constraint that the enclosure(s) be convex.

One of the challenging issues that arises in our geometric version of the knapsack problem is that in contrast with the classical knapsack problem, the "cost" of including an object in the knapsack is not purely a function of the object, since it depends on the choice of the other objects that are included. Here "cost" refers to the amount of capacity (e.g., extra perimeter or area) that must be expended in order to include an item within the enclosure. The difficulty is that in trying to change the shape of the enclosure to include one object, we affect the costs of including other objects. Since this phenomenon does not arise in the classical knapsack problem, it is not clear how to apply existing techniques to our class of problems.

Summary of Results: Our results are summarized below. Many of the proofs and algorithms are omitted from this shortened version of the paper. They all appear in the full paper [AKM 90].

(1) We show that when there is an upper bound on the perimeter or the area of the allowed enclosure (i.e., if $L < \infty$ or $A < \infty$), then the problem of enclosing a maximum-value subset of points with values $v_i > 0$ is NP-hard. This we establish by a reduction from the classical knapsack problem. The tricky part of the construction is to create a scenario in which the relative cost of including objects is not significantly affected by the choice of including other objects. We provide a pseudo-polynomial algorithm for the case in which all the values v_i are integral. We also provide an efficient algorithm that works well when L is small compared to the distances between the points themselves and does not depend on the values of the points.

As a by-product of our psuedo-polynomial time algorithm, we obtain efficient solutions to the following specific problems:

(a). Find the smallest perimeter polygon that encloses k points of a set of n points S. Our solution requires time $O(kn^3)$ and space $O(kn^2)$. (Alternatively, we can find the smallest *area* convex polygon enclosing k points.)

(b). Find the smallest perimeter polygon that (fully) encloses k segments of a set of n pairwise-disjoint segments S. (Segments are allowed to cross the boundary of the polygon.) Our solution requires time $O(n^6)$ and space $O(E)$ (where E is the size of the visibility graph). (Alternatively, we can find the smallest *area* convex polygon enclosing k segments.)

(2) We also show that when the values v_i are unrestricted in sign, the minimum-length enclosure problem is NP-hard. In fact, even the simple problem of finding a shortest cycle that separates two given sets of points (e.g., one set with values very large and positive, one with values very negative) is NP-hard, as we argue from the Euclidean traveling salesman problem.

Because of the above hardness results, we focus the remainder of our attention on the cases in which all values $v_i > 0$, and there is no upper bound on the perimeter or the area of the enclosure.

(3) If there is no bound on the length of fence available, but there is a cost $c > 0$ per unit length of fence, we solve various versions of the optimal single ($K = 1$) enclosure problem in polynomial time. The corresponding version of the classical knapsack problem allows one to purchase knapsack capacity at a given cost per unit and is solved easily by choosing to include only those objects whose value per unit capacity exceeds the cost per unit

capacity. The solution in our geometric setting is far less obvious, since costs of including an object depend on the other objects that are to be included. The objective function that we are trying to maximize is defined as the sum of the values of the objects enclosed minus the cost of the fence. We obtain the following results:

(a). (Section 2) For the case of enclosing n points, we give an an $O(n^3)$ time sequential algorithm. This algorithm also parallelizes efficiently. We also give an alternative sweep line algorithm to solve the problem that has the same complexity.

(b). (Section 4) We then consider the more general problem when there are polygons in the plane that model the objects to be included. We will also assume that objects have to be either entirely included or entirely excluded. The algorithm gets more complex, since we lose the convexity property that holds in the previous case. The running time of this algorithm is $O(En^2)$, where E is the size of the visibility graph, and it is also parallelizable efficiently.

(c). We further consider the case, when we are allowed to build fences through the polygons. (This arose in a real application as well, when there are some fields that have some value and we can build a fence through the field. The value we get is proportional to the area of the fence that we actually enclose.) This permits us to model objects that can be broken into parts and partially included. For this problem we obtain an $O(n^3)$ time algorithm. The other model we consider is when, the objects (line segments) are penetrable but the value is obtained only when we include them completely. Surprisingly, this turns out to be a much harder problem to solve. The corresponding algorithm that we provide has a running time of $O(n^6)$, and clearly has a lot of room for improvement. (These results are presented in the full paper [AKM 90].)

(4) In Section 5 we give some polynomial-time algorithms for fixed $K > 1$; however, we conjecture that problems in which K is part of the input are NP-hard.

Related Work: Problems that ask for an optimal polygon according to some objective function and subject to some set of constraints are prevalent in computational geometry. For example, [BDDG 85] solved the problem of finding a maximum area/perimeter k-vertex convex polygon whose vertices come from a given set S of n points. They give an $O(kn \log n + n \log^2 n)$ time algorithm that uses $O(n)$ space. These results were improved using "matrix-searching" techniques to yield a time bound of $O(kn + n \log n)$ [AKMSW 87]. [KL 85] consider the problem of finding a (single) convex enclosure of a subset of S that maximizes the sum of the values of points enclosed, where values v_i are unrestricted in sign. They give a dynamic programming algorithm that requires time $O(n^5)$. Most recently, an independent effort by [ORW 89] has examined the problem of finding a minimum/maximum area/perimeter convex enclosure that has vertices among a set S of n (possibly weighted) points and (1) is a convex k-gon; or (2) is an empty convex k-gon; or (3) encloses exactly k points of S. For these problems, [ORW 89] provide an $O(kn^3)$ time ($O(kn^2)$ space) algorithm, based on dynamic programming methodologies similar to ours, which matches our time bound for this special case. We know of no previous work on the more general versions of the knapsack problem treated here; most notably, in the much harder case of S being a set of polygonal objects, we know of no previous results.

2 Optimally Enclosing Point Objects

In this section we assume that S is a set of n points, each with value v_i. We are searching for a single ($K = 1$) enclosure that maximizes the total net profit, which is the sum of the values of the enclosed points, minus the cost of the enclosure. We assume that $L = \infty$, so that there is no upper bound on the perimeter of the allowed enclosure. First, we show that if values v_i are unrestricted in sign, then our problem is hard:

Theorem 1 *Let S be a set of n points in the plane with values v_i unrestricted in sign, and assume that fence costs c per unit length. Then the problem of finding a fence that maximizes net profit (sum of values enclosed minus cost of fence) is NP-complete.*

Proof: First, consider the following problem:

Red-Blue Separation: Given a set B of blue points, and a set R of red points in the plane, find a shortest cycle that separates B from R.

This problem is easily seen to be NP-complete by a reduction from the Euclidean traveling salesman problem ([GJ 78]). (This was shown independently by [ER 91].) Map each point in the TSP instance to a pair of very close points – one red and the other blue.

The proof of our theorem is completed by noting that the red-blue separation problem can be written as an instance of the optimal enclosure problem: put a value of M on each red point, and a value of $-M$ on each blue point, where M is a very large positive number. Then an optimal enclosure must surround all red points while excluding all blue points. □

Hence, in the remainder of the paper, we will assume that all values v_i are nonnegative.

2.1 Simple Properties of Optimal Fences

We prove a few simple properties of optimal fences; here, we only state:

Lemma 1 *Optimal fences in the plane are convex (assumes $v_i \geq 0$).*

Remark: Note that in three dimensions, the *convexity* property does not hold. Consider four non-coplanar points in $\Re^3$. The optimal solution (minimum surface area "sheet") is not convex.

2.2 Triangle Queries

In later sections of this paper, we will need a solution to the following problem: Given a set S of n points in the plane, with values $v_1, v_2, \ldots, v_n$ we wish to preprocess the points so that we can efficiently answer queries of the following form: Given a triple of points $(p, q, r)(p, q, r \in S)$ compute the sum of the values of the points in $\triangle pqr$.

To our knowledge, this problem has not been solved previously in the literature. We give an algorithm which does the preprocessing in time $O(n^2)$ and uses $O(n^2)$ space so that the queries can be answered in $O(1)$ time. The details are omitted in this abstract. We provide a preprocessing algorithm which answers "wedge" and "half-plane" queries in $O(1)$ time, and then show how to use these queries to answer the triangle queries in $O(1)$ time. The final result is:

Theorem 2 *We can answer triangle queries of the type $\triangle pqr$ to compute the sum of the values of the points in the triangle defined by p, q, r in $O(1)$ time. The preprocessing takes $O(n^2)$ time and uses $O(n^2)$ space.*

2.3 Shortest Path formulation

In this section we consider the two-dimensional version of the problem when the fence is unrestricted in length and costs c per unit length. We wish to maximize the "value" of the enclosure (which is the value of the points enclosed minus the cost of the fence).

In the full paper, we begin with a sweep line algorithm that solves the problem in time $O(n^3)$. We omit this here, and instead outline how to reduce the optimal enclosure problem to an *all-pairs shortest path* problem in directed graphs. This yields an $O(n^3)$ time sequential algorithm to solve the problem. Since shortest paths in directed graphs can also be obtained in parallel on a PRAM, this also yields a parallel algorithm to solve the fence problem.

Given the set S of n points x_i in the plane (we will assume that the points are sorted by their x-coordinates), define the complete weighted directed graph $G(V, E)$, with $V = \{i \mid x_i \in S\}$ and $E = \{(i,j)\}$. Let $U_{i,j} = \{x_k \mid (x_k \text{ "above" } \overline{x_i x_j}) \wedge (i < k < j)\}$ and $L_{i,j} = \{x_k \mid (x_k \text{ "below" } \overline{x_i x_j}) \wedge (i < k < j)\}$. Let $L_i = \{x_k \mid k < i\}$ and $R_i = \{x_k \mid k > i\}$. We define $cost(L_i) = \sum_{k \in L_i} v_k$. We define $cost(R_i)$ similarly.

We define

$$cost(i,j) = \begin{cases} cl_{ij} + \sum_{k \in U_{i,j}} v_k & \text{if } i < j \\ cl_{ij} + \sum_{k \in L_{i,j}} v_k & \text{if } i > j \end{cases}$$

Now compute the shortest paths $sp(i,j)$ between all pairs of vertices i, j. The optimum solution is obtained as follows (assuming $i < j$):

$$V^* = \min_{i,j}[sp(i,j) + sp(j,i) + cost(L_i) + cost(R_j)].$$

$$\text{Value of } S^* = \sum_{x_k \in S} v_k - V^*$$

The term in the minimization is essentially summing the cost of the enclosure (in terms of fence length) together with the sum of the values of the points *not* enclosed by the fence.

This algorithm computes the best enclosure with at least two points in it. We can compare the obtained solution with all the enclosures containing only a singleton point.

The construction of the graph can be done in $O(n^3)$ time very simply. The shortest path computation also takes $O(n^3)$ time (for an all-pairs shortest path computation) [PS 82].

3 Bounded Length Fence

We prove that the following version of the two-dimensional fence enclosure problem is NP-hard by showing a polynomial time reduction from the *Knapsack* problem, which is known to be NP-complete [GJ 78].

Fixed Length Fence Enclosure: Given a set S of n points $x_i \in \Re^2$, for each point x_i a value v_i, and real numbers V and L. Is there a subset S' of the points such that $\sum_{j \in S'} v_j \geq V$ and the perimeter of the convex hull of the points in S' is $\leq L$?

Knapsack Instance: Given a set U, and for each element $i \in U$, a weight w_i, and a value v_i, and positive integers W and C. Is there a subset $U' \subseteq U$ such that $\sum_{j \in U'} w_j \leq W$ and $\sum_{j \in U'} v_j \geq C$?

Theorem 3 *The Fixed Length Enclosure problem is NP-hard.*

Proof: Assume that the set U has cardinality n (in the instance of the knapsack problem). We create a set S of $2n$ points in $\Re^2$ as follows ($S = A \cup B$): Set A consists of n points placed on the vertices of a regular convex n-gon. Assume that the points are labeled $p_1, p_2, ..., p_n$ in clockwise order on the boundary of the n-gon. The value of each point $p_i \in A$ is M (which is a very large constant). Assigning very large values to points in A ensures that all of them are chosen to be in the enclosure (provided a sufficient amount of fence is available). These points effectively "fix" the general outline of the optimal enclosure. The value of L is such that we can include all the points of set A in the enclosure.

The points in set B (of cardinality n) correspond to items in set U. Point $q_i \in B$ (corresponding to $i \in U$) is placed off the side $\overline{p_i p_{i+1}}$ of the n-gon (see Figure 1). We draw the bisector of the angle $\angle p_i c p_{i+1}$ (where c is the center of the n-gon), and extend it beyond the intersection with segment $\overline{p_i p_{i+1}}$. The distance of q_i from c is determined by w_i (described later). The value of q_i is taken to be the value of the item i, i.e., v_i.

We would like the items that are chosen to be in the Knapsack in the optimal solution to correspond exactly to the items of B that are chosen to be in the optimal enclosure. To achieve that we set $V = nM + C$. (The value is obtained by choosing the n points in set A and a subset of the other points of value at least C.) We set $L = \text{perim}(\text{ConvHull}(A)) + sW$ (where s is an appropriate scale factor).

The points in set B are placed so that the boundary of the enclosure (and not the interior) contains all the points in A; the scale factor s is chosen to ensure this property. The points q_i are placed at a distance d_i from the side $\overline{p_i p_{i+1}}$, so that the extra fence required to include q_i is sw_i (see Figure 1).

Notice that the coordinates of some of the points may be irrational. It is easy to see that the points can be perturbed slightly to place them at rational coordinates. This is done by perturbing them all away from the center and providing a small amount of extra fence to take into account the perturbation. It is easy to see that the reduction can be performed in polynomial time. $\square$

3.1 Pseudo-Polynomial Time Algorithms

We present two algorithms to solve the version of the problem when the fence is of a bounded length L. The first algorithm runs in polynomial time when the values of the points are small integers. The second algorithm performs well when L is not very large compared to the smallest distance between two points (the values of the points may be large). In this abstract, we omit the second algorithm.

Theorem 4 *There is a pseudo-polynomial time algorithm to solve the fixed length enclosure problem.*

Proof: The algorithm is based on dynamic programming and is outlined below. We wish to compute $f(x_j, x_k, k)$ which is the least length of fence required to enclose at least value k with a "supporting fence" to the left of $\overline{x_j x_k}$. We assume that the values of x_j and x_k are included in the supporting fence. We also assume that x_i is a point in the half-plane left of $\overline{x_j x_k}$.

$f(x_j, x_k, k)$ is computed as follows:

$$f(x_j, x_k, k) = \begin{cases} l_{jk} & \text{if } v_j + v_k \geq k \\ \min_{x_i}\{f(x_i, x_k, k - (\Delta ijk + v_j)) + l_{ij}\} & \text{for all } x_i \text{ to the "left" of } \overline{x_j x_k} \} \\ \infty & \text{otherwise} \end{cases}$$

Notice that to compute any $f(p, q, k)$ we need the value of $f(*, *, k')$ where $k' < k$. Thus the function $f(p, q, k)$ can be computed in increasing values of k for all pairs of points p and q. The first values to be computed are $f(p, q, 1)$, for each pair of points p, q.

To compute the optimal solution we wish to compute the largest value of k, such that for some pair of points x_j, x_k the following holds:

$$(l_{jk} + f(x_j, x_k, k)) \leq L.$$

The optimal solution could also be a singleton point, and thus we compare k with each singleton value to obtain the optimal solution. We call this value k^* and it represents the optimal solution to the problem. $\square$

The running time of the algorithm is $O(kn^3)$. The space complexity is $O(kn^2)$.

4 Optimally Enclosing Polygonal Objects

We now consider a more general version of the fence enclosure problem with obstacles in the plane. A formal description of the problem is given as follows: Given a set S of simple polygons (obstacles), with obstacle O_i having value v_i, we wish to find an enclosure which maximizes "value"; where "value" is again defined to be the sum of the values of the obstacles enclosed minus the cost of the fence (where fence costs c per unit length). We do not permit the fence to be in the interior of any simple polygon (obstacle) in the set S. Thus, an obstacle is either entirely included, or entirely excluded from the enclosure. All points with values will be treated as obstacles for convenience. Hence, S consists entirely of obstacles, and we use $x_1, x_2, \ldots, x_n$ to denote the vertices corresponding to the points defining the obstacle boundaries. The optimum enclosure may not have a convex shape due to the presence of obstacles.

We can show that the shape of the optimum solution is geodesically convex (i.e., there is a shortest path between any two points in the region enclosed by the optimum solution that is also entirely inside the region). We provide an algorithm to solve the fence enclosure problem in the presence of obstacles, which is based on a reduction to an *all-pairs shortest path* problem in directed graphs. This yields an $O(En^2)$ time sequential algorithm to solve the problem. (We let n denote the size of the input, i.e., the total number of points describing the boundaries of the polygons.) Since shortest paths in directed graphs can be obtained in parallel on a PRAM, this yields a parallel algorithm to solve the fence problem for the two-dimensional version of the problem in the presence of obstacles.

Outline of Algorithm: Assume that point x^* is on the boundary of the optimum enclosure. The algorithm begins by finding shortest geodesic paths P_i from each vertex x_i (belonging to an obstacle in S) to x^*. Now consider all the vertices in angularly sorted clockwise order about x^*. We sort them in the angular order of the geodesic paths P_i emanating from x^*. We define $x_i \prec x_j$ iff x_i is strictly before x_j in the angular order $\mathcal{P}$ (if one of the geodesic paths, say P_i, is a subpath of P_j, then there is no ordering between x_i and x_j, and they are said to coincide). For example, in Figure 2, the path $P_0 \prec P_1, P_0 \prec P_2, P_0 \prec P_3$, and $P_1 \prec P_3$. But there is no ordering between P_1 and P_2, and also between P_2 and P_3 (and we say that they coincide).

We show that the vertices on the boundary of the optimum enclosure S^* (considered in clockwise order) never "back up", in the order $\mathcal{P}$. (Intuitively to "back up" in $\mathcal{P}$ means to go from x to y, where $y \prec x$.)

Lemma 2 *Consider the optimum solution S^*, with point x^* on the boundary ∂S^* (of the enclosure). The clockwise ordering of the vertices on ∂S^* is consistent with the ordering $\mathcal{P}$.*

We now build a directed graph $G(V, E)$ as follows:

$$V = \{i \mid x_i \text{ is a vertex of an obstacle in } S\}$$

$$E = \{(i,j) \mid x_i \text{ is visible from } x_j, \text{ and it is not the case that } x_j \prec x_i\}$$

In other words, the edge set consists of all edges in the visibility graph of the set of points. The edges are directed in the following manner: edge (i,j) is directed from i to j if either $x_i \prec x_j$ in $\mathcal{P}$, or they coincide in $\mathcal{P}$.

We now show how to assign costs to the edges in such a way, that the cost of a cycle in G corresponds precisely to the sum of the values of the obstacles left out of the enclosure, together with the cost of constructing the fence. The shortest cycle in G corresponds to the optimal enclosure with x^* on the boundary of the enclosure.

We can assume that the boundary of the enclosure has at least three vertices, and it goes through x^* in a clockwise manner. Let (x^*, x^+) be the first outgoing edge from x^*, and (x^-, x^*) be the (last) incoming edge to x^*. For an illustration of the incoming and outgoing edges from x^* see Figure 2.

For each obstacle O_i (with value v_i), we associate its entire value with a point q_i on its boundary (q_i is chosen to be any point on the boundary of the obstacle). All the other points on O_i have zero value, and q_i has value v_i. We include the obstacle O_i in our enclosure, if and only if we include q_i in the enclosure.

The costs on the edges are assigned as follows: Consider an edge (x_i, x_j) such that x_j is visible from x_i. Consider the "triangle" (more precisely, geodesic triangle) formed by P_i, P_j and the line segment $\overline{x_i x_j}$. If $x_i \prec x_j$, then we define $Out(i,j)$ to be the set of obstacles O_k, such that the path from q_k to x^* goes through the segment $[x_i, x_j)$. Note that the interval is closed at one endpoint of the segment, to take into account the obstacles to whom the shortest paths go through x_i. The obstacles in $Out(i,j)$ are all excluded from the optimal enclosure.

$$cost(i,j) = cl_{ij} + \sum_{O_k \in Out(i,j)} v_k$$

In other words, the cost of the edge corresponds to the length of the edge together with the cost of the obstacles left out due to this edge. If x_i and x_j coincide, then the cost of the edge (i,j) is cl_{ij} (this is the case when P_i is a subpath of P_j, or vice versa).

Define the set $O(x^-, x^+)$ to be the set of obstacles O_k, such that the path from q_k to x^*, is angularly strictly in between $\overline{x^- x^*}$ and $\overline{x^* x^+}$. The obstacles whose shortest paths contain x^- are also included in $O(x^-, x^+)$. (Note that the obstacles whose shortest paths contain x^+ will be counted in the shortest path from x^+ to x^-.) We associate a cost $cost(x^+, x^-)$ with the pair x^+, x^- where

$$cost(x^+, x^-) = c(l_{x^- x^*} + l_{x^* x^+}) + \sum_{O_k \in O(x^-, x^+)} v_k$$

Let X^+ be the set of vertices that are visible from x^*. We now find the shortest path tree in G, from each vertex $x^+ \in X^+$ to all the vertices x^-, such that x^- is visible from x^* and it is not the case that $x^+ \prec x^-$ (i.e., either x^- is "behind" x^+ in $\mathcal{P}$ or coincides with x^+). The edges $(x^-, x^*), (x^*, x^+)$ together with the shortest path $sp(X^+, x^-)$ forms a cycle $C(x^*, x^+, x^-)$.

For a fixed x^* and x^+, we compute the following:

$$V(x^*, x^+) = \min_{x^-}[cost(x^+, x^-) + sp(x^+, x^-)]$$

Taking the minimum over all $x^+ \in X^+$ gives us the shortest cycle through x^*. This gives us the optimal enclosure passing through vertex x^*. Let $C(x^*)$ be the shortest cycle passing through x^*.

Theorem 5 *The cost of cycle $C(x^*)$ exactly equals the cost of constructing the fence, together with the values of the obstacles left out.*

Proof: To prove this theorem, we need to show that the value of an obstacle not enclosed by the fence is charged to exactly one edge in the cycle. Consider the shortest path of any obstacle O_k that is not inside the enclosure, to the vertex x^*. Either, the path cuts the boundary of the enclosure and enters the interior of $C(x^*)$ to reach x^*, or reaches x^* without ever entering the interior. In the latter case, it is easy to see that the value of obstacle O_k is charged to $cost(x^+, x^-)$. In the former case, we need to show that the path from q_k to x^* crosses the boundary of the fence exactly once. If it does cut the boundary more than once, it is easy to see that the fence must "backup" in the order $\mathcal{P}$, which is a contradiction to Lemma 2. The actual cost of the fence is charged separately to each edge of the cycle. $\square$

Analysis For a fixed x^* and x^+ the cost of computing the shortest path tree is $O(n^2)$. Once we enumerate over all all such pairs the running time of the algorithm is $O(En^2)$, where E is the number of edges in the visibility graph. The worst case running time of the algorithm is $O(n^4)$.

5 Multiple Enclosures

In this section we consider the version of the problem when multiple fences are permitted in the enclosure. Suppose we are allowed to use upto K fences to build the enclosure. It is easy to see that we wil use all K (assuming $n \geq K$). The optimal solution will consist of convex fences, all of which can be separated from each other by straight lines passing through two of the points in set S. Fences F_i and F_j are separated from each other by line l_{ij}. In fact all these separator lines come from the $O(n^2)$ lines defining the arrangement of the lines passing through a pair of points in S.

Thus the entire set of separators has size $K(K-1)/2$. We can choose $K-1$ lines from the lines defining the arrangement. These lines are our "guess" for the actual separators for fence F_1, from F_2 through F_k. These lines define a set of half planes that we intersect to obtain a single region that contains one fence of the enclosure. Now we find a single optimal fence for this point set, define this to be the fence F_1. We discard all the points in F_1 and continue the same procedure to successively find the other $K-1$ fences. Clearly, this algorithm runs in polynomial time for fixed K. In fact, a naive implementation has running time $O(n^{O(K^2)})$.

A faster algorithm can be obtained using the fact that the fences can be separated by a planar graph. This is due to a lemma in [ERS 87] (see also [CRW 89, W 90]) that states:

Lemma 3 *Let $F_1, \ldots, F_k$ be k convex, compact and pairwise disjoint sets in the plane. Then there is a planar graph $G(V, E)$ that has vertices corresponding to the K given sets, with the following properties:*

- *For each edge* $(F_i, F_j) \in E$, *there is a line which cuts the plane into two open half-planes* H_{ij} *and* H_{ji}, *such that* R_i *is contained in* H_{ij}, *and* R_j *is contained in* H_{ji}.

- *For each* R_i:

$$R_i \subseteq R_i' := \bigcap_{R_i, R_j \in E} H_{ij}.$$

- *The regions* R_i' *are disjoint.*

As in the naive algorithm, we can choose these lines from the set of $O(n^2)$ lines defining the arrangement of lines passing through a pair of points in S. Let the number of (non-isomorphic) planar graphs with K vertices be $f(K)$. Note that $f(K)$ is not a function of n. Thus we need only consider all planar graphs on K vertices. For each graph we generate possible edges from the set of $O(n^2)$ lines and compute the optimal fence within each region F_i'. This algorithm has running time $O(f(K)n^3 n^{O(K)})$ for constant K.

References

[AKM 90] E.M. Arkin, S. Khuller and J.S.B. Mitchell, "Optimal enclosure problems", *Technical Report TR 90-1111*, Dept. of Computer Science, Cornell University.

[AKMSW 87] A. Aggarwal, M. Klawe, S. Moran, P. Shor and R. Wilber, "Geometric applications of a matrix searching algorithm", *Algorithmica* 2 (1987), pp. 195–208.

[AMP 90] E.M. Arkin, J.S.B. Mitchell, and C. Piatko, "On Geometric bicriteria path problems", Manuscript, 1990.

[BDDG 85] J.E. Boyce, D.P. Dobkin, R.L. Drysdale and L.J. Guibas, "Finding extremal polygons", *SIAM Journal on Computing* 14 (1985), pp. 134–147.

[CRW 89] V. Capoyleas, G. Rote, and G. Woeginger, "Geometric clusterings", *Technical Report B-89-04*, Freie Universität, Berlin. To appear: *Journal of Algorithms*.

[Chan 86] J.S. Chang, "Polygon optimization problems", Ph.D. Dissertation, Technical Report No. 240, Robotics Report No. 78, August, 1986, Dept. of Computer Science, New York University.

[CY 86] J.S. Chang and C.K. Yap, "A polynomial solution for the potato peeling problem", *Discrete and Computational Geometry* 1 (1986), pp. 155–182.

[Chaz 83] B. Chazelle, "The polygon containment problem", *Advances in Computing Research*, JAI Press (1983), pp.1–33.

[Chv 83] V. Chvatal, *Linear Programming*, W.H. Freeman and Co., 1983, pp. 374-380.

[Da 63] G.B. Dantzig, *Linear Programming and Extensions*, Princeton University Press, Princeton, NJ, 1963.

[ER 91] P. Eades and D. Rappaport, "The complexity of computing minimum separating polygons", To appear: *Pattern Recognition Letters*.

[ERS 87] H. Edelsbrunner, A.D. Robison and X. Shen, "Covering convex sets with non-overlapping polygons", *Technical Report UIUCDCS-R-87-1364*, Dept. of Computer Science, University of Illinois at Urbana-Champaign.

[GJ 78] M.R. Garey and D.S. Johnson, "Computers and Intractability: A guide to the theory of NP-completeness", *Freeman*, San Francisco.

[GN 72] R.S. Garfinkel and G.L. Nemhauser, *Integer Programming*, John Wiley and Sons, New York, 1972.

[KL 85] B. Korte and L. Lovasz, "Polyhedral results on antimatroids", *to appear* in Proc. of the New York Combinatorics Conference.

[LC 85] D.T. Lee and Y.T. Ching, "The power of geometric duality revisited", *Information Processing Letters* 21 (1985), pp. 117–122.

[ORW 89] M.H. Overmars, G. Rote and G. Woeginger, "Finding minimum area k-gons", *Technical Report RUU-CS-89-7*, University of Utrecht, Utrecht, Netherlands.

[PS 82] C.H. Papadimitriou and K.Steiglitz, "Combinatorial Optimization: Algorithms and Complexity", *Prentice-Hall*, New Jersey.

[W 90] R. Wenger, "Upper bounds on geometric permutations for convex sets", *Discrete and Computational Geometry*, 5 (1990), pp. 27–33.

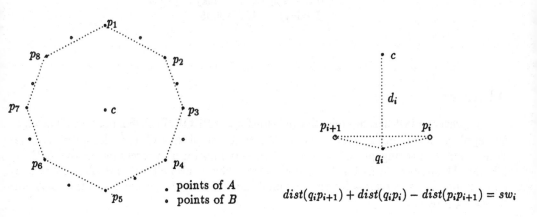

$$dist(q_i p_{i+1}) + dist(q_i p_i) - dist(p_i p_{i+1}) = s w_i$$

Figure 1: Figures for NP-hardness proof

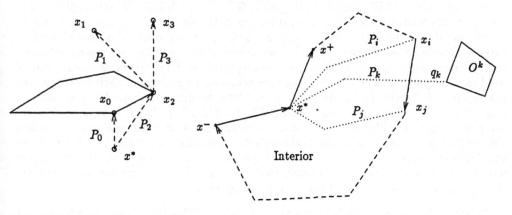

Figure 2: Figures for Obstacles case

A Fast Derandomization Scheme and Its Applications

Yijie Han

Department of Computer Science
University of Kentucky
Lexington, KY 40506

Abstract

We present a fast derandomization scheme for the PROFIT/COST problem. Through the application of this scheme we show the time complexity $O(\log^2 n \log \log n)$ for the $\Delta + 1$ vertex coloring problem using $O((m + n)/\log \log n)$ processors on the CREW PRAM. The power of this fast derandomization scheme also allows us to obtain fast and efficient parallel algorithms for the maximal independent set problem and the maximal matching problem.

1. Introduction

Recent progress in derandomization has resulted in several efficient sequential and parallel algorithms [ABI][BR][BRS][H1][HI][KW][L1][L2][L3][MNN][PSZ][Rag][Sp]. The essence of the technique of derandomization lies in the design of a sample space easy to search, in the probabilistic analysis showing that the expectation of a desired random variable is no less than demanded, and in the search technique which ultimately returns a good sample point. Raghavan[Rag] and Spencer[Sp] showed how to search an exponential sized sample space to obtain polynomial time sequential algorithms. Their technique can not be applied directly to obtain efficient parallel algorithms through derandomization. Alon *et al.*[ABI], Karp and Wigderson[KW] and Luby[L1][L2][L3] developed schemes using n random variables with limited independence on a small sample space, and thus obtained efficient parallel algorithms through derandomization. Berger and Rompel[BR] and Motwani *et al.*[MNN] presented novel designs in which $(\log^c n)$-wise independent random variables are used in randomized algorithms and then NC algorithms are obtained through the derandomization of these randomized algorithms.

To obtain linear processor NC algorithms, Luby[L2][L3] outlined an elegant framework in which pairwise independent random variables are designed on a sample space that facilitates binary search. His framework[L2][L3] consists of derandomization schemes for the bit pairs PROFIT/COST problem, the general pairs PROFIT/COST problem, and applications of these schemes to the three problems: the $\Delta + 1$ vertex coloring problem, the maximal independent set problem and the maximal matching problem. By applying his derandomization schemes he obtained a linear processor CREW algorithm for the

$\Delta + 1$ vertex coloring problem with time complexity $O(\log^3 n \log \log n)$. Although he put the three problems in a very nice setting, his derandomization scheme is not efficient enough to improve on parallel algorithms for the maximal independent set problem and the maximal matching problem obtained through ad hoc designs[GS][IS]. To illustrate his idea, Luby gave algorithms for the maximal independent set problem and the maximal matching problem with time complexity $O(\log^5 n)$ through derandomization[L3].

Recently Han and Igarashi[HI] gave a fast derandomization scheme for the bit pairs PROFIT/COST problem. Their scheme yields a fast CREW parallel algorithm for the bit pairs PROFIT/COST problem with time complexity $O(\log n)$ using a linear number of processors. Han then showed[H1] how to obtain an EREW algorithm with the same time and processor complexities. Their result improves the time complexity of Luby's $\Delta + 1$ vertex coloring algorithm to $O(\log^3 n)$. The most interesting feature in Han and Igarashi's scheme[HI] is the design of a sample space which allows redundancy and mutual independence to be exploited.

In this paper we give a new scheme to speed up the derandomization process of the general pairs PROFIT/COST problem. This scheme allows several bit pairs PROFIT/COST problems in one general pairs PROFIT/COST problem to be solved in one pass. We note that our scheme can not be constructed efficiently under the setting of previous derandomization schemes[L2][L3] because it would require more than a linear number of processors. The power of our derandomization scheme allows us to improve the time complexity for the $\Delta + 1$ vertex coloring problem, the maximal independent set problem and the maximal matching problem.

The previous best result for the $\Delta + 1$ vertex coloring is due to Luby[L2] which has time complexity $O(\log^3 n \log \log n)$ using $O(m+n)$ processors on the CREW model, and subsequent improvement of Han and Igarashi[HI] which reduces the time complexity to $O(\log^3 n)$. By applying our derandomization scheme we remove randomness from Luby's randomized parallel algorithm[L3] for the $\Delta + 1$ vertex coloring problem, improving the time complexity of the $\Delta + 1$ vertex coloring problem to $O(\log^2 n \log \log n)$ using $O((m + n)/\log \log n)$ processors on the CREW model.

Sophisticated applications of our fast derandomization scheme also yield fast parallel algorithms for the maximal independent set problem and the maximal matching problem[H2] which improve the best previous results. In particular, we are able to obtain a parallel algorithm for the maximal independent set problem with time complexity $O(\log^{2.5} n)$ using $O((m+n)/\log^{1.5} n)$ processors and a parallel algorithm for the maximal matching problem with time complexity $O(\log^{2.5} n)$ using $O((m+n)/\log^{0.5} n)$ processors. We are unable to present these algorithms here due to page limit.

2. Preliminaries

The *bit pairs PROFIT/COST* (BPC) and the *general pairs PROFIT/COST* (GPC) problems as formulated by Luby[L2] can be described as follows.

Let $\vec{x} = < x_i \in \{0, 1\} : i = 0, ..., n - 1 >$. Each point $\vec{x}$ out of the 2^n points is assigned probability $1/2^n$. Given function $B(\vec{x}) = \sum_{i,j} f_{i,j}(x_i, x_j)$, where $f_{i,j}$ is defined as a function $\{0, 1\}^2 \to \mathcal{R}$. The bit pairs PROFT/COST problem is to find a good point $\vec{y}$ such that $B(\vec{y}) \geq E[B(\vec{x})]$. B is called the bit pairs BENEFIT function and $f_{i,j}$'s are called the bit pairs PROFIT/COST functions.

Let $\vec{x} =< x_i \in \{0,1\}^q : i = 0, ..., n-1 >$. Each point $\vec{x}$ out of the 2^{nq} points is assigned probability $1/2^{nq}$. Given function $B(\vec{x}) = \sum_{i,j} f_{i,j}(x_i, x_j)$, where $f_{i,j}$ is defined as a function $\{0,1\}^q \times \{0,1\}^q \to \mathcal{R}$. The general pairs PROFIT/COST problem is to find a good point $\vec{y}$ such that $B(\vec{y}) \geq E[B(\vec{x})]$. B is called the general pairs BENEFIT function and $f_{i,j}$'s are called the general pairs PROFIT/COST functions.

The size m of the problem is the number of PROFIT/COST functions present in the input. The input is dense if $m = \theta(n^2)$ and is sparse if $m = o(n^2)$.

Let $G = (V, E)$ be a graph with $|V| = n$ and $|E| = m$. The degree of $v \in V$ is denoted by $d(v)$. Let $\Delta = \max\{d(v)|v \in V\}$. The output of the $\Delta + 1$ vertex coloring is $color(v) \in \{1, ..., \Delta + 1\}$ for all $v \in V$ such that if $(i, j) \in E$ then $color(i) \neq color(j)$.

Han and Igarashi[HI] have formulated the BPC problem as a tree contraction problem[M]. Assuming without loss of generality n is a power of 2. n 0/1-valued uniformly distributed mutually independent random variables r_i, $0 \leq i < n$, are used. A *Random variable tree* T is built for $\vec{x}$. T is a complete binary tree with n leaves plus a node which is the parent of the root of the complete binary tree (thus there are n interior nodes in T and the root of T has only one child). The n variables x_i, $0 \leq i < n$, are associated with n leaves of T and the n random variables are associated with the interior nodes of T. The n leaves of T are numbered from 0 to $n - 1$. Variable x_i is associated with leaf i.

Variables x_i, $0 \leq i < n$, are randomized as follows. Let $\vec{r} =< r_i : i = 0, ..., n-1 >$ and let $r_{i_0}, r_{i_1}, ..., r_{i_{\log n}}$ be the random variables on the path from leaf i to the root of T. Random variable x_i is defined to be $x_i(\vec{r}) = (\sum_{j=0}^{\log n - 1} i_j \cdot r_{i_j} + r_{i_{\log n}}) \bmod 2$, where i_j is the j-th bit of i starting with the least significant bit. It can be verified that random variables x_i, $0 \leq i < n$, are uniformly distributed mutually independent random variables.

Due to the linearity of expectation and pairwise independence of random variables in $\vec{x}$, $E[B(\vec{x})] = \sum_{i,j} E[f_{i,j}(x_i, x_j)] = \sum_{i,j} E[f_{i,j}(x_i(\vec{r}), x_j(\vec{r}))] = E[B(\vec{x}(\vec{r}))]$. The problem now is to find a sample point $\vec{r}$ such that $B|_{\vec{r}} \geq E[B] = \frac{1}{4}\sum_{i,j}(f_{i,j}(0,0) + f_{i,j}(0,1) + f_{i,j}(1,0) + f_{i,j}(1,1))$.

Han and Igarashi's algorithm[HI] fixes random variables r_i (setting their values to 0's and 1's) one level in a step starting from the level next to the leaves (level 0) and going upward on the tree T until level $\log n$. Since there are $\log n + 1$ interior levels in T all random variables will be fixed in $\log n + 1$ steps.

Let random variable r_i at level 0 be the parent of the random variables x_i and $x_{i\#0}$ in the random variable tree, where $i\#j$ is a number obtained by complementing the j-th bit of i. r_i will be fixed as follows. Compute $f_0 = f_{j,j\#0}(0,0) + f_{j,j\#0}(1,1) + f_{j\#0,j}(0,0) + f_{j\#0,j}(1,1)$ and $f_1 = f_{j,j\#0}(0,1) + f_{j,j\#0}(1,0) + f_{j\#0,j}(0,1) + f_{j\#0,j}(1,0)$. If $f_0 \geq f_1$ then set r_i to 0 else set r_i to 1. All random variables at level 0 will be fixed in parallel in constant time using n processors. The fixing results in a smaller space with higher expectation for B. Therefore this smaller space contains a good point.

If r_i is set to 0 then $x_i = x_{i\#0}$, if r_i is set to 1 then $x_i = 1 - x_{i\#0}$. Therefore after r_i is set, x_i and $x_{i\#0}$ can be combined. The n random variables x_i, $0 \leq i < n$, can be reduced to $n/2$ random variables. PROFIT/COST functions $f_{i,j}$, $f_{i\#0,j}$, $f_{i,j\#0}$, and $f_{i\#0,j\#0}$ can also be combined into one function. It can be checked that the combining can be done in constant time using a linear number of processors.

During the combining process variables x_i and $x_{i\#0}$ are combined into a new variable $x^{(1)}_{\lfloor i/2 \rfloor}$, functions $f_{i,j}$, $f_{i\#0,j}$, $f_{i,j\#0}$, and $f_{i\#0,j\#0}$ are combined into a new function $f^{(1)}_{\lfloor i/2 \rfloor, \lfloor j/2 \rfloor}$. After combining a new function $B^{(1)}$ is formed which has the same form of B but has only $n/2$ variables. As we stated above, $E[B^{(1)}] \geq E[B]$.

What we have explained above is the first step of the algorithm in [HI]. This step takes constant time using a linear number of processors. After k steps the random variables at levels 0 to $k-1$ in the random variable tree are fixed, the n random variables $\{x_0, x_1, ..., x_{n-1}\}$ are reduced to $n/2^k$ random variables $\{x^{(k)}_0, x^{(k)}_1, ..., x^{(k)}_{n/2^k-1}\}$, functions $f_{i,j}$, $i,j \in \{0,1,...,n-1\}$, have been combined into $f^{(k)}_{i,j}$, $i,j \in \{0,1,...,n/2^k-1\}$.

After $\log n$ steps $B^{(\log n)} = f^{(\log n)}_{0,0}(x^{(\log n)}_0, x^{(\log n)}_0)$. The bit at the root of the random variable tree is now set to 0 if $f^{(\log n)}_{0,0}(0,0) \geq f^{(\log n)}_{0,0}(1,1)$, and 1 otherwise. Thus Han and Igarashi's algorithm[HI] solves the BPC problem in $O(\log n)$ time with a linear number of processors.

Let $n = 2^k$ and A be an $n \times n$ array. Elements $A[i,j]$, $A[i,j\#0]$, $A[i\#0,j]$, $A[i\#0,j\#0]$ form a *gang* which is denoted by $g_A[\lfloor i/2 \rfloor, \lfloor j/2 \rfloor]$. All gangs in A form array g_A.

When visualized on a two dimensional array A (as shown in Fig. 1), a stage of Han and Igarashi's algorithm can be interpreted as follows. Let function $f_{i,j}$ be stored at $A[i,j]$. Setting the random variables at level 0 of the random variable tree is done by examining the PROFIT/COST functions in the diagonal gang of A. Function $f_{i,j}$ then gets the bit setting information from $g_A[\lfloor i/2 \rfloor, \lfloor i/2 \rfloor]$ and $g_A[\lfloor j/2 \rfloor, \lfloor j/2 \rfloor]$ to determine how it is to be combined with other functions in $g_A[\lfloor i/2 \rfloor, \lfloor j/2 \rfloor]$.

A *derandomization tree* D can be built which reflects the way the BPC functions are combined. D is of the following form. The input BPC functions are stored at the leaves, $f_{i,j}$ is stored in $A_0[i,j]$. A node $A_l[i,j]$ at level $l > 0$ is defined if there exist input functions in the range $A_0[u,v]$, $i * 2^l \leq u < (i+1) * 2^l$, $j * 2^l \leq v < (j+1) * 2^l$.

Tree D can be constructed[H1][HI] by first sorting the input by the file-major indexing and then building the tree bottom-up. The derandomization tree helps to reduce the space requirement for the BPC problem to $O(m)$. Han and Igarashi's algorithm[HI] has time complexity $O(\log n)$ using a linear number of processors.

The algorithm given by Han and Igarashi[HI] is a CREW algorithm. Recently, Han[H1] give an EREW algorithm for the BPC problem with the same time complexity.

3. A Scheme for the General Pairs PROFIT/COST Problem

In this section we present a scheme to speed up the derandomization process of the GPC problem.

In [L2][L3] Luby presented the following derandomization scheme for solving the GPC problem.

Let $\vec{y} = < y_i \in \{0,1\}^p : i = 0, 1, ..., n-1 >$. Let $\vec{x_u}, p \leq u < q$, be totally independent random bit strings, each of length n. Let $\vec{z}$ be a vector of n bits. Define $TB(\vec{y}) = E[B(\vec{x_{q-1}} \cdots \vec{x_{p+1}}\vec{x_p}\vec{y})]$. Then $E[TB(\vec{x_p}\vec{y})] = E[E[B(\vec{x_{q-1}} \cdots \vec{x_{p+1}}\vec{z}\vec{y})| \vec{z} = \vec{x_p}]]$ $= E[B(\vec{x_{q-1}} \cdots \vec{x_p}\vec{y})] = TB(\vec{y})$. Because in the GPC problem function B is the sum of GPC functions each depends on at most two variables, pairwise independent random variables can be used for bits in each $\vec{x_u}, p+1 \leq u < q$. Luby's algorithm for the GPC problem then uses his algorithm for the BPC problem to find a $\vec{z}$ satisfying $TB(\vec{z}\vec{y}) \geq E[TB(\vec{x_p}\vec{y})] = TB(\vec{y})$, thus fixing the random bits in $\vec{x_p}$.

Luby's solution[L2][L3] to the GPC problem can be interpreted as follows. Solving the GPC problem by solving q instances of the BPC problem, one for each $\vec{x_u}$. These instances are solved sequentially. After the instances for $\vec{x_u}, 0 \leq u < v$, are solved. BPC functions $f_{i_v, j_v}(x_{i_v}, x_{j_v})$ are evaluated based on the setting of bits $x_{i_u}, x_{j_u}, 0 \leq u < v$. Then the BPC algorithm is invoked to fix $\vec{x_v}$.

We present in this section a scheme which allows several $\vec{x_u}$'s to be fixed in one pass.

First we give a sketch of our approach. The incompleteness of the description in this paragraph will be fulfilled in the rest of this section. Let P be the GPC problem we are to solve. P can be decomposed into q instances of the BPC problem to be solved sequentially. Let P_u be the u-th instance of the BPC problem. Imagine that we are to solve $P_u, 0 \leq u < k$, in one pass, i.e., we are to fix $\vec{x_0}, \vec{x_1}, ..., \vec{x_{k-1}}$ in one pass, with the help of enough processors. For the moment we can have a random variable tree T_u and a derandomization tree D_u for $P_u, 0 \leq u < k$. In step j our algorithm will work on fixing the bits at level $j - u$ in $T_u, 0 \leq u \leq \min\{k-1, j\}$. The computation in each tree D_u proceeds as we have described in the last section. Note that BPC functions $f_{i_v, j_v}(x_{i_v}, x_{j_v})$ depends on the setting of bits $x_{i_u}, x_{j_u}, 0 \leq u < v$. The main difficulty with our scheme is that when we are working on fixing $\vec{x_v}, \vec{x_u}, 0 \leq u < v$, have not been fixed yet. The only information we can use when we are fixing the random variables at level l of T_u is that random variables at levels 0 to $l + c - 1$ are fixed in $T_{u-c}, 0 \leq c \leq u$. This information can be accumulated in the pipeline of our algorithm and transmitted on the *bit pipeline trees*. Fortunately this information is sufficient for us to speed up the derandomization process without resorting to too many processors. For the sake of a clear exposition we only describe a CREW derandomization algorithm. The conversion from the CREW algorithm to an EREW algorithm requires a more complicated scheme which we will present in the final version of the paper.

The scheme outlined in the last paragraph would require $O(mn^k)$ processors if it is to be designed under the setting of previous derandomization schemes[L2][L3]. In our setting we use only $O(c^k m)$ processors for a constant c.

Suppose we have $c \sum_{i=0}^{k} (m * 4^i)$ processors available, where c is a constant. Assign $cm * 4^u$ processors to work on P_u for $\vec{x_u}$. We shall work on $\vec{x_u}, 0 \leq u < k$, simultaneously in a pipeline. To be complete, there is a random variable forest F_u containing 2^u random variable trees for P_u. We are to fix the random bits on l-th level of F_v (for $\vec{x_v}$) under the condition that random bits from level 0 to level $l + u - 1, 0 \leq u \leq v$, in F_{v-u} have already been fixed. We are to perform this fixing in constant time. The 2^u random variable trees are not constructed before we start working on P_u, as this is the case for the BPC algorithm described in the last section. These trees are built bottom up as the derandomization process proceeds. Immediately before the step we are to fix the random bits on l-th level of F_v, the 2^u random variable trees are constructed up to the l-th level. The details of the algorithm for constructing the random variable trees are given at the end of this section.

Consider a GPC function $f_{i,j}(x_i, x_j)$ under the condition stated in the last paragraph. When we start working on $\vec{x_v}$ we should have the BPC functions $f_{i_v,j_v}(x_{i_v}, x_{j_v})$ evaluated and stored in a table. However, because $\vec{x_u}$, $0 \leq u < v$, have not been fixed yet, we have to try out all possible situations. There are a total of 4^v patterns for bits x_{i_u}, x_{j_u}, $0 \leq u < v$, we use 4^v BPC functions for each pair (i, j). By $f_{i_v,j_v}(x_{i_v}, x_{j_v})(y_{v-1}y_{v-2} \cdots y_0, z_{v-1}z_{v-2} \cdots z_0)$ we denote the function $f_{i_v,j_v}(x_{i_v}, x_{j_v})$ obtained under the condition that $(x_{i_{v-1}}x_{i_{v-2}} \cdots x_{i_0}, x_{j_{v-1}}x_{j_{v-2}} \cdots x_{j_0})$ is set to $(y_{v-1}y_{v-2} \cdots y_0, z_{v-1}z_{v-2} \cdots z_0)$.

For each pair $(w, w\#0)$ at each level l (this is the level in the random variable forest), $0 \leq l \leq \log n$, a *bit pipeline tree* is built (Fig. 2) which is a complete binary tree of height $2k$. Nodes at even depth from the root in a bit pipeline tree are selectors, nodes at odd depth are fanout gates. A signal *true* is initially input into the root of the tree and propagates downward toward the leaves. The selectors at depth $2d$ select the output by the decision of the random bits which are the parents of random variables $x_{w_d}, x_{w\#0_d}$ in F_d. There is one random variable corresponds to each selector. Let random variable r corresponds to the selector s. If r is set to 0 then s selects the left child and propagates the true signal to its left child while no signal is sent to its right child. If r is set to 1 then the true signal will be sent to the right child and no signal will be sent to the left child. If s does not receive any signal from its parent then no signal will be propagated to s's children no matter how r is set. The gates at odd depth in the bit pipeline tree are fanout gates and pointers from them to their children are labeled with bits which are conditionally set. Refer to Fig. 2 which shows a bit pipeline tree of height 4. If the selector at the root (node 0) selects 0 (which means that the random variable which is the parent of x_{w_0} and $x_{w\#0_0}$ in the random variable forest is set to 0), then $x_{w_0} = x_{w\#0_0}$, therefore the two random variables can only assume the patterns 00 or 11 which are labeled on the pointers from node 1. If, on the other hand, node 0 selects 1 then $x_{w_0} = 1 - x_{w\#0_0}$, the two random variables can only assume the patterns 01 or 10 which are labeled on the pointers of node 2. Let us take node 4 as another example. If node 4 selects 0 then $x_{w_1} = x_{w\#0_1}$, thus the pointers of node 9 are labeled with $\begin{smallmatrix} 1 & 1 \\ 0 & 0 \end{smallmatrix}$

and $\begin{smallmatrix} 1 & 1 \\ 1 & 1 \end{smallmatrix}$. This indicates that the bits for $(w_1 w_0, w\#0_1 w\#0_0)$ can have two patterns, $(01, 01)$ or $(11, 11)$.

The bit pipeline trees built for level $\log n$ have height k. No fanout gates will be used. This is a special and simpler case compared to the bit pipeline trees for other levels. In the following discussion we only consider bit pipeline tree for levels other than $\log n$.

Lemma 1: In a bit pipeline tree there are exactly 2^d nodes at depth $2d$ which will receive the true signal from the root.

Proof: Each selector selects only one path. Each fanout gate sends the true signal to both children. Therefore exactly 2^d nodes at depth $2d$ will receive the true signal from the root. $\square$

For each node i at even depth we shall also say that it has the *conditional bit pattern* (or *conditional bits, bit pattern*) which is the pattern labeled on the pointer from $p(i)$. The root of the bit pipeline tree has empty string as its bit pattern.

Define step 0 as the step when the true signal is input to node 0. The function of a bit pipeline tree can be described as follows.

Step t: Selectors at depth $2t$ which have received true signal selects 0 or 1 for $(w_t, w\#0_t)$. Pass the true signal and the bit setting information to nodes at depth $2t + 2$.

Now consider the selectors at depth $2d$. By lemma 1 a set of 2^d selectors at depth $2d$ receive the true signal. We call this set the *surviving set* $S_{w,d}^l$. We also denote by $S_{w,d}^l$ the set of bit patterns the 2^d surviving selectors have, where w in the subscript is for $(w, w\#0)$ and l is the level for which the pipeline tree is built. Let selector $s \in S_{w,d}^l$ have bit pattern $(y_{d-1}y_{d-2}\cdots y_0, z_{d-1}z_{d-2}\cdots z_0)$. s compares

$$f_{w_d,w\#0_d}^{(l)}(0,0)(y_{d-1}y_{d-2}\cdots y_0, z_{d-1}z_{d-2}\cdots z_0)+$$
$$f_{w_d,w\#0_d}^{(l)}(1,1)(y_{d-1}y_{d-2}\cdots y_0, z_{d-1}z_{d-2}\cdots z_0)+$$
$$f_{w\#0_d,w_d}^{(l)}(0,0)(z_{d-1}z_{d-2}\cdots z_0, y_{d-1}y_{d-2}\cdots y_0)+$$
$$f_{w\#0_d,w_d}^{(l)}(1,1)(z_{d-1}z_{d-2}\cdots z_0, y_{d-1}y_{d-2}\cdots y_0)$$

with

$$f_{w_d,w\#0_d}^{(l)}(0,1)(y_{d-1}y_{d-2}\cdots y_0, z_{d-1}z_{d-2}\cdots z_0)+$$
$$f_{w_d,w\#0_d}^{(l)}(1,0)(y_{d-1}y_{d-2}\cdots y_0, z_{d-1}z_{d-2}\cdots z_0)+$$
$$f_{w\#0_d,w_d}^{(l)}(0,1)(z_{d-1}z_{d-2}\cdots z_0, y_{d-1}y_{d-2}\cdots y_0)+$$
$$f_{w\#0_d,w_d}^{(l)}(1,0)(z_{d-1}z_{d-2}\cdots z_0, y_{d-1}y_{d-2}\cdots y_0)$$

and selects 0 if former is no less than the latter and selects 1 otherwise. Note that the selectors which do not receive the true signal (there are $4^d - 2^d$ of them) have bit patterns which are eliminated.

Let $LS_{w,d}^l = \{\alpha|(\alpha,\beta) \in S_{w,d}^l\}$ and $RS_{w,d}^l = \{\beta|(\alpha,\beta) \in S_{w,d}^l\}$.

Lemma 2: $LS_{w,d}^l = RS_{w,d}^l = \{0,1\}^d$.

Proof: By induction. Assuming that it is true for bit pipeline trees of height $2d - 2$. A bit pipeline tree of height $2d$ can be constructed by using a new selector as the root, two new fanout gates at depth 1, and four copies of the bit pipeline tree of height $2d - 2$ at depth 2. If the root selects 0 then patterns 00 and 11 are concatenated with patterns in $S_{w,d-1}^l$, therefore both $LS_{w,d-1}^l$ and $RS_{w,d-1}^l$ are concatenated with $\{0,1\}$. The situation when the root selects 1 is similar. $\square$

Now let us consider how functions $f_{i_d,j_d}^{(l)}(x_{i_d}, x_{j_d})(\alpha,\beta)$ are combined. Take the difficult case where both i and j are odd. By lemma 2 there is only one pattern $p_1 = (\alpha',\alpha) \in S_{i\#0,d}^l$ and there is only one pattern $p_2 = (\beta',\beta) \in S_{j\#0,d}^l$. If the selector having p_1 selects 0 then $x_{i_d} = x_{i\#0_d}$ else $x_{i_d} = 1 - x_{i\#0_d}$. If the selector having p_2 selects 0 then $x_{j_d} = x_{j \# 0_d}$ else $x_{j_d} = 1 - x_{j \# 0_d}$. In any case the conditional bit pattern is changed to (α', β'), i.e., $f^{(l)}_{i_d, j_d}(x_{i_d}, x_{j_d})(\alpha, \beta)$ will be combined into $f^{(l+1)}_{\lfloor i/2 \rfloor_d, \lfloor j/2 \rfloor_d}(x_{\lfloor i/2 \rfloor_d}, x_{\lfloor j/2 \rfloor_d})(\alpha', \beta')$. Note that $x_{\lfloor i/2 \rfloor_d}$ and $x_{\lfloor j/2 \rfloor_d}$ are new random variables and here we are not using superscript to denote this fact. The following lemma ensures that at most four functions will be combined into $f^{(l+1)}_{\lfloor i/2 \rfloor_d, \lfloor j/2 \rfloor_d}(x_{\lfloor i/2 \rfloor_d}, x_{\lfloor j/2 \rfloor_d})(\alpha', \beta')$.

Let $S = \{(\alpha', \beta') | (\alpha', \alpha) \in S^l_{i,d}, (\beta', \beta) \in S^l_{j,d}, \alpha, \beta \in \{0,1\}^d\}$.

Lemma 3: $|S| = 4^d$.

Proof: The definition of S can be viewed as a linear transformation. Represent $x \in \{0,1\}^d$ by a vector of 2^d bits with x-th bit set to 1 and the rest of the bits set to 0. The transformation $\alpha \mapsto \alpha'$ can be represented by a permutation matrix of order 2^d. The transformation $(\alpha, \beta) \mapsto (\alpha', \beta')$ can be represented by a permutation matrix of order 2^{d+1}. $\square$

Lemma 3 tells us that the functions to be combined are permuted, therefore no more than four functions will be combined under any conditional bit pattern.

We call this scheme of combining as *combining functions with respect to the surviving set*.

We have completed a preliminary description of our derandomization scheme for the GPC problem. The algorithm for processors working on $\vec{x_d}$, $0 \le d < k$, can be summarized as follows.

Step t ($0 \le t < d$): Wait for the pipeline to be filled.

Step $d + t$ ($0 \le t < \log n$): Fix random variables at level t for all conditional bit patterns in the surviving set. (* There are 2^d such patterns in the surviving set.*) Combine functions with respect to the surviving set. (* At the same time the bit setting information is transmitted to the nodes at depth $2d + 2$ on the bit pipeline tree. *)

Step $d + \log n$: Fix the sole remain random variable at level $\log n$ for the sole bit pattern in the surviving set. Output the good point for $\vec{x_d}$. (* At the same time the bit setting information is transmitted to the node at depth $2d + 2$ on the bit pipeline tree.*)

Theorem 1: The GPC problem can be solved on the CREW PRAM in time $O((q/k + 1)(\log n + \tau))$ and space $c \sum_{i=0}^{k}(m * 4^i)$ with $c \sum_{i=0}^{k}(m * 4^i)$ processors, where τ is the time for computing the BPC functions $f_{i_d, j_d}(x_{i_d}, x_{j_d})(\alpha, \beta)$.

Proof: The correctness of the scheme comes from the fact that as random bits are fixed a smaller space with higher expectation is obtained, and thus when all random bits are fixed a good point is found. Since k $\vec{x_u}$'s are fixed in one pass which takes $O(\log n + \tau)$ time, the time complexity for solving the GPC problem is $O((q/k + 1)(\log n + \tau))$. The space and processor complexities are obvious from the description of the scheme. $\square$

We have not yet discussed explicitly the way the random variable trees are constructed. The construction is implied in the surviving set we computed. We now give the algorithm RV-Tree for constructing the random variable trees. This algorithm will help understand better the whole scheme.

Procedure **RV-Tree**

begin

Step t $(0 \le t < u)$: Wait for the pipeline to be filled.

Step $u + t$ $(0 \le t < \log n)$:

(* In this step we are to build $T_u^{(t)}[i][j]$, $0 \le i < n/2^{t+1}$, $0 \le j < 2^u$. At the beginning of this step $T_{u-1}^{(t)}[i][j]$ has already been constructed. Let $T_{u-1}^{(t-1)}[i0][j]$ and $T_{u-1}^{(t-1)}[i1][j']$ be the two children of $T_{u-1}^{(t)}[i][j]$ in the random variable tree. $T_u^{(t-1)}[i0][j0]$ and $T_u^{(t-1)}[i0][j1]$ are the children of $T_{u-1}^{(t-1)}[i0][j]$, $T_u^{(t-1)}[i1][j'0]$ and $T_u^{(t-1)}[i1][j'1]$ are the children of $T_{u-1}^{(t-1)}[i1][j']$, in the bit pipeline tree for level $t - 1$. The setting of the random variable r for the pair $(i0, i1)$ at level t for P_{u-1}, i.e. the random variable in $T_{u-1}^{(t)}[i][j]$, is known. *)

make $T_u^{(t-1)}[i0][j0]$ and
$T_u^{(t-1)}[i1][j'r]$ as the children of
$T_u^{(t)}[i][j0]$ in the random variable forest for P_u;
(* jr is the concatenation of j and r. *)

make $T_u^{(t-1)}[i0][j1]$ and
$T_u^{(t-1)}[i1][j'\bar{r}]$ as the children of
$T_u^{(t)}[i][j1]$ in the random variable forest for P_u;
(* $\bar{r}$ is the complement of r. *)

make $T_u^{(t)}[i][j0]$ and
$T_u^{(t)}[i][j1]$ as the children of
$T_{u-1}^{(t)}[i][j]$ in bit pipeline tree for level t;

Fix the random variables in $T_u^{(t)}[i][j0]$ and $T_u^{(t)}[i][j1]$;

Step $u + \log n$:

(* At the beginning of this step the random variable trees have been built for T_i, $0 \le i < u$. Let $T_{u-1}^{(\log n)}[0][j]$ be the root of T_{u-1}. The random variable r in $T_{u-1}^{(\log n)}[0][j]$ has been fixed. In this step we are to choose one of the two children of $T_{u-1}^{(\log n)}[0][j]$ in the bit pipeline tree for level $\log n$ as the root of T_u. *)

make $T_u^{(\log n - 1)}[0][jr]$ as the child of
$T_u^{(\log n)}[0][jr]$ in the random variable tree;

make $T_u^{(\log n)}[0][jr]$ as the child of
$T_{u-1}^{(\log n)}[0][j]$ in the bit pipeline tree for level $\log n$;

fix the random variable in $T_u^{(\log n)}[0][jr]$;

output $T_u^{(\log n)}[0][jr]$ as the root of T_u;

end

Procedure RV-Tree uses the pipelining technique as well as the dynamic programming technique. These are some of essential elements of our scheme.

4. $\Delta + 1$ Vertex Coloring

Luby formulated the $\Delta + 1$ vertex coloring problem as $O(\log n)$ instances of the GPC problem[L3][1]. After each execution of his GPC algorithm a constant fraction of the *vertices* will be permanently colored and removed from the graph. It is not difficult, although necessary, for us to modify his formulation such that after each execution of a GPC algorithm a constant fraction of the *edges* will be removed from the graph. For the sake of completeness the details of this modification will be given in the final version of the paper.

Let $c^k m$ be the number of processors needed to compute k instances of the BPC problem in a pipeline. There will be $O(\log n)$ stages in the modified algorithm. Each stage will solve an instance of the GPC problem and therefore reduce the number of edges so that there will be no more than a fraction $1/c$ of the edges left. Thus during stage i there will be e edges in the remaining graph and $c^i e$ processors available. Because each stage has $O(\log n)$ instances of the BPC problem, the time complexity for stage i is $O(\log^2 n/i)$. Thus the time complexity of the whole algorithm becomes $O(\sum_{i=0}^{O(\log n)} \log^2 n/i) = O(\log^2 n \log\log n)$.

The number of processors used in the algorithm can be reduced to $O((m+n)/\log\log n)$. Because in stage i we can have $c^i/\log\log n$ processors for each edge under each conditional bit pattern. Therefore the tables for the BPC functions in stage i can be computed in time $O(\log^2 n \log\log n/c^i)$, the overall time for table construction for the whole algorithm is $O(\log^2 n \log\log n)$. The calculation for the time for constructing the derandomization trees is similar and can be shown to be $O(\log^2 n \log\log n)$ with $O((m+n)/\log\log n)$ processors. In the first $O(\log\log\log n)$ stages our GPC algorithm will be invoked with $k = 1$. The time complexity for these stages is

$$O\left(\sum_{i=0}^{O(\log\log\log n)} \frac{\log^2 n \log\log n}{c^i} \right) = O(\log^2 n \log\log n).$$

The rest stages take $O(\log^2 n \log\log n)$ time by the analysis in the last paragraph.

Theorem 2: There is a CREW algorithm with time complexity $O(\log^2 n \log\log n)$ using $O((m+n)/\log\log n)$ processors which computes $\Delta + 1$ vertex coloring for a graph. $\square$

References

[ABI]   N. Alon, L. Babai, A. Itai. A fast and simple randomized parallel algorithm for the maximal independent set problem. J. of Algorithms 7, 567-583(1986).

---

[1] The problem formulated[L2][L3] resembles a GPC problem. It is not a GPC problem in the strict sense. For our purpose it is alright to view it as a GPC problem because our GPC algorithm applies.

[BR]   B. Berger, J. Rompel. Simulating $(\log^c n)$-wise independence in NC. Proc. 1989 IEEE FOCS, 2-7.

[BRS]   B. Berger, J. Rompel, P. Shor. Efficient NC algorithms for set cover with applications to learning and geometry. Proc. 1989 IEEE FOCS, 54-59.

[GS]   M. Goldberg, T. Spencer. Constructing a maximal independent set in parallel. SIAM J. Dis. Math., Vol 2, No. 3, 322-328(Aug. 1989).

[H1]   Y. Han. A parallel algorithm for the PROFIT/COST problem. To appear in Proc. of 1991 Int. Conf. on Parallel Processing, (AUg. 1991).

[H2]   Y. Han. A fast derandomization scheme and its applications. Tech. Report, TR No. 180-90, Computer Science Dept., University of Kentucky, Lexington, Kentucky.

[HI]   Y. Han and Y. Igarashi. Derandomization by exploiting redundancy and mutual independence. Proc. SIGAL 1990, LNCS 450, 328-337.

[IS]   A. Israeli, Y. Shiloach. An improved parallel algorithm for maximal matching. Information Processing Letters 22(1986), 57-60.

[KW]   R. Karp, A. Wigderson. A fast parallel algorithm for the maximal independent set problem. JACM 32:4, Oct. 1985, 762-773.

[L1]   M. Luby. A simple parallel algorithm for the maximal independent set problem. SIAM J. Comput. 15:4, Nov. 1986, 1036-1053.

[L2]   M. Luby. Removing randomness in parallel computation without a processor penalty. Proc. 1988 IEEE FOCS, 162-173.

[L3]   M. Luby. Removing randomness in parallel computation without a processor penalty. TR-89-044, Int. Comp. Sci. Institute, Berkeley, California.

[MR]   G. L. Miller and J. H. Reif. Parallel tree contraction and its application. Proc. 26th Symp. on Foundations of Computer Science, IEEE, 291-298(1985).

[MNN]   R. Motwani, J. Naor, M. Naor. The probabilistic method yields deterministic parallel algorithms. Proc. 1989 IEEE FOCS, 8-13.

[PSZ]   G. Pantziou, P. Spirakis, C. Zaroliagis. Fast parallel approximations of the maximum weighted cut problem through Derandomization. FST&TCS 9: 1989, Bangalore, India, LNCS 405, 20-29.

[Rag]   P. Raghavan. Probabilistic construction of deterministic algorithms: approximating packing integer programs. JCSS 37:4, Oct. 1988, 130-143.

[Sp]   J. Spencer. Ten Lectures on the Probabilistic Method. SIAM, Philadelphia, 1987.

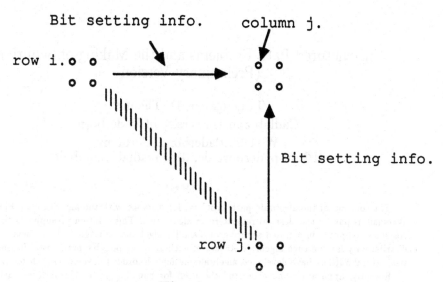

Fig. 1.

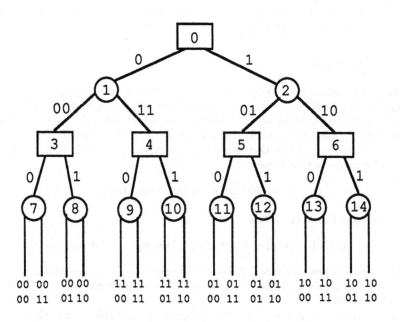

Fig. 2.

Unstructured Path Problems and the Making of Semirings (Preliminary Version)

T. Lengauer, D. Theune[1]

Cadlab and University of Paderborn

W–4790 Paderborn, Germany

tl@uni-paderborn.de, dirk@cadlab.cadlab.de

Abstract

The solution of the algebraic path problem, for instance with the algorithm by Floyd and Warshall, is part of the classical repertoire on algorithms. This solution presupposes that path costs are computed in a closed semiring or a similar algebraic structure. The associativity and distributivity laws in such algebraic structures exclude many possible path costs. In the seventies, several authors have developed algebraic methods in order to overcome such restrictions.

Recently, applications have created the need for handling such unstructured path costs. Motivated by such applications, we investigate the efficiency of algorithms solving path problems with unstructured costs. The resulting procedure allows for developing efficient algorithms for solving the algebraic path problem with respect to a wide variety of cost measures, including finding shortest paths with discounting, and counting paths or computing the expected length of paths with desired properties.

1 Introduction

The algebraic path problem is defined as follows. We are given a directed graph $G = (\{1,\ldots,n\}, E)$ with an edge labeling $\lambda : E \to R$ into a set R. Edge values can be composed with a binary operation $\odot$, in order to yield costs of paths of edges. Furthermore, an operation $\oplus : M(R) \to R$ can be used to combine costs of paths. Here, $M(R)$ is the set of all countable multisets that are composed of elements in R. (In the special case that $M = \{r_1, r_2\}$ we write $r_1 \oplus r_2 = \bigoplus M$.) The algebraic path problem asks for computing the values

$$d_{ij} = \bigoplus \{\lambda(p) \,|\, p \text{ is a path from } i \text{ to } j\}$$

where, for a path $p = (v_0, v_1, v_2 \ldots, v_k)$,

$$\lambda(p) = (\cdots (\lambda(v_0, v_1) \odot \lambda(v_1, v_2)) \odot \lambda(v_2, v_3)) \cdots) \odot \lambda(v_{k-1}, v_k)$$

The operation $\oplus$ has a neutral element $\bigoplus \emptyset = 0$. The operation $\odot$ has a neutral element 1 which is also the cost of the empty paths $p_{0,i}$ from each vertex i, $1 \leq i \leq n$, to itself. We call $C = (R, \oplus, \odot, 0, 1)$ an *algebraic cost structure*.

There are basically two approaches to solving algebraic path problems in the literature.

The first approach—pursued by several authors in different application areas, see e.g. [5, 6, 9, 10, 13, 15, 16, 19]—requires the algebraic cost structure $C = (R, \oplus, \odot, 0, 1)$ to fulfill the following additional requirements.

A1: $\odot$ is associative.

A2: 0 is absorptive over $\odot$, i.e., for all $r \in R$, we have $r \odot 0 = 0 \odot r = 0$.

---

[1]This research has been supported by the Project *EMC Simulation Systems* by the German Ministery of Research and Technology (BMFT) No. 0118/13AS0099

A3: $\oplus$ is associative and commutative over finite and countably infinite sets of values in R, i.e., for each $M \in M(R)$ and each partition $M = \bigcup_{i \in I} M_i$ into disjoint[2] subsets over a finite or countably infinite index set I we have

$$\bigoplus M = \bigoplus_{i \in I} (\bigoplus M_i)$$

A4: $\odot$ distributes over finite and countably infinite sums w.r.t. $\oplus$, i.e., for $M_1, M_2 \in M(R)$, we have

$$(\bigoplus M_1) \odot (\bigoplus M_2) = \bigoplus_{r_1 \in M_1} \bigoplus_{r_2 \in M_2} (r_1 \odot r_2)$$

If $C = (R, \oplus, \odot, 0, 1)$ has the properties A1 to A4 we call C a *closed semiring*.

The second approach has been presented by Tarjan [18]. It coaches all algebraic path problems in terms of one *generic* algebraic path problem, namely the problem of computing so-called *unambiguous path expressions*. This approach is more general than the approach via closed semirings, because it requires the relevant algebraic properties only on the elements in the closed semiring that are involved in possible computations, but the approach is also slightly more technical. We will restrict ourselves to semirings in this paper. Our methods can be extended to path expressions.

1.1 Algorithms

It is common knowledge that we can solve the algebraic path problem by the algorithm by Floyd and Warshall [9, 10, 19] also attributable to Kleene [13], or by Gaussian elimination [15, 17, 18], or with methods based on matrix multiplication [1, 7]. Rote [16] gives a recent overview over the algebraic path problem over closed semirings.

All of the algorithms mentioned above run in time $O(n^3(T_\odot + T_\oplus) + n^2 T_*)$ (plus an additional log factor, if matrix multiplication is used). Here $T_\odot$ and $T_\oplus$ are the worst-case run times for computing $r_1 \odot r_2$ and $r_1 \oplus r_2$, respectively, for $r_1, r_2 \in R$, and T_* is the worst-case run time for computing r^*, for $r \in R$. Rote [16] and Tarjan [17, 18] also discuss improvements for special graphs and algorithms for solving other path problems, such as single-pair path problems. The methods based on matrix multiplication can also be sped up, for special semirings [7].

1.2 Examples

Closed semirings can be quite diverse. Many examples of closed semirings appear in the literature. Despite this variety, there are many path-cost measures that cannot easily be coached in terms of closed semirings. Recently, applications in engineering, e.g., in circuit layout under electromagnetic compatibility (EMC) restrictions have created a need for solving path problems on cost structures that are not well behaved [12]. Other examples come from the area of artificial intelligence, where heuristic path search with diverse cost criteria has been a subject of study for a long time [8].

As an example of a cost structure that is not well-behaved, consider the problem of finding shortest paths with discounting [11]. Here, every edge obtains a cost value $\lambda(e) \in Q$. We define

$$\lambda(p) = \sum_{i=0}^{k-1} 2^{-i} \lambda(v_i, v_{i+1})$$

This cost measure violates A3, but it has all other properties of a closed semiring. To see, why A3 is violated, consider the paths p_1 and p_2 in Figure 1. p_1 from i to j has a higher cost (3) than p_2 (1). But, if we extend both paths with the segment p_3 to k, p_1 leads to a cheaper path (4) than p_2 (5). Classical algorithms for solving the algebraic path problem on the real numbers only remember the cost of the cheapest paths found so far. Therefore, they cannot be applied to this cost structure.

---

[2]Since we are concerned with *multisets* all unions ($\cup$) in this paper will be *disjoint*.

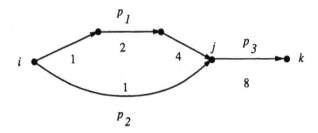

Figure 1: *Shortest paths with discounting is not distributive.*

1.3 The Transformation of Cost Structures

Strictly speaking, closed semirings are powerful enough to handle this example. So far, we just have been computing in the wrong cost structure C. This paper will detail how to turn a cost structure into a closed semiring, and illustrate the process by several examples.

Let us assume that we have a cost structure C given. If we inspect the axioms A1 to A4, we find that only A1 and A4 are critical restrictions. Since $\oplus$ tends to be some kind of census or tournament function for values in R, it is not restrictive to assume that $\oplus$ be associative and commutative (A3). Furthermore, the cost value 0 usually denotes the absence of any path, therefore (A2) is natural. In this paper we will discuss how to eliminate both A1 and A4.

The most direct way of doing so is to define a kind of homomorphism $g : C \rightarrow C'$ into a closed semiring C', and to provide a mapping $h : C' \rightarrow C$ that allows for interpreting the results computed in C' as costs back in C.

Formally, g and h need to have the following properties.

Definition 1 *Let* $C = (R, \oplus, \odot, 0, 1)$, $C' = (R', \oplus', \odot', 0', 1')$ *be cost structures. Let* $g : R \rightarrow R'$ *and* $h : R' \rightarrow R$ *be mappings. The pair* (g, h) *is called a* transformation *from* C *to* C' *if*

1. $g((\ldots(r_1 \odot r_2) \odot r_3) \ldots \odot r_n) = (\ldots(g(r_1) \odot' g(r_2)) \odot' g(r_3)) \ldots \odot' g(r_n))$ *for every finite sequence* $r_1, \ldots, r_n$ *in* R.

2. $\oplus M = h(\oplus'_{r \in M} g(r))$ *for every multiset* $M \subset R$.

By the following Theorem we can solve the algebraic path problem for C by replacing every edge weight $\lambda(e)$ in G by $\lambda'(e) = g(\lambda(e))$ and solving the algebraic path problem for C'.

Theorem 1 *For all* $i, j \in \{1, \ldots n\}$ *we have*

$$d_{ij} = h(\oplus'\{\lambda'(p) \mid p \text{ is a path from } i \text{ to } j\})$$

If C' is a closed semiring, all existing algorithms can be applied to solve the algebraic path problem on C'. Gondran and Minoux [11, Chapter 3] pointed out this method for solving path problems with unstructured costs and exemplified the method with several examples. Wongseelashote [20] gives a very general and nice algebraic framework for this sort of transformation of the cost structure, in the case that C meets A1. In general, however, finding an appropriate pair g and h can be quite trying, and furthermore the complexity of the solution of the algebraic path problem critically depends on the choice of g and h.

There are affinities of this question to the area of finite transductions and rational languages [3, 4]. But this area has mainly addressed closure properties and decidability questions. Efficiency considerations in this context are given in [2].

The purpose of this paper is to help take the mystery out of the choice of this transformation and to discuss its ramifications concerning the complexity of the resulting algorithms.

Section 2 of the paper provides the intuitive and formal presentation of how to eliminate A4. Section 3 gives examples of how to apply this process to nondistributive cost structures. Section 4 describes how to eliminate A1. In Section 5 we present conclusions and open problems.

2 Eliminating Distributivity

In this section we assume that C is a cost structure that meets A1.

For an intuitive explanation of the extension, we need to recapitulate that all of the algorithms mentioned above solve the algebraic path problem by computing *path expressions*. A path expression π_{ij} is a regular expression (over $+$, $\cdot$, $*$) that represents a set of paths between the same pair of vertices i and j. The set of the corresponding cost values is denoted by $M(\pi_{ij})$. The solution algorithms start with the path expressions

$$\pi_{ij} := \begin{cases} (i,j) & (i,j) \in E \\ \emptyset & \text{otherwise} \end{cases}$$

Here (i,j) represents the set $\{(i,j)\}$ and $\emptyset$ represents the empty path set between any pair of vertices. During the solution of the path problem the path expressions are composed by executing the operations

$$\lambda(\pi_{ij}\cdot\pi_{jk}) := \lambda(\pi_{ij}) \odot \lambda(\pi_{jk}) \tag{1}$$
$$\lambda(\pi_{ij}+\pi'_{ij}) := \lambda(\pi_{ij}) \oplus \lambda(\pi'_{ij}) \tag{2}$$
$$\lambda(\pi^*_{ii}) := \lambda(\pi_{ii})^* \tag{3}$$

The first of these operations concatenates paths, the second joins sets of paths, and the third accounts for the repeated traversal of cycles.

In fact, we do not compute the path expressions π_{ij} themselves—they would become far too long—but the values

$$\lambda(\pi_{ij}) := \bigoplus \{\lambda(p) \,|\, p \text{ is a path represented by } \pi_{ij}\}$$

If C is a closed semiring then, indeed, we can compute and remember the single cost value $\lambda(\pi_{ij})$ for each path expression π_{ij} considered during the algorithm. The associativity and distributivity laws of the closed semiring enable us to construct the final path cost from the costs of the intermediately occurring path expressions. However, if the cost structure is nondistributive, then we cannot get by with computing and storing a single value for each path expression. In Figure 1, we have to remember both paths between i and j before we compose them with the path from j to k. Formally, this prohibits us from combining the costs for path p_1 and p_2 using $\oplus$—i.e., minimizing over them— before we use $\odot$ to compose them with the cost of p_3. What this means is, that we need to compute and store more than a single cost value for each path expression.

However, we do not have to store all path costs encountered. For instance, if we are interested in finding shortest paths with discounting, we can discard the more costly of two paths p_1 and p_2 from i to j, *if both paths have the same cardinality*. In general, we will be able to combine some of the costs for paths from i to j, while we have to keep others separate. In general, we want to combine as many cost values as possible. However, we have to be careful to maintain the distributivity among the values which we combine. Pairs of values that destroy distributivity in the sense of Figure 1 have to stay separate.

We will now make this process formal. In the following, we restrict $M(R)$ to contain only sets $M(\pi)$, where π is a path expression.

Our goal is to define a closed semiring $\hat{C}$, in which we can feasibly compute path costs, and a transformation (g, h) from C into $\hat{C}$. At the basis of our method is the notion of a *grouping function*. The grouping function takes care of joining path costs only if distributivity is not violated.

Definition 2 *A function $f : M(R) \to M(R)$ is called a* grouping function *if the following properties hold.*

F1: *For $M_1, M_2 \in M(R)$, let $M_1 \odot M_2$ be the multiset such that $M_1 \odot M_2 := \{r_1 \odot r_2 \mid r_1 \in M_1 \text{ and } r_2 \in M_2\}$. ($M_1 \odot M_2$ contains one element for each pair $(r_1, r_2) \in M_1 \times M_2$.) Then, for $M_1, M_2 \in M(R)$*

$$f(M_1 \odot M_2) = f(f(M_1) \odot f(M_2))$$

F2: *For all $M_i \in M(R)$, where $i \in I$, and I is a countable index set, we have*

$$f(\bigcup_{i \in I} M_i) = f\left(\bigcup_{i \in I} f(M_i)\right)$$

F3: *If $M \in M(R)$ then*

$$\bigoplus M = \bigoplus f(M)$$

The width of a grouping function is defined as

$$w(f) = \sup\{|f(M)| \mid M \in M(R)\}$$

Intuitively, if M is a set of path costs, then f groups costs together as long as this does not violate the distributivity of $\odot$ over $\oplus$. Property F1 states that the grouping process is compatible with $\odot$, i.e., grouping before an evaluation does not change the result of the evaluation. F2 requires a natural commutativity property of the grouping process. Finally, F3 states that the grouping process is compatible with $\oplus$.

Let us now assume that $C = (R, \oplus, \odot, 0, 1)$ is an algebraic cost structure that meets A1. Furthermore, let f be a grouping function on C. Then we can turn C into the closed semiring $\hat{C} = (\hat{R}, \hat{\oplus}, \hat{\odot}, \hat{0}, \hat{1})$ that is defined as follows.

- $\hat{R} := f(M(R))$

- For $\hat{M}_1, \hat{M}_2 \in \hat{R}$ we have

$$\hat{M}_1 \hat{\odot} \hat{M}_2 := f(\hat{M}_1 \odot \hat{M}_2)$$

- Let $\hat{M}$ be a countable multiset of elements in $\hat{R}$. Then

$$\hat{\bigoplus}\hat{M} := f(\bigcup \hat{M})$$

- $\hat{0} := f(\emptyset)$

- $\hat{1} := f(1)$

Theorem 2 *Let C be an algebraic cost structure that meets A1, and let f be a grouping function on C. Then $\hat{C}$ is a closed semiring, and $(f, \oplus)$ is a transformation from C into $\hat{C}$.*

By Theorem 2, we can run any of the classical algorithms for the algebraic path problem on $\hat{C}$. This solves our algebraic path problem on C.

From Theorem 2 we can also deduce the run time for the algorithm. Assume that $w(f)$ is finite, and let us assume, that the computation of $f(M)$ takes linear time in $|M|$. The operations $\hat{\oplus}$ and $\hat{\odot}$ executed by the algorithm reduce to complex operations $\odot$ and $\oplus$ on pairs of sets of size at most $w(f)$ each. Thus each such operation takes time at most $O(w(f)^2)$. The time for executing the $*$ operator on $\hat{C}$ cannot be bounded in general. Let us denote this term with T_*. Then the total time of the algorithm is $O(n^3 w(f)^2 + n^2 T_*)$. This analysis presupposes that $w(f)$ is finite—which need not be the case, in general. Moreover, for efficiency reasons, $w(f)$ should be as small as possible.

Even if $w(f)$ is infinite, we can often run classical path algorithms: As long as $f(\hat{M})$ and $f(\hat{M}^*)$ are finite for all *finite* sets $\hat{M} \in \hat{R}$, we never create infinite sets and the algorithm terminates. In general, the size of the sets can grow exponentially, but in many cases we can derive much better bounds on the run time. We see such an example in Section 3.1.

Wongseelashote [20] has independently introduced a very similar algebraic framework for ensuring distributivity. He calls the grouping function a *reduction*. Here we suggest to choose grouping functions with the idea in mind of doing the maximum possible amount of computation that does not interfere with distributivity. Such a choice should minimize $w(f)$ and thus render the computations as efficient as possible.

3 Examples

In this section we will apply the general framework of Section 2 to several examples.

3.1 Shortest Paths with Discounting

We have introduced this problem already in Section 1 as an example. Solving this problem amounts to carefully designing an algebraic cost structure, selecting an appropriate grouping function and, finally, rendering the computations efficient.

The following structural lemma is important in this context.

Lemma 1 *For every pair (s, t) of vertices either*

1. *there is a simple optimal path from s to t or*

2. *the cost of an optimal path from s to t is attained in the limit. In this case, there is a sequence of paths whose costs converge to the limit, such that all paths in the sequence start with the same simple path segment p_1, followed by the same simple cycle σ, traversed an increasing and unbounded number of times. All paths end with the same simple path segment p_2.*

3.1.1 The Algebraic Cost Structure

We represent each path cost by a pair (c, k). Here, c denotes the *cost* of the path, i.e., the sum of its weighted edge labels, and k is the *cardinality* of the path, i.e., the number of its edges. The original edge labels—which were assumed to be rational numbers—are transformed into this cost structure by turning the value $\lambda(e)$ into the pair $(\lambda(e), 1)$. In addition, we need values to represent limit points of infinite sets of path costs. All such sets are obtained by repeatedly traversing cycles on the path an arbitrarily large number of times. We will denote such values with pairs (c, ∞), where c is some limit point of the cost values. Finally, the value $(0, 0)$ represents the cost of the empty path. Thus

$$R := (Q \cup \{\infty\}) \times (N \cup \{\infty\})$$

We now detail the definition of $\odot$.

$$(c_1, k_1) \odot (c_2, k_2) := (c_1 + c_2 2^{-k_1}, k_1 + k_2) \tag{4}$$

if $k_1, k_2 \neq \infty$. Furthermore

$$(c_1, \infty) \odot (c_2, \infty) := (c_1, \infty) \tag{5}$$

Finally

$$(c_1, \infty) \odot (c_2, k_2) := (c_1, \infty) \tag{6}$$
$$(c_1, k_1) \odot (c_2, \infty) := (c_1 + c_2 2^{-k_1}, \infty) \tag{7}$$

Equation (4) adds path costs and cardinalities of two finite-length paths. Equation (5) applies to the situation that two infinite sets of paths are concatenated. Equations (6) and (7) apply to the situation that one of the path sets is infinite.

The operation $\oplus$ is defined as follows. Let M be a set of cost values. Then

$$\bigoplus M = \inf M$$

where the infimum is taken w.r.t. to the following reflexive and transitive relation $\preceq$:

$$(c_1, k_1) \preceq (c_2, k_2) :\Leftrightarrow c_1 < c_2 \text{ or } (c_1 = c_2 \text{ and } k_1 \leq k_2) \tag{8}$$

The lexicographic ordering in equation (8) implies, that an actually attained cost value is preferred over a cost value attained only in the limit.

The neutral elements are $0 = (\infty, \infty)$ and $1 = (0, 0)$.

This cost structure solves the problem of computing shortest paths with disounting. If the minimum cost is attained for a path, than a value of the type (c, k) is returned. If the minimum cost value is only attained in the limit, then a value of type (c, ∞) is returned.

The cost structure is associative, but it does not satisfy A4, as Figure 1 has shown.

3.1.2 The Grouping Function

A straightforward grouping function is the function

$$f(M) := f_1(M) \cup f_2(M) \cup \cdots \cup f_{i-1}(M) \cup f_i(M) \cup \cdots \cup f_\infty(M)$$

where

$$
\begin{aligned}
f_i(M) &= \min\{(c,k) \in M \,|\, k = i\} \qquad i > 0 \\
f_\infty(M) &= \inf\{(c,k) \in M \,|\, k = \infty\} \cup \{(\liminf M, \infty)\}
\end{aligned}
\tag{9}
$$

Here $\min \emptyset = \infty$. In (9), $\liminf M$ is the smallest limit point of the first components of all finite-cardinality cost values in M.

The intuition behind this definition of f is, that we can select the cheapest among a set of simple paths *of the same cardinality* right away without violating distributivity. The same holds for sets of limit points.

An inspection shows that the function f meets F1 to F3. Unfortunately, $w(f)$ is infinite. However, we can render the computations efficient with an application of Lemma 1. Indeed, by Lemma 1, either the minimum cost is attained with a simple path or it is attained in the limit with a very structured path set. Using concatenation, union, and starring—the primitive operations of all path algorithms—such path sets can only be constructed, if we first assemble the cycle σ, then star it, and then concatenate the other simple paths to it. For all of these operations, we do not need any of the $f_i(M)$ for $i > n$, if we compute $f(M)^*$ as follows

$$
\begin{aligned}
f(M)_i^* &= f_i(M) \qquad 0 < i \le n \\
f(M)_\infty^* &= \min(\{f_\infty(M)\} \cup \{(c\frac{2^k}{2^k-1}, \infty) \,|\, (c,k) \in f_i(M),\ 0 < i \le n\})
\end{aligned}
\tag{10}
$$

Thus effectively, we only compute $n+1$ elements of $f(M)$. Each execution of $\hat{\odot}$ and $\hat{\oplus}$ takes time $O(n^2)$ and $O(n)$, respectively. By (10), the computation of the * operator can be done in time $O(n)$, and the total run time is $O(n^3 w(f)^2) = O(n^5)$.

3.2 Counting Paths with Desired Properties

One of the interesting features of the algebraic path problem—in contrast to single-pair or single-source shortest-path problems—is that $\oplus$ need not entail a minimization. Instead, arbitrary composition operations can be used. In the classical case, this fact has established the connection between the algebraic path problem and matrix inversion.

Here we give a few examples of nondistributive algebraic cost structures that do not entail path minimization. The commonality between all these examples is that $\oplus$ counts paths that have some desired property.

We first present the common framework, and then we give a number of instances.

Let Z be a set of *properties* exactly one of which a path can have. Let $z(p)$ denote the property of path p. Assume that $Y \subset Z$ is the subset of *desired* properties. We want to count paths p from i to j that have properties $z(p) \in Y$.

In order to do so, we need to be able to compute properties of paths from the properties of their segments. Thus we postulate the existence of the mapping $g : Z \times Z \to Z$, such that

$$z(pq) := g(z(p), z(q))$$

In order to be able to use the method described in Section 2, we have to require that g be *associative*. Furthermore, we assume that the property $z_0 := z(\emptyset)$ of the empty path is a neutral element for g.

For technical reasons, we assume the existence of a property z_∞ that is the property of no path. We assume that

$$g(z_\infty, z) = g(z, z_\infty) = z_\infty$$

for all $z \in Z$.

Then we can construct the following algebraic cost structure $C = (R, \oplus, \odot, 0, 1)$ that counts

- $R = Z \times (\mathbb{N} \cup \infty)$. The first component denotes a property, the second component denotes the number of paths having this property.

- $(z_1, n_1) \odot (z_2, n_2) := (g(z_1, z_2), n_1 n_2)$, since each of the n_1 path prefixes with property z_1 can be composed with each of the n_2 path prefixes with property z_2 to yield a different path with property $g(z_1, z_2)$.

- If $R' \subset R$ and $R' \neq \emptyset$, we set

$$\bigoplus R' := (z_0, \sum_{(z,n) \in R', z \in Y} n)$$

  Furthermore, $\bigoplus \emptyset := (z_\infty, 0)$. $\bigoplus$ counts all paths with properties in Y.

- The element $(z_0, 0) \in R$ is neutral w.r.t. $\bigoplus$.

- The element $(z_\infty, 0) \in R$ is neutral w.r.t. $\odot$.

Since g is associative, C is also associative. C is not distributive, since $\bigoplus$ turns all path properties into z_0. Thus we have to look for grouping functions for C.

Theorem 3 *The function $f : M(R) \to M(R)$ that is defined as follows is a grouping function for C.*

$$f(M) = \{(z, \sum_{(z,n) \in M} n) \mid z \in Z\}$$

Furthermore, $w(f) = |Z|$.

f just counts up all paths with the same desired property. We now give a few specific examples of path properties.

3.2.1 Paths with less than h edges

We want to count paths that have less than h edges. We set $Z := \{z_0 = 0, 1, 2, \ldots, h, z_\infty\}$. All edges have labels $(1, 1)$. Paths with $i < h$ edges receive the property i. Paths with h or more edges receive the property h. We have $Y = \{0, 1, 2, \ldots, h - 1\}$, and

$$g(z_1, z_2) = \min\{h, z_1 + z_2\}$$

The framework outlined above applies. The $\overset{*}{}$ operator on $\hat{C}$ can be computed in time $O(h^2 \log h)$. Thus the run time of the path algorithms on this example is $O(n^3 h^2 \log h)$.

3.2.2 Monotonic Paths

We want to count the number of paths whose sequence of (real) edge labels increases along the path. Let I be the set of occurring edge labels. We choose

$$Z := \{(\ell, u) \mid \ell, u \in I \text{ and } \ell \leq u\} \cup \{z_0, z_\infty, z_{\text{out}}\}$$

If $\lambda(e)$ is the edge label given to us, we assign to e the cost $(\lambda(e), \lambda(e), 1)$. A monotonic path whose first edge has label ℓ and whose last edge has label u has the property (ℓ, u). A nonmonotonic path receives the property z_{out}. We have $Y = \{(\ell, u) \mid \ell, u \in I \text{ and } \ell \leq u\}$. The function g is defined as follows:

$$g(z_1, z_2) := \begin{cases} (\ell_1, u_2) & \text{if } z_1 = (\ell_1, u_1), z_2 = (\ell_2, u_2), u_1 \leq \ell_2 \\ z_{\text{out}} & \text{if } z_1 = (\ell_1, u_1), z_2 = (\ell_2, u_2), u_1 > \ell_2 \\ & \text{or } z_1 = z_{\text{out}} \text{ or } z_2 = z_{\text{out}} \\ z_1 & \text{if } z_2 = z_0 \\ z_2 & \text{if } z_1 = z_0 \end{cases}$$

We have $|Z| = \frac{|I|(|I|+1)}{2} + 3$. The $\overset{*}{}$ operator can be computed in time $O(|I|^2)$. Thus the run time is $O(n^3 |I|^4)$.

3.2.3 Expected Path Length

A whole different set of examples asks for expected path lengths under some suitable probability distribution. For instance, we can regard G as a time-homogeneous Markov chain. That is, each edge e receives a second label, the *probability* $w(e)$, such that the sum of the probabilities of all edges leaving a vertex v is 1. This labeling assigns to each path the probability that is the product of the probabilities of its edges. Computing the expected length of a path with respect to these probabilities, is a cost structure that is a closed semiring. We can represent the cost of a path by a pair (ℓ, w), where ℓ is the length of the path, and w is the probability that the path is taken. More generally, a set of paths has the cost (ℓ, w), where ℓ is the expected length of a path in the set and w is the probability that any path in the set is taken. The corresponding semiring operations are

$$(\ell_1, w_1) \odot (\ell_2, w_2) := (\ell_1 + \ell_2, w_1 w_2)$$

$$\bigoplus_{i \in I} (\ell_i, w_i) := \left(\sum_{i \in I} \ell_i w_i \middle/ \sum_{i \in I} w_i , \sum_{i \in I} w_i \right)$$

The method presented in Section 3.2 can be modified to compute the expected length of paths with certain properties. To this end we must, as in Section 3.2, keep track of the properties. The corresponding cost structure has values with three components (z, ℓ, w), where z is a property, ℓ is the expected path length, and w is the probability that a path in the set under consideration is taken. A construction that is completely analogous to the one in Section 3.2 leads to a cost structure, with which expected lengths of paths with given properties can be computed. The run times are the same as in Section 3.2.

4 Eliminating Associativity

We have mentioned, that the associativity requirement A1 is another critical property of a closed semiring that can be violated by a variety of cost structures.

We choose to eliminate nonassociativity separately from and before nondistributivity, mostly because this structures the transformation procedure.

Let us first describe the general situation: A cost structure usually is nonassociative because the cost value of a path prefix per se does not give enough information on how to compute the path cost when extending the prefix by attaching another path segment to it. The intuitive method of ensuring associativity is to augment the cost structure by providing this context information.

Formally, the context information is defined by an equivalence relation $\sim$ on the set R. Cost values that are related by $\sim$ are costs of paths that provide the same context. (This also means that R must be such that contexts can be inferred from cost values—a requirement that can always be met by suitably modifying R.) The relation $\sim$ induces a partition of R into equivalence classes, the so-called *contexts*. Let Γ denote the set of contexts. Elements of Γ are denoted with γ. Let $[r]$ denote the equivalence class of cost value r with respect to $\sim$. In order to be suitable for our purposes, $\sim$ has to meet the following requirement.

E1: For all $r_1, r_2, r_3 \in R$ we have $r_1 \sim r_2 \Rightarrow r_1 \odot r_3 \sim r_2 \odot r_3$ and $r_3 \odot r_1 \sim r_3 \odot r_2$

Let $\gamma_1 \odot \gamma_2$ denote the multiplication of two contexts γ_1 and γ_2, respectively. By E1, this definition is well-formed.

We want to tailor our computation to the context. This means that we have to modify our composition operation, depending on the context. Specifically, we need different composition operations $\odot_\gamma$ in different contexts. The neutral element 0 must be absorptive for all of these operations. Furthermore, we require that the following natural consistency condition be met for these operations.

E2: For $r_1, r_2, r_3 \in R$ and $\gamma \in \Gamma$, we have

$$(r_1 \odot_{\gamma \odot [r_1]} r_2) \odot_{\gamma \odot [r_1] \odot [r_2]} r_3 = r_1 \odot_{\gamma \odot [r_1]} (r_2 \odot_{\gamma \odot [r_1] \odot [r_2]} r_3)$$

E2 means, that using the appropriate composition operations in the appropriate context renders the composition associative.

Finally, we have to assure that we compute actual path costs. Therefore we require

E3: For $r_1, r_2 \in R$ such that r_2 can label an edge in G, we have

$$r_1 \odot_{[r_1]} r_2 = r_1 \odot r_2$$

Given Γ and the $\odot_\gamma$, we can form an associative cost structure $C' = (R', \oplus', \odot', 0', 1')$ as follows.

- $R' = R^\Gamma$. The γth component of a cost vector gives the cost of the path set in the γth context.

- $(r_{1,\gamma})_{\gamma \in \Gamma} \odot' (r_{2,\gamma})_{\gamma \in \Gamma}$ is the cost vector $(r_{3,\gamma})_{\gamma \in \Gamma}$, that is defined componentwise as follows:

$$r_{3,\gamma} = r_{1,\gamma} \odot_{\gamma \odot [r_{1,\gamma}]} r_{2,\gamma \odot [r_{1,\gamma}]} \tag{11}$$

- $\oplus'$ is defined by applying $\oplus$ componentwise.

- $0'$ and $1'$ are the zero-vector and the one-vector, respectively.

Theorem 4 *If E1, E2, and E3 hold, and C fulfills A2 and A3, then C' fulfills A1, A2 and A3.*

In the modified cost structure C', the edge labels of the graph are vectors all of whose components are the original edge label. By E3, the cost component for the context of the empty path denotes the actual path cost.

The choice of the set Γ is problem-specific. In fact, there are different possible choices for the same problem. In general, the size of Γ can be infinite. The procedure becomes efficient, if the size of Γ is small.

As an example we examine a modification of the shortest paths with discounting cost structure. Modify the operation $\odot$ to

$$(c_1, k_1) \odot (c_2, k_2) := (c_1 + c_2 2^{-\lfloor \frac{k_1}{2} \rfloor}, k_1 + k_2) \tag{12}$$

Intuitively, we discount the path costs only every *other* time. The resulting cost structure is not associative. For example we have

$$((1,1) \odot (1,1)) \odot (2,1) = (3,3) \neq (4,3) = (1,1) \odot ((1,1) \odot (2,1))$$

In this case, apparently there are three contexts for a path, namely that the cardinality of the path be *odd* and *even* and *infinite*, respectively. The corresponding binary operations $\odot_{\text{even}}$, $\odot_{\text{odd}}$, and $\odot_{\text{inf}}$ are defined as follows:

$$(c_1, k_1) \odot_{\text{even}} (c_2, k_2) := (c_1 + c_2 2^{-\lceil \frac{k_1}{2} \rceil}, k_1 + k_2)$$
$$(c_1, k_1) \odot_{\text{odd}} (c_2, k_2) := (c_1 + c_2 2^{-\lfloor \frac{k_1}{2} \rfloor}, k_1 + k_2)$$
$$(c_1, k_1) \odot_{\text{inf}} (c_2, k_2) := (c_1, k_1)$$

By inspection, it can be seen that E1 through E3 hold. The definition of the composition operations, as well as the check of E1 through E3, can be illuminated by noting that the operations $\odot_{\text{even}}$, $\odot_{\text{odd}}$, and $\odot_{\text{inf}}$ facilitate the computation of the following path costs for a path $p = (v_0, \ldots, v_k)$ in the different contexts.

$$f_{\text{even}}(p) = \sum_{i=0}^{n-1} 2^{-\lfloor \frac{i}{2} \rfloor} \lambda(e_i)$$

$$f_{\text{odd}}(p) = \sum_{i=0}^{n-1} 2^{-\lceil \frac{i}{2} \rceil} \lambda(e_i)$$

$$f_{\text{inf}}(p) = 0$$

For the purposes of complexity analysis, note that the execution of $\odot'$ requires computing path costs for each context $\gamma \in \Gamma$. In order to do so, we have to be able to quickly concatenate contexts. Let us assume that this can be done in constant time, as it is the case in the example. Furthermore, assume that the $\odot_\gamma$ can be applied in constant time. Then we can execute $\odot'$ in time $O(|\Gamma|)$, in general, and in constant time, in our example.

The cost structure C' can be turned into a closed semiring with a method analogous to the one described in Section 3.1. The resulting run times are the same as in Section 3.1, up to constant factors.

5 Conclusions and Open Problems

We have discussed the application and efficiency of a general framework for eliminating the requirements of associativity and distributivity in closed semirings, when solving algebraic path problems.

Our investigations extend and elucidate algebraic frameworks proposed by Gondran and Minoux [11] and Wongseelashote [20].

We transform cost structures into semirings in two steps:

1. To ensure associativity we extend R to maintain the necessary context information.

2. To ensure distributivity we process to multisets of cost values.

Although the transformation can be done in one step, this two-step procedure seems to be a natural way of arriving at transformations that lead to efficient algorithms.

As a result, we can efficiently solve algebraic path problems on a large set of cost structures. The run time of the respective algorithms increases with the square of the largest number of values that have to be memorized for each path expression. We have exemplified the new method at a number of examples that exhibit its wide applicability. A related method that applies to single-pair and single-source shortest-path problems and extends Ford's method is presented in [14].

This paper presents a method that applies to many problems. We do not claim, that the run times obtained for the example problems discussed here are optimal. Rather, we conjecture that some of these problems may be solvable more efficiently, if additional problem-specific considerations and "tricks" are incorporated. For instance, the run time of the shortest paths with discounting example could be reduced by using more efficient methods for computing a natural convolution operation over + (with scaling of the second operand) and min.

6 Acknowledgements

We are grateful to Michel Minoux and Günter Rote for helpful comments on an earlier version of the paper, and for pointing out to us several relevant references. Discussions with Anja Feldmann, Werner John, and Ralf Thiele on EMC cost functions have provided the motivation for this research.

References

[1] A. V. Aho, J. E. Hopcroft, and J. D. Ullman. *The Design and Analysis of Computer Algorithms*. Addison-Wesley Series in Computer Science and Engineering. Addison-Wesley, Reading, MA, 1974.

[2] R. C. Backhouse and B. A. Carré. Regular algebra applied to path-finding problems. *Journal of the Institute of Mathematics and its Applications*, 15:161–186, 1975.

[3] J. Berstel. *Transductions and Context-Free Languages*. Teubner Verlag, Stuttgart, Germany, 1979.

[4] J. Berstel and C. Reutenauer. *Rational Series and Their Languages*. EATCS Monographs on Theoretical Computer Science. Springer Verlag, New York, NY, 1984.

[5] B. A. Carré. An algebra for network routing problems. *Journal of the Institute of Mathematics and its Applications*, 7:273–294, 1971.

[6] B. A. Carré. *Graphs and Networks*. Oxford Applied Mathematics and Computing Science Series. Clarendon Press, 1979.

[7] T. H. Cormen, C. E. Leiserson, and R. L. Rivest. *Introduction to Algorithms*. McGraw-Hill, New York, 1990.

[8] R. Dechter and J. Pearl. Generalized best-first search strategies and the optimality of A^*. *Journal of the Association for Computing Machinery*, 32(3):505–536, 1985.

[9] J. G. Fletcher. A more general algorithm for computing closed semiring costs between vertices of a directed graph. *Communications of the Association for Computing Machinery*, 23(6):350–351, 1980.

[10] R. N. Floyd. Algorithm 97 — shortest path. *Communications of the Association for Computing Machinery*, 5(6):345, 1962.

[11] M. Gondran and M. Minoux. *Graphs and Algorithms*. Wiley-Interscience Series in Discrete Mathematics. John Wiley & Sons, Chichester, U.K., 1984.

[12] W. John. Remarks to the solution of EMC-problems on printed circuit boards. In *Proceedings of the Seventh International Conference on Electromagnetic Compatibility*, pages 68–72, York, U.K., 1990.

[13] S. C. Kleene. Representation of events in nerve nets and finite automata. In C. Shannon and J. McCarthy, editors, *Automata Studies*, pages 3–40. Princeton University Press, Princeton, NJ, 1956.

[14] T. Lengauer and D. Theune. Efficient algorithms for path problems with general cost criteria. In B. Monien and M. Rodriguez-Artalejo, editors, *18th International Symposium on Automata, Languages, and Programming*, pages 314–326. Springer Lecture Notes in Computer Science, No. 510 Springer Verlag, New York, 1991.

[15] G. Rote. A systolic array algorithm for the algebraic path problem (shortest path, matrix inversion). *Computing*, 34(3):191–219, 1985.

[16] G. Rote. Path problems in graphs. *Computing Supplement*, 7:155–189, 1990.

[17] R. E. Tarjan. Fast algorithms for solving path problems. *Journal of the Association for Computing Machinery*, 28(3):594–614, 1981.

[18] R. E. Tarjan. A unified approach to path problems. *Journal of the Association for Computing Machinery*, 28(3):577–593, 1981.

[19] S. Warshall. A theorem on boolean matrices. *Journal of the Association for Computing Machinery*, 9(1):11–12, 1962.

[20] A. Wongseelashote. Semirings and path spaces. *Discrete Mathematics*, 26:55–78, 1979.

Neighborhood Graphs and Geometric Embedding

Frances Yao

Xerox Palo Alto Research Center

Abstract

Given a set of points S in the plane, define $G_k = (S, E_k)$ to be the *k-nearest-neighbors graph* of S where $(u, v) \in E_k$ if u is one of the k nearest neighbors of v. In this talk, we will present some results on the graph properties of G_k. In particular, we derive bounds on the diameter of the graph G_k. This study is motivated by geometric embedding problems. For example, in the simulation of many-body systems, it is desirable to map a set of particles to a grid of processors so as to preserve neighborhood relations, i.e., to minimize the dilation of G_k in the embedding. (Joint work with Mike Paterson)

Finding Optimal Bipartitions of
Points and Polygons

Joseph S. B. Mitchell [*]
School of Operations Research
& Industrial Engineering
Cornell University
Ithaca, NY 14853

Erik L. Wynters [†]
Center for Applied Mathematics
Cornell University
Ithaca, NY 14853

Abstract

We give efficient algorithms to compute an optimal bipartition of a set of points or a set of simple polygons in the plane. We examine various criteria involving the perimeter and the area of the convex hulls of the two subsets.

1 Introduction

In the standard "bipartition problem", we are interested in partitioning a set S of n points into two subsets (S_1 and S_2) in such a way as to optimize some function of the "sizes" ($\mu(S_i)$) of the two subsets. Avis ([2]) gave an $O(n^2 \log n)$ time algorithm to find a bipartition that minimizes the maximum of the diameters of the sets S_1 and S_2. Asano, Bhattacharya, Keil and Yao ([1]) improved the bound on the time complexity of this problem, obtaining an optimal $O(n \log n)$ algorithm. Monma and Suri ([12]) gave an $O(n^2)$ time algorithm for minimizing the sum of diameters. Most recently, Hershberger and Suri ([7]) have considered the problem in which the measure of "size" $\mu(S_i)$ is (a). the diameter, (b). the area, perimeter, or diagonal of the smallest enclosing axes-parallel rectangle, or (c). the side length of the smallest enclosing axes-parallel square. They provide $O(n \log n)$ time algorithms to find a bipartition that satisfies $\mu(S_i) \leq \mu_i$ ($i = 1, 2$) for two given numbers μ_1 and μ_2.

Here, we consider the version of the bipartition problem in which $\mu(S_i)$ is the perimeter or area of the convex hull, $conv(S_i)$, and we desire a partition that minimizes the sum $\mu(S_1) + \mu(S_2)$ or the maximum $\max\{\mu(S_1), \mu(S_2)\}$. We also consider the generalization in which S is a set of disjoint polygons.

Funnel trees and hourglasses are fundamental to several of our algorithms, so we discuss their properties next.

---
[*]Partially supported by NSF grants IRI-8710858 and ECSE-8857642, and by a grant from Hughes Research Laboratories. Email: jsbm@orie.cornell.edu

[†]Supported by a grant from Hughes Research Laboratories. Email: elw@macomb.tn.cornell.edu

2 Funnel Trees and Hourglasses

We use the term *vertex* to mean either a point in S or a polygon vertex. Let n denote the total number of vertices.

The visibility graph of S is the graph whose vertices are the vertices contained in or determined by S and whose edges are pairs of vertices (u, v) such that the open line segment between u and v does not intersect any polygon in S. The visibility graph can be computed in time $O(E + n \log n)$ and space $O(E)$ where E is the number of edges in the graph. Note that E is $O(n^2)$ when S is a set of points, but can be smaller when S is a set of polygons.

A *visible chain* is a path in the visibility graph. It is well known that a shortest path between two points in the plane that avoids polygonal obstacles is a visible chain [8,9,10,11,15]; the subgraph of the visibility graph consisting of all visible chains that are shortest paths from a fixed source vertex s to some other vertex is called the *shortest path tree rooted at s*.

Consider a vertex a that lies to the left of a directed edge (u, v) and is visible from a point b in the interior of the edge. Define the *lower chain* of a determined by u, v, and b to be the unique convex visibile chain from a to u for which the open region bounded by the chain and the line segments $\overline{uv}$ and $\overline{ab}$ contains no vertices. Intuitively, think of taking a rubber band between a and b in Figure 1 and sliding the end at b along $\overline{uv}$ until it reaches u. Define the *upper chain* from a to v analagously. These two chains form a *funnel* with apex a and base (u, v).

We require the base of a funnel to be an edge on the convex hull of the set S of points or polygons we wish to partition. In this setting a funnel is completely determined by its apex and by the next vertex on its lower (or upper) chain, i.e., by the first directed edge on its lower (or upper) chain [6]. Therefore, the total number of funnels in the visibility graph is at most $2E$. Furthermore, we can find all of the directed edges that determine funnels, in time proportional to their number, by traversing part of the visibility graph as explained below.

Consider a funnel f with apex a and base (u, v), and suppose b is the first vertex after a on its lower chain. There is a unique funnel p with apex b and base (u, v) contained in funnel f; its left chain is obtained from that of f by deleting edge (a, b). If we think of p as the parent of f, we see that the funnels with base (u, v) form a tree in the visibility graph (taking the location of a funnel to be that of its apex) rooted at u. Call this the *lower funnel tree on (u, v)*. Note that the path from the root of this tree to any node is precisely the lower chain of the funnel corresponding to that node. In an analogous way, the upper chains of funnels with base (u, v) determine an *upper funnel tree on (u, v)* rooted at v. Figure 2 shows a lower funnel tree, and an upper funnel tree on the same base.

Using the representation of the visibility graph generated by their algorithm, Ghosh and Mount show that parents, clockwise and counterclockwise siblings, and extreme clockwise and counterclockwise children of a node in a funnel tree can be found in constant time [6]. This implies that clockwise and counterclockwise traversals of lower and upper funnel trees can be made in the visibility graph without storing these trees explicitly. Also, we see that, summing over all bases, the total time needed to traverse these trees is $O(E)$.

If f_1 is a funnel determined by directed edge (a, b) and f_2 is a funnel determined by its reversal (b, a), then f_1 and f_2 together form an *hourglass* as shown in Figure 3. Since it takes two funnels to form an hourglass and each funnel is part of at most one hourglass, there are at most E hourglasses contained in the visibility graph. We find hourglasses by traversing the lower (or upper) funnel trees a second time and checking whether the reversal of the directed edge that determines the current funnel also determines a funnel.

Suppose that some function of an hourglass can be computed in constant time if the result of applying the function to the two funnels that form it is known. For example, the length of one of an hourglass's convex chains or the area of the region enclosed by an hourglass is easily computed from the chain lengths or areas of the funnels that form it. If the function is first applied to funnels during the intitial funnel tree traversals, then the function can be applied to all hourglasses during a second iteration of funnel tree traversals in $O(E)$ time. Furthermore, if the function can be evaluated on funnels in amortized constant time, the whole process can be done in $O(E)$ time.

We show below that the lengths of a funnel's upper and lower chains and the area of the region enclosed by a funnel can be calculated in amortized constant time and space during funnel tree traversals. We calculate the length of a funnel's lower (or upper) chain in constant time during a preorder traversal of the lower (or upper) funnel tree by simply adding the length of its first edge to the length of the lower (or upper) chain of its parent.

Consider the funnel shown in Figure 4. We compute the area of the region enclosed by the funnel by partitioning it into three smaller regions. The central region is a triangle, whose area is easy to calculate. We focus on the region between the lower chain and the segment $\overline{b_1 u_1}$. Call the area of this region the *lower area* of the funnel. The lower area of the funnel is just the lower area of its parent funnel in the lower funnel tree plus the area of triangle $\triangle b_1 u_1 u_2$. The upper area of the funnel is calculated analogously.

Our data structure for directed visibility graph edges allows us to store with each instance of an edge, the result of applying these length or area functions to the funnel or hourglass it determines. Thus the total amount of space needed to store the results of all of the function evaluations is $O(E)$. In the algorithms that follow we also maintain a variable corresponding to the current "best" hourglass, i.e., the one that minimizes the value of the applied function. After completing the funnel tree traversals this variable holds a representation of the optimal hourglass according to one of several criteria.

3 Hourglasses, Funnelglasses, and Bipartitions

Now we show that optimal bipartitions of points correspond to hourglasses and that optimal bipartitions of polygons correspond to hourglasses or "funnelglasses".

Start with an hourglass in the visibility graph of a set of points S and consider the two convex polygons formed by the set of edges obtained as follows: start with the edges of $conv(S)$ and delete the base edges of the hourglass and insert the edges of its two convex chains. Since the two chains (and hence the two polygons) are disjoint and the interior of the hourglass is empty, the polygons are the convex hulls of S_1 and S_2 for some bipartition $S = S_1 \cup S_2$ of S.

Now consider a bipartition $S = S_1 \cup S_2$ of a set of points S. Note that it is always beneficial to have both S_1 and S_2 nonempty in the bipartition since letting S_1 be a single vertex of the convex hull $conv(S)$ implies $\mu(S_1) = 0$ and $\mu(S_2) < \mu(S)$. This also implies that we never want $conv(S_1) \subset conv(S_2)$ or vice versa. And it's easy to see that any bipartition in which the convex hulls overlap cannot be optimal because the points in the intersection can be swapped from one set of the bipartition into the other in such a way that a new bipartition is created that reduces the perimeter and the area of each convex hull.

Suppose that we have a bipartition of S satisfying $conv(S_1) \cap conv(S_2) = \emptyset$. Then an inner common tangent to these convex polygons will be a line lying along a visibility graph edge (u, v), and each of the (closed) half-spaces determined by this line will contain one of the sets (S_i).

Since the hourglass determined by (u, v) is empty and each of the half-spaces above contains one of its convex chains, it will determine two polygons, as mentioned above, that are precisely $conv(S_1)$ and $conv(S_2)$.

We have shown that in any candidate $S = S_1 \cup S_2$ for an optimal bipartition of a point set S, S_1 and S_2 must have disjoint convex hulls. And any bipartition for which the convex hulls are disjoint can be obtained from an hourglass. Furthermore, every hourglass determines such a partition. Therefore, the problem of finding an optimal bipartition of a set of points is equivalent (via a linear-time reduction) to that of finding an optimal hourglass.

We also discuss the problem of finding an optimal bipartition of a set of polygons. Intuitively, when we were bipartitioning point sets, we were using two disjoint closed fences to contain the points, and we were minimizing the perimeters or areas of the fenced-in regions. We generalize this idea to the polygon case by considering only those bipartitions $S = S_1 \cup S_2$ of a set of polygons S that have the property that the shortest closed fence containing the obstacles in S_1 and the shortest closed fence containing the obstacles in S_2 form simple polygons with disjoint interiors. Recall that when partitioning points, if a point p of S_2 was contained in the convex hull of S_1 (or vice versa), it was better to delete p from S_2 and insert it in S_1. This led us to conclude that the fences needed to be disjoint. With polygons, however, a polygon P belonging to S_2 may be partially contained in the convex hull of S_1 in such a way that changing its membership increases the lengths of the fences. This means that optimal fences need not be convex – they may bulge in around polygons not contained in the fence – and they need not be disjoint – they may coincide along some portion of their length.

Now we define the polygon partitioning problem more formally. We say that a bipartition $S = S_1 \cup S_2$ of a set of polygons S is valid if S_1 and S_2 can be separated by a simple path between distinct edges of the convex hull of S (note that when partitioning points, valid partitions were separable by a single line). Given a valid bipartition, the relative convex hull $rconv(S_1)$ with respect to S_2 is the minimum perimeter polygon that contains every polygon in S_1 but contains no interior point of a polygon in S_2. Figure 5 shows a valid bipartition and Figure 6 shows the resulting relative convex hulls.

If the relative convex hulls $rconv(S_1)$ and $rconv(S_2)$ of a valid bipartition are disjoint, then they are also convex and can be obtained from an hourglass as above. In general, however, they will share a sequence of edges, i.e., they will coincide along a visible chain. In this case, they form what we call a funnelglass. A funnelglass is a pair of funnels and a shortest path between their apices that together satisfy the following conditions: the funnels are disjoint or they intersect only in having a common apex, and the extension of each funnel chain along the inter-apex path is a locally shortest or "taut string" path. The funnel chains and bases and the shortest path form a weakly simple polygon (one formed by a path with repeated vertices or edges but no self-crossings) as shown in Figure 6.

The correspondence between funnelglasses and bipartitions with "touching" relative convex hulls is as follows. Start with a funnelglass in the visibility graph of a set of polygons S and consider the two "touching" polygons formed by the following set of edges: take the edges of $conv(S)$ and delete the base edges of the funnelglass and insert the edges of its convex chains and its inter-apex path. Since the interior of the funnelglass is empty and since the inter-apex path can be extended through the interior of each funnel to a point on the funnel's base, the funnelglass separates the set of polygons and thus corresponds to a valid bipartition $S = A_1 \cup S_2$ for some S_1 and S_2. Furthermore, the "taut-string" property ensures that the polygons formed from it have minimal perimeter, i.e., they are precisely $rconv(S_1)$ and $rconv(S_2)$.

Conversely, if we start with the optimal valid bipartition $S = S_1 \cup S_2$, and $rconv(S_1)$ and $rconv(S_2)$ coincide along a visible chain, then the edge set $E = conv(S) \triangle (rconv(S_1) \cup rconv(S_2))$

forms a funnelglass as follows. The intersection $rconv(S_1) \cap rconv(S_2)$ must be a shortest path between two vertices u and v by optimality of the bipartition, and also the "taut-string" property must hold. And where the edges diverge at u and v they form funnels since relative convex hulls are convex where they don't "touch".

We have shown that an hourglass or a funnelglass determines a valid bipartition of a set of polygons. Furthermore, the optimal bipartition can always be obtained from an hourglass or a funnelglass. Therefore, the problem of finding an optimal bipartition is equivalent (via a linear-time reduction) to that of finding an optimal hourglass or funnelglass.

4 Partitioning Points

Now we consider the case in which S is a set of points in the plane, and the "sizes" $\mu(S_i)$ of the subsets are taken to be the perimeters or areas of their convex hulls. We seek a bipartition that minimizes the sum $\mu(S_1) + \mu(S_2)$ or the maximum $\max\{\mu(S_1), \mu(S_2)\}$. This gives us the following set of four problems: 1. the min-sum perimeter problem; 2. the min-max perimeter problem; 3. the min-sum area problem; and the min-max area problem.

Note that it is always beneficial to have both S_1 and S_2 nonempty in the bipartition since letting S_1 be a single vertex of the convex hull $conv(S)$ implies $\mu(S_1) = 0$ and $\mu(S_2) < \mu(S)$. We will show later that this is not always the case when partitioning polygons.

Theorem 1 *Given a set S of n points in the plane, a bipartition of S, $S = S_1 \cup S_2$, that solves any of problems 1 through 4 above can be found in time $O(n^3)$ using $O(n)$ space. Alternatively, problems 1 through 3 above can be solved in $O(n^2)$ time using $O(n^2)$ space.*

Proof: The algorithm for the first claim is based on the observation that there are only $O(n^2)$ distinct ways to partition n points, after which the convex hulls $conv(S_i)$ can be examined in a straightforward manner:

(0). Sort the points S by x-coordinate. Compute the convex hull of S.

(1). Consider each pair of points $p, q \in S$ ($p \neq q$) in turn. Each pair (p, q) such that $\overline{pq}$ does not lie on the boundary of the convex hull of S defines a line that partitions S into two nonempty subsets: Use the line $L_{p,q}$ obtained by taking the line through p and q and rotating it clockwise by an infinitesimal amount about the midpoint of the segment $\overline{pq}$.

    (a). For a given pair (p, q), we march through the list of points S (in x order), marking each point as being on the left or the right side of the oriented line through $L_{p,q}$. This yields each of the two sets S_1 and S_2 (for this pair (p, q)) in sorted x order.

    (b). In time $O(n)$, we can then find the convex hulls of S_1 and S_2.

    (c). Once we have the convex hulls of each set S_1 and S_2, we can easily compute the sum $\mu(S_1) + \mu(S_2)$ or the maximum $\max\{\mu(S_1), \mu(S_2)\}$ in linear time, where $\mu(S_i)$ denotes either perimeter or area (or any other function on a convex polygon that can be computed in linear time).

By keeping track of the best partitioning so far, we will have an optimal partitioning of S by the time we have examined all pairs (p, q) (since an optimal partitioning must correspond to *some* pair (p, q)).

The above algorithm clearly requires linear time per pair (p, q), so $O(n^3)$ time overall. The space requirement is only linear.

To prove the second claim of the theorem, we rely on methods developed in Section 2. Each pair of points (p, q) determines an hourglass whose bases are edges on the convex hull of S. The two convex chains of the hourglass together with the edges remaining in the convex hull after deleting the two base edges form two convex polygons that correspond to some bipartition of S. Conversely, every bipartition of S corresponds to an hourglass.

The "cost" of a bipartition is easily calculated from the chain lengths of its hourglass in constant time when minimizing either the total or maximum perimeter of $conv(S_1)$ and $conv(S_2)$. Similarly, the total area of the two convex hulls arising from a bipartition is just the area of the entire convex hull minus the area of the hourglass corresponding to the partition. By generating all hourglasses and comparing each newly considered hourglass to the best hourglass seen so far, we obtain the optimal hourglass and hence the optimal bipartition for each of problems 1, 2, and 3 above.

We compute the cost of all $O(n^2)$ hourglasses in time $O(n^2)$, as in Section 2. The space required by this algorithm is quadratic, since we use the (complete) visibility graph of the point set S. ∎

5 Partitioning Polygons

Now we consider the case in which S is a set of (pairwise-disjoint) simple polygons with a total of n vertices. The "size" of S_1 (S_2) will now be the perimeter of the *relative* convex hull, $rconv(S_1)$ $(rconv(S_2))$, of S_1 (S_2) with respect to the set S_2 (S_1). Again we seek a bipartition that minimizes the sum $\mu(S_1) + \mu(S_2)$ or the maximum $\max\{\mu(S_1), \mu(S_2)\}$, giving us once more the following set of four problems: 1. the min-sum perimeter problem; 2. the min-max perimeter problem; 3. the min-sum area problem; and the min-max area problem.

Note that, in contrast to the case in which S is a set of points, it is *not* always beneficial to have both S_1 and S_2 nonempty here in problems 1 and 2: it may be that every nontrivial bipartition of S results in the perimeters of $rconv(S_1)$ and $rconv(S_2)$ each being greater than the perimeter of $conv(S)$. Therefore, for these problems we distinguish between the case in which we want to find only beneficial bipartitions and the case in which we are forced to find nontrivial bipartitions even if every one of them is non-beneficial. This distinction is not necessary in problems 3 and 4 because the best nontrivial bipartition for these problems will always be beneficial.

Theorem 2 *Given a set S of (pairwise-disjoint) simple polygons with a total of n vertices, one can find a beneficial bipartition of S, $S = S_1 \cup S_2$, that solves the min-sum perimeter problem (or report that none exists) in time $O(En)$, using $O(n)$ space. Alternatively, one can solve this problem in time $O(E + n \log n)$ using $O(E)$ space.*

Proof: The algorithm for the first claim is straightforward:

(0). Triangulate the polygon with holes formed by the convex hull of S and the polygons in S.

(1). Generate the edges of the visibility graph using only $O(n)$ working space and $O(E \log n)$ time [13]. As each edge (u, v) is found, do the following calculations:

(a). Extend edge (u, v) at both ends until a polygon or the convex hull of S is hit.

(b). If the convex hull was hit on each side, i.e., (u, v) determines an hourglass, then compute the lengths of the hourglass's chains by finding the two funnels that form the hourglass as follows. The funnel with apex u is found by propagating the extension of (u, v) through adjacent cells of the triangulation until the triangle containing the convex hull edge hit by the extension is reached. The shortest paths from u to the endpoints of the convex hull edge that lie within the simple polygon formed from the pierced triangles is found in $O(n)$ time, and these paths are the upper and lower chains of the funnel [9]. The funnel with apex v is found similarly. Calculate the "cost" of this hourglass, compare it to the current best one, and update if neccessary by storing the edge (u, v) and its cost.

The above algorithm requires linear time per edge (u, v), so $O(En)$ time overall. The space requirement is only linear.

To prove the second claim of the theorem, we again compute hourglasses, but we compute the entire visibility graph first in time $O(E + n \log n)$ using $O(E)$ space [6]. Then we traverse funnel trees as in Section 2 and find the best hourglass in $O(E)$ time. ∎

To handle the min-max perimeter problem, it is no longer sufficient to consider only bipartitions corresponding to hourglasses; we must also consider funnelglasses. A *funnelglass* is a pair of funnels connected by a shortest path between the two apices of the funnels. Funnelglasses correspond to partitions in which the two relative convex hulls coincide along some visible chain.

For each funnel in a funnelglass, the total length of the two chains of the funnel must be greater than the length of its base; that's why we didn't need to consider funnelglasses when we were only interested in beneficial solutions to the min-sum perimeter problem. But for the min-max perimeter problem, a funnelglass may correspond to a beneficial partition, so it's no easier to find beneficial partitions than it is to find forced ones. We find beneficial partitions by finding the best nontrivial partition first. If this funnelglass has a lower cost then that of the entire convex hull, it is the optimal beneficial solution; otherwise, no beneficial solution exists.

Theorem 3 *Given a set S of (pairwise-disjoint) simple polygons with a total of n vertices, one can find a nontrivial bipartition of S, $S = S_1 \cup S_2$, that solves the min-max perimeter problem in time $O(E^2n)$, using $O(n)$ space. Alternatively, one can solve this problem in time $O(E^2 + n^2 \log n)$ time, using $O(E)$ space.*

Proof: The algorithm for the first claim appears below.

Generate the edges of the visibility graph using only $O(n)$ working space and $O(E \log n)$ time [13]. As each edge (u, v) is found, do the following:

(0). Extend edge (u, v) from u through v until it hits a polygon or an edge of the convex hull of S.

(1). If an edge of the convex hull was hit, i.e., if (u, v) determines a funnel, then

    (a). Determine the chain lengths of the funnel determined by (u, v) in linear time and space as in the proof of Theorem 2.

    (b). Build a shortest path map rooted at u in $O(n^2)$ time and $O(n)$ space [14].

(c). Generate the edges of the visibility graph using only $O(n)$ working space and $O(E \log n)$ time [13] while holding edge (u, v) from the outer loop fixed. As each edge (u', v') is found, if this edge determines a funnel, determine the chain lengths of the funnel, and calculate the cost of the funnelglass formed by this funnel, the funnel determined by (u, v), and the path from u' to u in the shortest path map. Compare the current funnelglass to the stored representation of the best funnelglass found so far and update if necessary.

The time needed by the above algorithm is dominated by the time needed to find all funnelglasses using one fixed funnel. This is an $O(En)$ time computation that needs to be done $O(E)$ times. The space requirement is only linear.

To prove the second claim of the theorem, we again find funnelglasses, but we compute the entire visibility graph first in time $O(E + n \log n)$ using $O(E)$ space [6]. Then we find all $O(E)$ funnels by traversing funnel trees as in Section 2. While doing this we create a list of all the funnels and a list at each vertex of the funnels having that vertex as their apex. Then for each vertex u we do the following:

(0). Find the shortest path tree rooted at u using Dijkstra's algorithm ($O(E + n \log n)$).

(1). For each vertex v and for each pair of funnels, one with apex u and one with apex v, compute the cost of the funnelglass formed by this pair of funnels and the shortest path from u to v. Update the representation of the best funnelglass found so far if necessary.

This algorithm requires time $O(En + n^2 \log n)$ for the n runs of Dijkstra's algorithm and $O(E^2)$ time to examine all pairs of funnels; thus the total running time is $O(E^2 + n^2 \log n)$ and the total space required for the visibility graph and funnel representations is $O(E)$. ∎

When we are forced to consider non-beneficial partitions in the min-sum perimeter problem, we have to look for funnelglasses as in the min-max case. But there is a difference: in the min-max case, any of the $O(n)$ funnels on each vertex could be part of the optimal funnelglass, but in the min-sum case there is a unique "best" funnel on each apex (taking the cost of a funnel to be the sum of its chain lengths minus the length of its base). This lets us find an optimal funnelglass more efficiently.

Theorem 4 *Given a set S of (pairwise-disjoint) simple polygons with a total of n vertices, one can find a nontrivial bipartition of S, $S = S_1 \cup S_2$, that solves the min-sum perimeter problem in time $O(n^3)$, using $O(n)$ space. Alternatively, one can solve this problem in time $O(En + n^2 \log n)$ time, using $O(E)$ space.*

Proof: The algorithm for the first claim is as follows:

(0). Find the best hourglass (if one exists) as in the proof of Theorem 2.

(1). Find all funnels in $O(En)$ time and $O(n)$ space as in the proof of Theorem 2 maintaining a representation of the best funnel found at each vertex together with its cost.

(2). Now for each vertex u build the shortest path map rooted at u and as each vertex v is labeled with its distance from u, calculate the cost of the funnelglass formed from the best funnel on u and the best funnel on v in constant time, updating a representation of the best funnelglass found so far if necessary.

(3). Output the bipartition corresponding to the hourglass or funnelglass with minimum cost.

The running time is dominated by the n shortest path map constructions which take a total of $O(n^3)$ time. The total space requirement is linear.

The algorithm for the second claim appears below:

(0). Calculate the visibility graph in $O(E + n \log n)$ time and $O(E)$ space [6].

(1). Generate all funnels and hourglasses using funnel tree traversals and store with each vertex of the visibility graph a representation of the best and the second best funnel having that vertex as apex. Also store a representation of the best hourglass found.

(2). For each edge e on the convex hull of S do the following:

(a). Make a copy VG_e of the visibility graph.

(b). Form an augmented visibility graph VG'_e as follows:
- Create an additional "funnel" node fn for each funnel f that was determined to be one of the "top two" funnels on its apex.
- Connect each funnel-node fn to the node v corresponding to the apex of funnel f. Let the length of edge (fn, v) be $c(f)/2$ where $c(f)$, the cost of a funnel is the sum of its chain lengths minus the length of its base.
- Create a "source" node s, linking it with an edge of length 0 to every funnel-node corresponding to a funnel with base e. Similarly, create and link a "sink" node t to every funnel-node not already linked to s.

(c). Find a shortest path from s to t in VG'_e. This determines the optimal funnelglass with one base on e.

(d). If necessary, update the stored representation of the best funnelglass found so far.

(3). Output the bipartition corresponding to the optimal funnelglass or hourglass. ∎

Funnelglasses are also used to solve the min-sum area problem. In this problem, the length of the path connecting the two apices is irrelevant; the cost of the funnelglass is just the area of the entire convex hull minus the sum of the two funnel areas. But there is a difficulty here that doesn't arise in the other problems: the funnelglass of minimal cost may not correspond to a feasible bipartition because its funnels may overlap.

Theorem 5 *Given a set S of (pairwise-disjoint) simple polygons with a total of n vertices, one can find a bipartition of S, $S = S_1 \cup S_2$, that solves the min-sum area problem in time $O(E^2 n^2)$, using $O(n)$ space.*

The algorithm for this claim is based on the algorithm for the first claim in Theorem 2. We again generate all pairs of funnels and calculate the cost of each pair, but this time a funnel's cost (or benefit) is the area of the region it encloses. Since a funnel can be triangulated in linear time, we can calculate its area in linear time and check that the two funnels in a pair don't intersect in linear time [3]. Therefore this algorithm takes $O(E^2 n)$ time and requires only linear space. ∎

Finally, we consider the min-max area problem. We show that this problem is NP-Hard by a polynomial-time reduction from the following NP-Complete problem [5]:

Partition: Given a set T of n objects with positive integer values $v(t)$, is there a partition $T = T_1 \cup T_2$ such that $\sum_{t \in T_1} v(t) = \sum_{t \in T_2} v(t)$.

Theorem 6 *The min-max area problem is NP-Hard.*

Proof: Given an instance of the partition problem, we generate in linear time a "stack" of rectangles as shown in Figure 7. Rectangle r_t corresponds to object $t \in T$ and has width 1 and length $v(t)$. The vertical separation between rectangles is infinitesimally small. Every bipartition $S = S_1 \cup S_2$ of this set S of rectangles can be separated by a simple path, and, for each subset in a bipartition, the area of its relative convex hull is just the total area of its rectangles. Therefore, the optimal solution of the min-max area problem would be a bipartition $S = S_1 \cup S_2$ satisfying

$$\sum_{r \in S_1} Area(r) = \sum_{r \in S_2} Area(r) = 1/2 \sum_{t \in T} v(t)$$

if and only if there exists a bipartition $T = T_1 \cup T_2$ of the objects satisfying $\sum_{t \in T_1} v(t) = \sum_{t \in T_2} v(t)$. And such a bipartition of rectangles can be transformed into the desired partition of objects in linear time. ∎

6 Conclusion

Our results are summarized in Table 1, where we list the time and space complexity of our algorithms for each problem in terms of the number of edges E in the visibility graph and the number n of points or polygon vertices in the partitioned set.

Several questions about these problems remain open. Perhaps the running times given for these algorithms can be improved. For several of these problems our best algorithm uses only linear space. Is there a faster algorithm using a quadratic amount of space? Also it might be interesting to find approximate solutions to the NP-Hard min-max area problem for polygons.

References

[1] T. Asano, B. Bhattacharya, M. Keil, and F. Yao. Clustering algorithms based on minimum and maximum spanning trees. In *Proceedings of the Fourth Annual ACM Symposium on Computational Geometry*, pages 252–257, 1988.

[2] D. Avis. Diameter partitioning. *Discrete and Computional Geometry*, 1(3):265–276, 1986.

[3] B. Chazelle. Triangulating a simple polygon in linear time. Technical Report CS-TR-264-90, Princeton University, 1990.

[4] M. R. Garey and D. S. Johnson. *Computers and Intractability: A Guide to the Theory of NP-Completeness*. W. H. Freeman and Company, New York, 1979.

[5] S. K. Ghosh and D. M. Mount. An output sensitive algorithm for computing visibility graphs. In *Proceedings of the 28th Annual IEEE Symposium on the Foundations of Computer Science*, pages 11–19, 1987. To appear: *SIAM J. Comp.*

	Points	Polygons	
		Beneficial Partitions	Forced Partitions
Min-Sum Perimeter	$O(n^2) / O(n^2)$ or $O(n^3) / O(n)$	$O(E + n \log n) / O(E)$ or $O(En) / O(n)$	$O(En + n^2 \log n) / O(E)$ or $O(n^3) / O(n)$
Min-Max Perimeter	$O(n^2) / O(n^2)$ or $O(n^3) / O(n)$	$O(E^2 + n^2 \log n) / O(E)$ or $O(E^2 n) / O(n)$	$O(E^2 + n^2 \log n) / O(E)$ or $O(E^2 n) / O(n)$
Min-Sum Area	$O(n^2) / O(n^2)$ or $O(n^3) / O(n)$	$O(E^2 n) / O(n)$	$O(E^2 n) / O(n)$
Min-Max Area	$O(n^3) / O(n)$	NP-Hard	NP-Hard

Table 1: The time and space complexity of our algorithms.

[6] J. Hershberger and S. Suri. Finding tailored partitions. In *Proceedings of the 5th Annual ACM Symposium on Computational Geometry*, pages 255–265, 1989.

[7] D. T. Lee. *Proximity and Reachability in the Plane*. PhD thesis, University of Illinois, 1978.

[8] D. T. Lee and F. P. Preparata. Euclidean shortest paths in the presence of rectilinear barriers. *Networks*, 14:393–410, 1984.

[9] T. Lozano-Pérez and M. A. Wesley. An algorithm for planning collision-free paths among polyhedral obstacles. *Communications of the ACM*, 22:560–570, 1979.

[10] J. S. B. Mitchell. A new algorithm for shortest paths among obstacles in the plane. *Annals of Mathematics and Artificial Intelligence*, 3:83–106, 1991.

[11] C. Monma and S. Suri. Partitioning points and graphs to minimize the maximum or the sum of diameters. In *Proceedings of the Sixth International Conference on the Theory and Applications of Graphs*, 1988.

[12] M. H. Overmars and E. Welzl. New methods for computing visibility graphs. In *Proceedings of the Fourth Annual ACM Symposium on Computional Geometry*, pages 164–171, 1988.

[13] J. H. Reif and J. A. Storer. Shortest paths in euclidean spaces with polyhedral obstacles. Technical Report CS-85-121, Brandeis University, Waltham, Mass., 1985.

[14] M. Sharir and A. Schorr. On shortest paths in polyhedral spaces. *SIAM J. Comp.*, 15:193–215, 1986.

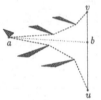

Figure 1: A funnel with base (u, v) and apex a.

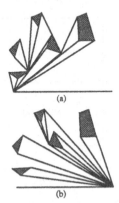

Figure 2: A lower funnel tree (a), and an upper funnel tree (b).

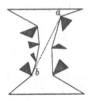

Figure 3: An hourglass formed from two funnels.

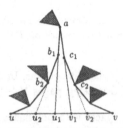

Figure 4: Partitioning the region enclosed by a funnel.

Figure 5: A simple path determines a valid bipartition of a set of simple polygons.

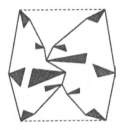

Figure 6: The relative convex hulls corresponding to the bipartition of Figure 5 and their corresponding funnelglass.

Figure 7: A "stack" of rectangles corresponding to a set of objects.

IMMOBILIZING A POLYTOPE

Jurek Czyzowicz[1], Ivan Stojmenovic[2], and Jorge Urrutia[2]

[1]Département d'Informatique, Université du Québec à Hull, Hull, Québec, Canada
[2]Computer Science Department, University of Ottawa, Ottawa, Ontario, Canada

Abstract

We say that a polygon P is immobilized by a set of points I on its boundary if any rigid motion of P in the plane causes at least one point of I to penetrate the interior of P. Three immobilization points are always sufficient for a polygon with vertices in general positions, but four points are necessary for some polygons with parallel edges. An O(n log n) algorithm that finds a set of 3 points that immobilize a given polygon with vertices in general positions is suggested. The algorithm becomes linear for convex polygons. Some results are generalized for d-dimensional polytopes, where 2d points are always sufficient and sometimes necessary to immobilize. When the polytope has vertices in general position d+1 points are sufficient to immobilize.

1. Introduction

The set of points I is said to immobilize a planar shape P if any rigid motion of P in the plane forces at least one point of I to penetrate the interior of P. By *shape* we mean a set bounded by a Jordan curve. Clearly, any minimal I contains only points belonging to the boundary of P. The disk is excluded from consideration since any number of points on its boundary leave it free to rotate.

Problems of immobilization of planar shapes were introduced by W. Kuperberg [K] and later reported in [O] where a number of open questions were presented:
- Do four points always suffice to immobilize any shape? Any convex shape?
- Find all the classes of convex shapes for which three points do not suffice.
- Do three points suffice for all smooth convex shapes?
- Design an algorithm for finding a set of immobilizing points for a given polygon.
- Extend to three (and higher) dimensions.

In this paper we concentrate on the problem of immobilizing a polytope. The results are applicable in robotics, especially in the problem of grasping. [MNP] and [MSS] study a related problem of closure grasp for piecewise smooth objects, i.e. ability to respond to any external force or torque by applying appropriate forces at the grasp fingers. [MP] found the position of the fingers so that the forces needed to balance the weight of the polygon are minimal. The ideas of using the inscribed circle and Voronoi diagram, exploited in our paper, were first used in [BFG], while the idea of normals to the boundary of a triangle meeting at a point appears in [MP], both in the context of an equilibrium grip.

[1,2] This research is partially supported by NSERC

A rigid motion of a set S on the plane is a mapping M from the set t×S (t represents time) to the plane, continuous with respect to its first coordinate, such that for every pair of points u, v∈S the distance between their images remains constant for all t and M(0, u)=u for every element of S.

In Section 2 we study first immobilization of a triangle. In sections 3 and 4 we extend considerations to immobilizing convex and simple polygons in the plane. It may be proved that any polygon in the plane can always be immobilized using four points. We describe a large class of polygons that require four points to immobilize. This class includes polygons other than parallelograms but any such polygon must always contain two parallel edges. For a given convex polygon and three given points on its boundary we have a criterion to check whether the points immobilize the polygon. When the points are not located at the vertices of the polygon we have a similar criterion for the class of simple polygons. An O(n log n) algorithm is obtained to find a set of three points that immobilize a polygon with vertices in general positions. In the case of a convex polygon the algorithm works in O(n) time.

In Section 6 some results are extended to higher dimensions. We prove that any d-dimensional polytope P can be immobilized by a set containing at most 2d points. Moreover, if some set of d+1 vectors normal to faces of P is linearly independent (as in the case of a polytope having vertices in general position), d+1 points suffice to immobilize P.

Some results of this paper are not obvious intuitively. Some other statements seem intuitively clear but often require involved and subtle proofs.

2. Immobilizing a triangle

In this section we study the problem of immobilizing triangles. In order to describe all possible sets containing three points that immobilize a given triangle, we first prove a lemma that is valid for any shape and will be also used later to obtain other results.

Suppose we are given a shape P which has moved from an initial position P to a new position P'. It is easy to show that the movement can be rerouted using the following two actions (see Fig. 1):

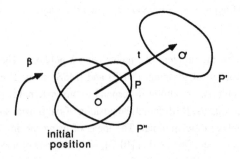

Figure 1

-rotation of P around any point O in the plane for an appropriate angle ß, to obtain the interim position P",

-translation of P" to the destination position P' for a vector t (t is equal to vector OO', where O' is the new position of O).

Since the movement is continuous, we may assume that we can always obtain the position P' such that both ß and |t| are less than an arbitrarily small value µ>0.

We will use the following method to build a set of points I that immobilize P. Each point W from I belonging to the boundary of P may restrict the movement of P to avoid W penetratating P. If P' is the position of P after an arbitrarily small movement, the movement can be rerouted as indicated and W should not become an interior point of P'. This will be used to prove the following lemma.

Lemma 2.1. Let OUV be a circle sector centered at O and determined by arc UV, and let W be an interior point of the arc UV that is also a point on the boundary of P (see Fig. 2). If the circle sector OUV lies entirely inside given shape P then any movement of P that brings the point O inside the circle w centered at W with radius WO implies that W must penetrate P.

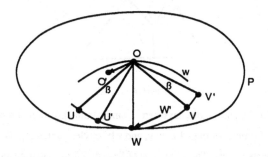

Figure 2

Proof. Suppose that O has moved to a point O' where O' is in the neighborhood of O and inside w. Reroute the movement of P as indicated above, with O being the center of rotation. The rotation for a small angle ß moves OUV to a new position OU'V' (see Fig. 2) such that W is still an interior point of the new arc U'V' and OU'V (or alternatively OUV') is still entirely inside the shape P. Translation moves O to O'. However, it is easy to note that the same translation moves a point W' from the interior of the circular sector OU'V' to the point W. This results in W penetrating P.

Theorem 2.1. Three points X, Y and Z immobilize a triangle T with vertices A, B and C if and only if the three orthogonal lines to the boundary of T at the points X, Y and Z are concurrent.

Proof: We prove first the necessity of the condition. Clearly each of X, Y and Z must lie on different sides of T. Let X, Y and Z belong to sides BC, AC, and AB, respectively. Suppose that the three orthogonals at the points X, Y and Z do not meet at a single point. Let O be a point in the interior of the triangle determined by these orthogonals. Then the three angles OXB, OYC and OZA are all acute (or all obtuse) (see Fig. 3). Therefore the triangle T may be rotated counterclockwise (or clockwise) around

O by an $\varepsilon > 0$ angle and the points X, Y and Z will remain outside the interior of T. ◆

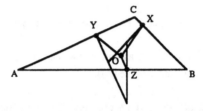

Figure 3

Suppose now that the three orthogonals intersect at a point O. To prove that X, Y and Z immobilize T, we show that any movement of T will force one of X, Y or Z to penetrate the interior of the image of T. We first consider the case when O is *inside* T (see Fig. 4).

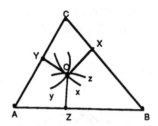

Figure 4

We show that the point O cannot move anywhere from its initial position. Suppose that O moves to a point O' within an $\varepsilon > 0$ distance from O. Let x, y, and z be circles containing O on their boundaries and centered at X, Y, and Z, respectively. The conditions of Lemma 2.1 are satisfied and O cannot move inside any of the circles x, y, and z. But, it is easy to note that any point from the neighborhood of O is inside at least one of circles x, y, or z; thus if O'≠O then at least one of the points X, Y, or Z will penetrate T. Therefore O=O' and the only allowable movement for T is a rotation around O. This is however impossible because in this case some interior points of T will move to X, Y and Z (causing the penetration of X, Y, and Z).

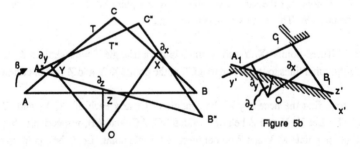

Figure 5a

Figure 5b

Consider now the case when O is outside T. Suppose without loss of generality, that the straight line passing through A and B separates O from C (see Fig. 5a). Then OZ is the shortest among the segments OX, OY, and OZ, and OZ is completely outside T while OY and OX intersect T. Suppose that T can move to a new position T'. We reroute the movement by a rotation around O by an appropriate angle ß and translation by corresponding vector t. Let T"=A"B"C" be the rotated position of T. It is easy to show that X and Y are interior points of T" while Z is the exterior one. Therefore the second step, translation of T" to destination T', should be chosen such that X and Y "escape" from T" while Z stays outside T". Let ∂_x, ∂_y, and ∂_z be the distances of X, Y, and Z from B"C", A"C", and A"B", respectively. Then $\partial_z < \partial_x$ and $\partial_z < \partial_y$ since OZ is the shortest among OZ, OY, and OX ($\varphi_x = |OX|(1-\cos \beta)$), and similarly for other two distances). Let x', y' and z' be straight lines parallel to B"C", A"C", and A"B", and with distances ∂_x, ∂_y, and ∂_z from O, respectively, such that O does not lie between any two corresponding parallel lines (see Fig. 5b). Observe that triangle $A_1B_1C_1$ in Fig. 5b is similar to ABC. X and Y can "escape" from T" only if the translation vector brings O to a point O' lying in both half-planes determined by x' and y' that do not contain O. To keep Z outside T", point O' must be located within the half-plane determined by z' containing O. However, as $\partial_z < \partial_x$ and $\partial_z < \partial_y$ the three mentioned half-planes have empty intersection.

The above argument holds also for the case of O on the boundary of T.♦

Corollary 2.1. Given two points X and Y on two different sides of T, it might not be possible to find a third point Z on the remaining side such that X, Y and Z immobilize T (see Fig. 6). This happens only for obtuse T.

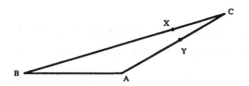

Figure 6

3. Immobilizing a convex polygon

We are now ready to give necessary and sufficient conditions under which three given points immobilize the convex polygon.

Given a convex polygon P we say that three of its sides x, y, and z *enclose* P if the triangle T(x,y,z) determined by the three lines containing them contains P.

Theorem 3.1. A convex polygon P can be immobilized by three points X, Y and Z if and only if:
   a) each of them belongs to the interior of a different side, say x, y and z of P respectively such that x, y and z enclose P, and
   b) the orthogonals to x, y and z at the points X, Y and Z respectively meet at a common point.

219

Proof. It is clear that x, y and z must enclose P, otherwise we can translate it away (see Fig. 7(a)). We prove now that each of X, Y and Z must belong to the interior of an edge of P. Suppose that one of them, say X, is a vertex of P (see Fig. 7(b)).

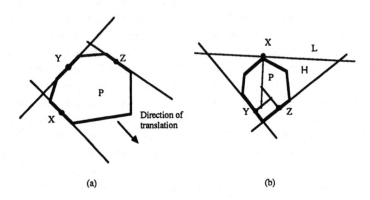

Figure 7

Then we can take a triangle H that encloses P and is formed by the two sides of P containing Y and Z and any line L that intersects P exactly at X. The line L can be chosen in such a way that the orthogonals at X, Y and Z do not meet. Then X, Y and Z do not immobilize H and therefore do not immobilize P.

To prove the sufficiency of our conditions, observe that by Theorem 2.1, the points X, Y, and Z immobilize triangle T(x,y,z) and therefore they also immobilize P.♦

4. Immobilizing a simple polygon

When the three immobilizing points are known not to be located at the vertices of the polygon the last result may be generalized for all simple polygons. First we generalize the definition of enclosing sides to any simple polygon P in the following way. Assign to each edge x of P the halfplane containing x on its boundary and containing the points from the interior of P that are in the proximity of x. The sides x, y, and z of P *enclose* P if the intersection of three halfplanes assigned to x, y, and z is nonempty and bounded (i.e. a triangle).

Generalizing the argument presented in Fig. 3 we will prove first the following

Lemma 4.1. If the three points X, Y and Z different from vertices of a given polygon P immobilize P, the orthogonals at X, Y and Z to its respective sides x, y and z must meet at a common point.

Proof: Suppose that the orthogonals to x, y, and z at X, Y, and Z do not meet at a common point. Obviously no two of these orthogonals may be parallel otherwise the translation along this parallel direction would have been possible. Take the orthogonals to X and Y. They partition the plane into four regions. One of them, R_1, is such that for any point O located in R_1 a small clockwise rotation of P around O would leave both X and Y outside P (see Fig. 8).

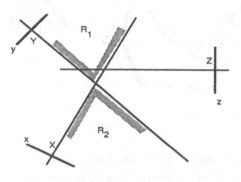

Figure 8

Similarly, the opposite region R_2 allows centers of the counterclockwise rotation. Clearly the orthogonal to Z must intersect one of these regions (R_1 in Fig. 8) partitioning it into two parts, one of which allowing a small rotation without any of X, Y, and Z penetrating P. ◆

Theorem 4.1. A polygon P can be immobilized by three points X, Y and Z different from vertices of P if and only if:

a) the orthogonals at the points X, Y and Z to its respective sides x, y and z meet at a common point, and

b) x, y and z enclose P.

Proof: Lemma 4.1 proves the necessity of the first condition. Suppose that x, y, and z do not enclose P. This may happen for one of two reasons: either the intersection of the halfplanes assigned to x, y, and z forms an unbounded region and then, as in the convex case (see Fig. 7a), P may be translated away, or this intersection is empty. In the latter case, if the orthogonals meet inside T(x,y,z) P may be rotated around the point of their intersection. In the remaining nontrivial case the orthogonals meet at a point O that is outside T(x,y,z) (as indicated on Fig. 9a). A repeated analysis as performed in the proof of Theorem 2.1 (refer to Fig. 5b) leads now to a different conclusion: any translation vector OU, where U is in the interior of the triangle T(x',y',z') (see Fig. 9b) sets the points X, Y, and Z outside P' (the new position of P). It is easy to see that this may be done for any small value of the rotation $\beta > 0$ and so that the corresponding translation $t(\beta)$ is a continuous function of β. As a consequence there exists a continuous motion that is the composition of the rotation around O and the translation by vector OU that does not cause any of X, Y, or Z to penetrate P. An interested reader may verify that example of such motion is when P moves ("slides") touching two of the three points X, Y and Z.

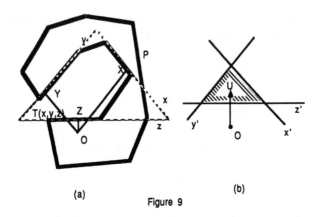

(a)　　　　　　　　　(b)

Figure 9

The sufficiency of both conditions can be proved along the similar lines as in Theorem 3.1. Triangle T from Fig. 5a stands now for the intersection of halfplanes assigned to x, y, and z. Existence of the nonempty intersection T of these halfplanes implies empty intersection of their complements (using similarity of triangles $A_1B_1C_1$ and ABC in Fig. 5b and 5a). This is valid for any location of the point of intersection O of three orthogonals in plane (Fig. 4 and Fig. 5b show two out of three cases of the location of O in the arrangement of three lines). ◆

Obviously, any convex polygon P needs at least three points to immobilize it. We will see soon that three points will suffice, also for simple polygon P, when there is no two parallel sides in P. Before that we have to turn our attention to polygons which may be immobilized using two points only. Clearly, at least one of these two points will have to be located at the reflex vertex of P.

Theorem 4.2. Two points X and Y immobilize a simple polygon P if and only if segment XY forms an angle at least $\pi/2$ with four adjacent sides of P and if two of these four sides are parallel they must lie on opposite sides of XY.

Proof. To prove the necessity observe that if one of the four angles (say XYZ on Fig. 10a) is $<\pi/2$ then P may rotate around X. If the two sides on the same side of XY are parallel (as XT and YZ in Fig. 10b) then P may be translated perpendicularly to XY.

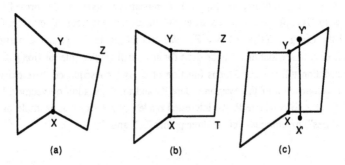

(a)　　　　　　(b)　　　　　　(c)

Figure 10

Sufficiency follows from the fact that for any pair of points X', Y' not in the interior of P, X' from the neighborhood of X and Y' from the neighborhood of Y, we have |X'Y'|>|XY| (see Fig. 10c). ◆

Theorem 4.3. Any polygon containing no parallel sides can be immobilized using three points.

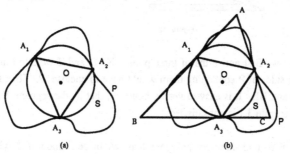

(a)   (b)

Figure 11

Proof. Let S be the largest circle contained inside given polygon P and let O be the center of S. If among the points at which S touches P we cannot chose three points A_1, A_2, and A_3 such that O is in the interior of the triangle $A_1A_2A_3$ then S must touch P in two endpoints of a diameter of S. As P has no parallel sides it is easy to see by Theorem 4.2 that these two points immobilize P. In the other case S touches P in three points A_1, A_2, and A_3 such that O is an interior point of triangle $A_1A_2A_3$. We call such a circle S a 3-type circle (see Fig. 11a). The conditions of Lemma 2.1 are now satisfied; the point O is the intersection of three orthogonals to P at points A_1, A_2, and A_3. Therefore we can repeat the proof given in Theorem 2.1. The triangle ABC is determined by the tangent lines to S at the points A_1, A_2, and A_3 as seen on Fig. 11b and Fig. 4. Applying Theorem 2.1 gives a straightforward result that the points A_1, A_2, and A_3 immobilize ABC and therefore immobilize P (triangle ABC is generated by three sides of polygon P that contain A_1, A_2, and A_3 respectively as interior points). ◆

It is not true, however, that all polygons can be immobilized using three points. For example, any parallelepiped with four vertices cannot be immobilized with three points (see [K]). In fact examples given by Kuperberg might suggest that each convex figure needing four points to immobilize is an intersection of two objects each one being either a strip or a disk. We show, however, that there are convex polygons other than parallelepipeds that also cannot be immobilized with three points.

Theorem 4.4. For every n>3 there are convex polygons with n vertices for which exactly four points are needed to immobilize them.

Proof. An example of a quadrilateral, but not a parallelepiped, for which four points are needed to immobilize it can be obtained as follows. Consider a triangle T with vertices A, B and C such that the angle at B is obtuse. Then the quadrilateral P with vertices A, B, E and D, such that the side DE is parallel to AB and close enough to it, cannot be immobilized by using three points (see Fig. 12).

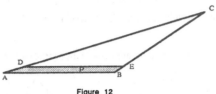

Figure 12

To prove this result, we first notice that if three points X, Y and Z were to immobilize P, then by Theorem 2.2, the three sides of P containing them would have to enclose P. Then these sides would be the segments AB, AD and BE. It is easy to verify, however, that if the segment DE is close enough to AB condition (b) of Theorem 3.1 is not satisfied.

To prove that for every n>3 there are polygons that cannot be immobilized with three points, it is sufficient to notice that we can substitute the side AD in P by a convex chain of edges close enough to AD and the same argument holds. ♦

It may be easily seen now that any polygon may be immobilized using only four points. As pointed out by an anonymous referee this may be also proved using some known facts from grasping. In [MSS] it was proved that for every polygon P there exists so-called force-torque closure grasp using a minimal set S of four finger points. From a discussion in [MS] it follows that any rigid velocity of P causes at least one of the points of S to have an instantaneous velocity strictly directed towards the interior of P. In consequence S immobilizes P. It was proved in [MSS] that the set S may be found in O(n) time. However, as the definitions of immobilization and force-torque grasp are not equivalent, in most cases this set S is not minimal for the purpose of immobilization. We know that three points are usually sufficient to immobilize a polygon.

Theorem 4.5. Let P be a polygon with vertices in general positions. In O(n log n) time (O(n) time if P is convex) we can find a set of three points immobilizing P.

Proof: As P does not have parallel edges, the largest circle inscribed in P touches its boundary in three points immobilizing it. Such circle is a vertex of a Voronoi diagram V(P) of the segments being edges of P. By [F, Ki, Y] V(P) may be constructed in O(n log n) time (O(n) time if P is convex following [AGSS]). It is then sufficient to check in linear time all vertices of V(P). ♦

5. Generalizations to higher dimensions

In this section we will generalize some results on immobilization of polygons in the plane to high-dimensional case.

Consider the largest inscribed sphere S (centered at O) of a given d-dimensional polytope P. Suppose S touches P in points $A_1, A_2, ..., A_t$. Let $T=CH(A_1, A_2,..., A_t)$ denote the convex hull of these points.

Lemma 5.1. $O \in CH(A_1, A_2,..., A_t)$.

Proof. If O is located outside T, O must be a vertex of $CH(O, A_1, A_2,..., A_t)$, and therefore there exists a (d-1)-dimensional hyperplane C passing through O such that $A_1, A_2,..., A_t$ are all on the same side of C (and not on C). Let OX be a vector normal to C such that all angles XOA_i ($1 \le i \le t$) are obtuse. Then, when we move point O in the direction OX it may be a center of a sphere larger than S. ♦

Lemma 5.2. Let $A_1 A_2...A_{d+1}$ be a d-dimensional simplex containing O in its interior. Then $\{A_1, A_2,..., A_{d+1}\}$ immobilizes P.

Proof. For any motion keeping O in place, the final position of this movement may be described as a composition of d-1 rotations around O. Some points among $A_1, A_2,..., A_{d+1}$ (all which move at all) will then penetrate the interior of P. Therefore, any possible motion must move O to a new position $O' \ne O$, and S moves to S'. Let b be the (d-1)-dimensional hyperplane that is bisector of OO'. Because $A_1 A_2...A_{d+1}$ contains O, when O is close enough to O', on each side of b there are some points among $\{A_1, A_2,..., A_{d+1}\}$. All points of S that lie on the opposite side of b than O are then inside S', the new position of S, (once more when O and O' are close enough), and thus penetrate P. For d=2 result follows from Lemma 2.1. ♦

Now we will turn our attention to the question of the upper bound for the number of points necessary to immobilize a polytope. Before we pass to the general d-dimensional case, let us consider, as a more intuitive illustration, the case of 3-dimensional polyhedra.

Theorem 5.1. Six points suffice to immobilize any polyhedron.

Proof. By Lemma 4.1, O is inside or on the boundary of $CH(A_1, A_2,..., A_t)$. Let m be the minimal number such that there exists an m-dimensional simplex T' with m+1 vertices taken from $\{A_1, A_2,..., A_t\}$ containing point O in its interior. Let these points be named $A_1, A_2,..., A_{m+1}$. Thus $T'=CH(A_1, A_2,..., A_{m+1})$. Consider the following cases:

Case 1) m=1. Then O is in the interior of a segment, say, $A_1 A_2$, and $A_1 A_2$ is a diameter of S. We will include A_1 and A_2 into the set of points to immobilize P. $A_1 A_2$ is the minimal distance between corresponding faces containing A_1 and A_2 (there may be, in case of non-convex polyhedron, several faces containing A_1 or A_2). This distance is exactly the distance between parallel planes that are tangent to S at A_1 and A_2, respectively. The points from these planes that are in the neighborhood of A_1 or A_2 are inside or on the boundary of P, and thus the only motion (if any) that does not cause A_1 or A_2 to penetrate P must be the motion within the plane normal to $A_1 A_2$. In other words, for any point $p \in P$, its motion remains within the plane containing p and normal to $A_1 A_2$. P intersects any such plane in a simple polygon, and that polygon can be immobilized in that plane with four points (Theorem 4.5). Thus P can be immobilized with six points.

Case 2) m=2. Then O is inside a triangle, say, $A_1A_2A_3$. We include A_1, A_2, and A_3 in the set of points to immobilize P. Consider the tangent planes to S at A_1, A_2, and A_3. P is obviously the superset of these planes in the neighborhood of touching points and will have restricted motion as the figure that is formed by the three tangent planes. The only possible motions of P are now translations along the line normal to the plane $A_1A_2A_3$. The translations can be prevented by chosing two more points, one for each direction of translation, thus giving a total of five points for immobilization.

Case 3) m=3. O is the interior point of the tetrahedron, say, $A_1A_2A_3A_4$. A_1, A_2, A_3, and A_4 will then immobilize P by Lemma 5.2. So, in this case four points suffice to immobilize P.♦

Theorem 5.2. 2d points are always sufficient and sometimes necessary to immobilize a given d-dimensional polytope P.

Proof. The proof is by induction on d. It is already proved for d=2 and d=3. For d=1 it is trivially sufficient to immobilize a segment on a line with two points. Suppose that the statement is true for any dimension smaller than d. We prove that the statement is then true for dimension d as well.

According to Lemma 5.1, the center O of the largest inscribed sphere S must be located inside or on the boundary of $CH(A_1, A_2,..., A_t)$. Let m be the minimal number $m \geq 1$ such that there exists a m-dimensional simplex T' with m+1 vertices taken from $\{A_1, A_2,..., A_t\}$ containing point O in its interior. Let these points be named $A_1, A_2,..., A_{m+1}$. Thus $T'=CH(A_1, A_2,..., A_{m+1})$. Consider the following cases:

Case 1) m=d. Then by Lemma 5.2, $\{A_1, ..., A_{d+1}\}$ immobilizes P.

Case 2) $1 \leq m < d$. Include the points $A_1, A_2,..., A_{m+1}$ in the set to immobilize P. Analogously as in Theorem 5.1, there is no motion of P within the m-dimensional space determined by the points A_1, $A_2,..., A_{m+1}$. Thus each of the possible motions, so far, must be within a (d-m)-dimensional space that is orthogonal to the above m-dimensional one. Since d-m<d, by induction hypothesis, this motion can be prevented by 2(d-m) additional points. Therefore $2(d-m)+m+1=2d-m+1 \leq 2d$ points suffice to immobilize P.

The necessity follows from the obvious fact that a d-dimensional cube (or parallelepiped) requires 2d points to immobilize it. ♦

The following theorem is a generalization of Theorem 4.3.

Theorem 5.3. Let P be a polytope in d-dimensional space. If there does not exist a linearly dependent set of d vectors v_1, v_2, ... ,v_d, such that each v_i is orthogonal to some face of P then P may be immobilized with d+1 points.

Proof. Let m and T' be defined as in the proof of Theorem 5.2. Vectors OA_1, OA_2, ..., OA_{m+1} then form a linearly dependent set of m+1 vectors (m+1 vectors in m-dimensional space). These vectors are indeed normals to some faces of P. According to the condition of the theorem it follows that m+1>d.

Therefore m=d and the result follows from Lemma 5.2. ◆

Corollary 5.1. Any d-dimensional simple polytope needs at least d points to immobilize it.

Proof. The proof is obvious by noting that in d dimensions for any d-1 points there exists an axes of rotation keeping these d-1 points in place. ◆

The reader may verify that there exist d-dimensional non-convex simple polytopes for which d points suffice to immobilize. From Lemma 5.2 (the conditions of the lemma are satisfied given a random polytope) and Corollary 5.1 follows

Corollary 5.2. For a simple d-dimensional polytope having vertices in general position the number of points needed to immobilize it is always equal to d or d+1.

In the case of convex P, however, d points will not be sufficient to immobilize P. The region delimited by the hyperplanes tangent to P at these d points must be unbounded and P may be translated away (similarly as in Fig. 7(a) for the planar case). As a consequence we have

Corollary 5.3. For a d-dimensional convex polytope having vertices in general position the number of points needed to immobilize it is always equal to d+1.

6. Conclusions and open problems.

In this paper we studied the problems of immobilization of polygons (polytopes). A number of interesting open problems follow from this work.

Theorem 4.1 gives a characterization of immobilization of a polygon by three points not located at its vertices. An interesting question is to extend this characterization to cover the placement of immobilization points anywhere on the boundary of the polygon.

Theorem 4.5 gives an O(n log n) algorithm finding three points immobilizing a given polygon having vertices in general positions. However, for some polygons four points are needed. For convex polygons, we can find out whether four points are actually needed and eventually output the optimal solution but it will take an $O(n^3)$ time following Theorem 3.1. It is an open problem to reduce the complexity of the algorithm finding the optimal number of immobilization points. For the case of non-convex polygons, it remains an open problem to recognize those that need four points to immobilize them. The problem of finding the optimal immobilizing set may be solved also by giving first the full answer to the question stated in [K] about the characterization of the class of polygons (convex polygons) needing four points to immobilize. The result from Theorem 4.4 is not a complete solution of this problem.

Another question worth pursuing is to propose the algorithm deciding whether a given set of n points immobilize a given polygon. Is it possible to chose a small subset of these points which will also immobilize the polygon? We believe that this question may have a positive answer only when the points are not allowed to be placed at the vertices of the polygon.

It seems that the theorem 3.1 (and theorem 4.1) may be fully extended to the case of d-dimensional polytope P. In particular, d+1 points should immobilize a convex polytope if and only if the (d-1)-dimensional hyperplanes tangent to P in these points enclose P, and the lines orthogonal to the hyperplanes at the points of immobilization meet at a common point. Another extension to higher dimension was suggested in [K] where instead of using points, immobilization by lines, planes, etc... may be considered.

References

[AGSS] A. Aggarwal, L.J. Guibas, J. Saxe, and P.W. Shor, A linear-time algorithm for computing the Voronoi diagram of a convex polygon, Discr. Comput. Geom., 4, 591-604, 1989.

[BFG] B. S. Baker, S. Fortune, E. Grosse, Stable Prehension with Three Fingers, Proc. 17th Symp. on Theory of Computing, 1985, pp.114-120.

[CSU] J. Czyzowicz, I. Stojmenovic, J. Urrutia, Immobilizing a shape, Rapport de recherche RR 90/11-18, Dept. Informatique, Université du Québec à Hull, November 1990 (earlier version available as Technical Report TR-90-37, Dept. of Computer Science, Univ. of Ottawa, July 1990).

[E]   H. Edelsbrunner, Algorithms in Combinatorial Geometry, Springer-Verlag, 1987.

[F]   S. Fortune, A sweep-line algorithm for Voronoi diagrams, Proc. 2nd ACM Symp. on Computational Geometry, 1986, 313-322.

[Ki]   D.G. Kirkpatrick, Efficient computation of continuous skeletons, Proc. 20th IEEE Symp. on Found. of Comp. Sci., 1979, 18-27.

[K]   W. Kuperberg, DIMACS Workshop on Polytopes, Rutgers University, Jan. 1990.

[MP]   X. Markenscoff and Ch. H. Papadimitriou, Optimal grip of a polygon, Int. J. Robotics Research, 8, 2, 1989, 17-29.

[MNP] X. Markenscoff, L. Ni and Ch. H. Papadimitriou, The Geometry of Grasping, Int. J. Robotics Research, 9, 1, 1990, 61-74.

[MSS] B. Mishra, J.T. Schwartz and M. Sharir, On the Existence and Synthesis of Multifinger Positive Grips, Algorithmica (1987) 2: 541-548.

[MS]   B. Mishra and N. Silver, Some Discussions of Static Gripping and Its Stability, IEEE Transactions on Systems, Man and Cybernetics, 19, 4, July/August, 1989, pp. 783-796.

[O]   J. O'Rourke, Computational geometry column 9, SIGACT News, 21, 1, 1990, 18-20, Winter 1990, #74.

[Y]   C.K. Yap, An O(n log n) algorithm for the Voronoi diagram of a set of simple curve segments, Discr. and Comput. Geom., 2, 1987, 365-393.

WHAT CAN WE LEARN ABOUT SUFFIX TREES FROM INDEPENDENT TRIES?

Philippe Jacquet*
INRIA
Rocquencourt
78153 Le Chesnay Cedex
France

Wojciech Szpankowski[†]
Department of Computer Science
Purdue University
W. Lafayette, IN 47907
U.S.A.

Abstract

A suffix tree of a word is a digital tree that is built from suffixes of the underlying word. We consider words that are random sequences built from independent symbols over a finite alphabet. Our main finding shows that the depths in a suffix tree are asymptotically equivalent to the depths in a digital tree that stores independent keys (i.e., independent digital trees known also as tries). More precisely, we prove that the depths in a suffix tree build from the first n suffixes of a random word are *normally distributed* with the mean asymptotically equivalent to $1/h_1 \log n$ and the variance $\alpha \cdot \log n$, where h_1 is the entropy of the alphabet, and α is a parameter of the probabilistic model. Our results provide new insights into asymptotic properties of compression schemes, and therefore find direct applications in computer sciences and telecommunications, most notably in coding theory, theory of languages, and design and analysis of algorithms.

1. INTRODUCTION

Periodicities, autocorrelations and related phenomena in words are known to play a central role in many facets of science, notably in coding theory, in the theory of formal languages, in the design and analysis of algorithms, and last but not least in molecular sequence comparison. Several efficient algorithms have been designed to detect and to exploit the presence of repeated subpatterns and other kinds of avoidable or unavoidable regularities in words [LO]. In this paper, we investigate the lengths of substrings that can be recopied in a random word X.

Periodicities, autocorrelations and related phenomena can be equivalently studied on an associated digital tree called a *suffix tree* [AA, AHU, AS, WE]. A suffix tree is a digital tree that is built over a finite alphabet Σ. In general, a digital tree – that is also called a *trie* – stores a set of words (strings, keys) $\mathcal{W}$ built over the alphabet Σ. A trie consists of branching nodes, called also internal nodes, and external nodes that store the keys. In addition, we assume that every external node is able to store only one key. The branching policy at any level, say k, is based on the k-th symbol of a string (key, word). For example, for a binary alphabet $\Sigma = \{0, 1\}$, if the k-th symbol in a key is "0", then we branch-out left in the trie, otherwise we go to the right. This process terminates when for the first time we encounter a different symbol between a key that is currently inserted into the trie and all other keys already in the trie. Then, this new key is stored in a newly generated external node. In other words, the access path from the root to an external node (a leaf of a trie)

*This research was primary supported by NATO Collaborative Grant 0057/89.

[†]This research was primary done while the author was visiting INRIA in Rocquencourt, France. Support was provided in part by NATO Collaborative Grant 0057/89, in part by NSF Grants NCR-8702115 and CCR-8900305, and from Grant AFOSR-90-0107, and in part by Grant R01 LM05118 from the National Library of Medicine.

is the minimal prefix of the information contained in this external node; it is minimal in the sense that this prefix is not a prefix of any other keys. The *depth of a key* is the length of a path from the root to the external node containing this key. The height of a trie is the maximum over all such depths. For more information regarding tries the reader is referred to [AHU, KN].

A suffix tree is a special trie that is built from a *single* word X, that is, the keys in $\mathcal{W}$ consists of suffixes of the word X. We do not compress the trie as in PATRICIA (cf. [KN]), that is, in our construction of a suffix tree no substrings are collapsed into one node. There is a natural correspondence between lengths of substrings that can be recopied in a word X and depths of suffixes in the corresponding suffix tree. Therefore, parameters of interest are height of the suffix tree and depths of suffixes, but in this paper we only concentrate on the depth. This parameter is analyze in a probabilistic framework called *Bernoulli model*. In this model symbols of a string X are drawn independently from the alphabet Σ, however, it is possible to extend our analysis to some models with dependency between symbols (e.g., Markovian model, see [JS]). In passing, we note that a suffix tree has (statistically) *correlated* keys (subwords) which makes the analysis non-trivial. It should be compared with a trie that is built from a set of statistically *independent* keys. We coin a term *independent trie* for the latter digital trees, and we compare our results for suffix trees with the ones known for independent tries.

The literature on the analysis of suffix trees is very scare. To the best of our knowledge, the analysis of the height of the suffix tree was initiated by Apostolico and Szpankowski [AS], and recently Devroye, Szpankowski and Rais [DSR] have established exact asymptotics for the height. The size of a suffix tree was investigated by Blumer, Ehrenfeucht and Haussler [BEH] using a mixture of analytical and simulation tools (so some more work is needed here), and we present rigorous analysis for the average size of suffix trees. The limiting distribution of the depths in a suffix tree (which, as we shall argue below, is the hardest to analyze) was left open, and this paper is intended to fill this gap.

2. AUTOCORRELATION PARAMETERS IN WORDS

Let $X = x_1 x_2 x_3 ...$ be a string of unbounded length formed by symbols from an alphabet Σ of cardinality V, and let $S_i = x_i x_{i+1} ...$ be the i-th *suffix* of X. For every off-diagonal pair (i, j) of positions of X, we define the *self-alignment* C_{ij} as the length of the longest string that is a prefix of both S_i and S_j. We leave C_{ij} undefined when $i = j$. Thus, $C_{ij} = k$ iff S_i and S_j agree exactly on their first k symbols, but differ on their $(k + 1)$-st. Clearly, $C_{ij} = C_{ji}$ for all meaningful choices of i and j.

Let now n be any fixed integer. We define the *height* H_n of X and the *depth* $D_n(i)$ *of the i-th suffix* of X, as follows

$$H_n = \max_{1 \leq i < j \leq n} \{C_{ij}\}, \tag{2.1a}$$

$$D_n(i) = \max_{1 \leq j \leq n, j \neq i} \{C_{ij}\}. \tag{2.1b}$$

Furthermore, the *depth* D_n of a word X is defined as the depth of a randomly selected suffix among the first n suffixes of X (see Sec. 3.1 for more precise definition of the depth). Intuitively, H_n is the maximum possible length of a substring Z of X that has at least two occurrences in X, both starting within the first n positions of X. Thus, there are two positions i and j of X, $i < j \leq n$, such that the occurrence of Z starting at j can be fully recopied from the occurrence starting at i. The depth D_n represents the length – averaged over the first n suffixes of X – of the longest substring of X that can be recopied from the past. The height H_n and the depth D_n express structural correlations among the substrings of the word X. Such correlations play a crucial role in many combinatorial and algorithmic constructions, and our above definitions resemble the notions that have already appeared in the literature, most notably in [LZ, ZL,GO2, GO3].

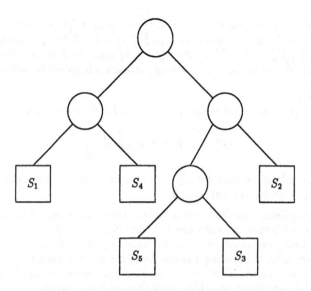

Figure 1: Suffix tree from Example 2.2

EXAMPLE 2.1. *Self-alignment matrix*

Let $X = abbabaa...$ and $n = 5$. Then $S_1 = X$, $S_2 = bbabaa...$, $S_3 = babaa...$, $S_4 = abaa...$ and $S_5 = baa...$. The corresponding self-alignment matrix $\mathbf{C} = \{C_{ij}\}_{i,j=1}^5$ is as follows.

$$\mathbf{C} = \begin{bmatrix} \star & 0 & 0 & 2 & 0 \\ 0 & \star & 1 & 0 & 1 \\ 0 & 1 & \star & 0 & 2 \\ 2 & 0 & 0 & \star & 0 \\ 0 & 1 & 2 & 0 & \star \end{bmatrix}$$

From $\mathbf{C}$ and the expressions (2.1), we obtain $H_n = 2$ and the average depth $D_n = (2+1+2+2+2)/5 = 9/5$. The associated suffix tree is presented in Figure 1, where circles represent branching nodes and squares are external nodes. ∎

We deal here with a probabilistic analysis of the depth under the *Bernoulli model* assumptions, that is: *the symbols of X are drawn independently from Σ, and the i-th symbol of Σ occurs in any position of X with probability p_i for $i = 1, 2, ..., V$, and $\sum_{i=1}^V p_i = 1$.*

Let us consider a depth of a fixed suffix, say the first one, and let us denote it as $D_n(1)$. According to (2.1b) $D_n(1) = \max_{2 \le j \le n}\{C_{1j}\}$. Note that the self-alignments $C_{1,j}$ are *strongly* dependent. In particular, to compute the distribution function $\Pr\{D_n(1) > k\}$ we need *all* joint distributions of the self-alignments. To be more precise, using *inclusion-exclusion formula* [BO] one immediately notes that

$$\Pr\{D_n(1) > k\} = \sum_{r=2}^{n} (-1)^r \sum_{i_1,...,i_r} \Pr\{C_{1,i_1} > k, ..., C_{1,i_r} > k\}, \qquad (2.2)$$

where i_j are distinct and $2 \le i_j \le n$ for every $1 \le j \le r$. An interesting fact is that, due to an alternating sum in (2.2), we *have to* take into account all terms of the above sum, and we need an *exact* formula for the joint distribution $\Pr\{C_{1,i_1} > k, ..., C_{1,i_r} > k\}$. This probability will be evaluated by some combinatorial method in Section 3, and this is a difficult part of our paper.

Now we are in a position to summarized our main results. The first – and the most important – finding deals with a comparison between independent tries and suffix trees. Let, for a moment, D_n^T, D_n^S denote the depths in an independent trie and a suffix tree with n keys, respectively. In addition, we define the appropriate distribution functions as $F_n^T(k) = \Pr\{D_n^T \leq k\}$ and $F_n^S(k)$, respectively. Note that for independent tries $\Pr\{D_n^T \leq k\} = \Pr\{D_n^T(i) \leq k\}$ for any key i, while for suffix tree we have $\Pr\{D_n^S \leq k\} = \frac{1}{n}\sum_{i=1}^n \Pr\{D_n^S(i) \leq k\}$ since the depth is defined for a randomly (and equally likely) selected suffix. The following proposition is proved in Section 3 .

PROPOSITION 1.
There exist $\beta > 1$ and $\epsilon > 0$ such that uniformly in k and n the below holds

$$|F_n^T(k) - F_n^S(k)| = O\left(\frac{1}{n^\epsilon \beta^k}\right). \tag{2.3}$$

In addition, all moments of the depth for suffix trees are in the same relationship to the appropriate moments of the depth for independent tries. ∎

Proposition 1 establishes a methodological tool to analyze some *dependent* data structures such as suffix trees. It basically says that suffix tree does not differ too much from independent tries. But, tries have been analysed extensively over last few years, and virtually we know almost everything about them. In particular, the limiting distribution of the depth is known, the average depth and the variance are also well known. Therefore, Proposition 1 and recent results of Jacquet and Régnier [JR, RJ], Pittel [PI2] and Szpankowski [SZ1] imply our next main result.

PROPOSITION 2.
(i) For large n the average ED_n depth of a suffix tree becomes for some $\epsilon > 0$

$$ED_n = \frac{1}{h_1} \cdot \{\log n + \gamma + \frac{h_2}{2h_1}\} + P_1(\log n) + O\left(\frac{1}{n^\epsilon}\right), \tag{2.4a}$$

and the variance $var D_n$ of the depth is

$$var D_n = \frac{h_2 - h_1^2}{h_1^3} \log n + C + P_2(\log n) + O\left(\frac{1}{n^\epsilon}\right), \tag{2.4b}$$

where $h_1 = -\sum_{i=1}^V p_i \log p_i$ and $h_2 = \sum_{i=1}^V p_i^2 \log p_i$, and $P_1(x), P_2(x)$ are fluctuating periodic functions with small amplitudes, and an explicit formula for the constant C can be found in [SZ1]. In the symmetric case, i.e., $p_1 = p_2 = ... = p_V = 1/V$, the variance becomes

$$var D_n = \frac{\pi^2}{6\log^2 V} + \frac{1}{12} + O\left(\frac{1}{n^\epsilon}\right), \tag{2.4c}$$

(ii) For the asymmetric model of suffix trees $(D_n - ED_n)/\sqrt{var D_n}$ is asymptotically normal with mean zero and variance one, that is, for all $x \in R$

$$\lim_{n\to\infty} \Pr\{D_n \leq ED_n + x\sqrt{var D_n}\} = \frac{1}{\sqrt{2\pi}} \int_{-\infty}^x e^{-t^2/2} dt ,$$

and for all integer m

$$\lim_{n\to\infty} E\left[\frac{D_n - ED_n}{\sqrt{var D_n}}\right]^m = \begin{cases} 0 & \text{when } m \text{ is odd} \\ \frac{m!}{2^{m/2}(\frac{m}{2})!} & \text{when } m \text{ is even} \end{cases}$$

(see Jacquet and Régnier [JR], and Pittel [PI2]). For the symmetric case, one proves that uniformly in $x \geq 0$

$$\lim_{n\to\infty} \sup_x |\Pr\{D_n \leq \log_V(n) + x\} - e^{-V^{-x}}| = 0 \tag{2.4d}$$

(see Pittel [PI2]). ∎

In some applications size of a suffix tree plays a more dominant role than the depth of the tree. By size of a digital tree we mean the number of (internal) nodes needed to build the tree. Most notably, size of a suffix tree determines space requirements, and therefore space-complexity of any algorithm based on the suffix tree. The next proposition presents one result in this direction, namely the average size EL_n of suffix trees built from n suffixes.

PROPOSITION 3.
There exist such $\epsilon > 0$ that the average size EL_n^S of suffix tree and the average size EL_n^T of regular tries satisfy the following relationship

$$|EL_n^S - EL_n^T| = O(n^{1-\epsilon}) . \tag{2.5a}$$

In particular, this implies that

$$EL_n^S = \frac{n}{h_1}(1 + P_3(\log n)) + o(n) \tag{2.5b},$$

where h_1 is the entropy of the alphabet, and $P_3(\log n)$ is a periodic function with a small amplitude (cf. [JR]). ∎

Finally, to get some idea about accuracy of our asymptotics (in particular Proposition 1) we have performed some simulation studies which are presented in the table below. The table compares *theoretical* values of the average depth ED_n^T, the variance $var D_n^T$ and the average size EL_n^T of independent tries with simulation results of ED_n^S, $var D_n^S$ and EL_n^S respectively for suffix trees in the range of n varying from 700 to 10,000. These results confirm, as expected, our theoretical findings presented above, and in addition they show good accuracy of the asymptotics even for small values of n.

COMPARISON OF TRIES AND SUFFIX TREES
Asymmetric alphabet $p = 0.1$ and $q = 0.9$

n	ED_n^T	ED_n^S	$var D_n^T$	$var D_n^S$	EL_n^T	EL_n^S
700	24.483	23.861	112.09	110.85	2153.3	2064.8
800	24.894	24.340	113.78	110.95	2460.9	2385.7
900	25.256	24.766	115.27	113.12	2768.5	2691.1
1000	25.579	25.080	116.61	114.84	3076.1	3014.1
2000	27.720	27.457	125.37	124.81	6152.3	6069.0
3000	28.959	28.802	130.50	130.04	9228.4	9125.3
4000	29.844	29.652	134.14	132.85	12304.6	12155.1
5000	30.531	30.348	136.96	135.87	15380.7	15227.6
10000	32.663	32.574	145.73	144.40	30761.4	30763.2

Some consequences of our findings for the design and analysis of algorithms on words are discussed in [AS]. Therefore, here we restrict our discussion to one example, namely *compression*. Consider the following natural compression technique. Instead of sending a whole code we may discover some repetitions in the code that can be recopied by transmitting pointers to the repeated pattern. In such a case we trade the lengths of repeated patterns for the length of two pointers. A natural question arises whether it is worth to do it, say in a *random code*. To answer it, we introduce *the compression coefficient C_V* as the ratio between the *average* length of repeated patterns and the length of overhead information, that is, in our case the length of two pointers. But, the average length of repeated patterns is just equal to the depth of the random word X (or the depth of associated suffix tree), while the length of two pointers that can carry information regarding n possible positions is simple $2\log_V n$. Therefore, we just proved that

$$C_V = 2h_1/\log V . \tag{2.6}$$

Since the entropy reaches its maximum for uniform distribution (symmetric case) from (2.6) one immediately shows that $C_V \leq 2$. But, for uniform distribution $C_V = 2$ (and $C_V > 1$ for distributions not too far away from the uniform distribution). Therefore, compression – as defined above – is *not* worth to do it. However, if one sends only one pointer to the beginning of the repeated pattern, and the length of the compressed pattern (encoded in a special compact form; cf [RPE]), then such a scheme is asymptotically optimal, that is, $C_V = h_1 / \log V$. There are other compression schemes that can achieve optimal compression. The interested reader is referred to works of Lampel and Ziv [LZ, ZL] (see also [RPE]).

3. ANALYSIS AND AUTOCORRELATIONS OF STRINGS

In this section we provide a sketch of the proof of our main result presented in Proposition 1. Detailed proof can be found in [JS1]. Our approach to solve the problem seems to be new, and it only resembles some similarities with the work of Guibas and Odlyzko [GO1, GO2, GO3] (see also [PI1]).

Before we plug into detailed analysis let us give an overview of our approach. In Section 2 we have shown that any analysis of the depth D_n of a digital tree, in particular, suffix tree needs an *exact* evaluation of the joint distribution of the self-alignments, as for example shown in (2.3) for independent tries. Such an evaluation for suffix tree is very complicated due to strong correlations of overlapping suffixes. Therefore, to circumvent the alternating sum problem (in fact, to hide it in a generating function form), we suggest a different, more combinatorial approach. We consider a finite string σ that is used as a "ruler" to measure the length of strings, in particular, to estimate overlapping between any two suffixes. For example, to evaluate the self-alignment between the i-th suffix S_i and the j-th suffix S_j we first compute the alignment between S_i and σ, and then the alignments between S_j and σ. These measures can be used to evaluate the self-alignment C_{ij} between S_i and S_j *with respect to* the ruler string σ. Finally, considering all possible ruler-strings σ we evaluate the self-alignments C_{ij}. This – although it looks more complicated than necessary – is the right approach as we shall prove below.

Using the above idea we shall compute the generating function of the average depth for suffix trees (difficult !) and independent tries (easy). These two generating functions are asymptotically compared to show that they do not differ too much for large n. This will lead to our main result Proposition 1.

It might be worth to point out that along the lines of our proof, we in fact explore the problem of autocorrelation properties of random strings. This finds many other applications in the combinatorics on words e.g., in squares of strings, in biprefix strings, and so forth [AA, AS, LO].

3.1 Some Notations and Preliminary Results

By X we denote a random unbounded string over a finite alphabet, generated according to the Bernoulli model introduced in Section 2. However, for simplicity of the analysis we restrict our attention to a binary alphabet $\Sigma = \{a, b\}$ with p (resp. q) denoting the probability of a (resp. b) occurrence in X. Our interest lies in evaluating the correlation between the first n symbols of X. To measure it we consider another (non-random !) string, say σ, which is further called a *ruler string*. The length of σ is denotes $|\sigma|$, and $0 < |\sigma| < \infty$. Let $\langle X, \sigma \rangle$ represent the subset of positions in X that σ and X agree, that is, σ overlaps with X starting from any position of $\langle X, \sigma \rangle$. At this time, we point out that some properties of $\langle X, \sigma \rangle$ were studied by Guibas and Odlyzko [GO1, GO2, GO3]. We shall use $\langle X, \sigma \rangle$ to investigate the depth $D_n(i)$ of the i-th suffix of X. There is a simple relationship between the depth and properties of $\langle X, \sigma \rangle$. Indeed, one immediately notes the following

$$\{D_n(i) > k\} \iff \exists \sigma \, \exists j \leq n \, : \, |\sigma| = k \text{ and } j \neq i \, : \, \{i, j\} \subset \langle X, \sigma \rangle \tag{3.1a}$$

and

$$\{D_n(i) \leq k\} \iff \exists \sigma \, : \, |\sigma| = k \text{ and } \langle X, \sigma \rangle \cap \{1, \cdots, n\} = \{i\} \, . \tag{3.1b}$$

Consider now all strings σ of fixed length k. Note that the events $\langle X, \sigma \rangle = \{i\}$ and $\langle X, \sigma' \rangle = \{i\}$ are disjoint for distinct σ and σ' such that $|\sigma| = |\sigma'| = k$. Hence, we can write

$$\Pr\{D_n(i) \leq k\} = \sum_{|\sigma|=k} \Pr\{\langle X, \sigma \rangle \cap \{1, \cdots, n\}\} = \{i\}\} \tag{3.1c}$$

The example below illustrates what we have done so far.

EXAMPLE 3.1 *Depth of the i-th suffix and* $\langle X, \sigma \rangle$.

Let $X = baabbabaaa...$ and we are interested in the depth $D_{10}(1)$ of the first suffix, that is, we would like to find such k that $\{D_{10}(1) \leq k\}$ holds. Let us first consider all σ such that $|\sigma| = 1$. Naturally, in this case $|\langle X, \sigma \rangle| > 1$, and $D_{10}(1) > 1$. Next all σ with $|\sigma| = 2$ and $|\sigma| = 3$ are investigated. But, for $\sigma = ba$ we have $\langle X, \sigma \rangle = \{1, 5, 7\}$, and for $\sigma = \{baa\}$ we find that $\langle X, \sigma \rangle = \{1, 7\}$, so $D_{10}(1) > 3$. Finally, string-rulers with $|\sigma| = 4$ are studied, and for $\sigma = \{baab\}$ we notice that $\langle X, \sigma \rangle = \{1\}$, so by (3.1) $\{D_{10}(1) = 4\}$. ∎

Our purpose is to derive an expression for the generating function of the depth $E[u^{D_n}] = \sum_{k=0}^{\infty} \Pr\{D_n = k\} u^k$. To recall, the depth D_n is defined as the length of a *randomly* (and equally likely) selected suffix in a *random* string X. If $E[u^{D_n(i)}]$ denotes the generating function of the depth of the i-th suffix, then our definition requires that[1]

$$E[u^{D_n}] = \frac{1}{n} \sum_{i=1}^{n} E[u^{D_n(i)}] . \tag{3.2}$$

It is more convenient, however, to work with the bivariate generating function $D(z, u)$ of the $E[u^{D_n}]$ defined as

$$D(z, u) = \sum_{n=0}^{\infty} n\, E[u^{D_n}] z^n . $$

To express $D(z, v)$ in terms of easy-to-compute generating functions we proceed as follows. Define an event $A_j = \{j \in \langle X, \sigma \rangle\}$ and the complementary event $\bar{A}_j = \{j \notin \langle X, \sigma \rangle\}$. Then $\Pr\{\langle X, \sigma \rangle \cap \{1, \ldots, n\} = \{i\}\} = \Pr\{\bigcap_{j \neq i} \bar{A}_j \cap A_i\}$. Noting that $\Pr\{\bigcap_{j \neq i} \bar{A}_j \cap A_i\} + \Pr\{\overline{\bigcap_{j \neq i} \bar{A}_j} \cap A_i\} = \Pr\{A_i\}$, and applying the inclusive-exclusive formula [BO] to the second probability of the LHS of the previous expression, we obtain

$$\Pr\{\bigcap_{j \neq i} \bar{A}_j \cap A_i\} = \Pr\{A_i\} - \sum_{j=1}^{n-1} (-1)^{j+1} \sum_{\{i_1, \ldots, i_j\}} \Pr\{\bigcap_{k=1}^{j} A_{i_k} \cap A_i\} , \tag{3.3a}$$

where $(i_1, ..., i_j)$ is a j-tuple of *distinct* elements from $\{1, \cdots, n\} - \{i\}$. But $\Pr\{\bigcap_{k=1}^{j} A_{i_k} \cap A_i\} = \Pr\{(i_1, ..., i_j, i) \subset \langle X, \sigma \rangle\}$. To simplify the above, let $\mathcal{L}$ be a finite set of integers. We define $P(\mathcal{L}, \sigma)$ as the probability of the event "$\mathcal{L} \subset \langle X, \sigma \rangle$". By denoting $m(\mathcal{L})$ as the greatest element of $\mathcal{L}$ and $|\mathcal{L}|$ as the size of $\mathcal{L}$, it is easy to see that for $\mathcal{L} = \{i, i_1, \ldots, i_{j-1}\}$ (3.3a) implies the following

$$\Pr\{\langle X, \sigma \rangle \cap \{1, \ldots, n\} = \{i\}\} = \sum_{j=1}^{n} (-1)^{j+1} \sum_{\substack{|\mathcal{L}|=j \\ m(\mathcal{L}) \leq n,\ i \in \mathcal{L}}} P(\mathcal{L}, \sigma) . \tag{3.3b}$$

To simplify the above notation, hereafter we consider only such sets $\mathcal{L}$ that $m(\mathcal{L}) \leq n$.

We can generalize (3.3b) to include empty string σ which is further denoted as $*$. Since $\Pr\{\langle X, * \rangle \cap \{1, \ldots, n\} = \{i\}\} = 0$, we adopt convention that $P(\mathcal{L}, *) = 1$ for every set $\mathcal{L}$. Then (3.3b) holds with the first sum starting from $\{j = 0\}$. Indeed, this is a direct consequence of the following simple combinatorial identity $\sum_{j=0}^{n} (-1)^j j \binom{n}{j} = 0$.

---

[1] Note that for independent tries all depths of keys $D_n(i)$ are equidistributed, that is, $E[u^{D_n(i)}] = E[u^{D_n(j)}]$ for all $i, j \in \{1, 2, ..., n\}$. Hence the depth D_n has the same distribution of the depth of a key. For suffix trees every depths might have different distribution, so D_n is a measure of the *mean* depth, and it is rather a parameter of a whole tree than a particular suffix.

The rest is easy. From (3.1b) we obtain

$$n \cdot \Pr\{D_n \leq k\} = \sum_{i=1}^{n} \Pr\{D_n(i) \leq k\} = \sum_{i=1}^{n} \sum_{|\sigma|=k} \Pr\{\langle X, \sigma \rangle = \{i\}\} \, ,$$

where the first equality is a simple consequence of our definition (3.2) of the depth D_n. But,

$$\sum_{\substack{|\mathcal{L}|=j \\ i \in \mathcal{L}}} P(\mathcal{L}, \sigma) = j \cdot \sum_{|\mathcal{L}|=j} P(\mathcal{L}, \sigma) \, ,$$

hence the above becomes

$$n \cdot \Pr\{D_n \leq k\} = \sum_{|\sigma|=k} \sum_{j=1}^{n} (-1)^{j+1} j P_j(\sigma) \tag{3.4}$$

where $P_j(\sigma) = \sum_{|\mathcal{L}|=j} P(\mathcal{L}, \sigma)$. Define two more generating functions

$$P_{n,\sigma}(v) = \sum_{j=1}^{n} P_j(\sigma) v^j \quad , \quad P_\sigma(z, v) = \sum_{n=1}^{\infty} P_{n,\sigma}(v) z^n \, .$$

We note that for $\sigma = *$ we have $P_j(*) = \binom{n}{j}$, and consequently $P_{n,*}(v) = (1+v)^n$ as well as $P_*(z, v) = (1+v)z/(1-(1+v)z)$.

After recognizing in the RHS of (3.4) the partial derivative of $P_{n,\sigma}(v)$ with respect to v at $v = -1$, we finally obtain our main result of this subsection.

THEOREM 1.
For $n > 1$ we have the identity

$$E[u^{D_n}] = \frac{(1-u)}{n} \sum_{\sigma} u^{|\sigma|} \frac{d}{dv} P_{n,\sigma}(v)|_{(v=-1)}$$

for $|u| < 1$, where $|\sigma|$ is the length of the string σ and $\sum_\sigma$ means the summation over all possible finite strings σ, that is, $\sum_\sigma f(\sigma) = \sum_{k=0}^{\infty} \sum_{|\sigma|=k} f(\sigma)$ for any function $f(\cdot)$ defined on strings. ∎

Moreover, since $E[u^{D_1}] = 1$ (indeed, $D_1 = 0$), we easily get the following corollary.

COROLLARY 2
When $|u| < 1$, we also have the identity

$$D(z, u) = (1-u)z + (1-u) \sum_{\sigma} u^{|\sigma|} \frac{\partial}{\partial v} P_\sigma(z, v)|_{(v=-1)} \, ,$$

for $|z| < 1$. ∎

3.2 Computation of $P_\sigma(z, v)$

From Theorem 1 we know that $D(z, u)$ depends on the generating function $P_\sigma(z, v)$. This generating function is, on the other hand, a function of a string-ruler σ. More formally, it is a function of the probability $P(\mathcal{L}, \sigma) = \Pr\{\mathcal{L} \subset \langle X, \sigma \rangle\}$. This implies that $P_\sigma(z, v)$ depends on the structure of $\mathcal{L}$ as well as on autocorrelation properties of σ itself. Indeed, let us consider a set of string-rulers with fixed length k. Then, as easy to see, the probability $P(\mathcal{L}, \sigma)$ for $|\sigma| = k$ relies on the fact whether suffixes of X with length smaller equal than k overlap or not. In other words, whether the set $\mathcal{L}$ consists of positions separated by less than k or not. The former case is the most difficult to treat, and it further depends on an autocorrelation structure of σ itself as illustrated in the example below.

EXAMPLE 3.2. *Autocorrelation of σ and k-clusters*

Let $X = bbabaabaabaababbbabaaba...$ and $\sigma = abaaba$ so $|\sigma| = k = 6$. Note that $\mathcal{L} = \langle X, \sigma \rangle = \{3, 7, 10, 19\}$ and the autocorrelation set denoted as $\langle \sigma, \sigma \rangle$ becomes $\langle \sigma, \sigma \rangle = \{1, 4\}$. There is a relationship between $\langle X, \sigma \rangle$ and $\langle \sigma, \sigma \rangle$, namely those positions of $\langle X, \sigma \rangle$ that are separated by less than $k = 6$ positions – we call it further a k *cluster* – are inherently correlated to $\langle \sigma, \sigma \rangle$. Indeed, in our case a k-cluster is $C_k = \{3, 7, 10\}$. This cluster is a direct consequence of the fact that the autocorrelation set $\langle \sigma, \sigma \rangle$ includes the position $\{4\}$, so a k-cluster of X with respect to σ can be created if and only if $\langle \sigma, \sigma \rangle - \{1\}$ is nonempty. It should be also obvious that all difficulties in evaluating the probability $P(\mathcal{L}, \sigma)$ arise from the necessity of taking into account k-clusters. ∎

In order to investigate k-clusters (for formal definition see below) we need to study autocorrelation property of a string-ruler. Throughout this section we assume that $|\sigma| = k$. Then for any finite string σ, let $p(\sigma)$ be defined as the product $p^{|\sigma|_a} q^{|\sigma|_b}$ where $|\sigma|_a$ is the number of a in σ and $|\sigma|_b$, the number of b in σ. Let $\mathcal{F}_\sigma = \langle \sigma, \sigma \rangle - \{1\}$ be the autocorrelation set of a string σ, that is, $i \in \mathcal{F}_\sigma$ if and only if σ overlaps with itself from position i. For simplicity we omit the trivial position $i = 1$. For instance, in Example 3.2 $\mathcal{F}_\sigma = \{4\}$. Furthermore, we define the *autocorrelation polynomial* $a_\sigma(z)$ of σ as

$$a_\sigma(z) = \sum_{i \in \mathcal{F}_\sigma} p(\sigma_{i-1}) z^{i-1} ,$$

where σ_{i-1} denotes the prefix of σ of length $i - 1$. In particular, in Example 3.2 the autocorrelation polynomial for $\sigma = abaaba$ becomes $a_\sigma(z) = p^3 q^2 \cdot z^3$. In passing, we note that although σ is not a random string the weight $p(\sigma)$ in the definition of the autocorrelation polynomial $a_\sigma(z)$, can be viewed as the probability of the occurrence of σ in the random string X. This is simple consequence of the fact that $\sum_{|\sigma|=k} p(\sigma) = 1$. Although this notation may cause some confusions, we decided to adopt it because of the latter property.

Using the autocorrelation polynomial, and after some tedious and rather complicated algebra, we finally obtain the following representation for the generating function $D(z, u)$ (for details consult [JS1]).

THEOREM 3.
The generating function for the depth D_n of suffix trees becomes

$$D(z, u) = z(1 - u) + (1 - u) \sum_\sigma u^{|\sigma|} \frac{p(\sigma)z}{[(1 - z)(1 + a_\sigma(z)) + p(\sigma)z^{|\sigma|}]^2} \qquad (3.5)$$

for every $|u| < 1$ and $|z| < 1$. ∎

Remark 2. We can define generating functions $E[u^{D_n}]$ and $D(z, u)$ for independent tries built over n independent strings $X_1, ..., X_n$, in the same manner as above. This case is much simpler since strings are not suffixes of any initial word, and do not suffer mutual correlations. We will use the superscript S for suffix trees and superscript T for regular trie when it will be necessary to point out a distinction. The generating function $D_n^T(z, u)$ for independent tries can be derived in the same fashion as before. However this time, we define $\langle \mathbf{X}, \sigma \rangle$, where $\mathbf{X} = (X_1, ..., X_n)$, as a subset of the strings that agree with σ. Then, the depth is evaluated as in (3.1) and Theorem 1, provided $P_\sigma(z)$ and $P_\sigma(z, v)$ can be computed. But independence assumption leads immediately to $P_j(\sigma) = \binom{n}{j} p^j(\sigma)$ and therefore

$$P_\sigma(z, v) = \frac{1}{1 - z[1 + p(\sigma)v]} - \frac{1}{1 - z},$$

and finally by Corollary 2 we obtain

$$D^T(z, u) = (1 - u)z + (1 - u) \sum_\sigma u^{|\sigma|} \frac{p(\sigma)z}{[1 - z + p(\sigma)z]^2} . \qquad (3.6)$$

We shall use this formula to compare regular tries with suffix trees. ∎

3.3 Asymptotics

In this section we present an asymptotic analysis of the depth D_n through a careful evaluation of the generating function $D(z,u)$ around its singularities. The asymptotics of $D(z,u)$ is carried out in three steps. *At first*, we prove that the generating function $D(z,u)$ can be analytically continued to $|u| < 1 + \epsilon$. This strengthens our results in the sense that not only convergence *in distribution* but also convergence *in mean* can be established (we use a well known fact that every analytical function is differentiable). In the *second step*, we prove that the expended generating function has only a single pole that completely determines the asymptotics . Finally, the *third step* consists of applying the celebrated Cauchy's theorem [HE] to prove asymptotics. However, to simplify our analysis we do not determine directly the asymptotics of the suffix tree, but rather compare the asymptotics of suffix trees with independent tries to take advantage of many well established results for tries (cf. [KN, JS, RS, PI2, SZ1]). In this conference version of the paper, we omit most of the derivations, and we only discuss the third step of the above plan.

We now compare asymptotics of suffix trees with asymptotics of independent tries, and conclude that they do not differ too much. Let us define two new generating functions $Q_n(u)$ and $Q(z,u)$ which represent difference between the *probability distribution functions* of the depth in the suffix tree and in the regular trie built over independent strings, that is

$$Q_n(u) = \frac{1}{1-u}\left(E[u^{D_n^S}] - E[u^{D_n^T}]\right)$$

$$Q(z,u) = \sum_{n=0}^{\infty} n\, Q_n(u) z^n = \frac{1}{1-u}\left(D^S(z,u) - D^T(z,u)\right) .$$

Then, by Theorem 3 and Remark 2 we find

$$Q(z,u) = \sum_{\sigma} u^{|\sigma|} p(\sigma) z \left(\frac{1}{[(1-z)(1+a_\sigma(z)) + p(\sigma)z^{|\sigma|}]^2} - \frac{1}{(1-z+p(\sigma)z)^2} \right) .$$

It is not difficult to establish asymptotics of $Q_n(u)$ by appealing to the Cauchy theorem. This is done in the following lemma.

LEMMA 4
There exists $B > 1$, such that the following evaluation holds for all $|u| \leq \beta$ such that $\beta > 1$:

$$Q_n(u) = \frac{1}{n} \sum_{\sigma} u^{|\sigma|} p(\sigma) \left(A_\sigma^{-n}(\frac{n}{C_\sigma^2 A_\sigma} + \frac{D_\sigma}{C_\sigma^3}) - n(1-p(\sigma))^{n-1} \right) + O(B^{-n}) .$$

PROOF : To prove this we apply Cauchy

$$n Q_n(u) = \frac{1}{2i\pi} \oint Q(z,u) \frac{dz}{z^{n+1}} ,$$

where the integration is done along a loop contained in a unit disk that encircles the origin. Details can be found in [JS1]. ∎

Finally, the main theorem of this section (and the paper) follows. The theorem below is our Proposition 1 from Section 2 rephrased in terms of generating functions rather than in probability distribution functions. It says that independent tries approximate very closely suffix trees (in fact, not only from the depth view point; see Proposition 3 and [DSR]).

THEOREM 5
For all $1 < \beta < \delta^{-1}$, there exists $\epsilon > 0$ such that uniformly for $|u| \leq \beta$: $E[u^{D_n^S}] - E[u^{D_n^T}] = (1-u)O(n^{-\epsilon})$.
PROOF : We use Mellin transform technique applied to RHS of the asymptotic expansion of $Q_n(u)$ presented in Lemma 4. The interested reader is referred to [JS1] for details. ∎

REFERENCES

[AA] A. Apostolico, The Myriad Virtues of Suffix Trees, *Combinatorial Algorithms on Words,* pp. 8596, Springer-Verlag, ASI F12 (1985).

[AHU] A.V. Aho, J.E. Hopcroft and J.D. Ullman, *The Design and Analysis of Computer Algorithms,* Addison-Wesley (1974).

[AS] A. Apostolico, W. Szpankowski, Self-alignments in Words and Their Applications, Purdue CSD-TR-732 (1987); *Journal of Algorithms,* to appear.

[BEH] A. Blumer, A. Ehrenfeucht and D. Haussler, Average Size of Suffix Trees and DAWGS, *Discrete Applied Mathematics,* 24, 37-45 (1989).

[BO] B. Bollobás *Random Graphs,* Academic Press, London (1985).

[DE] L. Devroye, A Note on the Average Depth of Tries, *Computing,* 28, 367-371 (1982).

[DSR] L., Devroye, W. Szpankowski and B. Rais, A note of the height of suffix trees, Purdue University, CSD TR-905 (1989); *SIAM J. Computing,* to appear.

[FL] P. Flajolet, On the Performance Evaluation of Extendible Hashing and Trie Searching, *Acta Informatica,* 20, 345369 (1983).

[FRS] P. Flajolet, M. Regnier and R. Sedgewick, Some Uses of the Mellin Transform Techniques in the Analysis of Algorithms, in *Combinatorial Algorithms on Words,* Springer NATO ASI Ser. F12, 241-254 (1985).

[GO1] L. Guibas and A. Odlyzko Maximal Prefix-Synchronized Codes, *SIAM J. Appl. Math,* 35, 401-418 (1978).

[GO2] L. Giubas and A. Odlyzko, Periods in Strings *Journal of Combinatorial Theory,* Series A, 30, 19-43 (1981).

[GO3] L. Guibas and A. W. Odlyzko, String Overlaps, Pattern Matching, and Nontransitive Games, *Journal of Combinatorial Theory,* Series A, 30, 183-208 (1981).

[HE] P. Henrici, *Applied and Computational Complex Analysis,* John Wiley & Sons (1977).

[JR] P. Jacquet and M. Regnier, Trie Partitioning Process: Limiting Distribution, *Proc. CAAP'86,* Lecture Notes in Computer Science **214,** 194-210 (1986).

[JS] P. Jacquet and W. Szpankowski, Analysis of Tries With Markovian Dependency, Purdue University, CSD TR-906, 1989; *IEEE Trans. Information Theory,* to appear.

[JS1] P. Jacquet and W. Szpankowski, Autocorrelation on Words and Its Applications. Analysis of Suffix Trees by String-Ruler Approach, INRIA TR-1106, 1989.

[KN] D. Knuth, *The Art of Computer Programming. Sorting and Searching,* Addison-Wesley (1973).

[LO] M. Lothaire, *Combinatorics on Words,* Addison-Wesley (1982).

[LZ] A. Lempel and J. Ziv, On the Complexity of Finite Sequences, *IEEE Information Theory* 22, 1, 75-81 (1976).

[MC] E.M. McCreight, A Space Economical Suffix Tree Construction Algorithm, *JACM,* 23, 262272 (1976).

[PI1] B. Pittel, Asymptotic growth of a class of random trees, *The Annals of Probability*, 18, 414 - 427 (1985).

[PI2] B. Pittel, Paths in a Random Digital Tree: Limiting Distributions, *Adv. Appl. Prob.*, 18, 139-155 (1986).

[RJ] M. Regnier and P. Jacquet, New Results on the Size of Tries, *IEEE Trans. Information Theory*, 35, 203-205 (1989).

[RPE] M. Rodeh, V. Pratt and S. Even, Linear Algorithm for Data Compression via String Matching, *Journal of the ACM*, 28, 16-24 (1981).

[SZ1] W. Szpankowski, Some Results on V-ary Asymmetric Tries, *Journal of Algorithms*, 9, 224-244 (1988).

[SZ2] W. Szpankowski, The Evaluation of an Alternating Sum with Applications to the Analysis of Some Data Structures, *Information Processing Letters*, 28, 13-19 (1988).

[SZ3] W. Szpankowski, On the Height of Digital Trees and Related Problems, *Algorithmica*, 6, 256-277 (1991).

[WE] P. Weiner, Linear Pattern Matching Algorithms, *Proc. of the 14-th Annual Symposium on Switching and Automata Theory*, 111 (1973).

[ZL] J. Ziv and A. Lempel, A Universal Algorithm for Sequential Data Compression, *IEEE Information Theory*, 23, 3, 337-343 (1977).

Competitive Algorithms for the Weighted List Update Problem [*]

Fabrizio d'Amore[1]   Alberto Marchetti–Spaccamela[2]

Umberto Nanni[2]

Abstract

In this paper we present some deterministic and randomized algorithms for the *Weighted List Update Problem*. In this framework a cost (weight) is associated to each item. The algorithms consist in modifying the well known Move-To-Front heuristic by adding randomness or counters in order to decide whether moving the accessed item. We prove that *Random Move-To-Front* and *Counting Move-To-Front* are 2-competitive against any *static* adversary, and that deterministic Move-To-Front does not share this property. We apply this approach to the management of non-modifiable trees by means of lists of successors proving that 2-competitivity property still holds.

1   Introduction

The *List Update Problem* has been extensively studied in the literature (see, for example, [12,2,1,6,13,7]). It consists in maintaining a dictionary as an unsorted linear list. While processing a sequence of requests, the list may be rearranged in order to minimize the access cost of subsequent operations. The cost of accessing an item is given by the position in the list the item is currently occupying.

In other cases the cost of visiting an item (i.e. comparing it with the searched item) may depend on the item itself. For example, such a cost could depend on the size of the element (an example is shown in section 4) or in many applications the weight may not necessarily correspond to a computational resource. So, we introduce the *Weighted List Update Problem*, which differs from the previous one because the cost of visiting an item only depends on the item itself, and not on its position in the list [13].

In this paper we consider online algorithms for the weighted version of the problem, comparing the well known Move-To-Front (MTF) [12,2,1,13,7], with the Random Move-To-Front (RMTF), and the Counting Move-To-Front (CMTF).

---

[*]Work supported by the ESPRIT II Basic Research Actions Program Project no. 3075 ("ALCOM") and by the Italian MPI National Project "Algoritmi e Strutture di Calcolo".

[1]Dipartimento di Informatica e Sistemistica, Università di Roma "La Sapienza", via Salaria 113, I-00198 Roma, Italia.

[2]Dipartimento di Matematica Pura ed Applicata, Università di L'Aquila, via Vetoio, Coppito I-67100 L'Aquila, Italia.

For the (unweighted) list update problem it has been shown that MTF is 2-competitive against any static [1] and dynamic algorithm [13,8,7]. On the other hand, Karp and Raghavan noted that no deterministic algorithm for this problem can be better than 2-competitive, so MTF is as good as any deterministic online algorithm [7].

In this paper we prove that, for weighted list problem RMTF and CMTF are both 2-competitive against any static algorithm, while MTF does not share this property, indeed it is not k-competitive against a static algorithm, for any fixed k.

The paper is organized as follows. In the next section the concepts of adversary and c-competitivity are discussed, the weighted list update problem is formally stated and two cost functions for this problem are considered.

In section 3 we present two algorithms, CMTF and RMTF, and prove that both of them are 2-competitive against a static algorithm. CMTF is a deterministic algorithm which makes use of auxiliary memory (one counter per item), while RMTF is a randomized variant which uses no extra space and has expected cost equal to that of CMTF. We also prove that MTF is not k-competitive in the same situation.

Section 4 describes the application of the weighted list update problem to the maintenance of trees in which the structure is not modifiable, and no total order is defined among its vertices. The only possible update operation is rearranging the lists of children of the vertices. If we associate to each vertex the size of its subtree, we can handle the list of the children of any vertex by using any weighted list update heuristic in order to reduce the overall cost of processing a sequence of searches over the tree. For example AND-OR trees and problem solving [10,5], diagnosis [11] are some of the areas which could exploit efficient solutions for this problem [4].

2 Preliminaries

The general framework to analyze the behavior of an online algorithm A while processing a request sequence σ consists in using an *adversary* [3] who generates the sequence σ and is charged the cost of processing the sequence by means of his own algorithm B. His goal is to maximize the ratio of the processing costs $c = A(\sigma)/B(\sigma)$.

Furthermore, we consider two kinds of adversaries: the *oblivious* adversary, who must generate the sequence without knowing the choices of the algorithm A, and the *adaptive* adversary, who can choose the next request on the basis of the previous answers given by the algorithm A. Thus an adversary is characterized on the basis of both the knowledge he has about the behavior of the algorithm A and the performance of his own algorithm B. Hence there are, for example, an *oblivious offline* adversary (often called *week* adversary [3]), an *adaptive online* (or *medium* [3]) adversary, an oblivious or adaptive *static* adversary, meaning that they are charged the cost of processing the generated sequence by means of an offline, an online and a static algorithm respectively.

In general, an algorithm A is c-*competitive* against an oblivious (adaptive) adversary using a given algorithm B if for each weighted list $\mathcal{L}$ and for each request sequence σ:

$$A(\sigma) \leq c \cdot B(\sigma) + f(\mathcal{L})$$

where f does not depend on the sequence choosen by the adversary but only on the handled list $\mathcal{L}$.

Oblivious and adaptive adversaries have the same power against a deterministic algorithm: in the (unweighted) list update problem any deterministic algorithm cannot be better than 2-competitive.

Randomization cannot help against an adaptive offline adversary. In general it is not known whether an adaptive online adversary is actually weaker: for the unweighted list update problem the answer is no [7]. Any c-competitivity result about randomized algorithms refers to oblivious adversaries and, of course, it is intended to provide an estimation of the expected behavior of the randomized algorithm, where the expectation is taken with respect to its random choices.

In [8,7] randomized algorithms for the list update problem are presented with a competitive ratio less than two.

In the weighted version of the problem the cost of visiting an item of the list depends on the element itself. To the best of our knowledge this problem has never been studied. In [13] a cost model was considered in which the cost of an element depends on its position.

Suppose we are given a pair $\langle S, w \rangle$, where S is a set $S = \{e_1, e_2, \ldots, e_n\}$ of elements and $w : S \to R^+$ is a total function mapping the elements of S into the set of positive reals R^+. In what follows we will denote by w_i the value of $w(e_i)$.

The elements of S are stored in a sequential unsorted list $\mathcal{L}$. Since $\mathcal{L}$ is modified during the processing of the sequence σ we will denote by $\mathcal{L}^t$ the list after request σ_t has been processed and by $\mathcal{L}^0$ the initial arrangement of $\mathcal{L}$.

It is easy to verify that for weighted lists the optimal static ordering is obtained by sorting the items by decreasing values of n_i/w_i, where n_i is the number of occurrence of e_i in σ. Such an ordering meets the one pointed out in [1] for non-weighted sequential lists.

In this paper the algorithms are analyzed by using two cost models (analogous to the $i - 1$ and i cost functions used in [7]). In the *wasted work* function the cost of accessing the i-th item of $\mathcal{L}^t$ is given by

$$\sum_{e_j \prec^t e_i} w_j \,,$$

where $e_i \prec^t e_j$ means that e_i is stored in $\mathcal{L}^t$ before e_j. In the *total cost* function the summation is extended to the elements $e_j \preceq^t e_i$ (i.e. including e_i itself).

3 Algorithms for weighted lists

The *Random Move-To-Front* (RMTF) heuristic works as follows: after a searched item e_j has been found, we toss a coin in order to decide whether move e_j to the front of the list or not. Such coin has a probability p_j of answering "yes". We will prove that the choice $p_j = w_{\min}/w_j$, where $w_{\min} = \min_i w_i$, makes this algorithm 2-competitive against an adaptive static adversary. We will also prove that Move-To-Front does not share this property in the case of weighted lists. It was proven that Move-To-Front has this property in the case of unweighted lists [1], and that it is 2-competitive also against an adaptive adversary [13].

The *Counting Move-To-Front* (CMTF) heuristic is completely deterministic and uses counters instead of coins. Any element e_j has associated a counter c_j with real values in $[0, 1)$. After a searched item e_j has been found, c_j is increased by w_{min}/w_j and if it reaches 1 or more, e_j is moved to the front of the list and the counter is decreased by 1. Also this deterministic algorithm is 2-competitive against any adversary using a static algorithm.

The following theorems prove the assertions about RMTF and CMTF.

Theorem 3.1 *For any weight function w, RMTF is 2-competitive in the wasted cost model against an adaptive static adversary.*

Proof. The expected number of times e_i is examined while searching for e_j starting from a given initial arrangement of the list depends only on the relative ordering of e_i and e_j in σ. This is the *pairwise independence property*, pointed out in [1]. In order to compute such contributions we can consider, for any i and j, the compressed sequence $\sigma_{i,j}$ obtained from σ by deleting all the elements out of e_i and e_j.

The theorem will be proved by showing that the expected cost of RMTF does not exceed twice the cost of the static algorithm SA, for *any* arrangement of its list. Namely we will prove that for each cost function w, for each sequence σ and for each static algorithm SA:

$$RMTF(w, \sigma) \leq 2 \cdot SA(w, \sigma) + f(w) \,,$$

where the costs refer to the wasted cost model, and $f(w) < 2 \cdot \sum_{1 \leq i < j \leq n}(w_i w_j)$.

Without loss of generality we assume that the elements in S are numbered according to the ordering of the static algorithm, that is $i < j$ if and only if $e_i \prec e_j$ in the static list.

In order to prove the theorem, we will separately examine (by virtue of the pairwise independence property) the contribution of any pair e_i, e_j to the global cost while processing the entire sequence σ.

The compressed sequence $\sigma_{i,j}$ can be subdivided into s *intervals* on the basis of the behavior or RMTF, namely on the basis of the presence of e_i before e_j (and vice versa) in $\mathcal{L}$ (the list handled by RMTF). Each of the resulting intervals starts when e_i is moved to the front of the list (and then before e_j) and consists of two portions: in the first $e_i \prec e_j$, and in the second $e_j \prec e_i$.

However it is worth observing that after $\sigma_{i,j}$ has been subdivided, its first last portions in general do not constitute an actual interval according to the above definition, but only a partial fragment. In order to keep account of this fact we assume that the SA pays nothing on the fragment and RMTF pays an expected cost $C_{i,j}$, where all the $C_{i,j}$, for $i < j$, sum to C.

Hence we have:

$$RMTF(w, \sigma) = \sum_{1 \leq i < j \leq n} RMTF(w, \sigma_{i,j})$$

$$SA(w, \sigma) = \sum_{1 \leq i < j \leq n} SA(w, \sigma_{i,j}) \,,$$

and the theorem can be restated as

$$\sum_{1 \leq i < j \leq n} RMTF(w, \sigma_{i,j}) + C \leq 2 \cdot \sum_{1 \leq i < j \leq n} SA(w, \sigma_{i,j}) \,. \tag{1}$$

Let $a_{i,j}(j)$ be the expected number of times that RMTF encounters e_i in a single interval of $\sigma_{i,j}$ while searching for e_j because $e_i \prec e_j$, and let $x_{j,i}(j)$ be the expected number of times that the searched item is e_j and $e_j \prec e_i$. It can be easily proven that $a_{i,j}(j) = 1/p_j$.

On the basis of the introduced notation we can write

$$RMTF(w, \sigma) = \sum_{1 \leq i < j \leq n} RMTF(w, \sigma_{i,j}) =$$

$$= \sum_{1 \leq i < j \leq n} \{s[a_{i,j}(j)w_i + a_{j,i}(i)w_j] + C_{i,j}\} ,$$

and

$$SA(w, \sigma) = s \sum_{1 \leq i < j \leq n} [a_{i,j}(j)w_i + x_{j,i}(j)w_i] .$$

Note that $C_{i,j} < a_{i,j}(j)w_i + a_{j,i}(i)w_j$.

A sufficient condition for inequality (1) to hold is:

$$a_{i,j}(j)w_i + a_{j,i}(i)w_j \leq 2[a_{i,j}(j)w_i + x_{j,i}(j)w_i] \tag{2}$$

and this is satisfied if

$$a_{j,i}(i)w_j \leq a_{i,j}(j)w_i ,$$

which is equivalent to

$$p_j w_j \leq p_i w_i . \tag{3}$$

By choosing $p_h = 1/w_h$ this always holds. Furthermore such choice yields $C_{i,j} < 2w_i w_j$, that is

$$C < 2 \cdot \sum_{1 \leq i < j \leq n} (w_i w_j) .$$

$\square$

Theorem 3.2 *For any weight function w, CMTF is 2-competitive against the adaptive static adversary:*

$$CMTF(w, \sigma) \leq 2 \cdot OSA(w, \sigma) + f(w) ,$$

where $f(w) < 2 \cdot \sum_{1 \leq i < j \leq n}(w_i w_j)$.

Proof. The proof is analogous to that of theorem 3.1 and therefore we omit it. We only note that in this case the quantity $a_{i,j}(j)$ is equal to the average number of increments which are needed to make the counter c_j greater or equal to 1. $\square$

As a remark to the proof of theorem 3.1 we observe that the neglect of term $x_{j,i}(j)$ in the right member of inequality (2) is equivalent to eliminate possible locality components in σ, thus disfavouring RMTF when compared to a static algorithm. On the other hand, possible strong locality properties in σ would favour RMTF with respect to any static algorithm: for example, if σ contains a subsequence of k consecutive identical queries (different from the first item of the static list) then RMTF pays the h-th among such queries with probability $(1 - p)^{h-1}$, where $1/p$ is the weight of the searched item, while a static algorithm *certainly* pays each query.

Also, it is worth observing that any choice of p_h proportional to $1/w_h$ would satisfy inequality (3). So, we could choose $p_h = w_{min}/w_h$ for maximizing the probability of moving items to the front of their lists, in order to obtain a faster convergence to the asymptotic behavior of the heuristic.

In the next theorem we prove that MTF is not k-competitive for any k with respect to the optimal static algorithm.

Theorem 3.3 *For each list of $n \geq 2$ items and for any k there exist a weight function w and a sequence σ of requests such that*

$$MTF(w, \sigma) > k \cdot OSA(w, \sigma).$$

Proof. On the basis of what seen in the proof of theorem 3.1 we can restate this theorem as:

$$\sum_{1 \leq i < j \leq n} MTF(w, \sigma_{i,j}) > k \sum_{1 \leq i < j \leq n} OSA(w, \sigma_{i,j}),$$

that is to say

$$s \sum_{1 \leq i < j \leq n} [a_{i,j}(j)w_i + a_{j,i}(i)w_j] > ks \sum_{1 \leq i < j \leq n} [a_{i,j}(j)w_i + x_{j,i}(j)w_i]. \tag{4}$$

Note that in the DMTF we have $a_{i,j}(j) = 1$ for any i, j; hence by choosing the following sequence $\sigma = \gamma, \gamma, \ldots, \gamma$, where $\gamma = e_1, e_2, \ldots, e_n$, we have $x_{j,i}(j) = 0$, for any $i < j$. Hence, inequality (4) becomes:

$$\sum_{1 \leq i < j \leq n} (w_i + w_j) > k \sum_{1 \leq i < j \leq n} w_i.$$

Note that term w_n appears in the left side of the inequality, while lacks in the right one. So, in order to make true the inequality, it suffices to make $w_n > \overline{w}_n$, where $\overline{w}_n$ can be easily computed as a function of $k, w_1, w_2, \ldots, w_{n-1}$. $\square$

Unfortunately, in the wasted work model, CMTF and RMTF are not c-competitive, for any constant c, against an oblivious adversary that uses the optimal offline algorithm. In order to show this, it suffices to show that CMTF is not c-competitive against an adversary who uses MTF (which is not better than the optimal offline algorithm).

Consider the case in which there are only three items and both CMTF and MTF start working on two initially identical lists, containing the items in the following order: $\langle e_1, e_2, e_3 \rangle$. Moreover suppose that $w_1 \geq w_2 \geq w_3$. Thus, counters c_1, c_2 and c_3 will be respectively incremented by the quantities w_3/w_1, w_3/w_2 and 1. For the sequence $\sigma = e_3 \cdot e_2^{\lceil w_2/w_3 \rceil} \cdot e_1^{\lceil w_1/w_3 \rceil}$, we have the following costs:

$$CMTF(\sigma) = (w_1 + w_2) + (w_3 + w_1)\frac{w_2}{w_3} + (w_2 + w_3)\frac{w_1}{w_3}$$

$$MTF(\sigma) = 2(w_1 + w_2 + w_3).$$

It is immediate to verify that the ratio $CMTF(\sigma)/MTF(\sigma)$ can be made greater than any fixed c.

It is interesting to note that such counterexample does not work in the case of the total cost model. On the contrary, it works for RMTF in the case of the wasted cost model.

4 Algorithms for Non-Modifiable Trees

A tree T is said to be non-modifiable if the father-child relationships cannot be modified. Such a structure is interesting when the father-child relationships capture relevant aspects of reality. For example, in decision trees the father-child relationships have a precise meaning and the order of the children of a vertex can affect only the complexity of discovering something, not the result of the search.

If we assume that the tree is searched by means of leftist depth first searches (ldfs's) then it makes sense considering heuristics which modify the relative ordering among siblings so that it results useful having the most frequently successfully accessed vertices as the leftmost ones among a set of siblings. A vertex u is *successfully accessed* if the searched vertex v belongs to the subtree rooted at u. Since the search for a vertex v univocally characterizes a path $\pi(v)$ from the root of the tree to vertex v itself, all $u \in \pi(v)$ are successfully accessed.

Let $T(v)$ be the subtree rooted at v, $w(v)$ be the number of vertices belonging to $T(v)$, and $f(v)$ be the frequency of v in σ. The optimal static ordering for non-modifiable trees is such that for two siblings v' and v'' we say that $v' \prec v''$ if and only if $f(v')/w(v') \geq f(v'')/w(v'')$.

The tree can be managed by means of any algorithm for the weighted list update problem. For example, in the case of RMTF, we perform the following steps. For any request v in σ we search the tree by means of a ldfs. Once v has been found we consider the path $\pi(v)$. For any $u \in \pi(v)$ we toss a coin (whose probability is proportional to $1/w(u)$) in order to decide whether u must be moved before all its siblings. In the case of CMTF, for any $u \in \pi(v)$ we increment its counter by a quantity proportional to $1/w(u)$, and if the counter reaches or exceeds 1 then we subtract 1 to it and move u before all its siblings.

After the searched vertex v has been found, for each $u \in \pi(v)$, we consider the list $\mathcal{L}(u)$ containing u and all its siblings. So, we have $|\pi(v)|$ lists, corresponding to as many weighted list update problems. We can apply one heuristic A for the weighted list update problem to update each of these lists. In the following, $A(T,\sigma)$ denotes the cost of processing a sequence σ of requests of vertices of T by means of A.

In order to estimate the involved wasted cost, we make use of the following theorem.

Theorem 4.1 *The cost of processing a sequence σ of search requests on a non-modifiable tree in the wasted cost model, $A(T,\sigma)$, is equal to the sum of all the wasted costs $A(\mathcal{L}(v),\sigma)$ for all the successfully accessed vertices v during the processing of σ.*

Proof. For each vertex v the weight function is defined as $w(v)$, namely the size of $T(v)$, the subtree rooted at v. The theorem follows from the fact that rearranging the children of a vertex v can only affect the cost of finding vertices in $T(v)$. □

Theorem 4.1 allows us to easily obtain the performance of RMTF and CMTF.

Theorem 4.2 *In the wasted cost model, for any non-modifiable tree T and any σ:*

$$RMTF(T,\sigma) \leq 2 \cdot OSA(T,\sigma) ,$$

$$CMTF(T,\sigma) \leq 2 \cdot OSA(T,\sigma) .$$

Proof. What asserted follows by theorems 3.1, 3.2 and 4.1. □

On the base of what seen we can extend the result on the non-competivity of MTF.

Theorem 4.3 *In the wasted cost model, for any $k \geq 0$ there exists a non-modifiable tree T_k and a sequence σ of requests such that*

$$MTF(T_k,\sigma) > k \cdot OSA(T_k,\sigma) .$$

Proof. The thesis holds by virtue of theorem 3.3 and theorem 4.1. □

5 Conclusions

In order to gain advantage against an oblivious adversary, CMTF can be modified by introducing a random initialization of the counters: this makes it similar to the *COUNTER* algorithm [7].

Besides, we observe that RMTF can be used for unweighted lists by moving accessed items to the front of the list with constant probability p. Using $p = 1/2$ we conjecture this algorithm to be as good as the *BIT* algorithm, which is 1.75-competitive against any oblivious offline adversary [7].

The question of finding c-competitive algorithms for the weighted list update problem against any adaptive offline adversary is still open.

References

[1] J. L. Bentley, and C. McGeogh, Amortized Analyses of Self-Organizing Sequential Search Heuristics, *Communications of the ACM* **28**, 4 (April 1985), 404–411.

[2] J. R. Bitner, Heuristics that Dynamically Organize Data Structures, *SIAM J. of Computing* **8**, 1 (February 1979), 82–110.

[3] S. Ben-David, A. Borodin, R. Karp, G. Tardos, and A. Wigderson, On the Power of Randomization in Online Algorithms, in *Proceedings of the 20th ACM Annual Symposium on Theory of Computing*, May 1990, 379–386.

[4] F. d'Amore, U. Nanni, and A. Marchetti-Spaccamela, Robust Algorithms for Diagnosis, Technical Report, Dipartimento di Informatica e Sistemistica, Univ. of Roma "La Sapienza", 1991.

[5] S. Gnesi, U. Montanari, and A. Martelli, Dynamic programming as graph searching: An algebraic approach, *Journal of ACM* **28**, (1981), 737–751.

[6] J. H. Hester, and D. S. Hirschberg, Self-Organizing Linear Search, *ACM Computing Surveys* **17**, 3 (September 1985), 295–311.

[7] S. Irani, N. Reingold, J. Westbrook, and D. D. Sleator, Randomized Competitive Algorithms for the List Update Problem, in *Proceedings of the 2nd ACM-SIAM Annual Symposium on Discrete Algorithms*, San Francisco, CA, January 1991, 251–260.

[8] S. Irani, Two Results on the List Update Problem, Technical Report TR-90-037, Computer Science Division, U. C. Berkeley, California, August 1990.

[9] M. S. Manasse, L. A. McGeoch, and D. D. Sleator, Competitive Algorithms for On-line Problems, in *Proceedings of the 18th ACM Annual Symposium on Theory of Computing*, May 1988, 322–333.

[10] N. J. Nilsson, *Principles of Artificial Intelligence*, Springer Verlag, (1982).

[11] R. Reiter, A Theory of Diagnosis from First Principles, *Artificial Intelligence* **32**, (1987), 57–95.

[12] R. Rivest, On Self-Organizing Sequential Search Heuristics, *Communications of the ACM* **19**, 2 (February 1976), 63–67.

[13] D. D. Sleator, and R. E. Tarjan, Amortized Efficiency of List Update and Paging Rules, *Communications of the ACM* **28**, 2 (February 1985), 202–208.

[14] R. E. Tarjan, Amortized Computational Complexity, *SIAM J. Alg. Disc. Meth.* **6**, 2 (April 1985), 306–318.

An Optimal Algorithm for the Rectilinear Link Center of a Rectilinear Polygon[*]

Bengt J. Nilsson[†]   Sven Schuierer[†]

Abstract

The problem of finding the center of an area has been studied extensively in recent years. $O(n \log n)$ time upper bounds have been given for the link center and the geodesic center of a simple polygon.

We consider the rectilinear case of this problem, and give a linear time algorithm to compute the rectilinear link center of a simple rectilinear polygon. As a consequence we also obtain a linear time solution for the rectilinear link radius problem. To our knowledge this is the first optimal algorithm which solves a non-trivial center problem.

1 Introduction

In many versions of motion planning problems the large cost involves making turns. An example is broadcasting a radio signal through a beam. At particular points relay stations must be erected to reflect the beam in a new direction. This leads to a cost measure for paths which is commonly known as the *link metric*.

The problem of finding link optimal paths between points inside a simple polygon has been solved by Suri [Sur87]. He gives a linear time algorithm for this problem. In addition he shows that the *link diameter*, the furthest link distance between any two points in a polygon, can be computed in $O(n \log n)$ time and linear storage (n will in the following always denote the number of polygon edges). A similar result is obtained by Ke [Ke89].

The *link center* problem is to find the set of points in a simple polygon for which the link distance to the furthest neighbour is minimum. It was first studied by Lenhart *et al* [LPS*87]. They give a quadratic time algorithm for this problem which was later improved to $O(n \log n)$ by Djidjev *et al* [DLS89] and Ke [Ke89].

In many cases we may want to restrict the paths to be rectilinear, i.e., all segments of the path are axis parallel. This particular restriction has been studied for simple rectilinear polygons by de Berg [B89]. He devises a data structure which allows you to compute, given two query points inside the polygon, the shortest path between the points in $O(\log n + l)$ time, where l is the number of links of the path. The data structure can be constructed in $O(n \log n)$ time and requires $O(n \log n)$ storage. Furthermore, he shows how to compute the rectilinear link diameter in $O(n \log n)$ time.

In the more general case when the obstacles do not form the boundary of a simple polygon but instead are just rectilinear line segments in the plane, a generalization of the shortest path problem has been studied by de Berg *et al* [BKNO90].

In this paper we give a linear time algorithm to compute the rectilinear link center of any simple rectilinear polygon. To our knowledge this is the first optimal algorithm found for the center problem of non-trivial classes of polygons.

The basic idea to solve the link center problem is as follows. First, we compute a chord c which splits the polygon into two parts such that all the points of the link center are "close to" c. For our purposes close means that any point in the link center is at most two links away from c. c can be

---

[*]This work was supported by the Deutsche Forschungs Gemeinschaft under Grant No. Ot 64/5–4.

[†]Institut für Informatik, Universität Freiburg, Rheinstr. 10–12, D-7800 Freiburg, Fed. Rep. of Germany.

constructed by computing a diametral pair of vertices v_1 and v_2 and then extending the middle link of the shortest path between v_1 and v_2 until the end points touch the boundary of the polygon. The chord obtained in this way is a is a good candidate for c though it may have to be shifted a bit in order to conform with our requirements. The second step of the algorithm consists of subdividing the region of points that can reach c with at most two links into histograms. This subdivision will consist of either three histograms or of one histogram plus a (possibly linear) number of histograms that are independent in a way which will be described later. The last part of the algorithm is to apply a subroutine to each of the histograms found. The subroutine computes the part of the link center contained in a histogram in linear time.

The paper is organized as follows. In the next section we state our definitions and give some preliminary results for simple rectilinear polygons. Before we give the actual algorithm we present a necessary subroutine in Section 3 which shows how to compute the link center in linear time for a special case, i.e., the part of the link center contained in a maximal histogram having a rectilinear chord in the polygon as base. In Section 4 we present the actual algorithm to compute the link center given a chord c such that there are points with great distance to c on both sides of c. Since the algorithm we state needs a chord Section 5 deals with the problem of computing the appropriate chord.

2 Definitions

Let P be a Jordan curve consisting of n axis parallel line segments. We define a *simple rectilinear polygon* **P** to be $P \cup$ interior(P). A (*rectilinear*) *path* $\mathcal{P}$ is a curve that consists of (a finite number of) axis-parallel line segments inside **P**. The length of $\mathcal{P}$, denoted by $\lambda(\mathcal{P})$, is the number of line segments it consists of. From now on we will only consider rectilinear polygons and rectilinear paths in a polygon **P**. Hence, whenever we talk of polygons, we mean simple rectilinear polygons and whenever we talk of paths, we mean rectilinear paths in **P**.

Let p and q be two points in a polygon **P** and e be an axis-parallel line segment in **P**. The (*rectilinear*) *link distance* between p and q, denoted by $d(p, q)$, is defined as the length of the shortest path connecting p and q. We say a polygonal path $\mathcal{P}$ from p to e is *admissible* if it is rectilinear and the last link of $\mathcal{P}$ is orthogonal to e. We define the link distance of p and e, again denoted by $d(p, e)$, to be the length of the shortest admissible path from p to a point of e.

We define the (*rectilinear*) *link diameter* $D(\mathbf{P})$ as the maximum rectilinear link distance between any two points in **P**. The (*rectilinear*) *link radius* $R(\mathbf{P})$ is defined as the minimum integer k for which there is some point in **P** from which all other points can be reached with at most k links. When the polygon **P** is given we will refer to the radius as R. Finally, we define the (*rectilinear*) *link center* $lc(\mathbf{P})$ as the set of points in **P** that can reach all other points using at most R links.

In the rest of the paper the letter v with various subscripts and superscripts will always denote vertices of **P**. In the same way the letters p, q, and r denote points of **P**.

It can be shown that, for any point p in **P**, there is a vertex having the furthest distance to p. Hence, the diameter, radius, and link center definitions can be equivalently stated in terms of the vertices of **P**.

We can establish the following relation between the radius and the diameter of a polygon, $\lceil D(\mathbf{P})/2 \rceil \leq R(\mathbf{P}) \leq \lceil D(\mathbf{P})/2 \rceil + 1$ and these bounds are tight. Our link center algorithm assumes a priori knowledge of the the radius of the polygon. We can get around this problem by first running the algorithm with the value $R(\mathbf{P}) = \lceil D(\mathbf{P})/2 \rceil$. If the algorithm produces a non-empty region then we are done. Otherwise we run the algorithm again with the value $R(\mathbf{P}) = \lceil D(\mathbf{P})/2 \rceil + 1$.

The link diameter can be found in linear time; see [NS91]. The algorithm is based on a divide and conquer algorithm presented in [B89]. The idea is to split the polygon into two subpolygons of approximately equal size, compute the diameter of the two subpolygons recursively, and to compute the longest distance between any pair of vertices in different subpolygons. The maximum of these three values is the diameter. This algorithm runs in $O(n \log n)$ time. In order to reduce the running time to linear it is possible to show that one only needs to recur on one of the subpolygons. For the other subpolygon one can show that the diameter is either smaller than the diameter of the full

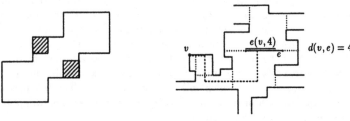

(a) The link center can be disconnected.   (b) A path from v to e of four links.

Figure 1:

polygon or the diameter of the subpolygon can be computed explicitly without recursion and in linear time.

An interesting fact is that the rectilinear link center may be disconnected; see Figure 1a. On the other hand, we can prove that the link center can have at most two components. The diameter for the example in Figure 1a is 3 and the radius is 2.

We will make heavy use of the following construction. For an axis-parallel line segment e and a point p in $\mathbf{P}$, let $e(p, d)$ denote the part of e that can be reached from p with an admissible path of length at most d. de Berg shows in [B89] that, for any axis parallel line segment e of $\mathbf{P}$, the sets of subedges $\{e(v, d(v, e)) \mid v \in \mathbf{P}\}$ and $\{e(v, d(v, e) + 1) \mid v \in \mathbf{P}\}$ can be computed in linear time. It is important to note that $e(p, d_p)$ is a connected set, for any point p and any integer d_p. Because of this property, we will often refer to the subedges as *intervals*. Note that, in particular, the sets of subedges have the property that their intersection is non-empty if and only if the pairwise intersection of any two subedges is non-empty.

Since we rely heavily on these results we give an informal argument as to the truth of these statements. de Berg shows in [B89] that, for any vertex v with $d(v, e) > 2$, there is vertex v_{next} such that the distance of v_{next} to e is one less than $d(v, e)$ and v and v_{next} have the same interval on e. The reason that such a vertex v_{next} exists is, of course, that the second link of a shortest path from v to e can be chosen such that it touches the boundary of $\mathbf{P}$. It can be shown that if we are given the horizontal and vertical vertex-edge visibility pairs computed in [Cha90], v_{next} can be found in constant time. Hence, the family of intervals $\{e(v, d(v, e)) \mid v \in \mathbf{P}\}$ only depends on the intervals of vertices that are exactly two links away and these can be computed in linear time. By keeping a reference at each vertex v to v_{next} we, thus, also obtain the intervals of vertices with $d(v, e) > 2$. For the family $\{e(v, d(v, e) + 1) \mid v \in \mathbf{P}\}$ a similar argument applies. Furthermore, it is not difficult to see that the intervals can be computed in sorted order according to any of the end points.

To see that the subedges are intervals it is enough to consider the histogram having the chord through e as base. A path from a point outside the histogram to e enters the histogram through a *window* and can then make a turn to reach the base. This turning point must lie within the interval between the window and the "opposite" side of the histogram.

A *histogram* is a monotone rectilinear polygon with one of the monotone chains forming a single straight line segment. The single straight line segment forming one of the monotone chains is called the *base* of the histogram. We define a *maximal histogram* $\mathbf{H}$ inside a rectilinear polygon $\mathbf{P}$ having an axis parallel chord c in $\mathbf{P}$ as its base to be the maximum area histogram interior to $\mathbf{P}$ with c as its base. The definition does not yield a unique maximal histogram since $\mathbf{H}$ can extend to either side of c. However, if we specify a direction orthogonal to c the ambiguity can be resolved.

A *window* is a maximal segment of the boundary of a histogram which is not also part of the boundary of $\mathbf{P}$. When the base is horizontal and all interior points of $\mathbf{H}$ lie above the base, we will use the term *west edge* to denote a vertical edge of $\mathbf{H}$ that has the interior to the right of the edge. An *east edge* is defined similarly as a vertical edge having the interior to the left.

3 The Solution of a Special Case

In this section we will establish a solution to a special case of the link center problem. Suppose we are given the radius R of the polygon $\mathbf{P}$ and a maximal histogram $\mathbf{H}$ along an axis parallel chord of the polygon. We consider the question of how to output the part of the link center contained in $\mathbf{H}$.

Let c be a horizontal chord in $\mathbf{P}$ and $\mathbf{H}$ a horizontally monotone maximal histogram in $\mathbf{P}$ with base c. We assume that all interior points of $\mathbf{H}$ lie above c. Let $\mathbf{P}_w$ be the subpolygon cut off by window w of $\mathbf{H}$.

DEFINITION 3.1 The region $\mathbf{A}_w(d)$ associated with window w is the region in $\mathbf{H}$ which can reach all points in $\mathbf{P}_w$ using at most d links.

It is easy to see that for every window w of $\mathbf{H}$, $\mathbf{A}_w(R)$ contains $lc(\mathbf{P}) \cap \mathbf{H}$. Furthermore, we can prove that the intersection of all the regions $\mathbf{A}_w(R)$ is exactly the part of the link center contained in $\mathbf{H}$.

LEMMA 3.1 Let $w_1, \ldots, w_m$ be the windows of $\mathbf{H}$.
If $R = 2$, $lc(\mathbf{P}) \cap \mathbf{H}$ is equal to the intersection $\bigcap_{1 \le i \le m} \mathbf{A}_{w_i}(R)$ intersected with $lc(\mathbf{H})$, the link center of $\mathbf{H}$.
If $R \ge 3$, $lc(\mathbf{P}) \cap \mathbf{H}$ is equal to $\bigcap_{1 \le i \le m} \mathbf{A}_{w_i}(R)$.

PROOF: Assume first that $R \ge 3$. This means that we can disregard the vertices of $\mathbf{H}$, since any such vertex can reach any point in $\mathbf{H}$ with at most three links. Hence, we only need to concern ourselves with the vertices contained in the subpolygons cut off by the windows of $\mathbf{H}$. Take a point $p \in \bigcap_{1 \le i \le m} \mathbf{A}_{w_i}(R)$. From p it is possible to reach the furthest vertex in each subpolygon $\mathbf{P}_{w_i}$ using at most R links. Hence, $\bigcap_{1 \le i \le m} \mathbf{A}_{w_i}(R) \subseteq lc(\mathbf{P}) \cap \mathbf{H}$. To see the reverse inclusion, take a point, $q \in \mathbf{H} \setminus \bigcap_{1 \le i \le m} \mathbf{A}_{w_i}(R)$. This means that q is not interior to some $\mathbf{A}_{w_\ell}(R)$ and therefore the shortest path from q to a furthest vertex in $\mathbf{P}_{w_\ell}$ uses more than R links. Hence, $\bigcap_{1 \le i \le m} \mathbf{A}_{w_i}(R) \supseteq lc(\mathbf{P}) \cap \mathbf{H}$.

When $R = 2$, $lc(\mathbf{H})$ is the area below the lowest horizontal edge of the top chain. From any point in this area it is possible to reach any other point in $\mathbf{H}$ using two links. Thus, $lc(\mathbf{H}) \supseteq lc(\mathbf{P}) \cap \mathbf{H}$ and by a similar argument as in the first part of the proof we get $\bigcap_{1 \le i \le m} \mathbf{A}_{w_i}(R) \cap lc(\mathbf{H}) = lc(\mathbf{P}) \cap \mathbf{H}$. $\square$

The previous lemma shows that in order to find $lc(\mathbf{P}) \cap \mathbf{H}$ we need to compute the intersection $\bigcap_{w \in \mathbf{H}} \mathbf{A}_w(R)$ efficiently.

We will in the following solve a more general problem. We show how to compute, for an arbitrary polygon $\mathbf{P}$ and an arbitrary value k, the intersection $\bigcap_{w \in \mathbf{H}} \mathbf{A}_w(k)$ in linear time (in the size of $\mathbf{P}$). This intersection is the set of points in $\mathbf{H}$, $\mathbf{H}$ being contained in $\mathbf{P}$, which can reach all the vertices of $\mathbf{P}$ with at most k links. In this section we will denote the intersection $\bigcap_{w \in \mathbf{H}} \mathbf{A}_w(k)$ as $\mathbf{R}(\mathbf{H})$. Note that the polygon $\mathbf{P}$ can be a subpolygon of the original polygon for which we want to compute the link center.

To each window w in $\mathbf{H}$ we associate the value $\delta_w = k - d_w$ where $d_w = \max_{v \in \mathbf{P}_w} d(v, w)$. The value δ_w tells us how many free links there can be in $\mathbf{H}$. So when $\delta_w = 0$, the region $\mathbf{R}(\mathbf{H})$ can be at most one link away from the window w. In general, a point in $\mathbf{R}(\mathbf{H})$ can be at most $\delta_w + 1$ links away from w.

If $\delta_w < 0$ then the region $\mathbf{A}_w(k)$ is empty and, hence, $\mathbf{R}(\mathbf{H})$ is also empty. On the other hand, if $\delta_w \ge 3$, then $\mathbf{A}_w(k) = \mathbf{H}$. This means that our algorithm only needs to concern itself with the windows for which $0 \le \delta_w \le 2$ since these are the only cases when the area $\mathbf{A}_w(k)$ is non-trivial.

We will divide the windows into different classes according to their value of δ_w and whether they lie on east or west edges of $\mathbf{H}$. This gives six different classes (in fact seven, since the base c of $\mathbf{H}$ is considered a window. But this class has only one member, c itself). We will show how to find a constant number of regions associated to each class such that the intersection of these regions is $\mathbf{R}(\mathbf{H})$. To achieve this we need to characterize the areas $\mathbf{A}_w(k)$ properly according to the class the window w belongs to. To this end we will use the notion of the d-area.

Let $\mathbf{P}$ be a simple rectilinear polygon and let e be an edge (or part of an edge) of $\mathbf{P}$.

DEFINITION 3.2 We define the d-area (or visibility region) $\mathbf{V}_{\mathbf{P}}(e, d)$ of $\mathbf{P}$ with base e recursively as

$d = 1$ the maximal histogram in $\mathbf{P}$ with e as its base,

$d > 1$ the 1-area of $\mathbf{P}$ with e as its base and the $(d-1)$-areas of the subpolygons, $\mathbf{P}_{w_1}, \ldots, \mathbf{P}_{w_k}$, cut off by the windows, $w_1, \ldots, w_k$, of $\mathbf{V}_{\mathbf{P}}(e, 1)$ with these windows as their bases.

It is easy to show that the d-area $\mathbf{V}_{\mathbf{P}}(e, d)$ contains exactly the points that can reach some point on e with at most d links of an admissible path.

In the following the d-area will always be interior to the maximal histogram we are looking at. This allows us to find nice properties of these areas which make them simple to work with.

Let I_w be the intersection of all the intervals $w(v, d_w)$, where $v \in \mathbf{P}_w$, and let I_w^1 be the interval which has the top end point of w as its upper end point and the topmost of the bottom end points of the intervals $w(v, d_w)$ as its lower end point. Also, let I_w^2 be the interval on w which has the top end point of w as its upper end point and the topmost of the bottom end points of the intervals $w(v, d_w + 1)$, for $v \in \mathbf{P}_w$, as its lower end point.

We state the following result, which characterizes the regions $\mathbf{A}_w(k)$, as a lemma without proof.

LEMMA 3.2 *Let w be a vertical window in $\mathbf{H}$. We have for the associated area $\mathbf{A}_w(k)$ that*

1. $\mathbf{A}_w(k) = \emptyset$ *if $I_w = \emptyset$ and $\delta_w = 0$,*

 $\mathbf{A}_w(k) = \mathbf{V}_{\mathbf{H}}(I_w, 1)$ *if $I_w \neq \emptyset$ and $\delta_w = 0$,*

2. $\mathbf{V}_{\mathbf{H}}(I_w^1, 2) \subseteq \mathbf{A}_w(k) \subseteq \mathbf{V}_{\mathbf{H}}(I_w^1, 2) \cup \mathbf{V}_{\mathbf{H}}(w, 1)$ *if $\delta_w = 1$,*

3. $\mathbf{A}_w(k) = \mathbf{V}_{\mathbf{H}}(I_w^2, 3)$ *if $\delta_w = 2$.*

We can show that the region in each case can be computed in linear time in the size of $\mathbf{P}$.

We will in the following show that it is necessary to compute only a constant number of regions that have to be intersected in order to get the region $\mathbf{R}(\mathbf{H})$. To do this we first prove that some windows in the classes when $1 \leq \delta_w \leq 2$ will not further restrict the region $\mathbf{R}(\mathbf{H})$ and can therefore be disregarded.

LEMMA 3.3 *If w and w' are two vertical windows both lying on west (east) edges of $\mathbf{H}$, with $\delta_w = \delta_{w'} = 1$ or 2, such that w' lies on the boundary of $\mathbf{V}_{\mathbf{H}}(I_w^{\delta_w}, 2)$, then $\mathbf{A}_{w'}(k) \subseteq \mathbf{A}_w(k)$.*

PROOF: We assume w.l.o.g. that the windows w and w' are west edges. We show that

$$\mathbf{A}_{w'}(k) \subseteq \mathbf{V}_{\mathbf{H}}(w', \delta_w + 1) \subseteq \mathbf{V}_{\mathbf{H}}(I_w^{\delta_w}, \delta_w + 1) \subseteq \mathbf{A}_w(k).$$

We can prove the first inclusion using the definition of d-area and Lemma 3.2.

To prove the second inclusion take a point $p \in \mathbf{V}_{\mathbf{H}}(w', \delta_w + 1)$. p can reach some point q on w' with $\delta_w + 1$ links (i.e., 2 or 3) of an admissible path w.r.t. w' and since $q \in \mathbf{V}_{\mathbf{H}}(I_w^{\delta_w}, 2)$, q can reach some point on the interval $I_w^{\delta_w}$ with two links (admissible w.r.t. w). Let p' be the last turning point of the path from p to q. Since the link from p' to q is horizontal we can use the property that $\mathbf{H}$ is a histogram and therefore it is possible to project the points p' and q to the base giving two new points. These four points define an empty rectangle within $\mathbf{H}$. Extend the last link of the path from q to $I_w^{\delta_w}$ until it reaches the other side of the rectangle where it connects to the extension of the second to last link of the path from p to q. Hence, we have constructed a path from p to $I_w^{\delta_w}$ of $\delta_w + 1$ links and, thus, $p \in \mathbf{V}_{\mathbf{H}}(I_w^{\delta_w}, \delta_w + 1)$. See Figure 2 for an example.

The third inclusion follows from Lemma 3.2. $\qquad\square$

The windows which do not help to restrict the region $\mathbf{R}(\mathbf{H})$ can be discarded in linear time by walking along the top boundary of $\mathbf{H}$, pushing windows (of the same class) onto a stack. Whenever the y-coordinate of the boundary reaches a value below the lower y-coordinate of the interval $I_w^{\delta_w}$, where w lies on the top of the stack, w is saved and the stack is emptied. The windows that are kept in this way are called the *relevant* windows. (Note that all windows for which $\delta_w = 0$ are also relevant.)

The next lemma shows that in order for $\mathbf{R}(\mathbf{H})$ to be non-empty, $\mathbf{H}$ can only have two relevant windows in each of the four classes when $\delta_w = 0$ or 1. We state this lemma without proof.

LEMMA 3.4 *If w, w', and w'' are three vertical relevant windows all lying on west (east) edges of $\mathbf{H}$, with $\delta_w = \delta_{w'} = \delta_{w''} = 0$ or 1, then the intersection $\mathbf{A}_w(k) \cap \mathbf{A}_{w'}(k) \cap \mathbf{A}_{w''}(k)$ equals the empty set*

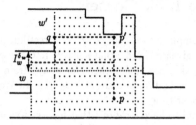

Figure 2: *p can reach both $I_w^{\delta_w}$ and w' with two links.*

This deals with the cases when $\delta_w \leq 1$. It remains to show how to handle the two cases when $\delta_w = 2$.

Let $w_1, \ldots, w_\ell$ be ℓ vertical relevant windows, $\ell \geq 2$, lying on west (east) edges of **H** (in the order of a scan of the boundary of **H**), with $\delta_{w_1} = \cdots = \delta_{w_\ell} = 2$, and let e be the lowest horizontal segment on the boundary of **H** such that w_1 lies before e (again in a scan along the boundary) and w_ℓ lies after e. We state the following lemma without proof.

LEMMA 3.5

$$\bigcap_{1 \leq i \leq \ell} \mathbf{A}_{w_i}(k) = \mathbf{V_H}(e, 2)$$

Computing the region $\mathbf{V_H}(e, 2)$ is easy, using a scan along the top boundary of **H** in order to find the edge e. Two more scans, one towards the right of e and one towards the left, give the top boundary of $\mathbf{V_H}(e, 2)$.

Finally, we look at the class $w = c$, the base of **H**. This class has only one member and, hence, we want to show properties that enable an algorithm to compute $\mathbf{A}_c(k)$ efficiently. It turns out that only the case when $\delta_c = 1$ is difficult. For this case we claim that we can partition the vertices in $\mathbf{P}_c$ into two sets $S_\mathbf{L}$ and $S_\mathbf{R}$ depending on whether the corresponding interval $c(v, d_c + 1)$ extends to the left or right of the interval $c(v, d_c)$. (We know that they always have one end point in common.) We associate one region to each of the two sets. The region **L** is the set of points in **H** from which we can reach all vertices in $S_\mathbf{L}$ with at most k links. The region **R** is defined similarly. We claim that we can compute these two regions in linear time, again using plane sweep as our main tool.

For the base case we state the following lemma without proof. (I_c is the intersection of all the intervals $c(v, d_c)$, where $v \in \mathbf{P}_c$.)

LEMMA 3.6 *Let c be the base of histogram **H**. The corresponding area $\mathbf{A}_c(k)$ is*

1. $\mathbf{A}_c(k) = \emptyset$ *if $I_c = \emptyset$ and $\delta_c = 0$,*
   $\mathbf{A}_c(k) = \mathbf{V_H}(I_c, 1)$ *if $I_c \neq \emptyset$ and $\delta_c = 0$,*
2. $\mathbf{A}_c(k) = \mathbf{L} \cap \mathbf{R}$ *if $\delta_c = 1$,*
3. $\mathbf{A}_c(k) = \mathbf{H}$ *if $\delta_c = 2$.*

For the vertical windows we get at most ten regions and for the base at most two regions that have to be intersected. Since the regions are all monotone with respect to the base of **H**, the intersection can be computed in linear time using a sweep line algorithm. This proves the main result of this section.

THEOREM 1 *Let $\mathbf{P}$ be an arbitrary rectilinear polygon and $\mathbf{H}$ a maximal histogram in $\mathbf{P}$ having c, a rectilinear chord in $\mathbf{P}$, as its base. The algorithm described above computes the region inside $\mathbf{H}$ from which all other points in $\mathbf{P}$ can be reached with at most k links. The algorithm runs in linear time in the size of $\mathbf{P}$.*

In particular we have,

COROLLARY 2 *The region $lc(\mathbf{P}) \cap \mathbf{H}$ can be computed in linear time.*

4 Computing the Link Center

We now turn to the task of computing the link center of a rectilinear polygon P. Throughout this section we assume that we are given a chord c that splits P into two parts P_1 and P_2 which contain vertices $v_1 \in P_1$ and $v_2 \in P_2$ that are *both* rather "far away" from c. For our purposes far away means that the distances from v_1 and v_2 to c are at least $R - 1$. We will show in the next section how to find such a chord in linear time.

To illustrate why it is helpful to have a chord c that has vertices on both sides that are far away we consider the area in P_2 that may contain the link center if we are given a vertex v in P_1 with $d(v, c) = d$. We claim that a point p in the part of the link center in P_2 has a distance of at most $R - d + 1$ to c. To see this, consider a shortest path $\mathcal{P}$ from v to p and let l be the first link of $\mathcal{P}$ that intersects c. Since p is in the link center of P, we have $\lambda(\mathcal{P}) \leq R$. l is at least the dth link of $\mathcal{P}$ as seen from v since $d(v, c) = d$. Hence, the part of $\mathcal{P}$ from c to p consists of at most $R - d$ links plus the part of l that is contained in P_2. Therefore, we have $d(p, c) \leq R - d + 1$. Hence, the bigger d, the more restricted is the area in P_2 that may contain the link center. Since we want to split the polygon only once, we choose c in such a way that the part of the link center of P that is contained in P_2 as well as the part that is contained in P_1 is sufficiently close to c. Since we require $d_1 = \max_{v \in P_1} d(v, c)$ and $d_2 = \max_{v \in P_2} d(v, c)$ to be both at least $R - 1$, "sufficiently close" means that the points in the link center of P have a distance of at most $R - (R - 1) + 1 = 2$ links to c.

It is the aim of this section to show that once we are given c, we can compute the link center of P in linear time. The remaining part of this section is dedicated to proving the following theorem.

THEOREM 3 *Let c be a chord of P with d_1 and d_2 defined as above. If $R \leq d_i + 1$, for $i = 1, 2$, then the parts of the link center in P_1 and P_2 can be computed in linear time.*

4.1 Where to Look for the Link Center

So suppose we are given a chord c in P that meets the conditions of Theorem 3. From now on throughout this section we assume that c is vertical and that P_1 is to the left of c and P_2 to the right. We distinguish two cases.

$$\text{(i)} \quad R \leq d_1 \qquad \text{and} \qquad \text{(ii)} \quad R = d_1 + 1$$

and the same for d_2. We will only show how to compute the part of the link center in P_2. The part of the link center in P_1 can, of course, be computed in the same way. Note that the part of the link center in P_2 is contained in the region R that can be reached from c with $R - d_1 + 1$ links. If $R - d_1 + 1 \leq 1$, i.e., $R \leq d_1$, R is contained in the maximal histogram H_c of c in P_2.

Since we know how to compute the link center of a polygon inside a histogram in linear time by Corollary 2, we can immediately solve Case (i). So consider Case (ii) and assume that $R = d_1 + 1$. From the above observation we have that the link center is contained in the 2-area of c in P_2 since $R - d_1 + 1 = 2$. Unfortunately, we cannot solve the link center problem for general 2-areas directly. So we have to restrict the region in P_2 that may contain the link center a bit more. To this end let $\mathcal{I}_1$ be the family of intervals $\{c(v, d_1) \mid v \in P_1\}$ and I_1 the intersection of all intervals in $\mathcal{I}_1$, i.e., $I_1 = \cap \mathcal{I}_1$.

First, we treat the case that $I_1 = \emptyset$. Let $\bar{I}_1$ be the smallest interval that intersects all the intervals in $\mathcal{I}_1$. If we consider $\bar{I}_1$ as a line segment, we can move it to the right in P_2 until we hit a point x on the boundary of P. Let h be the horizontal line segment from x to $\bar{I}_1$. Note that h is a chord in P_2. Let H_a be the maximal histogram in P_2 above h and H_b be the maximal histogram below h. We can show that if $R = d_1 + 1$ and $I_1 = \emptyset$, then the part of the link center in P_2 is contained in $H_a \cup H_b \cup H_c$ if we denote the maximal histogram of c in P_2 by H_c. We state this as a lemma.

LEMMA 4.1 *If $R = d_1 + 1$ and $I_1 = \emptyset$, then $lc(P) \cap P_2 \subseteq H_a \cup H_b \cup H_c$.*

PROOF: Let p be a point of the link center of P in P_2 outside $H_a \cup H_b$. We show that p is contained

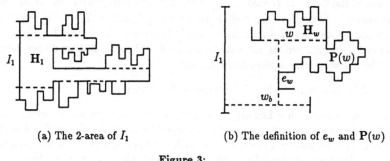

(a) The 2-area of I_1 (b) The definition of e_w and $\mathbf{P}(w)$

Figure 3:

W.l.o.g. assume that p is in $\mathbf{P}_a$. Let v be a furthest neighbour of c in $\mathbf{P}_1$ with $c(v, d_1)$ below $\overline{I}_1$, i.e., below h, and $\mathcal{P}$ a shortest path from p to v. Let l be the first horizontal link of $\mathcal{P}$ that intersects c as seen from v. If l intersects $c(v, d_1)$, then we need at least two more links to reach p since l is below h and p is above it and outside the 1-area of h which, in particular, implies that p cannot be reached from any horizontal link below h with one link. Since l is horizontal and intersects $c(v, d_1)$, it is at least the $d_1{}^{\text{st}}$ link of $\mathcal{P}$ as seen from v. Hence, $\mathcal{P}$ is at least $d_1 + 2$ links long which contradicts the assumption that p can reach v with a path of R links. Therefore, $\mathcal{P}$ does not intersect $c(v, d_1)$. But then l is at least the $(d_1 + 1)^{\text{st}} = R^{\text{th}}$ link of $\mathcal{P}$. By assumption, we have $\lambda(\mathcal{P}) \leq R$ and, thus, l connects to p. Since l intersects c orthogonally, p belongs to $\mathbf{H}_c$. $\square$

The above lemma implies that $lc(\mathbf{P}) \cap \mathbf{P}_2$ can be computed with three applications of the algorithm to compute the link center inside a histogram. By Corollary 2 this takes only linear time.

So we are left with the case that $R = d_1 + 1$ and I_1 is non-empty. It can be easily seen that in this case the part of the link center in $\mathbf{P}_2$ is contained in the union of the 2-area of I_1 in $\mathbf{P}_2$ and $\mathbf{H}_c$. Since we can compute the part of the link center in $\mathbf{H}_c$ efficiently, we will only concern ourselves with the 2-area of I_1 in the remaining part of this section. Unfortunately, it seems not to be possible to decompose the 2-area of I_1 into a constant number of histograms so that we cannot make immediate use of Corollary 2 but, instead, we have to take a different approach.

4.2 Computing the Link Center in the 2-Area of I_1

In the remaining part of this section we show how to compute the part of the link center that is contained in the 2-area of I_1. Recall that the 2-area of I_1 consists of $\mathbf{H}_1$ together with a number of maximal histograms that belong to the north and south windows of $\mathbf{H}_1$. An example is shown in Figure 3a.

Unfortunately, our algorithm—which is presented subsequently—to compute the part of the link center in the 2-area of I_1 does not work unless we are given a vertex v in $\mathbf{P}_2$ with $d(v, I_1) \geq R$. Now, although in "most" cases such a vertex exists, it may happen that we have $d(v, I_1) < R$, for all vertices v in $\mathbf{P}_2$. In this case it is possible to split I_1 into two subintervals I_{11} and I_{12} such that $I_1 = I_{11} \cup I_{12}$ and, for each interval I_{1i}, there is a vertex v_i with $d(v_i, I_{1i}) \geq R$. We then apply our algorithm to the 2-areas of I_{11} and I_{12} separately and, thus, obtain the part of the link center of $\mathbf{P}$ in the 2-area of I_1.

Since the only property of I_{1i} we make use of is that all points in I_{1i} can reach any point in $\mathbf{P}_1$ with d_1 links, we assume for simplicity that we do not have to split up I_1 and that there is a vertex v with $d(v, I_1) \geq R$. In the following we denote the maximal histogram of I_1 in $\mathbf{P}_2$ by $\mathbf{H}_1$. Let w_b be the window of that cuts off the part of $\mathbf{P}_2$ v belongs to. W.l.o.g. we assume that w_b is a south window.

Since we cannot compute the part of the link center of $\mathbf{P}$ in the 2-area of I_1 directly, we have to find out which parts do not have to be considered. w_b splits $\mathbf{P}_2$ into two polygons, the polygon that is below w_b which we denote by $\mathbf{P}_{w_b}^b$ and the polygon above w_b which we denote by $\mathbf{P}_{w_b}^a$. Note that the only part of the 2-area of I_1 that in $\mathbf{P}_{w_b}^b$ is the maximal histogram $\overline{\mathbf{H}}$ below w_b. Since the part of the link center in $\overline{\mathbf{H}}$ can be computed in linear time by Corollary 2, we only consider $\mathbf{P}_{w_b}^a$.

In the following we show that a number of the histograms of the 2-area of I_1 cannot intersect the link center of $\mathbf{P}$. So what are the parts of the 2-area of I_1 above s_b that we do not have to take into account? The following lemma gives a first answer.

LEMMA 4.2 *The link center in $\mathbf{P}_{w_b}^a$ is contained in the 2-area of w_b.*

An immediate consequence Lemma 4.2 is that we can disregard all the south windows in the 2-area of I_1 and also all the north windows that cannot be reached with one link from w_b.

So we only have to concern ourselves with north windows that intersect the maximal histogram $\mathbf{H}_{w_b}$ above w_b. Since we can compute the part of the link center in the maximal histogram $\mathbf{H}_{w_b}$ above w_b in linear time, we only concern ourselves with windows w that are not completely contained in $\mathbf{H}_{w_b}$. We say that w is *partially visible* from w_b in this case. We denote the leftmost vertical edge of $\mathbf{H}_1$ that is below w by e_w and the part of $\mathbf{P}_2$ that is below w and to the right of the line segment from the upper end point of e_w to w by $\mathbf{P}(w)$ (for illustration refer to Figure 3b). Since w is partially visible from w_b, e_w has two reflex vertices and can be reached from w with one link. Now consider a window w' that is below e_w and above w_b. Let $\mathbf{P}_{w'}$ be the part of $\mathbf{P}_2$ that is cut off by w'. The following lemma shows that $\mathbf{H}_w$ contains a part of the link center only if the vertices in $\mathbf{P}_{w'}$ are close enough to w'.

LEMMA 4.3 *A point p in the intersection of $\mathbf{H}_w \setminus \mathbf{H}_{w_b}$ and the 2-area of w_b can reach all the vertices in $\mathbf{P}_{w'}$ with R links if and only if $d(v', w') \le R - 3$, for all $v' \in \mathbf{P}_{w'}$.*

PROOF: We first show that if $p \in \mathbf{H}_w \setminus \mathbf{H}_{w_b}$ is in the link center of $\mathbf{P}$, then any vertex v in $\mathbf{P}_{w'}$ can reach w' with $R - 3$ links. To see this let $\overline{t_w}$ be the maximal chord through e_w and t_w the part of $\overline{t_w}$ above e_w. Since we can reach all the points in $\mathbf{H}_w$ to the left of t_w with one link from w_b, p is to the right of t_w. Furthermore, w' is also to the right of $\overline{t_w}$ since we chose w' to be below e_w. Let h be the horizontal line segment from some point on e_w to c. Clearly, w' is below h. Therefore, a shortest path from p to vertices in $\mathbf{P}_{w'}$ has two horizontal links that intersect $\overline{t_w}$; hence, an admissible path from p to w' consists of at least four links. Now let v be a vertex in $\mathbf{P}_{w'}$. Since $d(p, w') \ge 4$ and $d(p, v) \le R$, this implies that $d(v, w') \le R - 3$.

On the other hand, suppose v can reach w' with $R - 3$ links. We show that $d(p, v) \le R$ in this case. p belongs to the 2-area of w_b; hence, there is path $\mathcal{P}$ of length 2 from a point on w_b to p. Since p is to the right of t_w and cannot reach w_b with one link, the last link of $\mathcal{P}$ is horizontal and intersects t_w in a point p'. Now let v' be some point on w' that v can reach with $R - 3$ links. We project v' orthogonally onto $\overline{t_w}$ (which is always possible since we are in a histogram) and call the projection v''. The path from v' to v'' to p' to p consists of three links; hence, p can reach v with R links. $\square$

So let w' be the lowest window of $\mathbf{H}_1$ with $d(v', w') > R - 3$, for some vertex $v' \in \mathbf{P}_{w'}$. By the above lemma we can disregard all the windows w where e_w is above w'. Since we can find all the edges e_w in linear time, the north windows w with e_w above w' can also be computed in linear time. These will not be considered further.

In the following let s_w be the maximal line segment in $\mathbf{P}_2$ that contains window w. In our next step we show that not only the vertices below e_w may lead to the removal of $\mathbf{H}_w$ from the possible locations of points of the link center of $\mathbf{P}$ but also that the vertices above s_w may yield that no point in $\mathbf{H}_w$ can belong to $lc(\mathbf{P})$. So in the next lemma we deal with the vertices in $\mathbf{U}_w$ where $\mathbf{U}_w$ denotes the part of $\mathbf{P}_2$ above s_w.

LEMMA 4.4 *A point p in the maximal histogram $\mathbf{H}_w$ of north window w can reach all the points in the part $\mathbf{U}_w \setminus \mathbf{P}_w$ with R links if and only if all the intervals $c(v, R-1)$, with $v \in \mathbf{U}_w$, contain the left end point of s_w.*

PROOF: Let v be some vertex above s_w but not in $\mathbf{P}_w$. First assume that $c(v, R-1)$ contains p_l. Hence, there is an admissible path $\mathcal{P}$ of length $R - 1$ from v to p_l. Let l_1 be the first vertical link that intersects s_w as seen from v. Since the last link of $\mathcal{P}$ is horizontal, l_1 is at most the $(R - 2)^{\text{nd}}$ link of

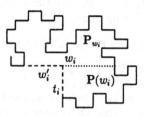

(a) $\mathbf{P}_{w_i} \cup \mathbf{P}(w_i)$ contains the remaining vertices.

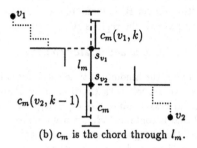

(b) c_m is the chord through l_m.

Figure 4:

$\mathcal{P}$. Now, p can reach any point on s_w with two links. Therefore, we can connect p to $\mathcal{P}$ yielding a path of length $R - 2 + 2 = R$.

Now suppose that there is a path $\mathcal{P}$ of length R from p to v. At least two vertical links l_1 and l_2 of $\mathcal{P}$ intersect s_w. Hence, the first vertical link l_1 that intersects s_w (as seen from v) is again at most the $(R-2)^{\text{nd}}$ link of $\mathcal{P}$. Since the intersection point of l_1 and s_w can be connected to p_l with one more horizontal link, we have $d(v, p_l) \leq R - 1$ and $p_l \in c(v, R-1)$. □

It can be easily seen that the north window that satisfy the conditions of Lemma 4.4 can be found in time linear in the number of vertices of $\mathbf{P}_2$. Thus, we obtain a set of north windows $w_1, \ldots, w_k$ with the following properties. The north windows $w_1, \ldots, w_k$ are partially visible to w_b, a point in the maximal histogram $\mathbf{H}_{w_i}$ above w_i can reach all the vertices that belong to the part of $\mathbf{P}_2$ that is below e_{w_i} and above w_b with R links and the same is true for all the vertices that belong to $\mathbf{U}_{w_i} \setminus \mathbf{P}_{w_i}$.

The link center can be viewed as the region that all vertices can reach with R links. Above we have shown that "most" vertices can reach the points in $\mathbf{H}_{w_i}$, $1 \leq i \leq k$ with R links. Which vertices have we not yet considered? First of all, all the vertices that are below w_b and secondly, the vertices that are between s_{w_i} and e_{w_i}. These are exactly the vertices that belong to $\mathbf{P}(w_i)$. Finally, we also have to consider the vertices of $\mathbf{P}_{w_i}$. So in the remaining part of this section we have to show how to compute the regions of points in $\mathbf{H}_{w_i}$ that can be reached by all the vertices in $\mathbf{P}(w_i) \cup \mathbf{P}_{w_i}$ and all the vertices below w_b with R links. Taking the intersection of these two regions then yields the part of the link center in $\mathbf{H}_{w_i}$.

We first consider the vertices below w_b. So let v_1 be a vertex in the part $\mathbf{P}_{w_b}$ of $\mathbf{P}_2$ below w_b whose interval $w_b(v_1, R-1)$ has the leftmost right end point and v_2 a vertex in $\mathbf{P}_{w_b}$ whose interval $w_b(v_2, R-1)$ has the rightmost left end point of all intervals $w_b(v, R-1)$ with v in $\mathbf{P}_{w_b}$. We have the following result.

LEMMA 4.5 A point p in the intersection of the 2-area of w_b and $\mathbf{H}_{w_i} \setminus \mathbf{H}_{w_b}$ can reach all the vertices below w_b with R links if and only if it can reach $w_b(v_1, R-1)$ and $w_b(v_2, R-1)$ with two links.

We now describe how to compute the 2-area of $w_b(v_1, R-1)$ in $\mathbf{H}_{w_i}$. Clearly, the parts w_i' of w_i that can be reached from $w_b(v_1, R-1)$ with one link can be computed in linear time using the information the edges e_{w_i}. All the points in $\mathbf{H}_{w_i}$ that can reach $w_b(v_1, R-1)$ with two links belong to the 2-area of w_i' in $\mathbf{H}_{w_i}$ which can be computed in time proportional to the size of $\mathbf{H}_{w_i}$. Since the 2-area of a subedge of the base in a histogram is monotone, we can compute the intersection of the 2-areas of $w_b(v_1, R-1)$ and $w_b(v_2, R-1)$ in $\mathbf{H}_{w_i}$ also in time proportional to the number of vertices in $\mathbf{H}_{w_i}$. We denote the region in $\mathbf{H}_{w_i}$ that can be reached by all vertices below w_b with R links by $\mathbf{R}_i$. Hence, the total amount of time spent for computing the regions $\mathbf{R}_i$ in all $\mathbf{H}_{w_i}$, $i = 1, \ldots, k$, is $O(\sum_{i=1}^{k} |\mathbf{H}_{w_i}|) = O(n)$.

Finally, we have to consider the vertices in $\mathbf{P}(w_i) \cup \mathbf{P}_{w_i}$, for each w_i, $i = 1, \ldots, k$. Let t_i be the chord from the upper end point of e_{w_i} to w_i and w_i' the part of w_i from t_i to the left. The part $\mathbf{P}_i'$ of $\mathbf{P}_2$ that is cut off by $t_i \cup w_i'$ just contains the vertices of $\mathbf{P}(w_i) \cup \mathbf{P}_{w_i}$ (see Figure 4a).Hence, if we view $\mathbf{P}'$ as a simple polygon, we can apply Theorem 1 and compute the part $\mathbf{R}_i'$ of $\mathbf{H}_{w_i}$ that can be reached by all vertices in $\mathbf{P}(w_i) \cup \mathbf{P}_{w_i}$ with R links in time proportional to $|\mathbf{P}(w_i) \cup \mathbf{P}_{w_i}|$. Again the resulting region $\mathbf{R}'_i$ is monotone and of size $O(|\mathbf{P}(w_i) \cup \mathbf{P}_{w_i}|)$. In order to estimate the time needed to compute all the regions $\mathbf{R}'_i$, $i = 1, \ldots, k$ we need the following lemma.

LEMMA 4.6 *If w and w' are two partially visible windows, then $\mathbf{P}(w)$ and $\mathbf{P}(w')$ are disjoint.*

Therefore, the amount of time spent for this step is $O(\sum_{i=1}^{k} |\mathbf{P}(w_i) \cup \mathbf{P}_{w_i}|) = O(|\mathbf{P}_2|) = O(n)$.

So we are left with the task of intersecting the regions $\mathbf{R}_i$ and $\mathbf{R}'_i$ both of which are monotone w.r.t. to x-axis and of size $O(|\mathbf{P}(w_i) \cup \mathbf{P}_{w_i}|)$, for $i = 1, \ldots, k$. This yields the part of the link center of $\mathbf{P}$ that is contained $\mathbf{H}_{w_i}$. Since intersecting two monotone polygons requires only time linear in the size of the polygons, this lasts step again takes time $O(\sum_{i=1}^{k} |\mathbf{P}(w_i) \cup \mathbf{P}_{w_i}|) = O(n)$ which completes the algorithm and shows that the part of the link center in the 2-area of I_1 can be computed in linear time.

5   Finding a Splitting Chord

In this section we show how to find a chord c in $\mathbf{P}$ that splits $\mathbf{P}$ into two parts $\mathbf{P}_1$ and $\mathbf{P}_2$ such that $\max_{v \in \mathbf{P}_1} d(v, c)$ and $\max_{v \in \mathbf{P}_2} d(v, c)$ differ by at most one from the radius R of $\mathbf{P}$. Given c Theorem 3 of the previous section enables us to find the part of the link center in $\mathbf{P}_1$ and $\mathbf{P}_2$ in linear time.

The rest of this section is concerned with showing that we can find a chord with the above mentioned property in linear time. A good candidate for c is the chord c_m through the middle link l_m of a path between two vertices that span the diameter D of $\mathbf{P}$. So let v_1 and v_2 be two vertices with $d(v_1, v_2) = D$ and let $\mathcal{P}$ be a shortest path between them. We will only concern ourselves with the case that D is even. The case that D is odd can be treated in essentially the same way but turns out to be a little more involved. So assume from now on that $D = 2k$.

It is our aim to show that we can find a chord c with $d(v_1, c) = k$ and $d(v_2, c) = k$. Note that once we have found c it automatically satisfies the conditions of Theorem 3 since we have that $R \le \lceil D/2 \rceil + 1 = k + 1$ as mentioned in Section 2 and, hence, $R - d_1 \le 1$ and $R - d_2 \le 1$. Recall that d_1 (d_2) is the maximal distance of a vertex in the part of $\mathbf{P}$ to the left (right) of c to c.

Let l_m be the $(k+1)^{\text{st}}$ link of $\mathcal{P}$ as seen from v_1 and c_m the maximal chord through l_m. We have $d(v_1, c_m) = k$ and $d(v_2, c_m) = k - 1$. Suppose that c_m is vertical, that v_1 is to the left of c_m and, furthermore, that $c_m(v_1, k)$ is above $c_m(v_2, k-1)$ (see Figure 4b). Clearly, $c_m(v_1, k)$ and $c_m(v_2, k-1)$ do not intersect for otherwise we have that $d(v_1, v_2) \le k + k - 1 < D$. c_m nearly satisfies the above conditions except that v_2 is still to close to it. The idea is now to move c_m further away from v_2 without decreasing the distance to v_1.

To this end let s_{v_1} be the bottommost point that v_1 can reach with k links on c and s_{v_2} the topmost point v_2 can reach with $k - 1$ links. We now move the line segment l from s_{v_1} to s_{v_2} to the left until it intersects some vertical edge e on the boundary of $\mathbf{P}$. We denote the new line segment we obtain by l' and its lower end point by s'_{v_2}. Let c be the chord from the top vertex v_t of e upward to the next boundary point. In order to avoid degeneracies we can assume that c is a little bit to the left of v_t. It is easy to see that v_1 is still to the left of c and that we have $d(v_1, c) = k$. Clearly, v_2 is to the right of c. To see that $d(v_2, c) = k$ also holds note that v_t is above s'_{v_2} since we chose e to intersect l'. Now s_{v_2} is the topmost point on c_m v_2 can reach with an admissible path of $k - 1$ links. Since we do not encounter a boundary point when moving l to the left, v_2 cannot reach any point above s'_{v_2} on l' with an admissible path of length $k - 1$. Hence, this is also true for all points on c and we have that $d(v_2, c) = k$.

We have shown that by choosing c as a splitting chord we obtain a chord that satisfies the conditions of Theorem 3. Clearly, we can compute c in linear time. This completes the algorithm.

6   Conclusions

In this paper we have studied the concept of rectilinear link distance within simple rectilinear polygons. We have given an optimal linear time algorithm to compute the rectilinear link center. This algorithm computes a partial cover of histograms in which we can efficiently compute the link center. To our knowledge this is the first optimal algorithm for a non-trivial center problem.

An open problem is of course to generalize the technique for polygons and path links that come from some fixed set of orientations. It may be possible to improve on the $O(n \log n)$ time upper bound which exists for general polygons. Here, one problem that arises is that the regions that have to be intersected may no longer be monotone with respect to a common line.

In a rectilinear world consisting of axis parallel line segments you can also define the link center. There exists a simple $O(n^4)$ time algorithm to solve the link center problem in this case and a lower bound of $\Omega(n^2)$ time since the link center may consist of a quadratic number of regions. It would be of interest to tighten these two bounds.

References

[B89]   MARK DE BERG. *On Rectilinear Link Distance.* Technical Report RUU-CS-89-13, Department of Computer Science, University of Utrecht, P.O.Box 80.089, 3502 TB Utrecht, the Netherlands, May 1989.

[BKNO90]   M.T. DE BERG, M.J. VAN KREVELD, B.J. NILSSON, M.H. OVERMARS. Finding Shortest Paths in the Presence of Orthogonal Obstacles Using a Combined L_1 and Link Metric. In *Proc. 2nd Scandinavian Workshop on Algorithm Theory*, Lecture Notes in Computer Science 447, pages 213–224, 1990.

[Cha90]   BERNARD CHAZELLE. Triangulating a Simple Polygon in Linear Time. In *Proc. 31th Symposium on Foundations of Computer Science*, pages 220–230, 1990.

[DLS89]   H.N. DJIDJEV, A. LINGAS, J.-R. SACK. An $O(n \log n)$ Algorithm for Computing a Link Center in a Simple Polygon. In *Proc. STACS*, Lecture Notes in Computer Science 349, pages 96–107, 1989.

[Ke89]   YAN KE. An Efficient Algorithm for Link-distance Problems. In *Proceedings of the Fifth Annual Symposium on Computational Geometry*, pages 69–78, ACM, ACM Press, Saarbrücken, West Germany, June 1989.

[LPS*87]   W. LENHART, R. POLLACK, J.-R. SACK, R. SEIDEL, M. SHARIR, S. SURI, G. TOUSSAINT, S. WHITESIDES, C. YAP. Computing the Link Center of a Simple Polygon. In *Proceedings of the Third Annual Symposium Computational Geometry*, pages 1–10, ACM, ACM Press, Waterloo, Ontario, Canada, June 1987.

[NS91]   B.J. NILSSON, S. SCHUIERER. Computing the Rectilinear Link Diameter of a Polygon. In *Proc. Workshop on Computational Geometry, CG91*, Lecture Notes in Computer Science, 1991, To Appear.

[Sur87]   SUBHASH SURI. *Minimum Link Paths in Polygons and Related Problems.* PhD thesis, Johns Hopkins University, Baltimore, Maryland, August 1987.

Geometric Searching and Link Distance
(Extended Abstract)

Gautam Das   Giri Narasimhan

Mathematical Sciences Department
Memphis State University
Memphis, TN 38152

ABSTRACT

Given n orthogonal line segments on the plane, their *intersection graph* is defined such that each vertex corresponds to a segment, and each edge corresponds to a pair of intersecting segments. Although this graph can have $\Omega(n^2)$ edges, we show that breadth first search can be accomplished in $\mathcal{O}(n \log n)$ time and $\mathcal{O}(n)$ space. As an application, we show that the *minimum link rectilinear path* between two points s and t amidst rectilinear polygonal obstacles can be computed in $\mathcal{O}(n \log n)$ time and $\mathcal{O}(n)$ space, which is optimal. We mention other related results in the paper.

1   Introduction

Given a set S of n orthogonal line segments on the plane, the *intersection graph* of S is defined as follows: each segment corresponds to a vertex in the graph, and an edge connects a pair of vertices if the two corresponding segments intersect. Clearly this graph is bipartite with the horizontal and vertical segments forming two independent sets.

This graph can potentially have $\Omega(n^2)$ edges. Given an initial segment h, the main result of this paper is an efficient algorithm to label every segment by its shortest distance from h in the graph. This is equivalent to breadth first search. Our algorithm does not explicitly generate all the edges, and in fact runs in $\mathcal{O}(n \log n)$ time and $\mathcal{O}(n)$ space. It uses elementary data structures such as binary trees and priority queues. Using this algorithm we solve several other problems efficiently.

Asano and Imai have shown how to perform breadth first and even depth first search on such graphs in $\mathcal{O}(n \log n)$ time and $\mathcal{O}(n \log n)$ space ([IA86], [IA87]). They use fairly complex data structures. However their techniques have wider applications.

The primary application of our result is motivated by a motion planning problem. Suppose we are given a collection of disjoint rectilinear polygonal obstacles inside a rectilinear polygonal room on the plane, and a pair of points s and t in the free space. Let the input size be n. Consider the problem of computing the *minimum link rectilinear path* between s and t that avoids all the obstacles. Such a path is composed of a minimum number of line segments, each of which is parallel to either the x or the y axis. The number of line segments on the path is known as the *link distance* between the pair of points. We first reduce the problem to performing *two* breadth first searches on an intersection graph of $\mathcal{O}(n)$ orthogonal line segments. The technique in [IA86] and [IA87] can accomplish these searches in $\mathcal{O}(n \log n)$ time and $\mathcal{O}(n \log n)$ space. Instead, we show how to use our algorithm to solve the problem in $\mathcal{O}(n \log n)$ time and linear space. We also solve several other related problems.

Before we summarize our results, we introduce the following definitions. Given some rectilinear polygonal obstacles inside a polygonal room, let H (resp. V) be the set of horizontal (vertical) line segments formed by extending each horizontal (vertical) polygonal edge within the free space (possibly in both directions) until it hits an obstacle or the room. Our results are listed below, and assume rectilinear geometry.

Results

1. Given a collection of polygonal obstacles inside a room, and their H and V sets, we show that breadth first search on the intersection graph of $H \cup V$ can be accomplished in $\mathcal{O}(n \log n)$ time and $\mathcal{O}(n)$ space, using elementary data structures.
2. Given *any* collection S of line segments, we can perform breadth first search on their intersection graph in $\mathcal{O}(n \log n)$ time and $\mathcal{O}(n)$ space. This is a generalization of the previous result.
3. Given a collection of polygonal obstacles inside a room, and points s and t, the minimum link path can be computed in $\mathcal{O}(n \log n)$ time and $\mathcal{O}(n)$ space, which is optimal.
4. Given a collection of polygonal obstacles in a room, and a point s, we can preprocess the input in $\mathcal{O}(n \log n)$ time and $\mathcal{O}(n)$ space such that given a query point t, its minimum link path from s can be reported in $\mathcal{O}(\log n + k)$ time, where k is the link distance. Its link distance from s can be reported in $\mathcal{O}(\log n)$ time.

Remarks

Our results are interesting for several reasons. Firstly, as was noted by Imai and Asano, extending their scheme in [IA86] to perform breadth first search in $\mathcal{O}(n)$ space seems difficult, although we note that their technique is quite general and easily allows other searching schemes such as depth first search. Secondly, to achieve our bounds, we employ several interesting ideas. To start with, we show that if the line segments were the same as those generated by extending the edges of a group of polygons, then the claimed bounds can be achieved. The algorithm uses a key idea of searching the horizontal and vertical lines *separately*, which allows us to exploit various geometric properties. We then generalize this result for *any* set S of line segments by showing that there exists a group of polygons such that $H \cup V = S$. Given S, these polygons can be computed in $\mathcal{O}(n \log n)$ time and $\mathcal{O}(n)$ space. Interestingly, this is related to the problem of computing *many faces* in an *arrangement* of line segments [EGS88].

Next, consider the applications to link distance problems, which is motivated by motion planning. Here a robot has to navigate amidst obstacles inside a cluttered workspace, where translational motion is considered cheap while directional changes are considered expensive. Thus we seek to minimize the links along the path. Several results have appeared for the nonrectilinear versions of the problem. Suri has studied the problem of minimum link distances inside a simple polygon without holes, and has obtained linear time optimal algorithms [S86]. Several people have also studied related concepts such as the *link diameter* and the *link center* of the region ([S90], [LPS87], [K89], [DLS89]). The rectilinear link distance problem has been studied for the case of a simple polygon without holes ([B91]). The problem becomes more complex when we allow multiple obstacles. Recently Mitchell et al. [MRW90] have designed an algorithm for computing nonrectilinear link distances amidst multiple polygons, and their algorithm runs in $\mathcal{O}(n^2 \alpha(n) \log^2 n)$ time and $\mathcal{O}(n^2)$ space, where $\alpha(n)$ is the inverse Ackermann's function.

The above result is suboptimal, and in fact this is generally true for most existing algorithms for shortest path problems amidst polygonal obstacles under various distance metrics. For example, consider the shortest path problem under the *Euclidean metric*. The best known algorithm runs in $\mathcal{O}(n^2)$ time and $\mathcal{O}(n^2)$ space [GM87]. Under rectilinear geometry, this problem can be solved in $\mathcal{O}(n \log^2 n)$ time and $\mathcal{O}(n \log n)$ space [CKV87]. Optimal algorithms have only been obtained for very restricted inputs. For example, if the input is a collection of rectangles, then the rectilinear Euclidean shortest path can be computed in $\mathcal{O}(n \log n)$ time [DLW89]. This is possible by a plane sweep because it can be shown that such paths have to be monotone in some direction.

In our case, we allow rectilinear polygons of arbitrary shapes, and the resulting minimum link paths need not be monotone in any direction. Yet our algorithm runs in optimal time and space.

The rest of this paper is organized as follows. In Section 2, we discuss the breadth first search algorithm (Result 1). Section 3 discusses how to generalize this algorithm where the input is an arbitrary set of orthogonal line segments (Result 2). Several problems related to minimum link distances are discussed in Section 4 (Results 3 and 4). We conclude with some open problems.

2   Labelling Algorithm

For this algorithm we are given a collection of disjoint rectilinear polygonal obstacles inside a polygonal room in the plane along with their H and V sets (as described in the Introduction). We are also given a distinguished line segment h in $H \cup V$. Without loss of generality we assume that h is a horizontal line segment. Consider the intersection graph I obtained by all these horizontal and vertical line segments. In this section we show how to perform an efficient breadth-first search in this graph starting from h so that every line segment is labelled with its distance (in I) from h.

For the sake of simplicity, we will assume throughout this section that all the x-coordinates (resp. y-coordinates) of the vertical (resp. horizontal) sides of the polygons are distinct, although our algorithm can be modified to tackle the general case.

Outline

An efficient search in I is achieved by searching in two different planar partitions – the horizontal planar partition, and the vertical planar partition. The horizontal (vertical) planar partition, which is denoted by HPP (VPP), includes $H(V)$ and all the edges of the obstacles. The algorithm works in two parts, each of which is divided into many phases. In the first part, the algorithm searches in the HPP to compute the shortest path from h to all the other horizontal segments. In the second part, the algorithm performs a search in the VPP to compute the shortest path from h to all the vertical segments.

We first describe the search performed in the HPP. In each phase the algorithm determines the set of horizontal line segments that are at distance 2 away from the horizontal line segments discovered in the previous phase. We denote the set of new horizontal segments chosen in phase k by H_k with $H_0 = \{h\}$. One way to imagine the algorithm is by using ideas from Suri et al. [S90]. One could imagine that in phase k the algorithm determines all new horizontal segments that get illuminated by placing light sources on the horizontal segments in H_{k-1} and shining them in the upward or downward direction. Hence in each phase the algorithm effectively jumps two steps by going from a set of horizontal segments H_{k-1} (at distance $2(k-1)$ from h) to a set of horizontal segments H_k (at distance $2k$ from h).

In the second part, the algorithm does a search in the VPP to label all the vertical line segments. In phase k of the second part, the algorithm labels all vertical segments at distance $2k-1$ from h in I.

Intuitively, the search is made efficient by maintaining two "complementary" data structures (HPP and VPP) instead of one. Maintaining these two data structures separately reduces the information stored about all the intersections between the segments. However, there is enough information stored in each of the structures to perform the requisite search.

It is possible to alternate the phases of the search in the HPP with that of the search in the VPP. This will ensure that all segments labelled k will be processed before any of the segments labelled $k + 1$.

We now get into more details of the labelling algorithm.

Searching the HPP

The set H_0 consists only of the line segment h. In phase k the algorithm computes the set of new horizontal segments H_k that get illuminated by placing a light source on some segment from H_{k-1}.

We first describe an outline of phase k, which consists of *Initialization*, *UpSweep*, and *DownSweep* procedures. In the initialization procedure, the algorithm places light sources on the set of segments in H_{k-1}. It is clear that not all line segments in H_{k-1} need to be attached with a light source. We only need to attach light sources to portions of an "outermost" set of segments (as shown in Figure 2.1). The upsweep procedure directs all the light sources in the upward direction and labels (with label $2k$) all the new horizontal segments that get illuminated in the process. The horizontal segments are labelled in sorted order, sorted according to the y-coordinates of these segments. This is followed by downsweep, which directs all the light sources in the downward direction, and labels (with label $2k$) as many new segments as possible (in the order of decreasing y-coordinates). Figures 2.1, 2.2, and 2.3 illustrate the three procedures.

We now describe the upsweep procedure in more detail. At the start of an upsweep, the outermost line segments in H_{k-1} will be called *Fronts*, and beams of light are directed upwards from portions of each Front. Each such beam is called a *Window*. The upsweep consists of a sequence of upward advancements of Fronts. The next Front that is selected is the one that advances to the horizontal segment above with the least y-coordinate. This is facilitated by using a priority queue. These new Fronts get labelled $2k$. Effectively, the light sources are moved from the previous segment to the illuminated portions of these new segments. However, the original beams (i.e., windows) are likely to get modified (*narrowed, split* or *terminated*) if they encounter obstacle edges. Each Front is thus a dynamic collection of disjoint windows (see Figure 2.4 for examples). For a particular Front, when all its windows have terminated (which happens when all the light beams have been terminated by obstacles), the Front is also terminated. The upsweep terminates when all the Fronts terminate.

There are two main data structures used by the upsweep procedure. Since the set of windows in a Front are a set of disjoint intervals, the windows associated with a Front will be maintained as a balanced tree structure that maintains disjoint intervals. The balanced tree structure used here must allow the following operations: *Search, Insert, Merge, Split, Delete* (see Figure 2.4 for examples). Each of these operations can be performed in $\mathcal{O}(\log n)$ time, assuming any of the basic balanced tree structures like red-black trees. The second data structure is the priority queue, which is maintained as a heap. This facilitates the following operations: *DeleteMin, Insert*, both of which can be performed in $\mathcal{O}(\log n)$ time.

The following are some of the subtle problems that the upsweep correctly handles. Firstly, propagating a set of windows to the new Front can be achieved by a constant number of elementary operations on the windows data structure of the previous Fronts. Secondly, using a heap structure effectively eliminates the possibility of a new segment being processed by different advancing Fronts at different times. Thirdly, a Front could encounter a segment of a previous phase. If this happens, the Front is immediately terminated. The reason is that either everything above it has already been processed, or there will be another Front that will carry on the advancement in that direction. This is illustrated in Figure 2.2, where Front c is terminated when it advances and hits Front b, since Front a would illuminate any new segments that c would have.

The downsweep procedure is similar to the upsweep procedure. There is one situation that needs further explanation. It is possible that a Front encounters a segment that was processed by the upsweep. However, this is ignored, and the sweep continues. The reason is that the downward sweep may have more segments that need to be labelled that could not possibly have been processed by an upward sweep. Thus some set of line segments could get processed twice within a phase (but not more number of times). No line segment is processed once in two different phases. This is crucial for our analysis.

Searching the VPP

This part of the algorithm is also divided into phases. The first phase involves finding out all the vertical segments that intersect h. These set of segments can also be set up as a collection of Fronts that can illuminate other vertical segments by shining light in the right or the left direction. From this point onwards, the algorithm proceeds along the lines of the earlier part of searching the HPP, with the difference that the search is done in the VPP. It suffices to say that in phase k, it will determine the vertical segments at distance $2k - 1$ from h in I.

Analysis of Algorithm

Each line segment is processed in exactly one phase. Processing a line segment involves: 1) removing it from the heap structure, which can be performed in $\mathcal{O}(\log n)$ time; 2) modifying the Windows data structure, which can be performed in $\mathcal{O}(\log n)$ time; 3) finding the next segment to be inserted into the heap (which takes $\mathcal{O}(1)$ time), and inserting that segment into the heap (which again takes $\mathcal{O}(\log n)$ time). Within each phase, each line segment may be processed twice, once along an upward sweep, and once along a downward sweep. Consequently, it is clear that the algorithm runs in time $\mathcal{O}(n \log n)$.

The heap data structure uses $\mathcal{O}(n)$ space. At any instant in the algorithm there are at most n Windows active in all the Fronts put together. Hence the space complexity of the algorithm is $\mathcal{O}(n)$.

3 Arbitrary Sets of Orthogonal Line Segments

In this section we will show that given *any* set S of orthogonal line segments, breadth first search on their intersection graph can be accomplished in $\mathcal{O}(n \log n)$ time and linear space. To do this, we shall compute a collection of polygons inside a polygonal room such that $S = H \cup V$, where H and V are formed by extending each horizontal (vertical) edge of the polygons in the free space (possibly in both directions) until it hits the room or another polygon. After this is done, the algorithm in Result 1 can be used.

Let $S = H1 \cup V1$, where $H1$ ($V1$) is the set of horizontal (vertical) line segments in S. Consider the *arrangement* [EGS88] of the segments in S on the plane, as Figure 3.1 shows. Let us compute those faces of the arrangement (including the external face) that contain an endpoint of some segment in S. These *faces* will be our polygonal obstacles, with the room being the external face. We will shortly describe how to efficiently compute these faces. For now, let us imagine that these faces have been computed, and we are ready to run the algorithm in Result 1 on the intersection graph of their H and V segments. The following facts hold (some of them are trivial).

1. $H = H1$ and $V = V1$. Thus the intersection graph of $H \cup V$ is the same as the intersection graph of S.

2. If the intersection graph of S is disconnected then some of the obstacles may themselves have holes within which other obstacles may reside. However, this is not a problem for our algorithm in Result 1, because it will only restrict the search to one component.

3. The collection of polygons may have several horizontal (vertical) edges sharing the same y (x) co-ordinate. Again, this is not a problem because the algorithm in Result 1 can be modified to handle such degeneracies.

Thus, all we need to do first is to compute these faces efficiently. Let P be the set of $2n$ endpoints of all segments in S. Our problem is a particular instance of a more general problem, which is as follows. Given a set of arbitrarily oriented segments S and a point set P, compute all faces of the arrangement of S that contain some point in P. This problem has been solved in [EGS88], unfortunately it has a running time which is unacceptable for our purposes. However, we can exploit the orthogonality of our S and modify the algorithm to run in $\mathcal{O}(n \log n)$ time and linear space. For this it will help if we briefly review the algorithm in [EGS88].

The algorithm works roughly as follows. S is divided into two sets S_1 and S_2 of approximately equal size. Then the faces in the arrangement of S_1 (S_2) which contain points of P are recursively computed. Each face resembles a polygon with holes. The faces from S_1 (S_2) are known as the *red* (*blue*) polygons. The algorithm then performs what is known as a *red-blue merge*, which outputs every connected component of the intersections between the red and blue polygons that contain points of P. This merge is performed by a plane sweep, and if all the red (blue) polygons had r (b) edges the running time of the merge is $\mathcal{O}((r + b + |P|)\log(r + b + |P|))$ [EGS88]. The algorithm uses a priority queue as a data structure and requires $\mathcal{O}(r + b + |P|)$ space.

In our case we can considerably simplify the above algorithm. We can eliminate recursion and perform the red-blue merge only once. Partition S into $H1$ and $V1$. We know that horizontal (vertical) segments do not intersect among themselves. Thus the arrangement of $H1$ ($V1$) has *only one* multiply connected face which is the entire plane, with the horizontal (vertical) segments representing "holes". Let us call the face of $H1$ ($V1$) the red (blue) face. We run the red-blue merge only once with these two polygons and the point set P as input. The output is clearly the set of obstacles we are seeking. Since r, b, and $|P|$ are each $\mathcal{O}(n)$, the algorithm runs in $\mathcal{O}(n\log n)$ time and $\mathcal{O}(n)$ space.

In the next section we shall discuss some applications of our breadth first search algorithms.

4   Minimum Link Path Problems

In this section we consider the following problem. Suppose we are given a collection of rectilinear polygons inside a polygonal room, and points s and t within the free space. Let the total input size be n. We are required to compute the minimum link path from s to t which avoids all the obstacles. We have designed an algorithm for solving the problem in $\mathcal{O}(n\log n)$ time and $\mathcal{O}(n)$ space. In this version of the paper, we describe the algorithm for finding the link distance. We will only briefly outline how to modify the algorithm for computing the actual path.

Consider the VPP and HPP of the set of polygons, with s and t being treated as point obstacles. Both partitions can be computed in $\mathcal{O}(n\log n)$ time and $\mathcal{O}(n)$ space by a plane sweep algorithm described in [FM84]. It is easy to see that there exists a minimum link path which is confined to the *grid* formed by overlaying VPP on HPP. Of course we do not want to compute the grid as it will be too time consuming. Let S be the set of the horizontal segments of HPP and the vertical segments of VPP. Clearly s is associated with a horizontal (vertical) segment h_s (v_s), and similarly t is associated with a horizontal (vertical) segment h_t (v_t). The minimum link path has to start along either h_s or v_s, and end along either h_t or v_t.

We now make two copies of each partition, called HPP_h, HPP_v, VPP_h and VPP_v. The labelling algorithm is first run with h_s as the initial segment. In this run the partitions HPP_h and VPP_h are labelled. The algorithm is then run with v_s as the initial segment. In this run the partitions HPP_v and VPP_v are labelled. At this stage we have four partitions with each segment labelled. Consider for example VPP_h. The label associated with v_t, say k, tells us that the minimum link path from s to t which *starts* along h_s and *ends* along v_t has $k+1$ links. By examining all four partitions, we can find out the minimum link distance from s to t if the path originated along either h_s or v_s and terminated along either h_t or v_t. The minimum of all four values gives us the link distance between s and t. To compute the actual path, we have to modify the labelling algorithm to keep back pointers so that the actual path can be retrieved by following pointers. We omit the details in this version of the paper.

Clearly the algorithm runs in $\mathcal{O}(n \log n)$ time and $\mathcal{O}(n)$ space. Notice that if we had performed the breadth first searches as in [IA86], our space complexity would have been nonlinear.

Optimality Proof

By reducing integer sorting to the the minimum link path problem, we show that our algorithm is optimal.

Assume that you are given n integers $a_1, \ldots, a_n$. Construct a polygon, P_i for each a_i. P_i is a strip of width $\epsilon = 0.1$ connecting the following points: $(a_i, a_i - \epsilon), (a_i, -a_i), (-a_i, -a_i), (-a_i, a_i), (a_i, a_i)$. It leaves a gap of width ϵ in the top right corner of the square region that the strip encloses. Hence, if $a_i > a_j$ then P_i completely encloses P_j. Now consider the problem of determining the minimum link path from the origin to $(B, 0)$, where B is 1 more than the largest integer in the input. The minimum link path will have to extricate itself from each of the polygonal regions. In the process it must sort the numbers.

Link Distance Query Problem

We next consider the *query* version of the above problem. Suppose we are given a collection of polygons in a room, and a point s. We have to preprocess the input into a data structure such that given any query point t, its minimum link path from s can be reported efficiently.

We describe an algorithm which takes $\mathcal{O}(n \log n)$ preprocessing time and $\mathcal{O}(n)$ space, and answers each query in $\mathcal{O}(\log n + k)$ time where k is the link distance between s and t. In fact we describe in detail an $\mathcal{O}(\log n)$ algorithm for reporting the link distance, and briefly outline how that may be modified to extract the path.

We first compute the HPP and VPP of the free space, treating s as a point obstacle. The point s corresponds to two line segments h_s and v_s. We then make two copies of each partition, called HPP_h, HPP_v, VPP_h and VPP_v. The labelling algorithm is first run with h_s as the initial segment. In this run the partitions HPP$_h$ and VPP$_h$ are labelled. The algorithm is then run with v_s as the initial segment. In this run the partitions HPP$_v$ and VPP$_v$ are labelled. At this stage we have four partitions with each segment labelled. Finally we organize each partition into a data structure for *planar point location* queries [K83]. All of the above can be done in $\mathcal{O}(n \log n)$ time and $\mathcal{O}(n)$ space.

We are now ready for query processing. Given a t, for each of the four partitions we find out the respective rectangle that contains it. Consider for example VPP$_h$. Suppose t is contained in a rectangle, both of whose vertical sides were labelled k. This means that the minimum link path from s to t which starts along h_s and ends along a vertical segment through t has $k + 1$ links. But suppose the two vertical sides had different labels. Clearly they cannot differ by more than one, so let one be k and the other be $k + 1$. This means that the minimum link path from s to t which starts along h_s and ends along a vertical segment through t has $k + 2$ links.

Thus we can find out the minimum link distance from s to t if the path originated along either h_s or v_s and terminated along either a horizontal or vertical segment through t. The minimum of all four values gives us the link distance between s and t. Clearly all four point locations can be performed in $\mathcal{O}(\log n)$ time.

To compute the actual path, we have to modify the labelling algorithm to keep back pointers so that the actual path can be retrieved by following pointers. We omit the details in this version of the paper.

5   Conclusions

In this paper we show that breadth first search can be accomplished in $\mathcal{O}(n \log n)$ time and $\mathcal{O}(n)$ space in an intersection graph of n orthogonal line segments. The main idea behind the algorithm is that it searches the horizontal and vertical lines separately. We apply it to several link distance problems to obtain optimal algorithms. We conclude with some open problems.

1. Can depth first search be done on a set of orthogonal line segments in $\mathcal{O}(n \log n)$ time and linear space?
2. Are there other applications for the techniques used in this paper, namely that of maintaining separate data structures for the horizontal and vertical line segments?
3. Can the *Link Diameter*, and the *Link Center* problems [S90] be solved more efficiently in the rectilinear case?

6 References

[B91]   de Berg, On Rectilinear Link Distance, *Computational Geometry: Theory and Application*, to appear.

[CKV87]   Clarkson, Kapoor, and Vaidya, Rectilinear Shortest Paths through Polygonal Obstacles in $\mathcal{O}(n \log^2 n)$ time, *ACM Symposium on Comp. Geometry*, 1987.

[DLS89]   Djidjev, Lingas, and Sack, An $\mathcal{O}(n \log n)$ Algorithm for Finding a Link Center in a Simple Polygon, *Proceedings of Sixth STACS, Lecture Notes in Computer Science, Springer Verlag Series*, 1989.

[EGC88]   Edelsbrunner, Guibas, and Sharir, The Complexity of Many Faces in Arrangements of Lines and Segments, *ACM Symposium on Comp. Geometry*, 1988.

[FM84]   Fournier, and Montuno, Triangulating a Simple Polygon and Equivalent Problems, *ACM Trans. on Graphics*, 1984.

[GM87]   Ghosh, and Mount, An Output Sensitive Algorithm for Computing Visibility Graphs, *IEEE FOCS*, 1987.

[IA86]   Imai, and Asano, Efficient Algorithm for Geometric Graph Search Problems, *SIAM J. of Comp.*, 1986.

[IA87]   Imai, and Asano, Dynamic Orthogonal Segment Intersection Search, *J. of Algorithms*, **8** (1987), pp. 1-18.

[K89]   Ke, An Efficient Algorithm for Link Distance Problems, *ACM Symposium on Comp. Geometry*, 1989.

[K83]   Kirkpatrick, Optimal Search in Planar Subdivision, *SIAM J. of Computing*, 1983.

[LPS87]   Lenhart, Pollack, Sack, Seidel, Sharir, Suri, Toussaint, Whitesides, and Yap, Computing the Link Center of a Simple Polygon, *ACM Symposium on Comp. Geometry*, 1987.

[MRW90]   Mitchell, Rote, and Woeginger, Minimum Link Paths among Obstacles in the Plane, *ACM Symposium on Comp. Geometry*, 1990.

[RLW87]   de Rezende, Lee, and Wu, Rectilinear Shortest Path with Rectangular Barriers, *Discrete and Comp. Geometry*, 1987.

[S86]   Suri, A Linear Time Algorithm for Minimum Link Paths inside a Simple Polygon, *Computer Vision, Graphics, Image Processing*, 1986.

[S90]   Suri, On some Link Distance Problems in a Simple Polygon, *IEEE Trans. on Robotics and Automation*, 1990.

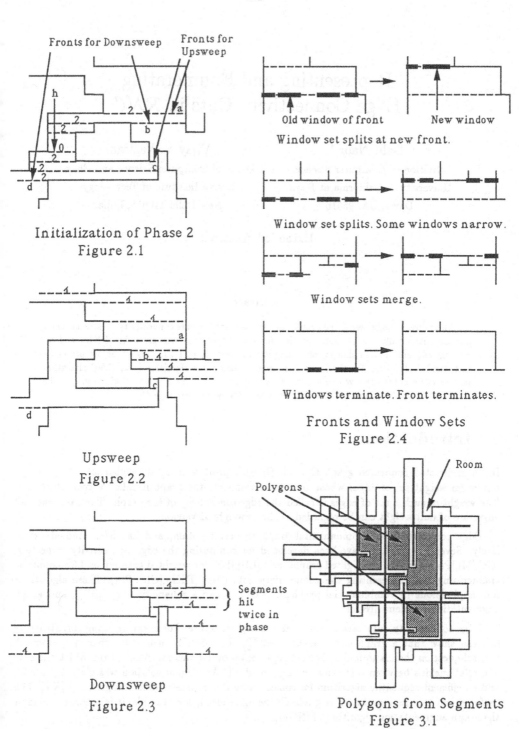

Fronts for Downsweep   Fronts for Upsweep

h

2
2
2
0
2
2
2

a
b

c

d

Initialization of Phase 2
Figure 2.1

Old window of front   New window

Window set splits at new front.

Window set splits. Some windows narrow.

Window sets merge.

Windows terminate. Front terminates.

Fronts and Window Sets
Figure 2.4

1
1
1
a
b 1
1
1
c

d

Upsweep
Figure 2.2

1
1
1

1

1   1

Segments
hit
twice in
phase

Downsweep
Figure 2.3

Room

Polygons

Polygons from Segments
Figure 3.1

Representing and Enumerating
Edge Connectivity Cuts in $\mathcal{RNC}$

Dalit Naor *
Division of Computer Science
University of California at Davis
Davis, CA. 95616

Vijay V. Vazirani
Dept. of Computer Science and Engg.
Indian Institute of Technology
New Delhi 110016, India.

Extended Abstract

Abstract

An undirected edge-weighted graph can have at most $\binom{n}{2}$ edge connectivity cuts. A succinct and algorithmically useful representation for this set of cuts was given by [DKL], and an efficient sequential algorithm for obtaining it was given by [KT]. In this paper, we present a fast parallel algorithm for obtaining this representation; our algorithm is an $\mathcal{RNC}$ algorithm in case the weights are given in unary. We also observe that for a unary weighted graph, the problems of counting and enumerating the connectivity cuts are in $\mathcal{RNC}$.

1 Introduction

In an undirected connected graph $G = (V, E)$ with positive integral weights $wt : E \rightarrow Z^+$, a minimum weight set of edges whose removal disconnects the graph is called a connectivity cut. The weight of such a set of edges is called the edge-connectivity of the graph. The case when all edges are of unit weight can be regarded as the unweighted version.

Edge connectivity is a fundamental graph theoretic notion, and has been studied extensively. Several algorithms have been developed for computing the edge connectivity value (e.g. [Ga],[M],[Po, ADK] for unweighted graphs, and [EH],[ET] for weighted graphs), and for counting, enumerating and representing the connectivity cuts ([BP],[DKL],[GN],[PS]). These algorithms find applications in combinatorial problems [PQ2], network reliability (e.g. [BP]), and graph augmentation problems [NGM].

Edge connectivity cuts have a rich combinatorial structure. For example, using this structure one can show that a graph can have at most $\binom{n}{2}$, i.e. $O(n^2)$, such cuts, where $|V| = n$. A natural question then is to find a succinct representation for this set of cuts, from which the cuts and relationships between them can be easily read off. An elegant solution was given by [DKL], and an efficient sequential algorithm for constructing this representation was given by [KT]. The usefulness of this representation is gradually being realized; for example, it is a corner stone in the graph augmentation algorithm of [NGM].

In this paper, we give a fast parallel algorithm for constructing this representation. For unweighted graphs, or if the weights are given in unary, our algorithm is an $\mathcal{RNC}$ algorithm; the only

*Research partially supported by a grant CCR-8803704 from the National Science Foundation.

step requiring randomization is the computation of max-flow in unary weighted graphs [KUW], [MVV]. Our algorithm uses new structural properties which may be of independent interest. Also, our algorithm may be a step towards parallelizing the graph augmentation algorithm of [NGM]. We also observe that the problems of counting and enumerating all connectivity cuts in an unweighted (or unary weighted) graph are in $\mathcal{RNC}$. There are not very many natural enumeration problems that have only polynomially many solutions and, to our knowledge, this is the only one that can be solved fast in parallel. Besides this, we know of only two counting problems that can be solved in $\mathcal{NC}$: counting the number of spanning trees in an undirected graph, and counting the number of Eulerian circuits in a directed graph.

[DKL] represent the set of connectivity cuts of G by an undirected edge-weighted graph $\mathcal{H} = (U, F)$ (also denoted by $\mathcal{H}(G)$). Each node in U consists of a subset (possibly empty) of distinct vertices from V, and the partition of V induced by a connectivity cut of $\mathcal{H}$ turns out to be a connectivity cut of G. Every connectivity cut of G is represented in $\mathcal{H}$ in this manner. Moreover, the structure of $\mathcal{H}$ is simple enough that from it these cuts can be enumerated in time linear in their number, and other information, such as whether two cuts are crossing or not, can also be easily read off.

In case G has no crossing cuts, the structure of $\mathcal{H}$ is particularly simple: it is a tree. This happens, in particular, if the edge connectivity is odd. In general, $\mathcal{H}$ can have cycles; however, two cycles can share at most one node. We call such a graph a cactus. In this paper, we first handle the case that $\mathcal{H}$ is a tree. To deal with the case of a cactus, we first restrict attention to a subset of the connectivity cuts of G; the cuts chosen are non-crossing, and the set is maximal with respect to this property. Moreover, the cuts are chosen in such a way that a representation (tree) for these cuts can be easily constructed, and easily modified to yield the cactus.

The works of [DKL] and [KT] have remained relatively unknown in the West. So, in Sections 2 and 3 we provide the background by defining the structure of connectivity cuts and their representation $\mathcal{H}(G)$, and in Section 4 we outline the main ideas behind the sequential algorithm of [KT]. Finally, the parallel algorithm is presented in Section 5.

2 Background - the Crossing Property

Consider an undirected graph $G = (V, E)$ with n node, m edges, and edge connectivity λ. A connectivity cut whose removal from the graph disconnects it into two sets of nodes, A and $\overline{A}$, is denoted by $(A, \overline{A})$. Since G may contain several connectivity cuts, it is natural to ask whether the connectivity cuts are structured in some nice manner. An important property of connectivity cuts, namely the Crossing Property, reveals that they are highly structured, and that their number is quite small. We define this property precisely.

Definitions - (1) The cuts $(X, \overline{X})$ and $(Y, \overline{Y})$ are Crossing Cuts if and only if all four sets $X \cap Y$, $\overline{X} \cap Y$, $\overline{X} \cap \overline{Y}$ and $X \cap \overline{Y}$ are non-empty. **(2)** For $U, W \subset V$, $d(U, W)$ is the sum of the weights on edges between vertices in U and vertices in W, and $d(U)$ is the sum of the weights on edges between vertices in U and vertices in $\overline{U}$, where $\overline{U} = V \setminus U$.

Lemma 2.1 [Crossing Cuts Lemma] *Let $(X, \overline{X})$ and $(Y, \overline{Y})$ be two crossing connectivity cuts. Then*

1. $d(X \cap Y, \overline{X} \cap Y) = d(\overline{X} \cap Y, \overline{X} \cap \overline{Y}) = d(\overline{X} \cap \overline{Y}, X \cap \overline{Y}) = d(X \cap \overline{Y}, X \cap Y) = \lambda/2$

2. $d(X \cap Y, \overline{X} \cap \overline{Y}) = d(X \cap \overline{Y}, \overline{X} \cap Y) = 0$

This lemma is proved in [B, Lemma 1], [DKL] and indirectly in [K1,K2]. Its immediate corollary is that if all weights are integral (in particular, if the graph is unweighted) and if λ is odd then there are no crossing connectivity cuts in G, since it is impossible to have $d(U) = \lambda/2$ for any subset U as required by the lemma. Another important corollary is the Circular Partition Lemma ([B, Lemma 2],[DKL]).

Lemma 2.2 [Circular Partition Lemma] *If there are crossing connectivity cuts in G then the set of vertices V can be partitioned into $k \geq 3$ disjoint subsets $V_1, \ldots, V_k$ that can be oredered on a cycle with the following property:*

(i) the number of edges between any two **adjacent** *sets V_i and $V_{(i+1)\bmod k}$ on the cycle is exactly $\lambda/2$.*

(ii) the number of edges between any two **non-adjacent** *sets of nodes on the cycle is 0.*

(iii) For any $1 \leq a < b \leq k$, if $A = \cup_{i=a}^{b-1} V_i$ then the cut $(A, \overline{A})$ is a connectivity cut, and for any connectivity cut $(B, \overline{B})$ which is not of the above form, $B \subset V_i$ for some i.

Note that the lemma does not imply that there is only one circular partition in the graph; there may be many, but they do not cross, i.e. one is a partition of a single subset, say V_i, of the other. A bound on the number of connectivity cuts can be derived from this lemma ([B], [DKL], [K1]). It can be shown by induction that the number of non-crossing cuts in G is linear in n, and that the number of crossing cuts is at most $\binom{n}{2}$, i.e. $O(n^2)$.

3 Representation $\mathcal{H}(G)$

Dinits, Karzanov and Lomosonov [DKL] derived a simple and elegant representation for the set of all connectivity cuts in a weighted graph. The set of all connectivity cuts can be compactly represented by an edge-weighted graph $\mathcal{H}(G)$ of $O(n)$ nodes and edges, and this representation can be constructed in $O(\lambda n^2) = O(nm)$ time, where $|E| = m$, if G is unweighted (if G is weighted, then the time is dominated by n max-flow computations) [KT].

Each vertex in G maps to exactly one node in $\mathcal{H}(G)$, so that any node in $\mathcal{H}(G)$ corresponds to a subset (possibly empty) of vertices from G. A cut $(S, \overline{S})$ in $\mathcal{H}(G)$ induces a cut $(A, \overline{A})$ in G where A consists of all the vertices from G that are mapped into nodes in S. The edge-connectivity of $\mathcal{H}(G)$ is also λ, and each connectivity cut in $\mathcal{H}(G)$ gives a partition that corresponds to a connectivity cut in G; each connectivity cut in G corresponds to one or more connectivity cuts in $\mathcal{H}(G)$. Hence, the connectivity cuts in $\mathcal{H}(G)$ compactly represent all connectivity cuts in G.

If G has no crossing cuts, in particular if λ is odd, then $\mathcal{H}(G)$ is particularly simple – it is a tree, all its edges are of weight λ, and any connectivity cut in $\mathcal{H}(G)$ is obtained by removing one of its edges. If there are crossing cuts in G then $\mathcal{H}(G)$ can contain cycles, but any two cycles can have *at most* a single node in common, so no edge is in more than one cycle. We call such a graph a cactus. Every edge in a cycle is called a *cycle-edge* and is given weight $\lambda/2$, and each of the other edges is called a *tree-edge* and is given weight λ. Hence, connectivity cuts in $\mathcal{H}(G)$ are of exactly two types: a cut of the first type is obtained by removing a tree-edge, and a cut of the second type is obtained by removing any pair of cycle-edges that lie on the same cycle. Every tree is a cactus with no cycles; hence, in order to unify our discussion we will always refer to $\mathcal{H}(G)$ as a cactus, unless we specify otherwise. Throughout the paper we use the term "vertex" to denote a vertex in

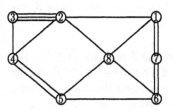

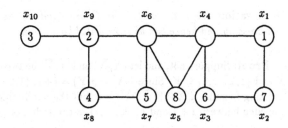

Figure 1: A graph G and its cactus representation $\mathcal{H}(G)$ ($\lambda = 4$)

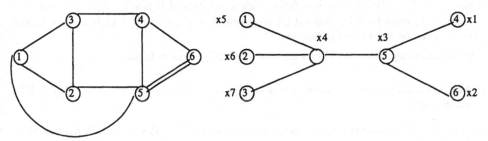

Figure 2: A graph G and its tree representation $\mathcal{H}(G)$ ($\lambda = 3$)

G, and the term "node" to denote a node in $\mathcal{H}(G)$. In Figures 1 and 2 we give two examples of a graph G and its connectivity cuts, represented by $\mathcal{H}(G)$: in one example $\mathcal{H}(G)$ is a cactus, and in the other it is a tree.

4 Sequential Construction of $\mathcal{H}(G)$

This section gives a high-level outline of the sequential algorithm of [KT]. Let $V = \{1, 2, \ldots, n\}$ be an arbitrary numbering of the vertices of G. For each connectivity cut $(A, \overline{A})$ assume that vertex 1 is in A and define $j(A)$ to be the smallest indexed vertex that is separated from 1 by the cut $(A, \overline{A})$, i.e. $\{1, \ldots, j(A) - 1\} \in A$ and $j(A) \in \overline{A}$. Clearly $j(A)$ is well defined for every cut $(A, \overline{A})$. For each i, $i = 1, 2, \ldots, n - 1$, define $S_i = \{(A, \overline{A}) \mid j(A) = i + 1\}$ (note that for some values of i, S_i may be empty).

A straightforward, but useful, observation ([GN]) is that the S_i's are disjoint, and that the set of all connectivity cuts is the union of all S_i's. Hence, one can construct $\mathcal{H}(G)$ by obtaining a representation for the S_i's and merging them appropriately into a single structure. There is a well known Directed Acyclic Graph (DAG) representation for all cuts in S_i due to Picard and Queyranne [PQ1]; this DAG can be computed from a maximum-flow between $\{1, 2, \ldots, i\}$ and $i + 1$ with an additional strongly connected component computation (see [PQ1] and [GN]). The method of [KT] adopts this approach but, in addition, uses another "trick" that simplifies matters: they choose a numbering scheme for the vertices so that the DAG representing S_i's is particularly simple – it is a chain! We denote by C_i the chain representation of the set of cuts S_i.

Suppose that the vertices of G are numbered so that for any i, vertex $i + 1$ is adjacent to at least one of the vertices from the set $\{1, 2, \ldots, i\}$. If G is connected, then one can always find such a numbering by, say, a BFS traversal. The following observation is what makes the representation of S_i particularly simple.

Observation 4.1 *If the vertices in G are numbered as described above, then all cuts in S_i are non-crossing and hence can be represented by a chain.*

Proof: Suppose not, and let $(X, \overline{X})$ and $(Y, \overline{Y})$ be two crossing cuts in S_i such that $\{1, 2 \ldots, i\} \subseteq X$ and $\{1, 2 \ldots, i\} \subseteq Y$. Since $j(X) = j(Y) = i+1$, $\{1, 2, \ldots, i\} \subseteq X \cap Y$ and $i+1 \in \overline{X} \cap \overline{Y}$. By the Crossing Lemma, $d(X \cap Y, \overline{X} \cap \overline{Y}) = 0$, but the numbering scheme guarantees that there is at least one edge between $i + 1$ and $\{1, 2, \ldots, i\}$, a contradiction. Hence, all cuts in C_i are non-crossing.

It is now easy to see that a set of non-crossing cuts, all of which contain $\{1, 2, \ldots, i\}$ on one side and $i + 1$ on the other, can be represented as a chain: simply order the cuts (e.g. the sides that contain $\{1, 2, \ldots, i\}$) by inclusion. This yields a partition of V into $V_1, V_2, \ldots, V_k$ such that $\{1, 2, \ldots, i\} \subseteq V_1$ and $i+1 \in V_k$. Any cut $(A, \overline{A})$, where $A = \cup_{j=1}^a V_j$ (for $a < k$), is a cut in S_i, and any cut in S_i can be represented in this way. $\square$

The sequential algorithm of [KT] for constructing $\mathcal{H}(G)$ works as follows:

(1) Enumerate the vertices such that there is an edge connecting $i+1$ and $\{1, 2, \ldots, i\}$ (e.g. BFS numbering).

(2) For each $i < n$, compute the maximum flow between $\{1, 2, \ldots, i\}$ and $i + 1$. If the flow value is λ, obtain the chain representation C_i of S_i (this requires an additional strongly connected component computation), otherwise C_i is the empty chain.

(3) Merge chains $C_{n-1}, \ldots, C_1$ one at a time (in reversed order) into a single cactus structure $\mathcal{H}(G)$.

The merging step (step 3) of the sequential algorithm is based on the following idea. Denote by G^i the graph obtained by identifying the vertices $\{1, 2, \ldots, i\}$ into a single vertex. Observe that the set of all λ-cuts in G^i is exactly the set of all cuts in $S_{n-1} \cup S_{n-2} \cup \ldots \cup S_i$. Since G^i is an undirected graph (possibly with parallel edges), its cuts of size λ can be represented by a cactus structure, call it $\mathcal{H}(G^i)$. (Note that the edge-connectivity of G^i is $\geq \lambda$, and if it is strictly greater, then $\mathcal{H}(G^i)$ is empty).

The algorithm in [KT] maintains the invariant that at the $(n - t)^{th}$ step, $t = n - 1, n - 2, \ldots, 1$, the representation $\mathcal{H}(G^t)$ of all cuts of size λ in G^t is produced; thus, after $n - 1$ iterations, the desired $\mathcal{H}(G^1) \equiv \mathcal{H}(G)$ is constructed. Since $S_{n-1} \cup S_{n-2} \cup \ldots \cup S_t$ is the set of all connectivity cuts in G^t, the $(n - t)^{th}$ step merges the chain C_t that represents S_t into the existing structure to create the next structure. The chains are merged one at a time, in reversed order. The details of the sequential merging step are described in the full version.

5 Parallel Construction of $\mathcal{H}(G)$

We first observe that since steps (1) and (2) are known to be in $\mathcal{NC}$ and $\mathcal{RNC}$ respectively, we get (step (2) requires $O(n^{4.5}m)$ processors [MVV]):

Theorem 5.1 *The connectivity cuts of an unweighted undirected graph can be enumerated in $\mathcal{RNC}^2$ using $O(n^{4.5}m)$ processors.*

For obtaining the representation, the approach taken by [KT] for implementing the merging step is highly sequential, so other ideas need to be employed. In particular, we need to derive additional properties of empty nodes in $\mathcal{H}(G)$, and also a relationship between sets of vertices occurring on the chains C_i and sets of vertices that are mapped to nodes in $\mathcal{H}(G)$. In this section we solve the parallel chain merging problem, i.e., the parallel implementation of Step (3). We first handle, in Section 5.1, the simpler case where $\mathcal{H}(G)$ is a tree; in Section 5.2 we consider the general case where $\mathcal{H}(G)$ is a cactus.

5.1 Parallel Chain Merging for Trees

In this section we assume that $\mathcal{H}(G)$ is a tree (i.e. all connectivity cuts in G are non-crossing). We also assume, without loss of generality, that in $\mathcal{H}(G)$ no more than one vertex is mapped into a node. This is not a real limitation, as we remark later. We start by showing (in Section 5.1.1) that the cuts in S_i correspond to a path in $\mathcal{H}(G)$, and that the paths that correspond to $S_1, \ldots, S_{n-1}$ are edge-disjoint and cover the entire tree $\mathcal{H}(G)$.

5.1.1 Sets of Cuts and Paths in $\mathcal{H}(G)$

Suppose that in $\mathcal{H}(G)$ all nodes that contain the vertices $\{1, 2, \ldots, i\}$ and the paths between them are contracted into a single node. Since the resulting graph is a tree, there is a unique path between the contracted node and the node that contains $i + 1$. Let this path be denoted by P_i, and let $x_1, \ldots, x_k$ be the nodes (from $\mathcal{H}(G)$) along P_i. For example, in Figure 2, $P_3 = x_4, x_3, x_1$.

Proposition 5.1 *There is a one-to-one correspondence between the edges on the path P_i and the cuts in S_i.*

Proof: From the definition of $\mathcal{H}(G)$, every cut $(A, \overline{A})$ obtained by the removal of an edge from the path P_i is a λ-cut, and it also separates $\{1, 2, \ldots, i\}$ from $i + 1$; hence $(A, \overline{A}) \in S_i$. Also, any cut $(B, \overline{B})$ obtained by the removal of an edge not on P_i either splits $\{1, 2 \ldots, i\}$ or puts $\{1, 2, \ldots i + 1\}$ on one side; hence, $(B, \overline{B}) \notin S_i$. $\square$

We now point out some relevant properties of the family of paths $P_1, \ldots, P_{n-1}$ in $\mathcal{H}(G)$.

Proposition 5.2 *1. $P_1, \ldots, P_{n-1}$ are edge-disjoint and cover the entire tree $\mathcal{H}(G)$.*

2. For any path $P_i = x_1, \ldots, x_k$, at least one of its endpoints must be a non-empty node in $\mathcal{H}(G)$.

3. There are no empty nodes in $\mathcal{H}(G)$ of degree less than 3.

4. Let y be an empty node in $\mathcal{H}(G)$, then of the paths containing y, exactly one goes through it, and the rest terminate at it.

Proof: Recall that $S_1, \ldots, S_{n-1}$ are disjoint sets of cuts whose union is the entire set of λ-cuts in G, and that every λ-cut in G is represented by an edge in $\mathcal{H}(G)$. Hence, by Proposition 5.1, (1) follows. (2) is clear since either x_1 or x_k must contain $i + 1$. If y is an empty node of degree one, then the edge attached to it does not represent any λ-cut. Also, if y's degree is two, then the two edges adjacent to y represent the same cut. In both cases, y is redundant but we assume that there no such nodes in $\mathcal{H}(G)$; hence its degree must be at least three, so (3) follows. To prove (4), let i be the smallest index such that i and $i + 1$ are in two *different* subtrees attached to y (there must be such i). Clearly, y is in the middle of P_i. Also note that if P_j contains y, $P_j \neq P_i$, then $j > i$ (from minimality of i), but then it must terminate at y due to the contraction of $\{1, \ldots, i, i + 1, \ldots j\}$. $\square$

$\mathcal{H}(G)$ can be easily constructed from the family of paths $P_1, \ldots, P_{n-1}$. However, the input to the chain-merging step is a set of *chains* $C_1, \ldots, C_{n-1}$ which are related to these paths. So, we establish a relationship between this family of paths and the chains.

Let $P = x_1, \ldots, x_k$ be a path in $\mathcal{H}(G)$. For any i, $1 \le i \le k$, contract into x_i all subtrees that are attached to x_i by edges that are not on P, and let $V_i \subset V$ be the vertices mapped into x_i after the contraction. We define the *chain C induced by P* as $C = V_1, \ldots, V_k$. Note that $\cup_{i=1}^{k} V_i = V$. For example, the chain induced by the path x_4, x_3, x_1 in Figure 2 is $\{1, 2, 3\}, \{5, 6\}, \{4\}$. Note that from the definition of P_i and Proposition 5.1, C_i is *exactly* the chain induced by the path P_i. Therefore, we can define the input to the *Chain-Merging* step as a family of $n - 1$ chains $C_1, \ldots C_{n-1}$ that are induced by a family of $n - 1$ paths $P_1, \ldots, P_{n-1}$ with properties (1) – (4) from Proposition 5.2. (If a chain is empty, i.e. $V_1 = V$, then we discard it as it does not represent any cut).

5.1.2 Chain Merging

This section describes the chain merging operation. The idea is to convert the chains into their corresponding paths, and then to merge the paths rather than the chains (an easy task to perform in parallel). Our solution assigns a unique label for every node in $\mathcal{H}(G)$, and for any chain C_i it finds the labels of the nodes along P_i. In other words, if $x_1, \ldots, x_k$ are the labels of the nodes on path P_i and $C_i = V_1, \ldots, V_k$ is the chain given by the input, then we develop a (parallelizable) method to find for each V_j what is the label x_j that corresponds to it. Since the labels are unique, this essentially transforms every chain C_i to its path P_i, so that it is straightforward to find whether and at which node two paths intersect. Since the paths are edge-disjoint and cover the entire tree, $\mathcal{H}(G)$ can be completely determined from the paths.

The unique labels of the nodes in $\mathcal{H}(G)$ are assigned as follows. Recall that by our assumption a node in $\mathcal{H}(G)$ is either empty or contains exactly one vertex from V. If a node y is not empty and contains $a \in V$, then its label is simply $\mathcal{A}$. Otherwise its label, denoted by Φ_y, consists of a subset of vertices in V, which will be defined precisely.

Labeling Non-Empty Nodes

We start with the case of a non-empty node, labeled $\mathcal{A}$ (i.e. $a \in V$ is mapped into $\mathcal{A}$). Let $P = x_1, x_2, \ldots, x_k$ be a path in $\mathcal{H}(G)$ and $C = V_1, V_2, \ldots, V_k$ be its induced chain. We denote by C^a the subset V_j that contains a. Since a is in a unique V_j, C^a is well defined. Note that if P goes through $\mathcal{A}$ then it *must* be that $x_j = \mathcal{A}$. Hence, given a chain C such that $C^a = V_j$, we want to answer the question: is $\mathcal{A}$ the label of jth node on the path that induces C? i.e., is $x_j = \mathcal{A}$? The next two lemmas characterize the paths that go through $\mathcal{A}$ and provide the answer to this question.

Lemma 5.1 *If P passes through $\mathcal{A}$ and C is the chain induced by P, then C^a is minimal for a, i.e. for any other chain D, $D^a \not\subset C^a$.*

Proof: Let l be the degree of $\mathcal{A}$ in $\mathcal{H}(G)$, and $e_1, \ldots, e_l$ be the edges attached to it. Suppose, w.l.o.g., that e_1 is on the path P. Also, let $T_i \subset V$ be the vertices that are mapped to the subtree attached to $\mathcal{A}$ by e_i. Clearly, C^a and T_1 are disjoint. We will now demonstrate that for any other induced chain D, D^a contains some vertices from T_1, and this proves the lemma. Take any other path Q and its induced chain D. If Q goes through $\mathcal{A}$, then since P and Q are edge-disjoint D^a must contain T_1. If Q does not go through $\mathcal{A}$, then it must be completely contained in some subtree attached at $\mathcal{A}$, and hence D^a must also contain T_1 or some vertices from it. Hence, $D^a \not\subset C^a$ in both cases. $\square$

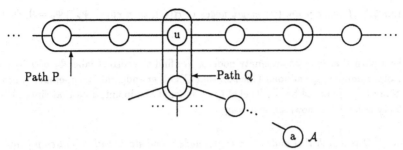

Figure 3: Exception of Lemma 5.2. P does not go through $\mathcal{A}$, but C^a is minimal

Lemma 5.2 *Let $a \in V$. If C^a is minimal for a, then P goes through $\mathcal{A}$, unless C^a is in the middle of C, and there is another chain D of length 2 such that $D^a = C^a$. (Figure 3)*

Proof Outline: Suppose that C^a is minimal for a but P does not go through $\mathcal{A}$, that is $x_j = u$ where $u \neq \mathcal{A}$ ($C^a = V_j$). $\mathcal{A}$ must be in a subtree that is attached to u by an edge e not on P. Let Q be the path that contains e (i.e., Q intersects with P at u on the way to $\mathcal{A}$), and D its induced chain. It can be shown that $D^a \subseteq C^a$, but from the assumption that C^a is minimal for a it must be that $D^a = C^a$. Also, we claim that this can occur if and only if C^a is in the middle of C and the length of D is 2. The claim follows. $\square$

Based on Lemmas 5.1 and 5.2, we can now state a simple characterization of all chains whose paths go through node $\mathcal{A}$: they are the chains C for which C^a is minimal, unless the exception of Lemma 5.2 holds. Hence, to answer the question: "is $\mathcal{A}$ the label of the jth node on the path that induces C, where $C^a = V_j$?" we have to check whether C^a is minimal for a. If it is, and the exception of Lemma 5.2 does not hold, then we label x_j with $\mathcal{A}$. To implement this in parallel, we need to order the sets C^a's of all chains by inclusion; as $|V_j| \leq n$, and there are $n - 1$ chains, this task can be done with $O(n^3)$ processors in constant time on a CRCW PRAM.

The above procedure is done, in parallel, for every $a \in V$. It identifies all non-empty nodes on the paths and assigns them their unique label. The remaining, unidentified, nodes on the paths correspond to empty nodes in $\mathcal{H}(G)$. Hence, at the end of this stage every chain $C = V_1, \ldots, V_k$ is converted into a path $P = x_1, \ldots, x_k$ such that its non-empty nodes are labeled: if x_j is a non-empty node that contains a, then its label is $\mathcal{A}$, and if x_j is empty then it has no label. The next stage uniquely labels the empty nodes on the paths.

Labeling Empty Nodes

We now handle the empty nodes of $\mathcal{H}(G)$. Since, in general, there can be many empty nodes in $\mathcal{H}(G)$, we first have to assign a unique label to every empty node in $\mathcal{H}(G)$, and then find the unique label for every empty node on the path. This completes the conversion of the chains into paths in $\mathcal{H}(G)$.

Proposition 5.2 (4) states that if y is an empty node in $\mathcal{H}(G)$ then there is *exactly one path P* that contains y and *does not terminate* at y; any other path that contains y must end at y. This asymmetry is used to assign to y its unique label Φ_y. Let $C = V_1, \ldots, V_k$ be the chain induced by path $P = x_1, \ldots, x_j = y, \ldots, x_k$. The label of y is defined to be $\Phi_y \equiv V_j$; i.e., if y is the jth node on P ($1 < j < k$) then its label is V_j, the *subset* of vertices that corresponds to x_j in C. This (trivially) uniquely labels all empty nodes which appear in the *middle* of paths. What is now left to show is how to find the label of an empty node that appears at the *end* of a path.

Observation 5.1 *If x and y are two empty nodes in $\mathcal{H}(G)$, then either $\Phi_x \cap \Phi_y = \emptyset$, $\Phi_x \subset \Phi_y$ or $\Phi_y \subset \Phi_x$.*

Let Q be a path that ends at an empty node x. We find the correct label Φ_x of x by using the label of Q's other *non-empty* endpoint (the fact that Q's other endpoint is non-empty is guaranteed by Proposition 5.2 (2)). Let $\mathcal{A}$ be the label of Q's non empty endpoint; note that since $\mathcal{A}$ is a label of a non empty node, it is known at that stage.

Lemma 5.3 *If Q is a path that ends at an empty node x and starts with a non-empty node labeled $\mathcal{A}$, then Φ_x is the minimal label among all labels Φ_z of empty nodes such that $a \in \Phi_z$.*

Proof: Clearly, Q ends at an empty node x only if $a \in \Phi_x$. Let x_1 and x_2 be two empty nodes such that $a \in \Phi_{x_1}$ and $a \in \Phi_{x_2}$. Since by Observation 5.1 two labels of empty nodes cannot properly intersect, assume $\Phi_{x_1} \subset \Phi_{x_2}$. In that case, it can be shown that the path in $\mathcal{H}(G)$ from $\mathcal{A}$ to x_2 must go through x_1. Hence, we can exclude the possibility that x_2 is the endpoint of Q, as this would imply that x_1 is in the middle of Q (so x_1's label is drawn from Q, which is impossible since $a \in \Phi_{x_1}$). Applying this argument repeatedly, we are finally left with the node x whose Φ_x is the minimal over all. $\square$

Based on Lemma 5.3, the label of an empty node that appears at the end of a path is found as follows: let $\mathcal{A}$ be the label of the other endpoint. Order by inclusion all empty-node labels Φ_x that contain a (the labels are subsets of vertices), and find the minimal label over all. By Observation 5.1, this label is unique. This can be done in parallel in constant time using $O(n^3)$ processors.

Summarizing the Merging Algorithm for Trees

Algorithm **Merge-Tree**: input $C_1, \ldots, C_{n-1}$, output $\mathcal{H}(G)$.

(A) Convert the chains into paths, find the non-empty nodes on the paths and assign them their unique labels:
For every vertex $a \in V$, order the subsets $C_1^a, \ldots, C_{n-1}^a$ by inclusion. If C_i^a is minimal, then label the corresponding node on path P_i with $\mathcal{A}$, unless the specific condition of Lemma 5.2 is met.

(B) Find a unique label for every empty node in $\mathcal{H}(G)$, and then assign all empty nodes on the paths their correct labels:
For the first task, every empty node y that appears in the middle of a path P (as the jth node of P) gets the label $\Phi_y \equiv V_j$, where V_j is the jth subset on the chain induced by P. Assign y its label Φ_y. Then, for any $a \in V$, order all empty-node labels Φ_y that contain a by inclusion. Finally, every path starting with a node labeled $\mathcal{A}$ and ending with an empty node assigns it the minimal label Φ_x that contains a among all labels of empty nodes.

(C) Merge the paths into a tree:
Any two paths that share a node-label intersect at this node; since the paths are edge-disjoint and cover the entire tree $\mathcal{H}(G)$, the construction is now obvious.

Remark - When one or more vertices from V are mapped to a node in $\mathcal{H}(G)$, then map to a node z all $Z \subset V$ such that C^a is minimal for a, for $a \in Z$ (with the additional condition of Lemma 5.2). Empty nodes are handled as before.

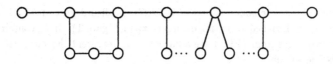

Figure 4: An "Almost" Simple Path in a Cactus

5.2 Parallel Chain Merging for Cacti

In this section we assume that $\mathcal{H}(G)$ is a cactus. We also assume, as before, that in $\mathcal{H}(G)$ no more than one vertex is mapped into a node. We show that the cuts in $\mathcal{S}_i$ form an "almost" simple path in the cactus $\mathcal{H}(G)$ (rather than a simple path as in the tree case) – it consists of a simple path with possibly few disjoint cycles that "hang off" it (see Figure 4). This path induces the chain C_i.

In Section 5.2.1 we study the properties of the family of these "almost" simple paths, and how they cover $\mathcal{H}(G)$. We then show, in Section 5.2.2, how to reduce the cactus case to the tree case: given $\mathcal{H}(G)$, we define a related structure $\mathcal{H}'(G)$ which is a *tree*. The family of "almost" simple paths (and hence their induced chains) that are inputs to our problem can be transformed into the family of simple paths (and their induced chains) that are inputs to the problem of constructing $\mathcal{H}'(G)$. Our algorithm transforms the input appropriately, constructs $\mathcal{H}'(G)$ (using algorithm **Merge-Tree**), and recovers $\mathcal{H}(G)$ from it.

5.2.1 Sets of Cuts and Paths in $\mathcal{H}(G)$

Let u and v be two arbitrary nodes in a cactus. If there are no cycles between u and v, then the path from u to v is unique and consists of only tree edges. Otherwise, the *order* of the cycles between u and v is unique (since cycles do not interleave), and each cycle can be traversed in two possible ways, clockwise or counterclockwise.

Consider now two nodes u and v such that, in G, there is an edge between the vertex that is mapped to u and the vertex that is mapped to v (for example, nodes x_1 and x_9 in Figure 1). Unless specified otherwise, we are only concerned with such pairs u, v. By the Circular Partition Lemma (Lemma 2.2), which states that there are edges only between *consecutive* subsets in the circular order, if there is a cycle with k edges along the path from u to v then the cycle can either be traversed by a *single* edge, or by $k - 1$ edges. Such a path is called an "almost" simple path (Figure 4). An "almost" simple path can be decomposed into two simple paths, the backbone path and the long path. We call the path that traverses *every* cycle through the *single* edge the backbone path, and the path that traverses *every* cycle through the *other* $k - 1$ edges the long path. Consider the paths between x_1 and x_9 in the example of Figure 1. The "almost" simple path between the two nodes is decomposed into the backbone path x_1, x_4, x_6, x_9 and the long path $x_1, x_2, \ldots, x_9$. Equipped with this terminology, we can now define the correspondence between the set of cuts $\mathcal{S}_i$ and the path they form in $\mathcal{H}(G)$.

Suppose that in $\mathcal{H}(G)$ all nodes that contain vertices $\{1, 2, \ldots, i\}$ and the paths between them are contracted into a single node. The resulting graph is a cactus. Let the *long* path between the contracted node and the node that contains $i + 1$ be denoted by P_i, and let $x_1, x_2, \ldots, x_k$ be the nodes along P_i.

We now relate the path P_i to the set of cuts $\mathcal{S}_i$. The following observation, which is the analogue of Proposition 5.1, can be verified in a similar manner: every edge on the path P_i corresponds to a cut in $\mathcal{S}_i$, and every cut in $\mathcal{S}_i$ is represented by one or two edges from P_i. The correspondence between edges on P_i and cuts in $\mathcal{S}_i$ is not one-to-one (as in the tree case) due to the existence of nodes which we call **degenerate nodes**: a node x in $\mathcal{H}(G)$ is called degenerate if x is an empty node with exactly two cycles attached to it (for example, x_4 in Figure 1). If P_i contains a degenerate node x_j then the two edges from P_i that are adjacent to x_j represent the same cut. Otherwise the correspondence is one-to-one.

Any "almost" simple path $P = x_1, x_2, \ldots, x_k$ induces a chain $C = V_1, \ldots, V_k$ in the following way: contract into x_j all cacti that are attached to x_j by edges *not* on P, and let $V_j \subset V$ be the vertices mapped into x_j after the contraction. Note that if x_j is degenerate, then $X_j = \emptyset$. If empty sets are removed from the chain, then the chain is in a **reduced form**. For example, the chain induced by path $x_4, x_5, x_6, x_7, x_8, x_9$ in Figure 1 is $\{1,6,7\}, \{8\}, \emptyset, \{5\}, \{4\}, \{2,3\}$, and its reduced form is $\{1,6,7\}, \{8\}, \{5\}, \{4\}, \{2,3\}$.

The input chains $C_1, C_2, \ldots, C_{n-1}$ are the chains induced by the paths $P_1, P_2, \ldots, P_{n-1}$ in their reduced form. As in the tree case, our goal is to convert C_i into P_i. Unlike the tree case, there are two issues to address: the first is to distinguish the **backbone** from the **long path**, i.e. to detect which edges along the path are cycle-edges, and which are tree-edges; the second is to reintroduce the empty sets that correspond to degenerate nodes into the chain (as C_i is in its reduced form). Since in a tree there are no cycle edges or degenerate nodes, these issues do not come up. We outline the idea of how this is done. If $P = x_1, \ldots, x_k$ is a path and $C = V_1, \ldots V_{k'}$ is the chain induced by P in its reduced form, then

Observation 5.2 *For any $1 \le i < j \le k$, $x_i, x_{i+1}, \ldots, x_j$ form a cycle in $\mathcal{H}(G)$ iff $\{\cup_{a=1}^{i} V_a\}, V_{i+1}, \ldots, V_{j-1}, \{\cup_{a=j}^{k} V_a\}$ is a Circular Partition (i.e. $\lambda/2$ edges between adjacent subsets, no edges otherwise). Also, if both $\{\cup_{a=1}^{i} V_a\}, V_{i+1}, \ldots, V_{j-1}, \{\cup_{a=j}^{k} V_a\}$ and $\{\cup_{a=1}^{j-1} V_a\}, V_j, \ldots, V_{l-1}, \{\cup_{a=l}^{k} V_l\}$ are Circular Partitions then there is a degenerate node between x_{j-1} and x_j on P in $\mathcal{H}(G)$.*

Given a chain C_i in its reduced form, we can use Observation 5.2 and find (in $\mathcal{NC}$) the edge types (tree or cycle) and degenerate nodes along the path P_i that induces it. For any interval (i, j), a set of $j - i$ processors "guesses" that (i, j) is a cycle; if the conditions are met and this interval is maximal, then (i, j) is declared as a *cycle interval*. Based on the intervals, new empty sets (degenerate nodes) are introduced and then for any detected cycle (i, j) $V_i, \ldots, V_j$ are connected by cycle edges. All other edges between consecutive nodes are tree-edges. Note that once the cycles and the degenerate nodes are identified, the backbone is well defined: it consists of all tree edges, along with all cycle edges between V_i and V_j if (i, j) is a cycle interval.

5.2.2 Chain Merging

Definition: Let $V_1, \ldots, V_k$ be a circular partition induced by a cycle $R = x_1, \ldots, x_k$ in $\mathcal{H}(G)$. Without loss of generality, assume that $1 \in V_1$, and let $i + 1$ be the smallest indexed vertex not in V_1. We also assume that $i + 1 \in V_2$ (from Lemma 2.2, $i + 1$ must be in a subset adjacent to i since there is an edge between $\{1, \ldots, i\}$ and $i + 1$). We say that $j(R) = i$. For example, for the cycle $R = x_6, x_7, x_8, x_9$ in Figure 1, $j(R) = 1$.

Proposition 5.3 *For any cycle $R = x_1, \ldots, x_k$ there is a unique path, $P_{j(R)}$, that contains the complete cycle R. We say that $P_{j(R)}$ reveals R.*

Proof: Take $l < j(R)$: R does not appear at all on P_l as there is no cut in S_l that is represented by edges from R. R is completely revealed by $P_{j(R)}$ since any edge from R together with the edge connecting x_1 and x_2 is a cut in $S_{j(R)}$. For any $l > j(R)$, since $\{1, \ldots, j(R), j(R) + 1, \ldots, l\}$ are contracted in P_l, the contraction results in collapsing x_1 and x_2 into a single node, so P_l either contains only parts of R, or does not contain nodes from R at all. □

Consider the tree $\mathcal{H}'(G)$ obtained from $\mathcal{H}(G)$ by removing a single edge from each cycle R in $\mathcal{H}(G)$: the removed edge is the edge between x_1 and x_2, where $\{1, 2, \ldots, j(R)\} \subseteq V_1$ and $j(R)+1 \in V_2$. $\mathcal{H}'(G)$ is a tree representation of a maximal subset of non-crossing cuts in G. Our algorithm transforms the chains of $\mathcal{H}(G)$ into the chains $C'_1, \ldots, C'_{n-1}$ which are the chains of $\mathcal{H}'(G)$, then to construct $\mathcal{H}'(G)$ from $C'_1, \ldots, C'_{n-1}$ (using algorithm **Merge-Tree**), and finally to restore $\mathcal{H}(G)$.

As a "preprocessing" step that is done on each chain C_i, we first introduce the degenerate nodes along P_i and then distinguish between the backbone and the long path of P_i. This is implemented as described in Section 5.2.1. We then transform every chain C_i into C'_i. The chain transformation requires that the following modifications are employed for *every* cycle R in $\mathcal{H}(G)$ (we state it without a proof and implementation details):

(i) Find the chain $C_{j(R)}$ such that R is completely revealed by $P_{j(R)}$, and remove from it the appropriate edge.

(ii) If R is partially contained in some chain C_l, remove it entirely from the C_l (i.e., collapse all of its nodes into a single one).

Claim 5.1 *The transformation correctly generates the input chains for $\mathcal{H}'(G)$.*

We now summarize the merging algorithm for any cactus:
Algorithm **Merge-Cactus**: input $C_1, \ldots, C_{n-1}$, output cactus $\mathcal{H}(G)$.

(A) Transform $C_1, \ldots, C_{n-1}$ into $C'_1, \ldots, C'_{n-1}$.

(B) Construct $\mathcal{H}'(G)$ using algorithm **Merge-Tree**.

(C) Restore $\mathcal{H}(G)$ by adding the missing edges into $\mathcal{H}'(G)$.

The number of processors required are upper-bounded by the step that requires n max flow computations, which is $O(n^{4.5}m)$ [MVV]. Therefore we get,

Theorem 5.2 *For an unweighted undirected graph G, the representation $\mathcal{H}(G)$ of all edge connectivity cuts can be constructed in $\mathcal{RNC}^2$, using $O(n^{4.5}m)$ processors.*

References

[ADK] G.M. Adelson-Velskii, E.A Dinits, A.V. Karzanov, *Flow Algorithms* [In Russian], Nauka, Moscow, 1976.

[B] R.E. Bixby, *The Minimum Number of Edges and Vertices in a Graph with Edge Connectivity n and m n-Bonds,* Networks, Vol. 5, pp. 253-298, 1975.

[BP] M.O. Ball, J.S. Provan, *Calculating Bounds on Reachability in Computer Networks,* Networks, Vol. 18, pp. 1-12, 1988.

[DKL] E.A. Dinits, A.V. Karzanov, M.V. Lomosonov, *On the Structure of a Family of Minimal Weighted Cuts in a Graph,* Studies in Discrete Optimization [In Russian], A.A. Fridman (Ed), Nauka, Moscow, pp. 290-306, 1976.

[EH] A.H. Esfahanian, S.L. Hakimi, *On Computing the Connectivities of Graphs and Digraphs,* Networks, Vol. 14 (1984), pp. 355-366.

[ET] S. Even, R.E. Tarjan, *Network Flow and Testing Graph Connectivity,* Siam J. Computing, Vol. 4, No. 4, pp. 507-518, 1975.

[Ga] H. Gabow, *A Matroid Approach to Finding Edge Connectivity and Packing Arborescences,* Proceedings of the 23rd Annual ACM Symposium on Theory of Computing, New Orleans, pp. 112-122, 1991.

[GN] D. Gusfield, D. Naor, *Extracting Maximal Information about Sets of Minimum Cuts,* Tech. Report CSE-88-14, UC Davis.

[K1] A. Kanevsky, *Graphs with Odd and Even Edge Connectivity are Inherently Different,* Tech. Report TAMU-89-10, June 1989.

[K2] A. Kanevsky, *On the Number of Minimum Size Separating Vertex Sets in a Graph and How to Find All of Them,* Proc. of the 1st Annual ACM-SIAM Symposium on Discrete Algorithms, San Francisco, January 1990.

[KT] A.V. Karzanov, E.A. Timofeev, *Efficient Algorithm for Finding all Minimal Edge Cuts of a Nonoriented Graph,* Cybernetics, (1986) pp. 156-162, Translated from Kibernetika, No. 2, pp. 8-12, March-April 1986.

[KUW] R.M. Karp, E. Upfal, A. Wigderson, *Constructing a Perfect Matching is in Random $\mathcal{NC}$,* Combinatorica, 6(1) 1986, pp. 35-48.

[M] D. Matula, *Determining Edge Connectivity in $O(nm)$,* Proc. of the 28th Annual IEEE Symposium on Foundations of Computer Science, Los-Angeles, pp. 249-251, 1987.

[MVV] K. Mulmuley, U.V. Vazirani, V.V. Vazirani, *Matching is As Easy As Matrix Inversion,* Combinatorica, 7(1) 1987, pp. 105-113.

[NGM] D. Naor, D. Gusfield, C. Martel, *A Fast Algorithm for Optimally Increasing the Edge-Connectivity,* Proc. of the 31th Annual IEEE Symposium on Foundations of Computer Science, St. Louis, pp. 698-707, 1990.

[PQ1] J.C.Picard, M. Queyranne, *On the Structure of All Minimum Cuts in a Network and Applications,* Mathematical Programming Study 13 (1980), 8-16.

[PQ2] J.C.Picard, M. Queyranne, *Selected Applications of Minimum Cuts in Networks,* INFOR - Can. J. Oper. Res. Inf. Process. 20 (1982), 394-422.

[Po] V.D. Podderyugin, *An Algorithm for Finding the Edge Connectivity of Graphs,* Vopr. Kibern., no. 2, 136 (1973).

[PS] J.S. Provan, D.R. Shier, *A Paradigm for Listing (s,t)-Cuts in Graphs,* Technical Report UNC/OR TR91-3, Department of Operations Research, University of North Carolina at Chapel Hill, February 1991.

Planar Graph Augmentation Problems*

(Extended Abstract)

Goos Kant   Hans L. Bodlaender

Dept. of Computer Science, Utrecht University

P.O. Box 80.089, 3508 TB Utrecht, the Netherlands

Abstract

In this paper we investigate the problem of adding a minimum number of edges to a planar graph in such a way that the resulting graph is biconnected and still planar. It is shown that this problem is NP-complete. We present an approximation algorithm for this planar biconnectivity augmentation problem that has performance ratio 3/2 and uses $O(n^2 \log n)$ time. An $O(n^3)$ approximation algorithm with performance ratio 5/4 is presented to make a biconnected planar graph triconnected by adding edges without losing planarity.

1 Introduction

Many problems concerning the planarity of graphs arise from the wish to draw the graph in an elegant way. Many planar graph drawing algorithms are known, but several of them work only on special classes of planar graphs. For example, the graph drawing algorithm of Tutte [13] to draw a graph with convex faces requires as input a triconnected graph, and the graph drawing algorithm of Woods [14] requires as input a biconnected graph. If the graph is not biconnected, several additional (dummy) edges are added, e.g., using an algorithm of Read [10]. However, there exist instances for which Read's algorithm adds $O(n)$ edges, even when only one edge would be sufficient to make the graph biconnected. (In this paper, n denotes the number of vertices and m denotes the number of edges.) When looking for elegant drawings of graphs, it is useful to preserve the structure of the original drawing as much as possible. For this reason, we are looking for the minimum number of edges, that when added to the input graph yield a biconnected or triconnected planar graph. To be precise, we are looking to the following optimization problem, and its decision variant:

[PLANAR BICONNECTIVITY AUGMENTATION] (PBA)
Instance: Planar graph $G = (V, E)$.
Question: Find a planar, biconnected graph $H = (V, F)$ with $E \subseteq F$, and $|F - E|$ as small as possible.

Eswaran & Tarjan [2] studied this augmentation problem without the requirement of planarity. They proved that this problem is NP-complete for weighted graphs and linear

---

*This work was supported by the ESPRIT II Basic Research Actions program of the EC under contract No. 3075 (project ALCOM).

solvable for unweighted graphs (see also Rosenthal & Goldner [11]). In [3], Frederickson & Ja'Ja give approximation algorithms for the augmentation problem on weighted graphs, working within two times optimal in $O(n^2)$ time. [2] and [3] include augmentation algorithms to meet edge-connectivity constraints as well, which formed the base for further research. Recently, Naor et al. [9] describe an $O(\delta^2 nm + nF(n))$ algorithm to find the smallest set of edges to increase the edge-connectivity of G by δ, where $F(n)$ is the time to perform one maximum flow on G. Naor's algorithm finds an optimal sequence of δ augmentations, where in each augmentation-step the edge-connectivity is increased by one, and adding a minimum number of edges. However, these algorithms cannot easily be modified to meet planarity requirements, but some of their techniques will be useful in our approach.

In this paper we prove that the Planar Biconnectivity Augmentation problem is NP-complete, even for the unweighted case. Therefore, we present an approximation algorithm, which has performance ratio 1.5 and runs in $O(n^2 \log n)$ time. Hereto we make use of matching techniques and incremental planarity testing (see [1]), which finds a nice application in this context. The same algorithms can be used to augment a planar graph such that the graph is bridge-connected and planar. In some special cases there exist optimal, polynomial time algorithms. Especially, when the inputgraph is outerplanar, linear algorithms can be obtained such that the augmented graph is bridge-connected, biconnected or triconnected and still planar, by adding a minimum number of edges [7].

We also consider the Planar Triconnectivity Augmentation problem. This is the problem to find the minimum number of edges, which when added to a planar graph yields a triconnected planar graph. It is unknown yet whether this problem is NP-complete. In this paper we describe an $O(n^3)$ approximation algorithm for biconnected graphs, which gives solutions within 5/4 times optimal.

This paper is organized as follows. In section 2 some definitions and preliminary results are given and we prove that deciding whether a planar graph can be made biconnected by adding $\leq K$ edges is NP-complete. In section 3 we present an $O(n^2 \log n)$ algorithm to make a planar graph biconnected, which works within 3/2 times optimal. In section 4 we describe the 1-2-matching problem, and present an approximation algorithm, which we will use in section 5 to augment a biconnected planar graph such that the resulted graph is triconnected and planar, which works only 5/4 from optimal. In section 6 some concluding remarks are made. Omitted proofs can be found in the full paper.

2 Preliminaries

Let $G = (V, E)$ be an undirected graph. Assume G has at least two biconnected components, also called *blocks*. A block of G is called *pendant* if it contains exactly one cutvertex. Let p be the number of pendant blocks of G. For each $v \in V$, let $d(v)$ denote the number of connected components of $G - \{v\}$, the graph obtained by removing v from G. Each of these components is called a v-block. Let $d = \max\{d(v)|v \in V\}$, and let $p(v)$ denote the number of pendants connected at v. Let q be the number of blocks in G, not connected with the remaining part of the graph. The *block-cutpoint graph* $bc(G)$ of G is defined as follows (cf. Harary [6]): every block and cutvertex of G is represented by a vertex of $bc(G)$ and two vertices v_1, v_2 of $bc(G)$ are adjacent if and only if the corresponding cutvertex of v_1 in G is contained in the corresponding block of v_2 in G or vice versa. An *outside* vertex of a block is an arbitrary vertex on the outerface of that block.

Now consider the following operation: take two vertices v_1, v_2 (not cutvertices) in

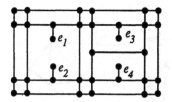

Figure 1: e_1 and e_2 can be matched with each other, but e_3 and e_4 can not.

two pendant blocks V_1, V_2 which lie in different v-blocks B_1, B_2 and connect them. This may destroy planarity. However, the new formed graph G' has one v-block less, and the number of pendant blocks is decreased by two, unless the new pendant containing v_1 and v_2 is an entire v-block, in which case the number of pendant blocks is decreased by one. This observation forms the basis for the following theorem.

Theorem 2.1 ([2]) $max\{d-1, q + \lceil \frac{p}{2} \rceil\}$ *edges are necessary and sufficient to make G biconnected.*

The proof of theorem 2.1 leads to an algorithm for finding a minimum augmenting set of edges to biconnect a graph, which runs $O(m + n)$ time [11] (without the requirement of planarity).

When we require that the augmented graph is also planar, then it is not always possible to connect pairs of pendant blocks. For example, in Figure 1, edges e_1, e_2, e_3 and e_4 each form a pendant block. e_1 and e_2 can be connected without losing planarity, but e_3 and e_4 cannot. Essentially, making a graph biconnected with the minimum number of edges corresponds to connecting as much as possible pairs of pendant blocks with each other without destroying planarity. The number of added edges equals the number of edges, that connect pairs of pendant blocks (hereafter called matching edges), plus the number of pendant blocks, not connected by a matching edge. Minimizing the number of extra edges is hard, as stated in the following theorem (using a transformation from 3-Partition):

Theorem 2.2 *The Planar Biconnectivity Augmentation problem (decision variant) is NP-complete*

In a special case, however, the problem becomes efficiently solvable:

Theorem 2.3 *If all cutvertices are part of one triconnected component, then we can find in $O(n^{2.5})$ time the minimum number of edges, which added to the graph gives a biconnected planar graph.*

Proof: Let G' be the triconnected component that contains all cutvertices of the inputgraph G. We construct a new graph H by representing every vertex $v \in G$ by $p(v)$ vertices in H. We add an edge (v_1, v_2) in H iff v_1 and v_2 share a common face in the unique embedding G', and v_1 and v_2 do not represent the same vertex in G. The planar biconnectivity augmentation problem is now equal to finding a maximum cardinality matching M in H. Let $P = \sum_{v \in G} p(v)$, then it follows that the number of required edges is equal to $P - |M|$. Crossings between added edges in one face can be removed easily. Constructing a maximum cardinality matching can be done in $O(\sqrt{|V_H|}.|E_H|)$ time [8], but since $|E_H|$ can be $O(|V_G|^2)$, the total running time of the algorithm is $O(|V_G|^{2.5}) = O(n^{2.5})$. □

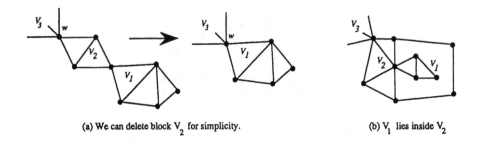

(a) We can delete block V_2 for simplicity.

(b) V_1 lies inside V_2

Figure 2: Reduce Chains.

3 Planar Biconnectivity Augmentation

As the Planar Biconnectivity Augmentation problem is NP-complete, we now look for approximation algorithms. We have two approximation algorithms, a fast and relatively easy one PBA_2OPT, which runs in $O(n \log n)$ time, and has approximation ratio 2, and a slower and more complicated one PBA_1.5OPT, which runs in $O(n^2 \log n)$ time, and has approximation ratio 1.5. The first algorithm is not included in this extended abstract. Now we describe PBA_1.5OPT.

When G has a pendant V_1, that is connected to exactly one other block V_2, and V_2 is connected to exactly one other block V_3, then we call $V_1, V_2, \cdots$ a *chain of blocks*. The first step in PBA_1.5OPT is to reduce chains of blocks. When we add an edge from V_1 to a block outside the chain, then all blocks $V_2, V_3, \cdots$ in the chain are in the same block as V_1. Hence we can discard the blocks $V_2, V_3, \cdots$, for simplicity (see Figure 2(a)). On the other hand, if V_1 lies *inside* V_2 then V_1 can only be connected with V_2 and we can union them into one pendant immediately (see Figure 2(b)). Inside means that the other cutvertex v_2 of V_2 shares a common face with the cutvertex v_1 of V_1. Therefore, we want to test whether adding (v_1, v_2) to G preserves the planarity. Let PLANAR(v_1, v_2) denote this test. Hereto we construct the PQRS-datastructure of Di Battista & Tamassia [1] to store the planar graph G such that we can test in $O(\log n)$ time amortized whether adding (v_1, v_2) to G preserves the planarity. The PQRS-datastructure can be built in $O(n \log n)$ time. Using the block-cutpoint graph $bc(G)$ and the incremental planarity testing, the algorithm REDUCE_CHAINS runs in $O(n \log n)$ time.

When two pendants V_1 and V_2 in a face F are matched by an edge, then all pendants left from this edge in F cannot be matched with pendants right from this edge in F without destroying the planarity. However, when such a crossing between added edges (v_1, v_2) and (v_3, v_4) occurs in a face, then we can always remove it: change (v_1, v_2), (v_3, v_4) into (v_1, v_3), (v_2, v_4) or into (v_1, v_4), (v_2, v_3). So, we allow this type of crossings. To model the situation and to facilitate planarity tests, we introduce *face vertices*.

Instead of adding an edge between two outside vertices v_1 and v_2 of pendants V_1 and V_2, respectively, we add an edge from v_1 and v_2 to the face vertex of F. This means that two pendants V_1 and V_2 can be matched with each other if PLANAR(v_1, F) and PLANAR(F, v_2) for a certain face vertex F holds, or if PLANAR(v_1, v_2) holds. Initially there are no face vertices. We now give the main step of PBA_1.5OPT. Recall that $p(v)$ denotes the number of pendants connected at v.

MATCH_PENDANTS
  while there is a matching between pendants possible **do**
    choose v_1, v_2, such that PLANAR(v_1, v_2) or there exists a face F with
      (PLANAR(v_1, F) and PLANAR(F, v_2)) with $p(v_1)$
      maximal, and for this choice of v_1, $p(v_2)$ maximal;
    if there is no F with (PLANAR(v_1, F) and PLANAR(F, v_2)) **then**
      introduce a new face vertex F;
    add an edge between outside vertices of the v_1- and v_2-pendant and F;
    Decrease $p(v_1)$ and $p(v_2)$ by 1
  od

If there are several cutvertices v_2 with the required property in MATCH_PENDANTS, then we take this vertex v_2, for which the coalesced pendant is as large as possible, hence for which v_1 and v_2 have the highest common ancestor in $bc(G)$.

Note that by matching two pendants in one face, the embedding of G is also restricted in a certain way. This may mean that several other pendants of the two corresponding cut vertices can only be matched with each other. Hence, in general, the $p(v)$ pendants of cutvertex v will be matched with pendants of less than $p(v)$ other cutvertices.

After completing MATCH_PENDANTS, we change the matching edges $(v_1, F), (F, v_2)$ of a matching between the blocks V_1 and V_2 into one edge (v_1, v_2). We delete the face vertices and remove crossings between the additional edges as described above.

At this point in the algorithm, there may still exist some labels $p(v) > 0$ in the graph, which correspond with pendants which cannot be matched anymore with another pendant. For each of these pendants, we have to add an additional edge, that only resolves the biconnectivity of that pendant, which we call an *extra edge*. Hence, if M is the sum of the unmatched labels then M extra edges are required. We distinguish two types of extra edges:

cheap edges = extra edges, which are also required in the optimal solution.

expensive edges = extra edges, which are not required in the optimal solution.

If $k \leq p(v)$ pendants, connected at cutvertex v, are matched with other pendants in a common face F or can only be embedded in a face F without destroying planarity, then we say that label k *belongs to* face F. To count the number of expensive edges, the following lemma is useful.

Lemma 3.1 *Let the labels $B_1, \ldots, B_k$ belong to a face F, with $\sum_{i=1}^{k} B_i = B$. If $max_i\{B_i\} < \lceil \frac{B}{2} \rceil$ then all pendants can be matched by $\lfloor \frac{B}{2} \rfloor$ matching edges, and if B is odd one extra edge, otherwise $B - max_i\{B_i\}$ matching edges and $2 \cdot max_i\{B_i\} - B$ extra edges are required.*

Since we are counting the expensive edges in worst-case situation and in the case $max_i\{B_i\} < \lceil \frac{B}{2} \rceil$ no extra edges are required, we assume in our analysis w.l.o.g. that $max_i\{B_i\} \geq \lceil \frac{B}{2} \rceil$ holds.

Lemma 3.2 $|expensive\ edges| \leq 2|matching\ edges|$.

Proof: Every matching edge has to *pay* for two expensive edges and initially no matching edge pays. Consider an optimal solution and our approximate solution of PBA_1.5OPT. Let $p(v) = A$ be a biggest label in MATCH_PENDANTS, matched with another label B (or some labels with total sum B), whereas in the optimal solution A is matched with labels $B_1, \ldots, B_k$. It follows by MATCH_PENDANTS that $\max_i\{B_i\} \leq B$. For suppose that $B_1 = \max_i\{B_i\} > B$, and A is not matched with B_1, then B_1 is matched before A, say with label A'. Then $B_1 > A$ or $A' > A$ holds and, hence, A was not the biggest label matched with another label, which yields a contradiction.

Notice that by the approximate matching $B' \leq B$ pendants remain unmatched, which are matched with B in the optimal solution. This leads to B' expensive edges, to be paid by the B matching edges (between labels B and A). In the optimal solution all A pendants are matched, so assume w.l.o.g. that $\sum_{i=1}^{k} B_i \geq A$. If in MATCH_PENDANTS the B_i-pendants are matched in a common face F, then we change the matching edges in F such that they appear between two B_i-pendants or between two other pendants. Since $\max_i\{B_i\} \leq B$ this means that at least $\sum_{i=1}^{k} B_i - B$ matching edges between B_i-pendants are possible, hence at most $2B - \sum_{i=1}^{k} B_i$ expensive edges. This leads to a total number of $B' + (A - B) + 2B - \sum_{i=1}^{k} B_i \leq 2B$ expensive edges and at least B matching edges. So assume further that labels $B_1, \ldots, B_p$ $(p \leq k)$ are matched in other faces and labels $B_{p+1}, \ldots, B_k$ are matched in F.

First suppose that labels $B_1, \ldots, B_p$ are matched with labels $C_1, \ldots, C_p$, with $C_i > B_i$ $(1 \leq i \leq p)$. Thus B_i matching edges can be made and $C_i - B_i$ pendants remain unmatched for $1 \leq i \leq p$. Since the B matching edges have to pay for the B' expensive edges, also the B_i matching edges have to pay for $B_i' \leq B_i$ expensive edges of pendants, which remained unmatched now, but are matched with B in the optimal solution. Hence we assign to each B_i a number of B_i expensive edges and we still have to pay for the following number of expensive edges: $B' + (A - B) + 2 \cdot \max_i\{B_i\} - (B_{p+1} + \ldots + B_k) - (B_1 + \ldots + B_p) \leq A + 2 \cdot \max_i\{B_i\} - \sum_{i=1}^{k} B_i \leq 2B$. These can be assigned to the B matching edges.
A similar argument holds when $C_i \leq B_i$.

From this it follows that although the matching of A and B may not lead to an optimal matching, all involved expensive edges can be paid by the involved matching edges, and some expensive edges are assigned at some matching edges, which are treated later. We can further ignore all those treated matching and expensive edges: delete these edges from the graph G, obtaining a reduced graph G'. Apply the same argument to G': find in G' the biggest label A, matched with another label B, whereas in the optimal solution A was matched with the labels $B_1, \ldots, B_k$. Suppose that some $B_i > B$ and B_i is now matched with A'. Then $B_i > A$ or $A' > A$ holds, which means that we have already treated the labels B_i and A', so they are already removed from the graph. Hence $\max_i\{B_i\} \leq B$ holds, and we can apply our argument to G'.

Since each time at least one edge is deleted from the graph, this certainly stops and every matching edge pays for at most two expensive edges, which proves the lemma. □

Our approximate solution consists of three types of edges: the matching edges (M_m), the cheap edges (M_{c1}) and the expensive edges (M_e). The optimal solution consists of matching edges (M_{opt}) and cheap edges (M_{c2}). From the observations that $M_e \leq 2M_m$ by lemma 3.2 and $M_{c2} \geq M_{c1}$ by definition, the following theorem follows:

Theorem 3.3 PBA_1.5OPT *gives solutions that use as most 3/2 times the optimal number of edges.*

For the implementation, we construct for each cutvertex v a binary balanced tree

$CV(v)$, containing those cutvertices w, for which in the original graph G PLANAR(v, w) holds. Furthermore we introduce for each cutvertex v another binary balanced tree $FV(v)$. $FV(v)$ contains these face vertices F, for which PLANAR(F, v) holds during the augmentation of G. So initially $FV(v)$ is empty for all cutvertices v of G. With bucketsort we sort the labels $p(v)$ in non-increasing order. In every step we take the biggest label $p(v_1)$ and search for the biggest label $p(v_2)$. Checking whether pendants v_1, v_2 can be matched is done as follows: (i) Check if $v_1 \in CV(v_2)$. If not then PLANAR(v_1, v_2) did not hold initially, so find another pair. (ii) Check if there is some F, with $F \in FV(v_1)$ and $F \in FV(v_2)$. (iii) If not, then check whether PLANAR(v_1, v_2) holds.

If there is some face vertex F with $F \in FV(v_1)$ and $F \in FV(v_2)$, then this means that there is a face vertex F, such that PLANAR(v_1, F) and PLANAR(F, v_2) holds. If there is no such F, but PLANAR(v_1, v_2) holds, then a new face vertex F must be created. After adding the edges (v_1, F) and (F, v_2), the trees FV are updated as follows: if F is new, then F is added to $FV(w)$ for every cutvertex w with $v_1, v_2 \in CV(w)$. If F is not new, then F is deleted from every $FV(w)$ for every cutvertex w with $F \in FV(w)$ and $v_1 \notin CV(w)$ or $v_2 \notin CV(w)$.

Using the algorithm of [1], it is possible to carry out all PLANAR(v_1, v_2) tests in $O(\log n)$ time per call, amortized. Analysis of the time complexity of our algorithm leads to the following result:

Theorem 3.4 *There exists an approximation algorithm for the Planar Biconnectivity Augmentation Problem with performance ratio 1.5 and time complexity $O(n^2 \log n)$.*

4 1-2-Matching

In this section we consider a problem on triconnected planar graphs, and give an approximation algorithm for this problem. The algorithm will be used as a 'tool' in section 5. Consider triconnected planar graphs $G = (V, E)$, in which every edge e has a set of one or more characteristics associated with it, denoted by $C(e)$, where each characteristic is one of the following: $0 \sim 0, 1 \sim 0, 1 \sim 1, 2 \sim 0, 2 \sim 1$ or $2 \sim 2$. If $|C(e)| \geq 2$ then for all $i \sim j \in C(e)$, $i + j$ is even or for all $i \sim j \in C(e)$, $i + j$ is odd. A 1-2-matching is a 'matching', that is obtained as follows:

1. We fix one characteristic $i \sim j \in C(e)$ for each edge e, and we assign i to one of the adjacent faces, and j to the other adjacent face.

2. In a face, it is possible to match a 1 with a 1 or with a 2, a 2 with a 2 or with two 1's, or three 2's with each other. (Every number may be matched at most once.)

The *cost* of a matching is the following: add together over all matched edges $e \in E$ the number $\max\{i + j \mid i \sim j \in C(e)\}$ plus two times the sum of all numbers in fixed characteristics that are not matched plus two times the number of matchings between a 2 and one 1. The problem is to fix the characteristics and assigning the characteristics to the faces such that the 1-2-matching has minimum cost, i.e, where as much as possible 1's and 2's are matched by the 1-2-matching. (The cost represents two times the number of edges, needed to make the necessary connections to obtain triconnectivity in section 5.) An interesting, but still open question is whether such a minimum 1-2-matching can be computed in polynomial time. In Figure 3 an inputgraph and a corresponding minimum 1-2-matching are given.

In this section we present an approximation algorithm APPROX_1-2-M, that yields a solution that is only a factor 5/4 larger than the optimal one.

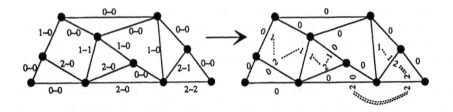

Figure 3: Example of an inputgraph and a 1-2-matching with total cost 18.

Definition 4.1 *An edge e is a 2-edge if there is a characteristic $2 \sim i \in C(e)$, otherwise it is a 1-edge.*

Note that if we can match a 2-edge with two 1-edges, then instead of it we could also match the two 1-edges with each other. Therefore, we first attempt to match a maximum number of 2-edges with each other, and then we try to match the remaining 2-edges with the present 1-edges and a maximum number of 1-edges with each other. This forms the basis for APPROX_1-2-M, which consists of constructing two maximum matchings, called the 2-matching and the 1-matching, on two to be constructed matching graphs, H_2 and H_1. H_2 is defined as follows:

If for a 2-edge e, $2 \sim 2 \in C(e)$, then e is represented by two vertices v_1 and v_2 in H_2, otherwise e is represented by one vertex in H_2. If two 2-edges e_i and e_j share a common face in G, then an edge between the two corresponding vertices v_i and v_j in H_2 is added. Notice that a triconnected planar graph has an unique embedding.

Every vertex v_i in H_2 receives an $a(v_i)$-value as follows: if the corresponding 2-edge e_i in G has $|C(e)| \geq 2$ or if e_i shares a common face with a 1-edge then $a(v_i) = 0$, otherwise $a(v_i) = 1$. We are now looking for a subgraph $H_2' = (V, E')$ of H_2 such that we first maximize the number of vertices v_i with $a(v_i) \leq deg_{H_2'}(v_i) \leq 1$ and then for this number of vertices, also maximize $|E'|$ (this is called the *degree constrained subgraph problem*, e.g. see [12]). In other words, if $a(v_i) = 1$, then v_i must be matched if possible.

For every unmatched vertex v in H_2 we test if there is a path $v = v_1, \ldots, v_k$ (k odd) in H_2 such that $(v_{2i}, v_{2i+1}) \in E'$, and v_k has two neighbours w_1, w_2, with $(w_1, w_2) \in E'$. If so, then we change this matching as follows: delete matching edges (v_{2i}, v_{2i+1}) and (w_1, w_2), add the matching edges (v_{2i-1}, v_{2i}), $1 \leq i \leq \lfloor \frac{k}{2} \rfloor$ and match the three vertices w_1, w_2 and v_k with each other. Notice that no two unmatched vertices can have a path to a common pair w_1, w_2, otherwise the matching E' was not maximal. The resulting matching is called a 2-matching.

We change in the characteristics the 2 into 0 of those 2-edges in G, whose corresponding vertices are matched in H_2'. Edges in G, that have a characteristic $0 \sim 0$ are ignored, but for the remaining 2- and 1-edges, we construct a second matching graph H_1. The edges e in G are represented in H_1 according to $C(e)$ by one of the matching components (assume e belongs to faces F_1 and F_2), as shown in Figure 4. This means that if $C(e) = \{1 \sim 1\}$ then e is represented by two vertices v_1 and v_2 in H_1. There is an edge from v_1 (v_2) to those vertices in H_1, whose corresponding edge in G belongs to face F_1 (F_2). All other characteristics are dealt with in a similar way. If e has except characteristic $2 \sim 2$ also characteristic $2 \sim 0$, then we can ignore characteristic $2 \sim 2$ for simplicity.

The matching components are chosen in such a way that there is a directed correspondence between the 1-2-Matching in G (except when these involve $2 \sim 0$ characteristics)

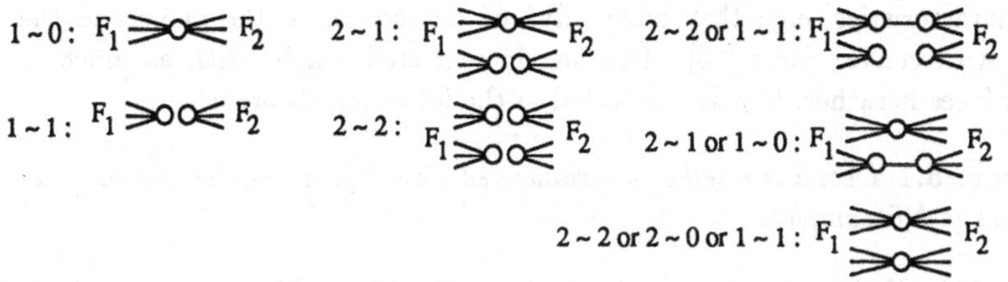

$1 \sim 0:$ $F_1 \gg\!\!\circ\!\!\ll F_2$ $2 \sim 1:$ $F_1 \gg\!\!\circ\!\!\ll F_2$ $2 \sim 2$ or $1 \sim 1:$ $F_1 \gg\!\!\circ\!\!-\!\!\circ\!\!\ll F_2$

$1 \sim 1:$ $F_1 \gg\!\!\circ\!\!\circ\!\!\ll F_2$ $2 \sim 2:$ $F_1 \gg\!\!\circ\!\!\ll F_2$ $2 \sim 1$ or $1 \sim 0:$ $F_1 \gg\!\!\circ\!\!-\!\!\circ\!\!\ll F_2$

$2 \sim 2$ or $2 \sim 0$ or $1 \sim 1:$ $F_1 \gg\!\!\circ\!\!\ll F_2$

Figure 4: Matching components for the characteristics.

and (ordinary) matchings in H_1. Note that for characteristic $2 \sim 0$ no corresponding matching component in H_1 is constructed. The problem is that both vertices must be matched in F_1 or both in F_2, which we are not able to represent by a matching component. These two faces F_1 and F_2 contain no other matched or unmatched 2-edges, otherwise the corresponding vertex was already matched by the 2-matching. If $k \geq 2$ 1-edges belong to face F_1 (or F_2), then we give every edge between the corresponding vertices in H_1 weight $1 + \frac{1}{k}$. If there is only one 1-edge in F_1 (or F_2) then we introduce a new vertex in H_1, with an edge of weight $\frac{3}{4}$ to the corresponding vertex. We now apply an algorithm to find a maximum weighted matching in H_1 [5]. The weights $1 + \frac{1}{k}$ mean that these edges are preferred in the maximum weighted matching. (The meaning of the values of these weights will become clear in the analysis.)

For each unmatched edge e in G, with $2 \sim 0 \in C(e)$, we remove one $(1+\frac{1}{k})$-edge in one of the two shared faces, and connect the two endpoints to two new vertices, representing e. If there is no such matched $(1 + \frac{1}{k})$-edge in H_1, then if the edge of weight $\frac{3}{4}$ is matched in H_1 (this occurs when F_1 or F_2 has only one 1-edge), then e is matched with the 1-edge, otherwise the 2-edge e remains unmatched. Call the resulting matching in H_1 the 1-matching. The required matching in G is obtained by taking together the 2-matching and the 1-matching. We call the resulting algorithm APPROX_1-2-M.

Similar to the proof of lemma 3.2 we count the number of extra edges and matching edges in APPROX_1-2-M and the number of extra edges and matching edges in an optimal solution of the 1-2-matching. It follows by complicated case analysis (see full paper) that the number of matching edges, involved by one extra edge is at least four, which leads to the following theorem:

Theorem 4.1 APPROX_1-2-M *works within 5/4 times optimal.*

The matching graphs can be constructed in $O(n^2)$ time. We can rewrite the degree constrained subgraph problem to finding a maximum cardinality matching [12], which can be solved in $O(\sqrt{|V_{H_2}|} \cdot |E_{H_2}|)$ time [8]. The 2-matching is equal to finding a maximum weighted matching, which can be implemented to run in $O(|V_{H_1}| \cdot |E_{H_1}| + |V_{H_1}|^2 \cdot \log |V_{H_1}|)$ time [5]. Since $|V_{H_1}| = O(|V_G|), |V_{H_2}| = O(|V_G|), |E_{H_1}| = O(|V_G|^2)$ and $|E_{H_2}| = O(|V_G|^2)$ worst-case, this yields an $O(n^3)$ running time for the algorithm APPROX_1-2-M.

5 Planar Triconnectivity Augmentation

In this section we deal with the question how to augment a biconnected graph such that the augmented graph is triconnected and still planar. Triconnected planar graphs have nice characteristics, e.g., they have only one embedding in the plane and they can be drawn with convex faces [13]. Drawing biconnected graphs with as much as possible convex faces is rather difficult, as stated in the following theorem.

Theorem 5.1 *Deciding whether a biconnected planar graph can be drawn with $\geq K$ convex faces is NP-complete.*

For this and theoretical reasons we consider the Planar Triconnectivity Augmentation Problem for biconnected planar graphs. In this section we present an approximation algorithm, in which we will use the approximation algorithm APPROX_1-2-M in section 4 for the 1-2-matching problem.

Assume G has at least two triconnected components (or, shortly, *tricomps*), and we want to add edges between pairs of tricomps to obtain a triconnected planar graph. Such a pair of tricomps is said to be matched by a *matching edge*. The other augmenting edges which do not go from one tricomp to another are called *extra edges*. A tricomp V_1 is connected at two cutvertices, say a and b, with the remaining part of the graph, $G - V_1$, and belongs to two faces of $G - V_1$. Other tricomps can be connected at a and b with the remaining part of the graph as well. We call this a *parallel chain* of tricomps. Another possibility is that a second tricomp V_2 is connected at b and c with $G - V_2$, a third tricomp at c and d, etc. We call this a *serial chain* of tricomps. Notice that a serial chain can be part of a parallel chain, which again can be part of a serial chain, etc.

To abstract the tricomps in a correct way, we introduce *characteristics*. A characteristic $i \sim j$ means that one side of V' between a and b must have i outgoing edges and the other side of V' must have j outgoing edges. We now initially represent every tricomp V' by an edge e between the corresponding two cutvertices a and b, with the characteristic $1 \sim 0$ associated with it, denoted by $C(e) = C((a, b)) = \{1 \sim 0\}$, and we delete V' from G (see Figure 5). $1 \sim 0$ means that this tricomp V' (possibly one vertex v with $deg(v) = 2$) must get one additional outgoing edge to become triconnected. All other edges receive characteristic $0 \sim 0$. Let a *2-subgraph* in this reduced graph G' be a serial or parallel chain or a tricomp. For a 2-subgraph V' we want to add matching edges between the edges with characteristics, which represents the number of outgoing edges of the corresponding tricomps. V' must also be triconnected with $G - V'$ and, hence, must also get characteristics associated with it. We try to add as few as possible (outgoing) edges between tricomps in V' such that V' is triconnected except some unmatched tricomps for the outgoing edges. These characteristics for V' are called *typical characteristics*. Notice that V' may contain several typical characteristics. Instead of characteristic $2 \sim 1$ one could have $1 \sim 0$ (see Figure 5), and similarly instead of $2 \sim 2$ one could have $2 \sim 0, 1 \sim 1$, or both.

The following lemma is crucial for our algorithm.

Lemma 5.2 *Every typical characteristic $i \sim j$ for a 2-subgraph V' can be changed such that $i \leq 2$ and $j \leq 2$, without increasing the number of extra edges.*

If the recognized 2-subgraph V' is not a serial or parallel chain, then we have to fix one characteristic for each edge in this tricomp, such that in a face of V', we can match (as much as possible) a 1 with a 1, a 2 with a 2 or with two 1's or three 2's with each other, while preserving planarity. Precise analysis shows that this exactly corresponds with the 1-2-matching problem as described in section 4. However, since V' must also have at least one outgoing edge to $G - V'$, not all characteristics must be matched inside. A small modification of APPROX_1-2-M is sufficient to take care of this. This leads to the following algorithm:

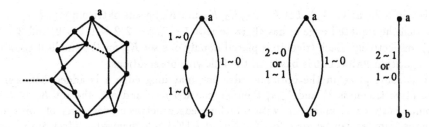

Figure 5: Example of typical characteristics for serial and parallel chains.

TRICONNECT

$G' := G$;

replace every tricomp V' by an edge e between the cutvertices with $C(e) := \{1 \sim 0\}$;
for the other edges $e, C(e) := \{0 \sim 0\}$; done := false;
while not done **do**
    if G' contains a serial chain V' **then** SERIAL(V') **else**
        if G' contains a parallel chain V' **then** PARALLEL(V') **else**
            do APPROX_1-2-M(V'), with V' a tricomp of G';
    if $G' = V'$ **then** done := true **else**
        replace V' by an edge (a, b) with $C((a, b)) :=$ the typical characteristic set of V'
od
Choose one typical characteristic for G' and from this, the unique typical
characteristics follow for all tricomps in G;
add corresponding edges between tricomps

We still must describe the algorithms SERIAL(V') and PARALLEL(V'), which compute
the typical characteristics for V', when V' is a serial or parallel chain. SERIAL(V') replaces a path of vertices with degree 2 in G' by one edge (a, b) with the set of typical
characteristics associated with it. Let $C(V') = \{C(e)|e \in V'\}$. Let $S(V')$ be the sum over
all edges $e \in V' \max\{i + j|i \sim j \in C(e)\}$. Then the following holds for V':
    If $S(V') = 1$ then $C((a, b)) := \{1 \sim 0\}$, if $S(V') = 2$ then if $1 \sim 0 \in C(V')$ then
$C((a, b)) := \{1 \sim 1, 2 \sim 0\}$, else $C((a, b)) := C(V')$, if $S(V') \geq 3$, then if $S(V')$ is odd, then
$C((a, b)) := \{2 \sim 1, 1 \sim 0\}$, else if $1 \sim i \in C(V')$ then $C((a, b)) := \{2 \sim 2, 2 \sim 0, 1 \sim 1\}$
else $C((a, b)) := \{2 \sim 2, 2 \sim 0\}$. This set $C((a, b))$ follows from the algorithm SERIAL(V')
for computing the matching edges between the tricomps, for which the following holds
(SERIAL(V') is not included in this extended abstract):

Theorem 5.3 SERIAL (V') *computes the set of typical characteristics in linear time for a*
serial chain V'.

Let now V' be a parallel chain, which essentially is a set of parallel edges $e_1, \ldots, e_k$,
with characteristics associated to them. We may permute these edges in any order and we
have to find an order such that adding a minimum number of matching edges between the
corresponding tricomps makes V' triconnected. If $e_1, \ldots, e_k$ is the optimal order then we
have to add augmenting edges between the corresponding tricomps V_i and $V_{i+1}, 1 \leq i < k$.
If both V_1 and V_k have outgoing edges, then for exactly one $j, 1 \leq j < k$, no augmenting
edges are required. This place is called a *gap* and if V_1 (V_k) has no outgoing edges, then the

gap is said to be at V_1 (V_k). Let K_1, K_2, K_3, K_4 and K_5 be sets of tricomps $\subseteq \{V_1, \ldots, V_k\}$ for which the replaced edge e_i has characteristic $1 \sim 1, 2 \sim 2, 2 \sim 1, 1 \sim 0$, and $2 \sim 0$ in $C(e_i)$, respectively. Each tricomp is placed in only one set K_i, for the smallest possible i. The algorithm PARALLEL is based on the following observations:

If at some place in the optimal order one matching edge e is added between two consecutive tricomps V_i and V_{i+1}, then we can delete e and add the set K_1 of $1 \sim 1$ tricomps between V_i and V_{i+1}, with matching edges between every pair of consecutive tricomps. Similar can be done for K_2. If $|K_4| + |K_5| > 2$ then extra edges are required, because otherwise there occurs more than one gap. Therefore, as long as $|K_4| > 2$ holds, we can add an extra edge between two $1 \sim 0$ edges, and treat this united tricomps as a $1 \sim 1$ tricomp, hence adding it to K_1. We can do similarly for K_5, adding the united tricomp to K_2, so we assume w.l.o.g. that $|K_4|, |K_5| \leq 2$. If $|K_4| + |K_5| > 2$ then we union a $1 \sim 0$ tricomp and a $2 \sim 0$ tricomp by an extra edge to a $2 \sim 1$ tricomp, thereby increasing the set K_3. If $|K_3|$ is even then we can union it optimally by matching edges to one $2 \sim 2$ or $1 \sim 1$ tricomp (adding it to K_1 or K_2), otherwise we can union it optimally to one $2 \sim 1$ tricomp. If $|K_4| + |K_5| = 2$ then one gap occurs inside the optimal order; if $|K_4| = 1$ or $|K_5| = 1$ then the gap occurs at V_1 or V_k.

The algorithm PARALLEL(V') is just a tedious case analysis of the different sizes of the sets K_i, and the following theorem holds:

Theorem 5.4 PARALLEL(V') *computes the optimal order of the tricomps and the set of typical characteristics for a parallel chain V' in linear time.*

Since the optimal characteristics can be computed in linear time for a serial and a parallel chain, and we can compute the characteristics for the 1-2-matching problem in $O(n^3)$ time within 5/4 times optimal, this leads to the following theorem.

Theorem 5.5 *There exists an approximation algorithm for Planar Triconnectivity Augmentation for biconnected planar graphs that has performance ratio 5/4 and uses $O(n^3)$ time.*

6   Final Remarks

In this paper we considered the problem of adding a minimum number of edges to a planar graph, such that the augmented graph is biconnected or triconnected. For both problems approximation algorithms are given, but the approximation algorithm for triconnectivity requires the graph to be biconnected, which can be done by APPROX_1-2-M. Unfortunately, it is not true that *every* optimal set of edges which increases the connectivity by one can be exended to an optimal solution to increase the connectivity by $k > 1$. See also Naor et al. [9] for this problem on graphs, without the requirement of planarity. Hence the performance ratios of 3/2 and 5/4 do not guarantee a performance ratio of 15/8 to make an arbitrary planar graph triconnected without losing planarity. This interesting open problem becomes efficiently solvable, when the inputgraph G is outerplanar. In this case, a linear algorithm can be obtained to augment G by a minimum number of edges such that the resulted graph is planar and triconnected (Kant [7]).

Instead of considering the planarity constraint for the augmentation problems, we can ask similar questions in which the augmented graph must satisfy some other specified properties, e.g., belonging to a class of outerplanar graphs, perfect graphs or partial k-trees. These open problems are an interesting field for further study and research, to come to a general technique for augmentation algorithms.

Acknowledgements

The authors wish to thank Marinus Veldhorst for some useful suggestions.

References

[1] Di Battista, G., and R. Tamassia, Incremental planarity testing, *Proc. 30th Annual IEEE Symp. on Found. on Comp. Science*, North Carolina, 1989, pp. 436–441.

[2] Eswaran, K.P., and R.E. Tarjan, Augmentation problems, *SIAM J. Comput.* 5 (1976), pp. 653–665.

[3] Frederickson, G.N., and J. Ja'Ja, Approximation algorithms for several graph augmentation problems, *SIAM J. Comput.* 10 (1981), pp. 270–283.

[4] Frank, A., Augmenting graphs to meet edge-connectivity requirements, *Proc. 31th Annual IEEE Symp. on Found. on Comp. Science*, St. Louis, 1990, pp. 708–718.

[5] Gabow, H.N., Data structures for weighted matching and nearest common ancestors with linking, in: *Proc. 1st Annual ACM-SIAM Symp. on Discrete Algorithms*, San Fransisco (1990), pp. 434–443.

[6] Harary, F., *Graph Theory*, Addison–Wesley Publ. Comp., Reading, Mass., 1969.

[7] Kant, G., *Optimal Linear Planar Augmentation Algorithms for Outerplanar Graphs*, in preparation.

[8] Micali, S., and V.V. Vazirani, An $O(\sqrt{V} \cdot E)$ algorithm for finding maximum matching in general graphs, in: *Proc. 21st Annual IEEE Symp. Foundations of Computer Science*, Syracuse (1980), pp. 17–27.

[9] Naor, D., D. Gusfield and C. Martel, A fast algorithm for optimally increasing the edge-connectivity, *Proc. 31st Annual IEEE Symp. on Found. of Comp. Science*, St. Louis, 1990, pp. 698–707.

[10] Read, R.C., A new method for drawing a graph given the cyclic order of the edges at each vertex, *Congr. Numer. 56* (1987), pp. 31–44.

[11] Rosenthal, A., and A. Goldner, Smallest augmentations to biconnect a graph, *SIAM J. Comput.* 6 (1977), pp. 55–66.

[12] Shiloach, Y., Another look at the degree constrained subgraph problem, *Inf. Proc. Lett.* 12 (1981), pp. 89–92.

[13] Tutte, W.T., Convex representations of graphs, *Proc. London Math. Soc.*, vol. 10 (1960), pp. 304–320.

[14] Woods, D., *Drawing Planar Graphs*, Ph.D. Dissertation, Computer Science Dept., Stanford University, CA, Tech. Rep. STAN-CS-82-943, 1982.

Parametric Search and Locating Supply Centers in Trees

Greg N. Frederickson*

Department of Computer Science
Purdue University
West Lafayette, Indiana 47907

gnf@cs.purdue.edu

Abstract. Linear-time and -space algorithms are presented for solving three versions of the p-center problem in a tree. The techniques are an application of parametric search.

1. Introduction

Parametric search is a powerful technique for solving various optimization problems [AP], [AASS], [C1], [C2], [CSSS], [CSY], [FJ1], [FJ2], [Gu], [HT], [LS], [M], [MT], [1], [R], [S], [SS], [Z]. To solve a given problem by parametric search, one must find a largest (or smallest) value that passes a certain feasibility test. In general, a set of candidate values is explicitly or implicitly identified, and the set is searched, choosing one value at a time upon which to test feasibility. As a result of the test, various values in the set will be discarded from further consideration. Eventually the search will converge on the optimal value.

In [M] and [C2], Megiddo and Cole have emphasized the role that parallel algorithms can play in the implementation of parametric search. Their techniques are quite general, and perhaps as a result do not lead to algorithms known to be optimal. In fact, rarely have algorithms for parametric search problems been shown to be optimal. (For several that are optimal, see [CSSS], [F2], and [LS].) In at least some cases the time required for feasibility testing matches known lower bounds for the parametric search problem. But in worst case $\Omega(\log n)$ values must be tested for feasibility. Is this extra factor of at least $\log n$ necessary? In this paper we show that for several problems of locating supply centers on trees, we show that this factor can be avoided, and give optimal algorithms.

---

*This research was supported in part by the National Science Foundation under grants CCR-86202271 and CCR-9001241, and by the Office of Naval Research under contract N00014-86-K-0689.

In [F2], we presented linear-time algorithms for solving the max-min tree k-partitioning problem [PeS] and the min-max tree k-partitioning problem [BPS]. In this paper we extend our techniques to handle the problem of finding a p-center in a tree [Hk]. Let T be an n-vertex tree in which there is a positive weight associated with each edge. Let $p \leq n$ be a positive integer. The problem is to designate as supply centers p points in the tree so that the maximum distance from a demand point to the nearest supply center is minimized. The simplest version of the problem $(V/V/p)$ allows supply centers to be placed only on vertices, and restricts demand points to be vertices. Other versions allow points on edges to also be demand points $(V/E/p)$, or points on edges to also be supply centers $(E/V/p)$, or both $(E/E/p)$. In these latter problems, the edge is viewed as a continuum of points, and the distance function is generalized to describe the distance of a point on an edge to each of the edge's endpoints. If we consider general graphs rather than trees the p-center problem is NP-hard [KH]. A polynomial-time approximation algorithm for $(V/V/p)$ on general graphs is given in [HS]. It generates a supply set that the maximum distance from a demand point to its nearest supply center is within a factor of 2 of minimum.

We survey previous results when the graph is a tree. For $p \leq 2$, $O(n)$-time algorithms have been presented in [Go1], [Go2], [Hn1], [Hn2], [KH]. For $p > 2$, the following results were known. Algorithms with $O(n^2 \log n)$ time and $O(n^2)$ space have been presented for $V/V/p$ and $E/V/p$ in [KH] and for $V/V/p$, $V/E/p$ and $E/V/p$ in [CT]. For the latter three problems, algorithms with $O(n(\log n)^2)$ time and $O(n \log n)$ space have been presented in [MTZC], and algorithms with $O(n \log n)$ time and space appear in [FJ1]. In this paper we give algorithms for these problems that use $O(n)$ time and space.

Beyond merely producing optimal algorithms, our results point up the need for carefully designed data structures and appropriate access to them. These supply location problems depend on having a suitable representation of the set of all intervertex distances in the tree. In [FJ1], the $\Theta(n^2)$ elements of this set were generated in a representation consisting of a set of $\Theta(n)$ implicitly stored matrices, which take a total of $\Theta(n \log n)$ time and space to generate. It was observed in [FJ1] that the tree itself constitutes a (not very convenient) representation of the set of all intervertex distances. The key point that we make is that some compromise between these two positions leads to an optimal algorithm. We do not represent all of the intervertex distances in matrices, but rather consider subsets of intervertex distances, never representing those distances that can be ruled out as the algorithm progresses. As in [F2] we manage the parametric search to produce intermediate information on which to base faster feasibility tests. To construct optimal algorithms for our problems, it appears to be necessary to interleave the narrowing of bounds on the optimal value with the construction of better feasibility tests.

Our paper is organized as follows. In section 2 we review basic methods for generating and searching sets of intervertex distances, and for testing feasibility of prospective values. In section 3 we motivate the path-like structure on which we search, which we call a stem. In section 4 we we present our approach for placing supply centers on a stem. In section 5 we we present our approach for placing supply centers on a tree.

2. Background

In this section we present background information that we shall use in designing our optimal algorithms. We first review straightforward tests of the feasibility of a search value in a tree. We then review how to represent all possible search values of a path within a sorted matrix. We next review an approach for searching that uses values from a collection of sorted matrices. Finally, we describe straightforward approaches for the path and the tree, which we will subsequently improve on. We concentrate on $V/V/p$, leaving the discussion of $E/V/p$ and $V/E/p$ to [F1].

We first describe a straightforward feasibility test for identifying the number of centers necessary so that no demand point is at distance greater than a certain threshold from the nearest supply center. The feasibility test takes a test value λ, and determines if at most p centers can be placed in the tree so that no demand point is at distance greater than λ from the nearest supply center.

Feasibility test $SFTEST0$ for $V/V/p$ is the following [KH]. Root the tree at a vertex of degree 1, and initialize $numctr$ to 0. We shall use two arrays $sup(\cdot)$ and $dem(\cdot)$, where $sup(v)$ will be the distance from v to the nearest descendant of v that is a supply center, and $dem(v)$ will be the distance from v to the farthest descendant of v that is an unsatisfied demand point. Explore the tree starting at the root. To explore the subtree rooted at vertex v, do the following. Set $dem(v)$ to 0 and $sup(v)$ to ∞, and explore the subtree rooted at each child w of v. Upon the return from child w to v, we do the following. If $dem(w) + sup(w) \le \lambda$, then no supply center is needed at w, and $dem(w)$ is reset to $-\infty$. Otherwise (when $dem(w) + sup(w) > \lambda$), if $dem(w) + c(v, w) > \lambda$, then the demand must be satisfied at w, by adding 1 to $numctr$ and resetting $sup(w)$ to 0 and $dem(w)$ to $-\infty$. In any event, reset $dem(v)$ to be $\max\{dem(v), dem(w) + c(v, w)\}$ and reset $sup(v)$ to be $\min\{sup(v), sup(w) + c(v, w)\}$. This completes the exploration at vertex v. When exploration at the root has completed, if $dem(root) + sup(root) > \lambda$, then the demand at the root is satisfied by adding 1 to $numctr$. When the traversal is complete, if $numctr \le p$, then λ is feasible ($\lambda \ge \lambda^*$), else λ is not feasible ($\lambda < \lambda^*$). Feasibility test $SFTEST0$ uses $O(n)$ time.

We next discuss a data structure that contains all candidate values for problems on a path. This structure is based on ideas in [FJ1] and [M]. Let the vertices on the path P

be indexed from 1 to n. For $i = 1, 2, \cdots, n$, let A_i be the distance from vertex v_1 to v_i. For any pair i, j with $i < j$, the distance from v_i to v_j is $A_j - A_i$. Let X_1 be the sequence of sums $A_n, A_{n-1}, \cdots, A_2$, and let X_2 be the sequence of sums $A_1, A_2, \cdots, A_{n-1}$. Then matrix $M(P)$ is the $(n-1) \times (n-1)$ Cartesian matrix $X_1 - X_2$, where the ij-th entry is $A_{n+1-i} - A_j$. Then $M(P)$ is a *sorted matrix*, in the sense that in every row the values are in nonincreasing order, and in every column the values are in nonincreasing order. Also, $M(P)$ can be stored in $O(n)$ space so that any matrix entry can be computed in constant time.

We next review a searching algorithm *MSEARCH*. The version for max-min problems was discussed specifically in [F2], but here we give the version for min-max problems. This algorithm combines and adapts the methods of [FJ1] and [FJ2]. Algorithm *MSEARCH* takes as its arguments a collection of sorted matrices, searching bounds λ_1 and λ_2, where λ_2 is feasible and λ_1 is not, and a stopping count. *MSEARCH* will produce a sequence of values one at a time to be tested for feasibility, with elements in the collection of matrices discarded as a result of each test. If a value λ is produced that is not in the range $\lambda_1 < \lambda < \lambda_2$, then it need not be tested. If λ is feasible, then λ_2 is reset to λ. Otherwise, λ_1 is reset to λ. *MSEARCH* will halt when the number of matrix elements remaining is no greater than the stopping count. We refer the interested reader to [F1] for further details.

The following theorem is taken from [F1] and [F2] and is similar in spirit to Lemma 5 in [FJ1] and Theorem 2 in [FJ2].

Theorem 2.1. ([F1], [F2]) Let $\mathcal{M}$ be a collection of N sorted matrices $\{M_1, M_2, \cdots, M_N\}$ in which matrix M_j is of dimension $m_j \times n_j$, $m_j \le n_j$, and $\sum_{j=1}^{N} m_j = m$. Let q be nonnegative. The number of feasibility tests needed by *MSEARCH* to discard all but at most q of the elements is $O(\max\{\log \max_j\{n_j\}, \log(m/(q+1))\})$, and the total time of *MSEARCH* exclusive of feasibility tests is $O(\sum_{j=1}^{N} m_j \log(2n_j/m_j))$. $\square$

In the remainder of this section, we provide first attempts at locating supply centers that motivate the introduction of further ideas in the next section. First suppose that the tree is limited to its simplest possible form, a path. We mimic the approach in [F2] that addressed the problem of tree partitioning. We describe a simple approach *PATH0* for a path P. Initialize λ_1 to 0 and λ_2 to ∞. Run algorithm *MSEARCH* on the set containing the single sorted matrix $M(P)$ with stopping count 0, using feasibility test *FTEST0*. When *MSEARCH* halts, $\lambda^* = \lambda_2$. By Theorem 2.1, *MSEARCH* will produce a sequence of $O(\log n)$ values to be tested. It follows that the total time is $O(n \log n)$.

From [F2], we have the following definition. For a tree rooted at a vertex of degree 1, let a *path-partition* be a partition of the edges of the tree into paths, where a vertex is an endpoint of a path if and only if it is of degree not equal to 2 with respect to the tree. Call any path in a path-partition that contains a leaf in the tree a *leaf-path*.

We next describe a simple approach for a tree, which unfortunately does not quite work. Initialize λ_1 to 0 and λ_2 to ∞. Initialize T to be the tree rooted at a vertex of degree 1. Repeat the following. Form a path-partition of T. For each leaf-path P_j in the path-partition, form sorted matrix $M(P_j)$. Run *MSEARCH* on this collection of matrices, with stopping count 0, using *FTEST0*. If T contains just one leaf, halt with $\lambda^* = \lambda_2$. Otherwise do the following. For each leaf-path P_j, infer the supply centers in P_j such that each vertex in turn going up in P_j is within λ_2 of either a supply center or the top of P_j, and as few as possible supply centers are used. Delete all vertices on the path supplied by these supply centers. This leaves a smaller tree. Reset T to be the smaller tree, and p to be the number of supply centers remaining to be placed in this tree. This concludes the description of the repeat loop.

This approach does not quite work, since not all vertices on a leaf path are necessarily deleted. Such a shortened leaf-path would then never be removed. We discuss how to overcome this problem in the next section.

3. Stems, thorns, twigs and buds

In presenting our algorithm for finding a p-center, we shall adapt and generalize the techniques introduced in the previous section and in [F2]. We first discuss the pruning strategy that we employed in partitioning trees [F2], and show how this strategy must be modified for locating p-centers. This basic strategy is to consider leaf-paths in the path-partition of the tree, and then use *MSEARCH* to determine the final position of supply centers on these paths. Once the position of supply centers on a leaf-path is known, we would like to remove these leaf-paths. Unfortunately, it appears that one cannot in general remove all of such a leaf-path. The reason is that the optimal placement of centers on such a leaf-path may leave some vertices on the leaf-path to be supplied by centers not on the leaf-path, or may place a supply center on a leaf-path which will supply some vertices not on that leaf-path. The best one can do in general is to replace a leaf-path by a path of constant size. Because of this complication, our algorithm does not deal with paths, but slightly more general structures, called stems. The stems will be searched and replaced as the algorithm progresses. Stems will be defined carefully after discussing the replacement strategy.

Let λ_1 and λ_2 be the current best lower and upper bounds, resp., on λ^*. Since we know that λ_2 is feasible, we handle replacement assuming that $\lambda^* < \lambda_2$. There are two cases for replacement. In the first case, there are some vertices on the stem that will not be supplied by supply centers placed on the stem. The stem is replaced by a single edge, whose cost is the distance from the farthest unsatisfied vertex on the stem up to the top of the stem. For any stem with no active values, the cost of the replacement edge will be less than λ_2. Since the stem has no active values, the cost of the replacement edge will be at most λ_1. In a path-partition, we call a leaf-path that is a single edge of cost at most λ_1 a *thorn*. We call the leaf endpoint of the thorn its *point*. In the second case, the stem will be replaced by a path of two edges, with the higher edge having cost equal to the distance from the top vertex to the highest supply vertex on the stem, and the lower edge having cost equal to the current value of λ_2 minus the cost of the upper edge. For any stem with no active values, the cost of the upper replacement edge will be at most λ_1, and the cost of both edges together will be at least λ_2. In a path-partition, we call a leaf-path that comprises two edges, with the upper edge of cost at most λ_1, and both together of cost at least λ_2, a *twig*. Since λ_2 has already been tested for feasibility, any subsequent test will have $\lambda < \lambda_2$, and thus a supply vertex will be placed at the middle vertex of the twig. We call this middle vertex a *bud*.

We now are ready to define a stem. Let T be a tree at any point in our algorithm, with the corresponding bounds on λ^*, $\lambda_1 < \lambda* \le \lambda_2$. Let $\tilde{T}$ be T with all thorns and twigs removed. A *stem* in T is a path in the path-partition of $\tilde{T}$, along with all thorns and twigs that attach to vertices in this path. We shall first show how to solve a p-center problem on a stem in a manner analogous to, but more complicated than, solving a p-center problem on a path. Let the *backbone* of the stem be the vertices on the path that remains when the edges of the thorns and twigs are removed from the stem.

We first show how to clean up a stem P so that it is simpler to handle. Let v_i, $i = 1, 2, \cdots, n$ be the vertices on backbone of P, listed in order along the backbone. If more than one thorn is attached to any vertex v_i, all but a longest thorn can be deleted, since its point will force a supply center to be closest to v_i. Thus we assume that there is at most one point u_i attached to any vertex v_i. Let u_i and u_j be points of thorns, with $i \ne j$. If $d(u_i, v_j) \le c(u_j, v_j)$, then u_i can be deleted, since any vertex that supplies u_j will supply u_i. A simple linear-time scan through P will delete any such points. If more than one twig is attached to any vertex v_i, all but the twig with the bud closest to v_i can be deleted, since the supply center located at its bud will cover at least all vertices not on these twigs that any other bud on these twigs would cover. Thus we assume that there is at most one bud w_i attached to any vertex v_i. Let w_i and w_j be buds of twigs, with $i \ne j$. If $d(w_i, v_j) \le c(w_j, v_j)$, then w_i can be deleted, since any vertex not on these twigs that is supplied by w_i will be supplied by w_j. A simple linear-time scan through P will delete any such twigs. Of course, the number p of supply centers to be placed should be decreased by 1 for each twig deleted.

We next describe how to form a sorted matrix that contains all distances of a stem that need to considered as search values in solving a $V/V/p$ problem on a stem. We generate a path P', called the *projected stem*, as follows. The idea is to take each point and each bud and "project" them onto a path containing the backbone vertices. Each point u_i will be projected to a new vertex u_i' at distance $c(v_i, u_i)$ in front of v_i, and a new vertex u_i'' at distance $c(v_i, u_i)$ beyond v_i. Similarly, each bud w_i yields new vertices w_i' and w_i''. Note that some new vertices may precede v_1 or follow v_n, but that the path is easily extended to contain these. Note that the leaf endpoint of a twig is ignored, as the cost of the edge to it is at most λ_1, and the length of the twig is at least λ_2. Thus the set of all intervertex distances in P' includes λ^*, unless $\lambda^* = \lambda_2$. Since λ_2 has already been tested, no difficulty results in this latter case. Projected stem P' will contain at most $5n$ vertices. The sorted matrix associated with the projected stem will be $M(P')$.

The projected stem can be determined in linear time as follows. For $i = 1, 2, \cdots, n$, let A_i be the distance from v_1 to v_i. Initialize P' to be the sequence of backbone vertices, with their associated sequence of distances from v_1. For each i for which (v_i, u_i) is a thorn, let B_i be the distance from v_1 to u_i, and let C_i be the distance from u_i to v_n. Note that if the indices i are in increasing order, then the values B_i are in increasing order, and the values C_i are in decreasing order. For the values B_i this follows since for any pair i and j with $i < j$, $d(v_i, u_i) < d(v_i, u_j)$, which implies $d(v_1, u_i) < d(v_1, u_j)$. The argument for the C_i is similar. Thus one can form the B_i values by scanning along the stem, and then perform a merge of A_i and B_i values to determine where to insert the vertices u_i'' in P'. Then let C_i' contain the values $d(v_1, v_n) - d(u_i, v_n)$, and generate these values from the C_i. Merge C_i' into the sequence of values currently associated with P' to determine where to insert the new vertices u_i'. This completes the projection of the points of thorns. The projection of buds is done similarly.

4. Locating supply centers on a stem

We present three schemes to perform parametric search on a stem of n vertices, with each scheme an improvement on the previous. The first scheme uses two phases of search to achieve $O(n \log \log n)$ time. The first phase gathers information with which subsequent feasibility tests can be performed in $o(n)$ time. The second phase then completes the parametric search using this faster feasibility testing. The second scheme applies recursively the idea of gathering information, and takes $O(n \log^* n)$ time.[1] The third scheme uses a variety of ideas to reduce the time to $O(n)$. The overall strategy is similar to that for path partitioning [F2]. We shall identify and emphasize those parts that are significantly different.

---

[1] The function $\log^* k$ is the iterated logarithm of k, defined by $\log^* 1 = \log^* 2 = 1$ and $\log^* k = 1 + \log^* \lceil \log k \rceil$ for $k > 2$.

As in [F2], we set up data structures that allow us to perform a feasibility test in less than linear time. These data structures are somewhat different from those in [F2]. Let P_j be a portion of a stem P formed by taking a subpath of the backbone of P, and attaching the thorns and twigs to the corresponding vertices of the subpath. We call this a *substem*. Let f_j be the index of the first backbone vertex in the substem, and t_j the index of the last. Assuming that a supply center is placed on every bud, $nctr(l)$ is the minimum number of supply centers on backbone vertices of P_j such that each vertex with index between l and t_j is within distance λ_1 of either the nearest supply center or v_{t_j}. Assume that the index of the backbone supply center of largest index is maximized, subject to using $nctr(l)$ supply centers for vertices with indices from l to t_j. Then $sply(l)$ will be the distance from v_{t_j} to its nearest supply center, and $dmnd(l)$ will be the distance from v_{t_j} to the farthest vertex with index at least l that is at least λ_2 from its nearest supply center. If there is no supply center (i.e., no bud in P_j and no supply center on a backbone vertex with index at least l), then let $sply(l)$ be ∞. If v_{t_j} is at most λ_1 from its nearest supply center, then let $dmnd(l)$ be any value.

We also need several other structures for efficient searching. Let $lastd(l)$ be the largest index such that all backbone and point vertices with indices from l to $lastd(l)$ can be supplied by a supply center at v_l. Let $fdem(l)$ be the longest distance from v_{f_j} to any backbone or point vertex whose index is between f_j and l. Let $succ(l)$ be the smallest index no smaller than l such that $u_{succ(l)}$ or $v_{succ(l)}$ is not supplied by a bud. For the whole stem, let $succh(l)$ be the smallest index no smaller than l, that is the index of a thorn. If no such thorn exists, let $succh(l) = l$. Let $fsup$ be the distance from the nearest bud in P_j to v_{f_j}, and $tsup$ the distance from the nearest bud in P_j to v_{t_j}. (If there is no bud in P_j, then take both of these quantities to be ∞.)

Our first scheme *STEM1* is similar to *PATH1* in [F2]. In *STEM1* we partition the vertices of the stem P into subsets that form substems. We perform a first phase of the parametric search to generate a data structure that allows us to test feasibility more quickly. As it narrows the bounds λ_1 and λ_2 on λ^*, *MSEARCH* will discard from consideration as λ^* almost all candidate values arising from the substems.

To make this scheme work, each substem P_j will have $\lfloor \log n \rfloor$ vertices. Generate $\mathcal{M}_1$, the set of sorted matrices for the substems. Call *MSEARCH* on set $\mathcal{M}_1$ and a stopping count $n/(\log n)^2$, using *SFTEST0*. From the at most $n/(\log n)^2$ candidate values remaining from $\mathcal{M}_1$, retain those that fall between λ_1 and λ_2. We call these candidate values *active values*. For each substem that contains no active values, we compute the functions $ncut(\cdot)$ and $rmdr(\cdot)$ using algorithm *SDIGEST1*.

By size of a substem we mean the number of edges in the backbone. For convenience, if there is no thorn attached to a backbone vertex v_i, we shall assume that $c(u_i, v_i) = 0$, and hence $d(u_i, v_l) = d(v_i, v_l)$ for any vertex v_l.

We discuss *SDIGEST*1, an algorithm to compute these data structures in linear time for a portion P_j of a stem P. First we note that $fsup$ and $tsup$ can be computed by simple linear scans of P_j. Also, $succ(\cdot)$ and $fdem(\cdot)$ can be computed by linear scans from the front of P_j. Similarly, $sucth(\cdot)$ for all P_j can be computed by a linear scan of P. A simple backward scan through P_j determines $lastd(\cdot)$. Then on a backward scan through P_j, $nctr(\cdot)$, $sply(\cdot)$ and $dmnd(\cdot)$ can be computed.

Lemma 4.1. Let P_j be a portion of stem P that contains r vertices and that has no active values. Then *SDIGEST*1 will compute $nctr(l)$, $dmnd(l)$, $sply(l)$, $lastd(l)$, $fdem(l)$, $fsup$ and $tsup$ for all backbone vertices v_l in P_j in $O(r)$ time. $\square$

We next describe a sublinear feasibility test *SFTEST*1, which is somewhat similar to *FTEST*1 in [F2]. Let λ be the value to be tested. We assume that $\lambda_1 < \lambda < \lambda_2$. Initially set $numctr$ to 0, $supdist$ to ∞ and $demdist$ to 0. Examine each substem in order starting with the first substem. To handle the next substem P_j, do the following. If there are active values for P_j, then scan P_j as in the straightforward feasibility test. Otherwise, do the following. Add the number of twigs in P_j to $numctr$. If $supdist + demdist \leq \lambda$, then the vertex at distance $demdist$ will be supplied by the supply center at distance $supdist$, and $demdist$ should be reset to 0. If $supdist + fdem(t_j) \leq \lambda$, then all vertices in P_j are supplied by the supply center at distance $supdist$, and $supdist$ should be reset to $\min\{tsup, supdist + length(P_j)\}$. Otherwise (if $supdist + fdem(t_j) > \lambda$), binary search among vertices on P_j to find a vertex v_l of smallest index such that $supdist + fdem(l) > \lambda$, and reset $numctr$ to be $numctr + nctr(l)$, $supdist$ to be $sply(l)$ and $demdist$ to be $dmnd(l)$. This concludes the discussion for the case when $supdist + demdist \leq \lambda$.

For the case when $supdist + demdist > \lambda$, we must focus on $demdist$ as follows. If $demdist + fsup \leq \lambda$, then reset $numctr$ to be $numctr + nctr(f_j)$, $supdist$ to be $sply(f_j)$, and $demdist$ to be $dmnd(f_j)$. Otherwise (if $demdist + fsup > \lambda$), do the following. Let z_1 be the index of the first point in P_j, or t_j if there is no thorn in P_j. Binary search for the maximum index s such that $demdist + d(v_{f_j}, v_s) \leq \lambda$ and $d(u_{z_1}, v_s) \leq \lambda$. If $s = t_j$, then postpone placement of the supply center to the next substem by resetting $demdist$ to $\max\{demdist + d(v_{f_j}, v_s), d(u_{z_1}, v_s)\}$. Otherwise, add 1 to $numctr$. If $lastd(s)$ is the last vertex in P_j, reset $supdist$ to be $\min\{tsup, d(v_s, v_{t_j})\}$. Otherwise, reset $numctr$ to be $numctr + nctr(lastd(s) + 1)$, $supdist$ to be $sply(lastd(s) + 1)$ and $demdist$ to be $dmnd(lastd(s) + 1)$. This completes the discussion when $supdist + demdist > \lambda$, and with it, the description of the examination of a substem P_j. When all substems have been examined, if $supdist > \lambda$, then add 1 to $numctr$. If $numctr \leq p$, then λ is feasible ($\lambda \geq \lambda^*$), else λ is not feasible ($\lambda < \lambda^*$). This completes the description of feasibility test

SFTEST1. Note that when all substems contain active values, then this test reduces in general to the straightforward linear-time feasibility test.

Lemma 4.2. Let P be a stem of n vertices partitioned into substems, each of size at most r. Let all but at most n/r^2 substems have the values $fsup$ and $tsup$ computed, and the functions $nctr(\cdot)$, $dmnd(\cdot)$, $sply(\cdot)$, $lastd(\cdot)$, and $fdem(\cdot)$, computed for all backbone vertices on them. Algorithm *SFTEST1* correctly determines the feasibility of a test value in $O((n/r)\log r)$ time.

Proof. There are $\lceil n/r \rceil$ substems to be examined. Consider the examination of substem P_j. The search for v_l will use $O(\log r)$ time. If there are no active values for P_j, then all other operations in examining P_j will take constant time. There are $\Theta(n/r)$ such substems, which will take $O((n/r)\log r)$ time in total. If there are active values for P_j, then examining the substem will take $O(r)$ time. But there are only $O(n/r^2)$ substems with active values, which will take $O(n/r)$ time in total. $\square$

The second phase of *STEM1* is the same basic idea as *PATH0*, except that *SFTEST1* is used and a stem is searched rather than a path, and the current values of λ_1 and λ_2 are not reinitialized to 0 and ∞.

Theorem 4.3. *STEM1* solves the p-center problem on an n-vertex stem in $O(n \log \log n)$ time.

Proof. By Theorem 2.1, the number of test values produced on the first call to *MSEARCH* is $O(\log \log n)$, and these will be produced in $O(n)$ time. Since each feasibility test takes $O(n)$ time, the total time for feasibility testing is $O(n \log \log n)$. By Lemma 3.1 the total time to compute the $nctr$, $sply$ and $dmnd$ values for $\Theta(n/\log n)$ substems is $O(n)$. By Theorem 2.1, the second call to *MSEARCH* will produce $O(\log n)$ test values in $O(n)$ time. By Lemma 4.2, each call to feasibility test *SFTEST1* will take $O(n \log \log n/ \log n)$, or $O(n \log \log n)$ for all tests. $\square$

We next apply the above strategy recursively to yield our second algorithm, *STEM2*, which runs in $O(n \log^* n)$ time. The approach is similar to that of *PATH2* in [F2]. Our basic idea is to start with the straightforward feasibility test, and repeatedly construct data structures for ever faster feasibility tests. In each phase of a sequence of phases, we generate these data structures as we search among sets of sorted matrices that represent larger and larger substems. Let $f(n)$ be the smallest power of 2 no smaller than $\log n \log \log n$. Let lev be the smallest integer such that $f^{lev}(n) = 8$. For each value of i from lev down to 0, do the following. Let the substem length $r_i = f^i(n)$. Form a set of sorted matrices, one for each substem of length r_i, and call *MSEARCH* on this set with stopping count n/r_i^2, using *FTEST1*. When *MSEARCH* terminates, for each substem P_j with no active values, compute $nctr(l)$, $sply(l)$ and $dmnd(l)$ for each vertex v_l in P_j.

Theorem 4.4. *STEM2* solves the p-center problem on an n-vertex stem in $O(n \log^* n)$ time.

Proof. There will be $lev + 1$ phases, with lev being $O(\log^* n)$. The call to *MSEARCH* with matrices of size $r_i \times r_i$ will use $O(n)$ time and generate $O(\log r_i)$ test values. Performing all feasibility tests on this phase will use $O((n/r_{i+1}) \log r_{i+1} \log r_i)$ time. Since r_{i+1} is $\Theta(\log r_i \log \log r_i)$, this time is also $O(n)$. The time to compute $nctr(\cdot)$, $sply(\cdot)$ and $dmnd(\cdot)$ values at the end of the phase will also be $O(n)$. Thus the time for each of the $O(\log^* n)$ phases is $O(n)$. □

Our third algorithm *STEM3* uses linear time. We shall use the same basic approach as in *STEM2*, but shall be more careful in three activities in which we spent $O(n)$ time per phase: feasibility testing, computing $nctr(\cdot)$, $sply(\cdot)$ and $dmnd(\cdot)$ values, and producing test values. First, we increase the value of the function $f(\cdot)$ somewhat, so that the feasibility tests will take less time in total on each subsequent phase. Second, we use the $nctr(\cdot)$, $sply(\cdot)$ and $dmnd(\cdot)$ data structures already generated to generate the $nctr(\cdot)$, $sply(\cdot)$ and $dmnd(\cdot)$ data structures on the next phase. Third, we call *MSEARCH* on carefully trimmed sets of submatrices to produce our search values more efficiently.

To reduce the total time for feasibility testing, we choose $f(n)$ to be the largest power of 2 no larger than $(\log n)^2$. Let lev be the smallest integer such that $f^{lev}(n) = 16$. For each value of i from lev down to 0, do the following. First, let r_i be $f^i(n)$. As before, we view the stem as divided into substems, each of length r_i.

We discuss the ideas behind an algorithm *SDIGEST2*, which updates the arrays containing $nctr(\cdot)$, $sply(\cdot)$, and $dmnd(\cdot)$ more efficiently than *SDIGEST1*. It is similar to *DIGEST2* in [F2]. To assist in computing these values, we shall keep track of $snext(l)$ and $dnext(l)$. Let $snext(l)$ be the index s of the backbone vertex or bud such that $sply(l)$ is the distance from this vertex to v_{t_j}, and $dnext(l)$ the index d of the backbone vertex or point such that $dmnd(l)$ is the distance from this vertex to v_{t_j}. We maintain $dlow(j)$, $slow(j)$, and $high(j)$. Let $dlow(j)$ be the largest index such that $fdem(dlow(j)) \leq \lambda_2$. Thus $dlow(j)$ is the largest index such that all vertices with index at most $dlow(j)$ can be supplied by a supply center at v_{f_j}. Let z_1 be the index of the first point in P_j. Let $slow(j)$ be the largest index such that $d(v_{f_j}, v_{slow(j)}) < \lambda_2$, and $d(u_{z_1}, v_{slow(j)}) < \lambda_2$ or $d(v_{f_j}, v_{z_1}) \geq \lambda_2$. Thus $slow(j)$ is the largest index such that a supply center placed at $v_{slow(j)}$ can supply all vertices with index at most $slow(j)$. Let $high(j)$ be the largest index such that $d(v_{high(j)}, v_{t_j}) \geq \lambda_2$. We shall also maintain $lastd(l)$ for any vertex v_l with $f_j \leq l \leq slow(j)$. Initially, $dnext(l) = l$ and $snext(l) = l$.

The approach of *SDIGEST2* uses several additional data structures and two queues. We first introduce the additional data structures. Let $tdem(l)$ be the longest distance from v_{t_j} to any backbone vertex or point whose index is between l and t_j. (Let $intd(l)$ be the index of this vertex.) The queues are Q_s, for the indices of backbone vertices that are potential supply centers, and Q_d, for the indices of backbone vertices to be supplied by a supply center. We initialize Q_s as follows. Insert a mark into Q_s. Then for each vertex v_l in turn while $l \le slow(j)$ and $fdem(l) < fdem(t_j)$ and $fdem(t_j) - d(v_{f_j}, v_l) > \lambda_1$, insert l into Q_s. Note that the second of the three conditions tests whether $fdem(t_j)$ is realized by some thorn whose index is less than l. If this were so, then there is no thorn between l and t_j, and since the thorn is of cost at most λ_1, $d(v_l, v_{t_j}) < \lambda_1$. Similarly, for the third condition, if $fdem(t_j) - d(v_{f_j}, v_l) \le \lambda_1$, then since the length of P_j is at most the distance from v_{f_j} to the vertex realizing $fdem(t_j)$, $d(v_l, v_{t_j}) < \lambda_1$. We initialize Q_d as follows. Insert a mark into Q_d. Then for each vertex v_l in turn while $l \le dlow(j)$ and $d(v_{succ(l)}, v_{t_j}) > \lambda_1$, insert l into Q_d.

We use pointers q_s and q_d in Q_s and Q_d, resp. Initialize q_s and q_d to the largest index initially in Q_s and Q_d, resp. Alternately, handle each of Q_s and Q_d until it contains only the mark, starting with Q_s. To handle Q_s, do the following. Extract the mark from the front of Q_s, rearrange Q_s to put it in sorted order, and insert the mark at the end. Then extract and handle each element v_l in turn from Q_s. If $lastd(l) = l$, first repeatedly add 1 to q_s while $d(v_l, u_{q_s+1}) < \lambda_2$, and then set $lastd(l)$ to q_s. This represents the case in which l is near the end of a substem of P_j, and we must move q_s to the next substem. Otherwise (if $lastd(l) > l$), reset q_s to $lastd(l)$. In either case, if $q_s + 1$ is not in Q_d and $tdem(succ(q_s + 1)) > \lambda_1$, then insert $q_s + 1$ into Q_d, else set $dmnd(q_s + 1)$ to $tdem(succ(q_s + 1))$, $dnext(q_s + 1)$ to $intd(succ(q_s + 1))$, $sply(q_s + 1)$ to $tsup$, and $snext(q_s + 1)$ to the index of the bud at distance $tsup$ from t_j. This completes the description of handling an item extracted from Q_s.

To handle Q_d, do the following. Extract the mark from the front of Q_d, rearrange Q_d to put it in sorted order, and insert the mark at the end. Then extract and handle each element v_l from Q_d up to the mark. We note that handling an element in Q_d may result in an element being inserted into either Q_s or Q_d. If $dnext(l) = l$ and $snext(l) = l$, then while $d(u_l, v_{q_d+1}) < \lambda_2$, increment q_d by 1. Otherwise (if $dnext(l) > l$ or $snext(l) > l$), if $sply(l) + dmnd(l) < \lambda_2$, then reset q_d to $snext(l)$ and m to "s", else reset q_d to $dnext(l)$ and m to "d". In either case, push the triple (m, l, q_d) onto the stack. If $m = $ "s", then do the following. If q_d is not in Q_s, $fdem(q_d) < fdem(t_j)$, and $fdem(t_j) - d(v_{f_j}, v_{q_d}) > \lambda_1$, then insert q_d into Q_s, else set $sply(q_d)$ to $\min\{d(v_{q_d}, v_{t_j}), tsup\}$ and set $snext(q_d)$ to the index of the vertex at this distance from v_{t_j}. If $m = $ "d", then do the following. If q_d is not in Q_d, and $d(u_{succ(q_d)}, v_{t_j}) > \lambda_1$, then insert q_d into Q_d, else set $dmnd(q_d)$ to $d(u_{succ(q_d)}, v_{t_j})$

and $dnext(q_d)$ to $succ(q_d)$. This completes the description of handling an item extracted from Q_d.

Once the queues are empty, we return back through P_j, using the stack. While the stack is not empty, do the following. Pop a triple (m, l, q_d) from the stack. If $m = $ "d", then set $sply(l)$ to $sply(q_d)$, $snext(l)$ to $snext(q_d)$, $dmnd(l)$ to $dmnd(q_d)$, $dnext(l)$ to $dnext(q_d)$, and $nctr(l)$ to $nctr(l) + nctr(q_d)$. If $m = $ "s", then set $sply(l)$ to $sply(lastd(q_d) + 1)$, $snext(l)$ to $snext(lastd(q_d) + 1)$, $dmnd(l)$ to $dmnd(lastd(q_d) + 1)$, $dnext(l)$ to $dnext(lastd(q_d) + 1)$, and $nctr(l)$ to $nctr(l) + 1 + nctr(lastd(q_d) + 1)$. When the stack is empty, all needed values $nctr$, $sply$, $dmnd$, $snext$, $dnext$, and $lastd$ for P_j have been updated. This completes the description of $SDIGEST2$.

Lemma 4.5. Over all phases, the time for all calls to $SDIGEST2$ is $O(n)$.

Proof. The proof is similar to that of Lemma 3.3 in [F2]. When a mark is deleted in Q_s, the contents of Q_d form a rotated list. When a mark is deleted in Q_d, the contents of Q_d form a concatenation of two rotated lists, since Q_d accepts items when both Q_s and Q_d are handled. Note that this can be reordered in time linear in the length of the queue. If the first entry in a triple (m, l, q_d) is "s", then it is assumed that a supply center is to be placed at v_l. When this triple is popped from the stack, the values of $sply(l)$, $dmnd(l)$, and $nctr(l)$ are set accordingly. If the first entry in a triple (m, l, q_d) is "d", then it is assumed that v_l is the first vertex following the last vertex supplied by a supply center. When this triple is popped from the stack, the values of $sply(l)$, $dmnd(l)$, $nctr(l)$, $snext(l)$ and $dnext(l)$ are set accordingly. $\square$

For $SDIGEST2$ to work correctly, it would appear that values of $fdem(l)$ and $tdem(l)$ must be updated on each phase. This can be simulated easily in the following way. For the whole stem, let $preth(l)$ be the largest index no larger than l, that is the index of a thorn. If no such thorn exists, let $preth(l) = l$. Note that a linear scan of the stem suffices to determine $preth(l)$ for all vertices v_l. Then for a substem P_j containing v_l, $fdem(l) = d(v_{f_j}, v_l)$ if $preth(l) < f_j$, and $fdem(l) = \max\{d(v_{f_j}, v_l), d(v_{f_j}, u_{preth(l)})\}$ otherwise. A similar approach using $succh(l)$ works for generating $tdem(l)$.

We do not discuss here how to generate sets of submatrices on which to produce test values. This is discussed in detail in [F1] and [F2] with respect to the path partitioning problem. Combining these ideas with what is discussed here gives our algorithm $STEM3$.

Theorem 4.6. Algorithm $STEM3$ solves the p-center problem $V/V/p$ on a stem of n vertices in $O(n)$ time.

Proof. The correctness follows from Lemmas 4.2 and 4.5, and Theorem 4.4, and the above discussion. The time complexity follows from a discussion similar to that in the proof of Theorem 3.3 in [F2]. □

5. Locating supply centers in a tree

We next discuss the changes that should be made to the algorithms in section 4 to give us algorithms for locating p-centers in trees. First, we discuss the formation of thorns and twigs, and how $TREE1$ from [F2] is modified into algorithm $STREE1$. Second, we discuss the replacement of the sorted matrix for the projected stem by a set of seven sorted matrices. This alternative presentation of candidate values appears necessary to enable $MSEARCH$ to run efficiently on the thin matrices. We discuss the dynamic representation of these matrices. Finally, we present data structures for implementing various tests efficiently in $SDIGEST1$, $SFTEST1$, and $SDIGEST2$, so that running time of $O(n)$ is achieved for our p-center algorithm $STREE3$ for a tree.

We first discuss the generation of thorns and twigs. Given a tree T, and a value $\lambda_1 < \lambda^*$, we have already defined a stem in T. Let the set of stems of T be called a *stem-partition* of T. Let a *leaf-stem* in T be a stem in T that contains a leaf that is a backbone vertex. The pruning strategy behind our algorithms is to reduce the number of leaf-stems. The first phase reduces the problem of placing supply centers in a tree to the problem of placing supply centers in a tree with fewer than $2n/(\log n)^2$ leaf-stems, and then generates a data structure for a fast feasibility test. Initialize λ_1 to 0 and λ_2 to ∞. While the number of leaf-stems in the tree is at least $2n/(\log n)^2$, do the following. Form a stem-partition of the tree. Fewer than $n/(\log n)^2$ leaf-stems are of length more than $(\log n)^2$. At least half of the leaf-stems are of length at most $(\log n)^2$. Generate the projected stem of each such leaf-stem. Let $\mathcal{L}$ be the set of sorted matrices defined on these leaf-stems that are of length at most $(\log n)^2$. Let n' be the number of vertices on these stems. Call algorithm $MSEARCH$ on set $\mathcal{L}$, with stopping count of $n'/(2(\log n)^2)$. Feasibility test $SFTEST0$ is used.

When $MSEARCH$ terminates, at most $n'/(2(\log n)^2)$ values will be active. Note that at most $n'/2$ vertices will be on stems with active values, so that at least $n'/2$ vertices will be on stems with no active values. Consider the leaf-stems that have no active values. For each such leaf-stem, infer the location of supply centers on the stem. If there is a vertex other than the top vertex of the stem that is not supplied by a supply center on the stem, do the following. Let y_d be the farthest such vertex from the top of the stem. Replace the stem (except for the top vertex) by a thorn whose edge cost equals the distance from y_d to the top of the stem. Reduce the number of centers remaining to be located in the tree by the number placed on the stem. Otherwise (if all vertices below the top vertex are supplied by supply centers on the stem), do the following. Let y_s be the supply center on the stem that is closest to the top of the stem. Let y_d be the farthest vertex no higher than y_s that is supplied by y_s. Delete the stem (except for the top vertex) and replace it by a twig. The higher of the two twig edges will be of cost equal to the distance from v_s to the top vertex. The lower of the two twig edges will be of cost equal to the distance from v_d to v_s. Reduce the number of centers remaining to be located in the tree by the number placed on the stem, minus 1 if the stem is replaced by a twig. For either of the above constructions, it is not hard to show that there is a one-to-one correspondence between a p-center in the original tree, and a p'-center in the new tree. Furthermore, each leaf-stem with no active values is either deleted or replaced by a thorn or a twig. This completes the description of one iteration of the while loop.

At least $(1/2)2n/(\log n)^2 - n/(2(\log n)^2) \geq n/(2(\log n)^2)$ stems will be removed in one iteration. The tree resulting at the end of an iteration will have at most 7/8 of the leaf-stems of the tree at the beginning of the iteration. Let T' be the tree resulting at the end of the while loop, which will contain fewer than $2n/(\log n)^2$ leaf-stems. Let p' be the number of supply centers remaining to be placed. The number of leaf-stems is reduced iteratively. Once the number of leaf-stems is less than $2n/(\log n)^2$, the data structures for a faster feasibility test can be formed. Form a stem-partition of T'. Then do the following, which is similar to the first phase of algorithm *STEM1*. Partition the stems into substems, each of length at most $(\log n)^2$. Note that this should result in a partition of the edges. Form a set $\mathcal{M}$ of sorted matrices based on the projected substems. Then call algorithm *MSEARCH* on $\mathcal{M}$, with stopping count $n/(\log n)^4$. Use *SFTEST0* for feasibility testing. Determine the active values, and for each substem P_j with no active value, compute $nctr(\cdot)$, $sply(\cdot)$ and $dmnd(\cdot)$ for P_j. Let T_c' be the *stem-tree*, a tree in which each node represents a substem in T', and node u in T_c' is a parent of node w in T_c' if the highest backbone vertex in the substem for w is the lowest backbone vertex in the substem for u. This completes the description of the first phase of *STEM1*.

The data structure generated in the first phase of *STREE1* is used in the second phase for a sublinear feasibility test, *SFTEST2*. The test of value λ is performed by *SFTEST2* as follows. As before we assume that $\lambda_1 < \lambda < \lambda_2$. Perform a traversal of T_c', starting at the root. At node v, representing substem P_j, do the following. Set $numctr(v)$ to 0, $supdist(v)$ to ∞ and $demdist(v)$ to 0. Then recursively explore each child w of v. Upon the return of the recursive call involving w, add $numctr(w)$ to $numctr(v)$, reset $supdist(v)$ to be $\min\{supdist(v), supdist(w)\}$ and reset $demdist(v)$ to be $\max\{demdist(v), demdist(w)\}$. After handling all children w of v, handle P_j in fashion identical to that in *SFTEST1*, except that $supdist(v)$ takes the role of $supdist$, $demdist(v)$ takes the role of $demdist$, and $numctr(v)$ takes the role of $numctr$. When P_j has been handled, $numctr(v)$ will have been increased to reflect the number of supply centers placed on P_j, and $supdist(v)$ and

$demdist(v)$ will be the distances from the top vertex in P_j to the nearest supply center and the farthest unsatisfied demand vertex, resp. When exploration at the root of T'_c has completed, if $demdist(root) + supdist(root) > \lambda$, then the demand at the root is satisfied by adding 1 to $numctr(root)$. If $numctr(root) \leq p$, then λ is feasible ($\lambda \geq \lambda^*$), else λ is not feasible ($\lambda < \lambda^*$). This completes the description of feasibility test $SFTEST2$.

Lemma 5.1. Let T be a tree of n vertices and at most n/r leaf-stems, whose stem-partition consists of $O(n/r)$ substems, each of size at most r. Let all but at most n/r^2 substems contain no active values. Algorithm $SFTEST2$ will correctly determine the feasibility of a test value in $O((n/r)\log r)$ time.

Proof. The proof is similar to that of Lemma 4.2. $\square$

We next describe the second phase of $STREE1$. Reset tree T to be T', and p to be p'. Generate a stem-partition of T'. While there is more than one leaf-stem, repeat the following. Consider the leaf-stems of T'. Let $\mathcal{M}$ be the set of sorted matrices defined on these stems. Call algorithm $MSEARCH$ on set $\mathcal{M}$, with stopping count of 0. Use the feasibility test $SFTEST2$ on the tree T with p. When $MSEARCH$ halts, all leaf-stems in T' will have no active value. If T' is a single stem, then $\lambda^* = \lambda_2$, and we are done. Otherwise, for each leaf-stem, identify the supply centers required on the stem, and either delete the stem or replace it by a twig or thorn as necessary. Reset T' to be the tree resulting from deleting these leaf-stems, and p' to be the number of cuts remaining to be made. This completes the description of one iteration of the while loop. This also completes the description of $STREE1$.

Theorem 5.2. Algorithm $STREE1$ solves the p-center problem $V/V/p$ on a tree of n vertices in $O(n(\log\log n)^2)$ time.

Proof. The proof is similar to the proof of Theorem 4.2 in [F2], but we include it for completeness. We first consider the generation of tree T'. Each call to $MSEARCH$ will take $O(n')$ time, produce $O(\log\log n)$ search values and result in the deletion of at least $n'/2$ vertices from the tree. Thus the total time to produce search values over all iterations will be $O(n)$. It will cost $O(n)$ time to test each search value. The number of iterations needed to reduce the tree to one with at most $2n/(\log n)^2$ leaves will be $O(\log\log n)$. Thus the total time to generate T' will be $O(n(\log\log n)^2)$.

Once T' has been generated, T'_c can be generated and partitioned, and the values of $nctr(\cdot)$, $sply(\cdot)$ and $dmnd(\cdot)$ can be computed in $O(n)$ time. By Lemma 5.1, any subsequent feasibility test will use $O(n\log\log n/(\log n)^2)$ time. Each vertex will participate in the formation of a sorted matrix for just one call to $MSEARCH$. Thus the total time to produce test values in all calls to $MSEARCH$ during the second phase is $O(n)$, and each call will produce $O(\log n)$ test values. Since the number of leaf-stems is at least halved on each iteration of the while loop, the number of calls to *MSEARCH* is $O(\log \dot{n})$. Thus the number of values tested is at most $O((\log n)^2)$. Since these can each be tested in $O(n \log \log n/(\log n)^2)$ time, the total time for feasibility testing in the second phase is $O(n \log \log n)$. $\square$

The above strategy can be applied recursively to yield an $O(n \log^* n)$-time algorithm *STREE2*. The basic idea is to start with the tree and a straightforward feasibility test, and bootstrap our computation by alternately pruning the tree to reduce the number of leaf-stems (and hence the number of stems in a stem-partition), and then constructing data structures for an ever faster feasibility tests. In each phase of a sequence of phases, we generate these data structures as we search among sets of sorted matrices that represent larger and larger substems. The structure of the algorithm is similar to *TREE2* in [F2].

Theorem 5.3. Algorithm *STREE2* solves the p-center problem $V/V/p$ on a tree of n vertices in $O(n \log^* n)$ time.
Proof. The proof is similar to that of Theorem 4.3 in [F2]. $\square$

We now discuss algorithm *STREE3*, which runs in linear time. This will be similar to *TREE3* in [F2], but will have some differences. First, we can compute *a priori* the distances between a vertex and any ancestor in the rooted tree. Simply let A_v be the distance from vertex v to the root, and then subtract two such values to compute any such distance. This is simpler than the corresponding task in *TREE3*, which had to be able to handle vertex weights that could increase. Second, we must be able to merge stems together as leaf-stems are replaced by thorns or twigs. This requires care, as our algorithms for handling stems required certain preprocessing of the stems, and it is too expensive to perform this preprocessing after each merge.

Instead of generating a projected stem, we generate a set of seven sorted matrices that contain all values under consideration on a stem. The set of values to be searched is the set of distances from backbone vertices and buds to backbone vertices and points of thorns. The data structure contains these values, plus others, in the form of seven sorted matrices that are succinctly represented and generated as follows. We assume that the vertices are indexed in the same preorder scheme as described in the previous section. Let B_i be the distance of vertex v_i from the root. Let D be the maximum value of B_i over all vertices v_i in the original tree. Let $A_i = D - B_i$. We shall generate sequences from which we derive our sorted matrices. For any sequence X, let X^R be the sequence X in reverse order. For any stem or substem P_j, let X_1 be the sequence of distances A_i in order by increasing i. Then sorted matrix $M_1(P)$ is the $n \times n$ matrix $X_1^R - X_1$. This matrix contains all distances between backbone vertices in P_j. Let X_2 be the sequence of distances B_i in order by increasing i. Let X_3 be the sequence of distances $A_i + c(v_i, u_i)$, for each i for which (u_i, v_i) is a thorn, in order by increasing i. Let X_4 be the sequence of distances $B_i + c(v_i, u_i)$, for each i for which (u_i, v_i) is a thorn, in order by increasing i. Note that if the indices i are in increasing order, then the values X_3 are in increasing order, and the values X_4 are in decreasing order. For the values X_3 this follows since for any pair i and j with $i < j$, $d(v_i, u_i) < d(v_i, u_j)$, which implies $d(v_1, u_i) < d(v_1, u_j)$. The argument for the X_4 is similar. Sorted matrices $M_2(P)$ and $M_3(P)$ are $X_3^R - X_1$ and $X_4 - X_2^R$, resp. These matrices contain all distances between backbone vertices and points. Let X_5 be the sequence of distances $A_i - c(v_i, w_i)$, for each i for which w_i is a bud, in order by increasing i. Let X_6 be the sequence of distances $B_i - c(v_i, w_i)$, for each i for which w_i is a bud, in order by increasing i. Note that if the indices i are in increasing order, then the values X_5 are in increasing order, and the values X_6 are in decreasing order. Sorted matrices $M_4(P)$ and $M_5(P)$ are $X_1^R - X_5$ and $X_2 - X_6^R$, resp. These matrices contain all distances between backbone vertices and buds. Sorted matrices $M_6(P)$ and $M_7(P)$ are $X_3^R - X_5$ and $X_4 - X_6^R$, resp. These matrices contain all distances between points and buds.

We use the same rearrangement scheme as in [F2] for copying stems when all of their siblings have been replaced by thorns or twigs. As stems are merged together, some thorns and twigs may need to be deleted in order to keep the resulting stems cleaned up. (If this is not done, then the resulting matrices would not necessarily be sorted.) This presents a problem for the call to *MSEARCH* involving the thin matrices. This call is intended to take less time than the size of the description of the thin matrices. Thus we cannot afford to copy the corresponding sequences into a concise array representation. We discuss how the sequences X_3 through X_6 are maintained so that the searching of matrices can be accomplished efficiently. We maintain these sequences both as doubly-linked lists, and as arrays stored in parallel with the values A_i and B_i. Any vertex v_i for which there is no thorn will have an entry "null" in location i of the arrays corresponding to X_3 and X_4. The same approach is applied for twigs. Imposed on top of the array is a binary tree index structure. (The structure is stored in a parallel array.) Each node in the index contains two pointers LN and SN. If there are non-null values in the portion of the sequence covered by the node, then LN and SN point to the largest and smallest such values, resp. If all values in the corresponding portion of the sequence are null, and no such node is an ancestor, then LN and SN are null. When a stem is copied in the rearrangement scheme, these arrays are also copied.

When two stems are merged, this structure can be updated in time proportional to the number of thorns and twigs removed, plus the logarithm of the size of the new stem.

As for searching the sequence, whenever we would have been testing a null value, we use either the next larger actual value or the next smaller actual value in the sequence. The time for handling the thin matrices will be within a constant factor of what it was previously.

As discussed in [F1], the values $fsup$, $tsup$, $sucth(l)$, $succ(l)$, $fdem(l)$, and $tdem(l)$ can be updated in a total of $O(n)$ time overall. In [F1] we survey all updates and accesses of these values in $SDIGEST1$, $SDIGEST2$, $SFTEST1$ and $SFTEST2$, and show how to represent the data so that all such updates and accesses take $O(n)$ time in total.

Lemma 5.4. The computation of all values of $fsup$, $tsup$, $prec(l)$, $succ(l)$, $fdem(l)$, and $tdem(l)$ will take all $O(n)$ time overall. □

Theorem 5.5. Algorithm $STREE3$ solves the p-center problem $V/V/p$ on a tree of n vertices in $O(n)$ time.
Proof. The correctness follows from the above discussion plus that in section 4. The time complexity follows from a discussion similar to that in the proof of Theorem 3.6 of [F2] and Theorem 4.4, and uses Lemma 5.4. □

The modifications necessary to handle the $E/V/p$ and $V/E/p$ versions of the p-center problem can be found in [F1].

References

[AASS] P. K. Agarwal, B. Aronov, M. Sharir, and S. Suri. Selecting distances in the plane. April 1990.

[AP] E. M. Arkin and C. H. Papadimitriou. On the complexity of circulations. *J. Algorithms*, 7:134–145, 1986.

[BPS] R. I. Becker, Y. Perl, and S. R. Schach. A shifting algorithm for min-max tree partitioning. *J. ACM*, 29:58–67, 1982.

[CT] R. Chandrasekaran and A. Tamir. Polynomially bounded algorithms for locating p-centers on a tree. *Math. Prog.*, pages 304–315, 1982.

[C1] R. Cole. Partitioning point sets in arbitrary dimensions. *Theor. Comput. Sci.*, 49:239–265, 1987.

[C2] R. Cole. Slowing down sorting networks to obtain faster sorting algorithms. *J. ACM*, 34:200–208, 1987.

[CSSS] R. Cole, J. S. Salowe, W. L. Steiger, and E. Szemeredi. An optimal-time algorithm for slope selection. *SIAM J. Comput.*, 18:792–810, 1989.

[CSY]   R. Cole, M. Sharir, and C. K. Yap. On k-hulls and related problems. *SIAM J. Comput.*, 17:61–77, 1987.

[F1]   G. N. Frederickson. Optimal algorithms for partitioning trees and locating p-centers in trees. Technical report CSD-TR-1029, Purdue University, 1990.

[F2]   G. N. Frederickson. Optimal algorithms for tree partitioning. In *Proc. 2nd ACM-SIAM Symposium on Discrete Algorithms*, pages 168–177, 1991.

[FJ1]   G. N. Frederickson and D. B. Johnson. Finding kth paths and p-centers by generating and searching good data structures. *J. Algorithms*, 4:61–80, 1983.

[FJ2]   G. N. Frederickson and D. B. Johnson. Generalized selection and ranking: sorted matrices. *SIAM J. on Computing*, 13:14–30, 1984.

[Go1]   A. J. Goldman. Optimal center location in simple networks. *Transport. Sci.*, 5:212–233, 1971.

[Go2]   A. J. Goldman. Minimax location of a facility in an undirected tree graph. *Transport. Sci.*, 6:407–418, 1972.

[Gu]   D. Gusfield. Parametric combinatorial computing and a problem of program module distribution. *J.ACM*, 30:551–563, 1983.

[Hk]   S. L. Hakimi. Optimum locations of switching centers and the absolute centers and medians of a graph. *Operations Res.*, 12:450–459, 1964.

[Hn1]   G. Y. Handler. Minimax location of a facility in an undirected tree graph. *Transport. Sci.*, 7:287–293, 1973.

[Hn2]   G. Y. Handler. Finding two-centers of a tree: the continuous case. *Transport. Sci.*, 12:93–106, 1978.

[HT]   R. Hassin and A. Tamir. Efficient algorithms for optimization and selection on series-parallel graphs. *SIAM J. Alg. Disc. Meth.*, 7:379–389, 1986.

[HS]   D. H. Hochbaum and D. B. Schmoys. Approximation algorithms for bottleneck problems. *J. ACM*, 33:533–550, 1986.

[KH]   O. Kariv and S. L. Hakimi. An algorithmic approach to network location problems. *SIAM J. Appl. Math.*, 37:513–538, 1979.

[LS]   C. Lo and W. Steiger. An optimal-time algorithm for ham-sandwich cuts in the plane. September 1990.

[M]   N. Megiddo. Applying parallel computation algorithms in the design of serial algorithms. *J. ACM*, 30:852–865, 1983.

[MT]   N. Megiddo and A. Tamir. New results on the complexity of p-center problems. *SIAM J. Comput.*, 12:751–758, 1983.

[MTZC] N. Megiddo, A. Tamir, E. Zemel, and R. Chandrasekaran. An $O(n \log^2 n)$ algorithm for the kth longest path in a tree with applications to location problems. *SIAM J. Comput.*, 10:328–337, 1981.

[PeS]   Y. Perl and S. R. Schach. Max-min tree partitioning. *J. ACM*, 28:5–15, 1981.

[1]   R. Pollack and M. Sharir. Computing the geodesic center of a simple polygon. Technical Report 231, Courant Institute, September 1986.

[R]   M. Reichling. On the detection of a common intersection of k-convex polyhedra. *Lect. Notes Comput. Sci.*, 333:180–187, 1988.

[S]   J. S. Salowe. L-infinity interdistance selection by parametric search. *Inf. Proc. Lett.*, 30:9–14, 1989.

[SS]   J. T. Schwartz and M. Sharir. Finding effective 'force-targets' for two-dimensional multifinger frictional grips. Technical Report 379, Courant Institute, June 1988.

[Z]   E. Zemel. A linear time randomizing algorithm for searching ranking functions. *Algorithmica*, 2:81–90, 1987.

On Bends and Lengths of Rectilinear Paths: A Graph-Theoretic Approach

C. D. Yang,   D. T. Lee[1]
Department of EE & CS
Northwestern University
Evanston, IL 60208

and

C. K. Wong
IBM Research Division
Thomas J. Watson Research Center
Yorktown Heights, NY 10598

Abstract

We consider the problem of finding a rectilinear path between two designated points in the presence of rectilinear obstacles subject to various optimization functions in terms of the number of bends and the total length of the path. Specifically we are interested in finding a minimum bend shortest path, a shortest minimum bend path or a least-cost path where the cost is defined as a function of both the length and the number of bends of the path. We provide a unified approach by constructing a *path-preserving graph* guaranteed to preserve all these three kinds of paths and give an $O(K + e \log e)$ algorithm to find them, where e is the total number of obstacle edges, and K is the number of intersections between *tracks* from *extreme point* and other tracks (defined in the text). K is bounded by $O(et)$, where t is the number of *extreme edges*. In particular, if the obstacles are rectilinearly convex, then K is $O(ne)$, where n is the number of obstacles. Extensions are made to find a shortest path with a bounded number of bends and a minimum-bend path with a bounded length. When a source point and obstacles are pre-given, queries for the assorted paths from the source to given points can be handled in $O(\log n + k)$ time after $O(K + e \log e)$ preprocessing, where k is the size of the goal path. The *trans-dichotomous* algorithm of Fredman and Willard[8] and the running time for these problems are also discussed.

1. Introduction

There have been many results [1, 3, 5, 10, 13, 14, 15, 17, 21, 26, 27, 28] about finding shortest collision-free paths among obstacles. The problem varies on the metric used, the types of obstacles and the path considered. In [1, 10, 15, 12, 17, 20, 22, 25, 26] Euclidean paths are sought, whereas [3, 5, 13, 21, 24, 27, 28] put emphasis on rectilinear paths. A typical problem in the latter case is to find a rectilinear shortest path among rectilinear obstacles. Two best known results to date are due respectively to Clarkson *et al.*[3] and Wu *et al.*[28]. The former runs in $O(e \log^{3/2} e)$ where e is the number of obstacle edges and the latter in $O((e + k) \log t)$, where t is the number of extreme edges, and k is the number of intersections among tracks emanating from extreme edges, which is bounded by $O(t^2)$.

On the other hand, the number of links(bends) on paths gains more attention recently [4, 6, 12, 16, 19, 23, 24, 25]. Ohtsuki [24] studied the problem of finding a rectilinear path with a minimum number of bends, or segments, referred to as the *minimum-bend rectilinear path* among obstacles and proposed an $O(e \log^2 e)$ algorithm. Using the data structure given by Imai and Asano [11], the problem of finding minimum-bend path can be solved in $\theta(e \log e)$ time. Instead of a minimum bend *rectilinear* path, Suri [25] considered

---

[1]Supported in part by the National Science Foundation under Grant CCR-8901815.

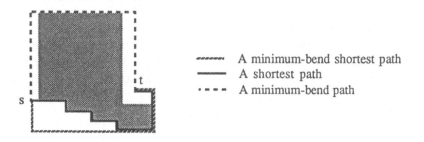

Figure 1: Paths with different criteria.

the problem of finding a minimum-bend path in a simple polygon, and presented a linear time algorithm for the problem when the polygon is triangulated.[2] The number of bends or links of the minimum-bend path between any two points is referred to as the *link distance* between these two points within the polygon. McDonald and Peters [19] presented a linear time algorithm for finding the *smallest path* in a simple polygon where *smallest* means the shortest both in length and in link-distance. As to find paths from a pre-given source to a queried destination, de Berg, *et al.*[4] can find the best path in $O(\log n + k)$ time with $O(n^2)$ preprocessing time, where a combined L_1 and link metric is used, and k is the size of the reported path .

We note that the existing algorithms which minimize the length of the path do not necessarily produce a path of minimum bends, and vice versa. For example, the algorithm due to Wu *et al.*[28] finds a shortest path but leaves the number of bends in the solution path uncertain. The shortest path obtained by Clarkson *et al.*[3] tends to find a shortest path with more bends than necessary, as shown in solid lines in Figure 1. On the other hand, the minimum-bend path obtained by Ohtsuki's algorithm is far from being the shortest. Therefore, we aim to take both the number of bends of the path and its length into consideration and give a graph-theoretical approach to solving these problems in a unified manner. We shall restrict ourselves to the rectilinear domain in which the objects, paths and obstacles, are rectilinear, and omit the term "rectilinear" from here on without confusion. More precisely we address a set of problems in which both the bends and lengths of paths are considered with different minimizing priorities. That is, we are interested in finding a shortest path with a minimum number of bends, a minimum-bend path whose length is the smallest, or a path with the least *cost* which is a function of both factors. In addition, we also consider the problem of finding an optimal path while one of the factors is bounded by a given limit and the problem of finding assorted paths to a queried destination while obstacles and source are pre-given.

2. Problems Description

We first define those problems discussed in this paper as follows.

---

[2]Simple polygon triangulation can be solved in linear time recently due to Chazelle[2]

Given a set of isothetic, rectilinear, non-intersecting obstacles, the starting point s and the destination point d, we consider the following problems:

1. Problem **MBSP**: Find among all the shortest paths from s to d the one with the minimum number of bends. Such a path is called the *minimum-bend shortest path*, denoted as $mbsp(s, d)$.

2. Problem **SMBP**: Find among all the minimum-bend paths from s to d the shortest path. Such a path is called the *shortest minimum-bend path*, denoted as $smbp(s, d)$.

3. Problem **MCP**: Find a minimum-cost path from s to d, where the cost is an increasing function f of the number of bends and the length of a path. Denote such a path $mcp(s, d)$.

4. Problem **BBSP**: Find among all paths from s to d with bends no greater than a given bound the shortest path. Such a path is called the *bounded-bend shortest path*, denoted as $bbsp(s, d)$.

5. Problem **BLMBP**: Find among all paths from s to d whose length is no greater than a given bound the one with the minimum number of bends. Such a path is called the *bounded-length minimum-bend path*, denoted as $blmbp(s, d)$.

Main Results

Using *graph-theoretic approach* we construct a *path-preserving* graph in which the desired objective paths, *mbsp, smbp, mcp, bbsp* and *blmbp* are guaranteed to be embedded.

Applying the shortest path searching algorithm of Fredman and Tarjan[7] to such a graph, an $O(K + e \log e)$ algorithm is derived for MBSP, SMBP and MCP, where K is the number of intersections between *tracks* from *extreme point* and other tracks, and is bounded by $O(et)$, where t is the number of *extreme edges* of the obstacles. The time complexity can be improved to $O(K + \frac{e \log e}{\log \log e})$, if we apply the *trans-dichotomous* algorithm by Fredman and Willard[8]. In the work by de Berg, *et al.*[4], segments, instead of obstacles, are given and the *mcp* path to a given destination from a fixed source is queried. Their algorithm spent $O(n^2)$ to compute one-to-all mcp paths for preprocessing. The same graph constructed can also be used to find the best assorted paths from s to all the vertices on obstacles in $O(K + e \log e)$ time. For BBSP, we present an algorithm which runs in $O(B(K + e \log e))$ time, where $B = min\{b^*, b\}$, b^* is the bend number of the $mbsp$ and b is the given bound for the number of bends. As for the BLMBP, we obtain an algorithm which runs in $O(b^*(K + e \log e))$ time.

3. The Path-Preserving Graph

A path-preserving graph $G(V, E)$ is a graph containing at least one of the objective paths described above. It can be shown that the *grid graph* obtained by drawing horizontal and vertical lines through the edges of all the obstacles and through points s and d without crossing any obstacles is a path-preserving graph. The grid graph is of quadratic size and can be constructed in $O(e^2)$ time and space, and $O(e^2 \log e)$ time suffices to obtain paths like *mbsp, smbp* and *mcp*. We aim to find a graph whose size is either smaller or dependent on a smaller factor, e.g., the number of extreme edges, so that a more efficient algorithm can be obtained.

Let V denote the set of vertices of all obstacles and CVX the set of convex vertices of all the obstacles.

Definition 1 An *extreme edge* of an obstacle is a boundary edge e with its two adjacent edges lying on the same side of the line containing e. An *extreme point* is an endpoint of an extreme edge. Let $EXTM$ denote the set of all extreme points.

Definition 2 A *horizontal track* $H(x)$, of x, where $x \in \{s, d\} \cup CVX$, is a horizontal collision-free maximal segment or ray emanating from x. A *vertical track* $V(x)$, of x, where $x \in \{s, d\} \cup CVX$, is defined similarly. The endpoints of $H(x)$ and $V(x)$ are called the *projection points* of x, denoted $PJ(x)$. Let T denote the set of horizontal and vertical tracks, *i.e.*, $T = \{H(x)\} \cup \{V(x)\}, x \in \{s, d\} \cup CVX$.

Let $BG(V_{BG}, E_{BG})$ denote the *boundary graph* in which $V_{BG} = V \cup \{PJ(x)|x \in \{s, d\} \cup CVX\}$ consists of all vertices on the boundary of obstacles and projection points and E_{BG} the edges connecting consecutive vertices in V_{BG} along the boundary(refer them as Type 0 edges). Let $TG(V_{TG}, E_{TG})$ denote the *track graph* in which $V_{TG} = \{s, d\} \cup CVX \cup \{PJ(x)|x \in \{s, d\} \cup CVX\}$ consists of projection points, convex vertices, and $\{s, d\}$ and E_{TG} is constructed as follows.

(1) Add into E_{TG} the edge connecting a vertex in $CVX \cup \{s, d\}$ to its projection points (referred to as Type 1 edges).

(2) Add into E_{TG} the edge connecting a convex vertex, u, to a extreme point, v, if there is intersection between $H(u)$ and $V(v)$ or between $V(u)$ and $H(v)$ (referred to as Type 2 edges).

The edge cost assigned to an edge is the rectilinear distance between its two end vertices.

The graph TG actually contains all those tracks between convex vertices and their projection points and edges connecting an extreme point to a convex vertex, if there is an intersection between the tracks of the extreme vertex and those of that convex vertex. We now define the path preserving graph PPG based on TG and BG.

Given the boundary graph and track graph, we obtain the *path-preserving graph*, $PPG(V_{PPG}, E_{PPG})$, where $V_{PPG} = V_{BG} \cup V_{TG}$ and $E_{PPG} = E_{BG} \cup E_{TG}$. That is, PPG is the union of BG and TG. An example of the graph PPG is given in Figure 2. Note that, the two graphs, BG and TG are glued at points in CVX. With the path-preserving graph described, we now prove that any of our goal paths is embedded in such a graph.

Definition 3 A three segment subpath, s_1, s_2 and s_3 such that s_1 and s_3 are on the same side with respect to a line containing s_2 is called a *U-shaped subpath*.

Lemma 1 *Given a set of obstacles, s and d, there exists at least one X-path, $X \in \{mbsp, smbp, mcp, bbsp, blmbp\}$, from s to d which can be decomposed into a sequence of maximal subpaths $\pi_i, i = 1 \ldots m$, such that all π_i's are paths containing no U-shaped subpaths and the two end points of each π_i are points in $\{s, d\} \cup EXTM$.*

Lemma 2 *There is at least one X-path, $X \in \{mbsp, smbp, mcp, bbsp, blmbp\}$, which is embedded in PPG.*

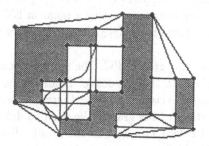

Figure 2: The graph PPG.

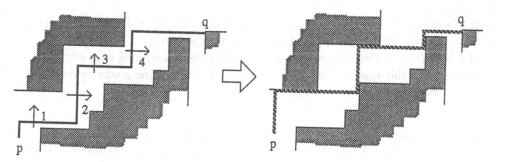

Figure 3: Generate the corresponding pushed path.

[**proof**] According to the previous lemma and the fact that EXTM $\cup\{s,d\}$ is in PPG, we can focus on a U-shape-free path whose end vertices are in EXTM $\cup\{s,d\}$. Let RP denote an X-path. Without loss of generality, assume that the two ends of RP are p and q satisfying $q.x \geq p.x$ and $q.y \geq p.y$. Let the segment sequence from p to q be $h_1, v_1, h_2, v_2, \ldots, h_m, v_m$ where h_i's are horizontal and v_i's are vertical and h_1 and v_m may be empty. We can perform the following dragging operations on RP starting from h_1 (or v_1) until v_m (or h_m) *without increasing the path length or the number of bends* (refer to Figure 3).

(1) Drag h_i upwards until it hits some obstacle, then

(2) Drag v_i rightward until it hits some obstacle.

While dragging a segment upwards or rightward, we adjust the length of its adjacent segments accordingly. Notice that while performing (1) and (2), we either reduce the length of the path or the number of bends if during the dragging operation of h_i (or v_i), h_i (or v_i) is aligned with h_{i+1} (or v_{i+1}) before hitting an obstacle. Therefore if RP is optimal with respect to length, then the segments h_i or v_i must be in contact with an extreme edge of some obstacle; if RP is optimal with respect to bends, no *alignment* would occur during the dragging operation. In the latter case, we have transformed RP into another path RP', preserving the number of bends, such that all segments on it are aligned with extreme edges of obstacles.

It is easy to see that in such a path, all the segments are either a boundary edge (type 0) in G_{BG} or a segment from a point in $CVX \cup \{s, d\}$ to one of its projection point (type 1), or two consecutive segments connecting a point in EXTM $\cup \{s, d\}$ to a point in $CVX \cup \{s, d\}$ (type 2). All of them are in E_{PPG}. This completes the proof. $\diamond$

We now have a path-preserving graph PPG. The remaining task is to devise algorithms to search in the graph for the path that we desire. Before doing so, note that there is no bend information on the edges. Actually whether there is a bend depends on the edges that are incident on the vertices of the graph. We need to define the *incidence* relation of an edge and its endpoint before we augment the graph. Recall that the PPG is the combination of TG and BG which contains three types of edges.

Definition 4 In PPG, *incidence(e,v)* of an edge e on one of its endpoint v is defined as the following:
(1) If e is of type 0 or 1, then e is a horizontal or vertical edge and incidence(e,v) equals a, b, l or r if e is above, below, left of or right of v respectively.
(2) If e is of type 2, then e represents a path along the intersecting tracks from its two end vertices. In this case, *incidence(e,v)* equals a, b, l or r if the involving track from v is above, below, left of or right of v, respectively.

Thus for a type 0 or type 1 edge $e = (v, w)$, we have *(incidence(e, v), incidence(e, w))*$\in$ $\{(X, Y), (Y, X)\}$ where $X = a, Y = b$ or $X = l, Y = r$, and for a type 2 edge $e = (v, w)$, we have *(incidence(e, v), incidence(e, w))*$\in \{(X, Y), (Y, X)\}$ where $(X, Y) \in \{a, b\} \times \{l, r\}$.
We define a new graph derived from PPG by expanding each vertex into a set of four *cell-vertices* with internal edges representing bend information as follows.

Definition 5 Define graph PPG* by the following construction method:
(1) Split each vertex v in PPG into four *cells*, denoted by $l(v), r(v), a(v)$ and $b(v)$.
(2) Replace each edge $e = (v, w)$ by $e' = (X(v), Y(v))$, where $(X, Y) = $ *(incidence(e, v), incidence(e, w))*.
(3) Change the edge cost, $c_{old}(e)$, of every edge e to be a vector cost $c(e) = (c_{old}(e), nb)$ for each edge e in PPG where nb represents the bend cost; $nb = 0$ if the e is of type 0 or 1, $nb = 1$ if e is of type 2.
(4) Add to the graph the following new edges:
$e = (l(v), r(v))$ with $c(e) = (0, 0)$,
$e = (u(v), d(v))$ with $c(e) = (0, 0)$ and
$e = (X(v), Y(v))$ with $c(e) = (0, 1)$ for all $(X, Y) \in \{(l, a), (l, b), (r, a), (r, b)\}$

We change the costs of all the edges to be vector costs because we attempt to search the graph and compare accumulated cost of path lexicographically, which reflects the priorities of those minimizing factors.

Theorem 1 *There is an optimal solution $X, X \in \{mbsp, smbp, mcp, bbsp, blmbp\}$, with length l and bend number b embedded as a path π on the graph PPG* and the total edge costs of π is (l, b).*

The algorithm *FindPPG** which generates PPG* is described as follows.
Algorithm *FindPPG**

1. Initially let V_{PPG^*} be $V \cup \{s, d\}$, where V is the set of vertices of all obstacles.

2. Sort all the horizontal edges and vertical edges respectively by merging those sorted boundary edge lists between consecutive extreme edges.

3. Sweep vertically and horizontally to find the projection point, $PJ(v)$ for each point v in V. Add those projection points to V_{PPG^*} and add the edges connecting v to $t \in PJ(v)$ to E_{PPG^*}. When a projection point is found on an obstacle edge e, insert it on a sorted list, $l(e)$ associated with e, which contains all projection points on e.

4. Traverse the boundary vertices of each obstacle with $l(e)$ being traversed in order between every two boundary vertices. Create an edge between every two consecutive points traversed, and let the cost be the rectilinear distance of them. Add all these edges into E_{PPG^*}.

5. Create an edge between a point in $EXTM \cup \{s, d\}$ and a point in $CVX \cup \{s, d\}$ if there is intersection between their tracks. Let the cost be the rectilinear distance between them. Add these edges to E_{PPG^*}. Here we adopt the line segment intersection reporting algorithm by Mairson and Stolfi [18].

6. Change every vertex in PPG* into four cells and replace the costs of edges by vector costs which reflect the bends and distances of paths as in the definition of PPG*.

Lemma 3 *The algorithm FindPPG* runs in $O(K + e \log t)$ time.*

Lemma 4 *The graph PPG* contains $O(e + K)$ edges and $O(e)$ vertices.*

With PPG*, one can apply different algorithms to find different objective paths. For *mbsp*, we apply a shortest path searching algorithm using the vector edge costs of the form *(distance, bend)* as the metric. For *smbp*, we apply the same searching algorithm but only change the metric to be *(bend, distance)* which is the reverse of the cost vector on edges. For *mcp*, we can precompute the metric of each edge to be the value of the cost function. Below is an algorithm for finding the minimum-bend shortest path. The algorithms to find the *smbp* and *mcp* are similar.

Algorithm FindMBSP:

- construct PPG* with edge costs to be vectors of form *(distance, bend number)*.

- apply shortest path algorithm by Fredman and Tarjan [7] to find the minimum bend shortest path on PPG* by comparing the vector cost *lexicographically.*

Theorem 2 *The algorithm FindMBSP finds the minimum bend shortest path in time $O(K + e \log e)$ time.*

[proof] The correctness follows Theorem 1. As for the run time, according to Lemma 3, the construction of PPG* needs $O(K + e \log t)$ time. The shortest path searching algorithm by Fredman and Tarjan [7] runs in $O(|E_G| + |V_G| \log |V_G|)$ time on a graph G, therefore, with Lemma 4, the searching on PPG* takes $O(K + e \log e)$ time.◇

Theorem 3 *The problems MBSP, SMBP and MCP can be solved in $O(K + e \log e)$ time.*

4. The problem BBSP and BLMBP

As mentioned in the previous section, the path *bbsp* and *blmbp* are also embedded in the graph PPG*. The general form of this problem, which is so called *shortest weight-constrained path* problem [9], is NP-complete. The problem here is polynomially solvable because the two keys (distance and number of bends) on the graph are very restricted. Consider the problem BBSP first, if the given bend bound b is greater than the bend number of a minimum-bend shortest path, denoted as b^*, we can simply find and report a *mbsp* as the result. So the number of bends of a *bbsp* we search on the graph is really bounded by b^*. This helps to limit the time complexity.

While searching a minimum path, for example, *mcp*, we are able to compare and discard paths except the best when we reach a graph node. The difficulty, however, in these bounded-factor problems is that we cannot discard paths when we reach a node without knowing the following situation of searching. For example, consider two paths, p_1 and p_2, from s to a node v. If the bend number and distance of p_1 are both smaller than, or one smaller than and the other equal to, the two corresponding factors of p_2 then we can discard p_2 since p_1 can replace p_2 in any further situation. We cannot determine which one is better when each of p_1 and p_2 has one factor better than the other. We have to keep all the necessary information efficiently. Since the bend number is discrete(integer) data, we are able to compare two paths with the same bend number while reaching a node v, i.e., keeping one path per number of bends suffices. Since the given bend bound is bounded by b^*, at each node, at most b^* paths are needed to be maintained while searching. We therefore obtain the algorithms as follows.

Algorithm FindBBSP
(We refer the reader to the paper by Fredman and Tarjan[7] for details of the Fibonacci Heap and the original searching algorithm)

1. Construct PPG*.

2. Find the *mbsp* with bend number b^*. If the given bend bound b is larger than b^*, return *mbsp* as the result. Continue the following otherwise.

3. Applying a modified Fredman and Tarjan's shortest path searching algorithm as follows:

   (a) Use a Fibonacci Heap FH for storing visited nodes. Initialize it to just contain the starting node. Each heap-node v stored in FH is uniquely identified by label $l(v) = (n_x, bd)$ where n_x is the index of the corresponding node in the graph and bd is the bend number of a shortest path, $\pi(v)$, from starting point to v. Heap-nodes are ordered in FH according to their distance fields, which are the distances of $\pi(v)$.

   (b) Pick and delete from FH the one, say v, with the least associated distance $d(v)$. Let $l(v) = (v_x, B)$. For each node u connecting to v_x by an edge e with metric (c, b), if there was a heap-node uu with label $(u, b + B)$ inserted in FH, we then check uu to see if its associated distance is longer than $c + d(v)$. If true, perform *decrease-key* operation to uu so that the new distance value of uu is $c + d(v)$. If not, do nothing (conceptually ignore this advancing). If there is no heap-node in FH with label $(u, b + B)$ and if $b + B$ does not exceed the given bound of bends, we then *insert* into FH a heap-node with label $(u, b + B)$ and distance $c + d(v)$.

Repeat this step until a heap-node z picked from FH has label (d, X) where d is the index of the destination node on the graph.

(c) Report the distance associated with z to be the length of *bbsp* and X to be the bends of *bbsp*. (One can store the historic advancing information on nodes in FH in order to trace back and report the actual path.)

Theorem 4 *The algorithm FindBBSP solves problem BBSP in $O(B(K + e \log e))$ time, where $B = min\{b, b^*\}$, b is the given bound for bend number and b^* is the bend number of the mbsp.*

[proof] For the correctness of the algorithm, we point out that while advancing to some node, u, from v, we label the node with the number of the bends of the path and we check both the label and the distance before we insert it to FH. This provides the room to consider paths with longer distance but less bends from s to u. The only time we do nothing, i.e. conceptually discard advancing our searching path from v to u, is when there has been a path discovered with the same number of bends but less distance. All the other advancing situations from v to u are considered by either decreasing the key in FH or inserting a new node to FH. Thus the correctness is guaranteed. As for the running time, according to the algorithm, no two nodes with the same label will appear in FH during the process, so at most $O(|V_{PPG^*}| * b)$ (or $O(e * b)$) nodes will be inserted into FH, where b is the given bend bound. Thus the we spend overall $O(eb \log(eb))$ time for $O(eb)$ insertions and deletions. Now consider the decrease-key operation. Since each edge e will be traversed when one of its end node v is picked from the FH and this can happen as many as b times in the worst case, there is a total of $O(|E_{PPG^*}| * b)$ edge-traversals. At each such edge-traversing, a decrease-key operation may be performed. Therefore, at most $O(|E_{PPG^*}| * b)$ (or $O((e + K)b))$ decrease-key operations will be performed during the process. Since each decrease-key operations take only $O(1)$ amortized time, overall $O((e + K)b)$ time is spent for decrease-key. Therefore the running time of algorithm Find-BBSP is $O(K + e \log e)$ if $b \geq b^*$ or $O((e + K)b + eb \log(eb))$ if $b < b^*$, or we can simply say it is of $O(B(K + e \log e))$, $B = min\{b, b^*\}$ since b is bounded by $O(e)$ obviously. ◇

Now we consider the problem BLMBP. One can see that if the length of a shortest minimum-bend path is within the given bound for length, then the *smbp* is also the solution of BLMBP. If not, we observe that if a minimum-bend shortest path has length l and which is greater than the given bound, then there is no solution, since l is the shortest possible distance of a feasible path. Otherwise, the bend number of the *blmbp* must have bend number b' less than or equal to b^* which is the bend number of *mbsp*. Again, b^* is a bound here to find the *blmbp*. The algorithm FindBLMBP is not much different from the one finding *bbsp* except that while advancing from a node to its neighbors, we will discard a path when either the number of bends exceeds b^* or the length exceeds the given bound.

Theorem 5 *The algorithm FindBLMBP solves problem BLMBP in $O(b^*(K + e \log e))$ time where b^* is the number of bend in mbsp.*

[**proof**] Similar to the previous one.◇

5. Finding Paths from a Fixed source to Arbitrary Destinations

Consider the following query problem: given a set of obstacles and a fixed source point, find the mcp, mbsp, smbp from the source to a query point. de Berg, *et al.*[4], study the query problem where segments, instead of obstacles, are given, and only the *mcp* problem is considered. There are two stages in their algorithm for preprocessing: firstly they spent $O(n^2)$ time to find all mcp's from the source to all endpoints of segments and secondly they construct a tree to partition the space in $O(n \log n)$ time to allow for $O(\log n)$ query time, where n is the number of segment. Adopting their algorithm directly for our problem would yield an $O(e^2)$ time preprocessing algorithm. With our graph, one can find all mcp's, mbsp's and smbp's from the source to all obstacle vertices in $O(K + e \log e)$ time. The same second-stage algorithm and data structure used by de Berg, *et al.*, can be reused for our case. Therefore, we can solve the query problem in $O(\log e + k)$ query time where k is the size of the path, using $O(K + e \log e)$ preprocessing time and $O(e \log e)$.

6. Applying Trans-dichotomous Algorithm

Regarding the practical performance while implementing the algorithms, one can adopt trans-dichotomous shortest path searching algorithm provided by Fredman and Willard[8]. The trans-dichotomous algorithm utilizes addressing of random access machine and speeds up computation in a way falling outside the framework of comparison based algorithm. The searching time, if applying their algorithm on PPG*, will be $O(K + \frac{v \log v}{\log \log v})$ for finding the *mbsp, smbp* and *mcp*. As for finding the *bbsp*, the time required becomes $O(B(K + \frac{e \log e}{\log \log e}))$ and for *blmbp*, $O(b^*(K + \frac{e \log e}{\log \log e}))$.

References

1. T. Asano, T. Asano, L. Guibas, J. Hershberger, and H. Imai. Visibility of disjoint polygons. *Algorithmica*, 1, 1986, pp.49-63.
2. B. Chazelle. Triangulating a simple polygon in linear time. Tech. Report CS-TR-264-90, Princeton Univ., 1990.
3. K. L. Clarkson, S. Kapoor, and P. M. Vaidya. Rectilinear shortest paths through polygonal obstacles in $O(n \log^{3/2} n)$ time. Submitted for publication.
4. M. de Berg, M. van Kreveld, B. J. Nillson, M. H. Overmars. Finding shortest paths in the presence of orthogonal obstacles using a combined L_1 and link metric. *Proc. of SWAT '90, Lect. Notes in Computer Science*, 447, Springer-Verlag, 1990, pp.213-224.
5. P. J. deRezende, D. T. Lee, and Y. F. Wu. Rectilinear shortest paths with rectangular barriers. *Discrete and Computational Geometry*, 4, 1989, pp. 41-53.
6. H. N. Djidjev, A. Lingas and J. Sack. An $O(n \log n)$ algorithm for computing the link center in a simple polygon. *Proc. of STACS '89, Lect. Notes in Computer Science*, 349, Springer-Verlag, 1989, pp.96-107.
7. M. L. Fredman and R. E. Tarjan. Fibonacci heaps and their uses in improved network optimization algorithms. *Proc. 25th IEEE Sympo. on Foundations of Computer Science*, 1984, pp.338-346.

8. M. L. Fredman and D. E. Willard. Trans-dichotomous algorithms for Minimum Spanning Trees and Shortest Paths. *Proc. 31th IEEE Sympo. on Foundations of Computer Science*, 1990, pp.719-725.

9. M. R. Garey and D. S. Johnson. Computers and Intractability. A Guide to the Theory of NP-Completeness. W. H. Freeman, San Francisco, 1979, pp.214.

10. L. J. Guibas and J. Hershberger. Optimal shortest path queries in a simple polygon. *Proc 3rd ACM Symp. on Computational Geometry*, 1987, pp.50-63.

11. H. Imai and T. Asano. Dynamic Segment Intersection Search with Applications. *Proc. 25th IEEE Symp. on Foundations of Computer Science*, Singer Island, Florida, 1984, pp.393-402.

12. Y. Ke. An Efficient Algorithm for Link-Distance Problems. *Proc. 5th ACM Symp. on Computational Geometry*, 1989, pp.69-78.

13. R. C. Larson and V. O. Li. Finding minimum rectilinear distance paths in the presence of barriers. *Networks*, 11, 1981, pp.285-304.

14. D. T. Lee. Proximity and reachability in the plane. PhD. Dissertation, University of Illinois, 1978.

15. D. T. Lee and F. P. Preparata. Euclidean shortest paths in the presence of rectilinear barriers. *Networks*, 14, 1984, pp.393-410.

16. Lenhart, R. Pollack, J. Sack, R. Seidel, M. Sharir, S. Suri, G. Toussaint, S. White-sides and C. Yap. Computing the link center of a simple polygon. *Proc. 3rd ACM Symp. on Computational Geometry*, 1987, pp.1-10

17. T. Lozano-Perez and M. A. Wesley. An algorithm for planning collision-free paths among polyhedral obstacles. *CACM*, 22, 1979, pp.560-570.

18. H. G. Mairson and J. Stolfi. Reporting line segment intersections in the plane. Tech. Rep., Dept. of Computer Sci., Stanford University, 1983.

19. K. M. McDonald and J. G. Peters. Smallest paths in simple rectilinear polygons. Submitted for publication.

20. J. S. B. Mitchell and C. Papadimitriou. Planning shortest paths. *SIAM conference* July 15-19, 1985.

21. J. S. B. Mitchell. Shortest rectilinear paths among obstacles. Technical report NO. 739, School of OR/IE, Cornell University, April 1987.

22. J. S. B. Mitchell. An optimal algorithm for shortest rectilinear paths among obstacles in the plane. *Abstracts of the First Canadian Conference on Computational Geometry*, 1989, pp.22.

23. J. S. B. Mitchell, G. Rote and G. Wöginger. Computing the Minimum Link Path among a Set of Obstacles in the Planes. *Proc. 6th ACM Symp. on Computational Geometry*, 1990.

24. T. Ohtsuki. Gridless Routers — New Wire Routing Algorithm Based on Computational Geometry. *International Conference on Circuits and Systems*, China, 1985.

25. S. Suri. A Linear Time Algorithm for Minimum Link Paths Inside a Simple Polygon. *Computer Vision, Graphics and Image Processing* 35, 1986, pp.99-110.

26. E. Welzl. Constructing the visibility graph for n line segments in $O(n^2)$ time. *Info. Proc. Lett.*, 1985, pp.167-171.

27. P. Widmayer. Network design issues in VLSI. Manuscript, 1989.

28. Y. F. Wu, P. Widmayer, M. D. F. Schlag, and C. K. Wong. Rectilinear shortest paths and minimum spanning trees in the presence of rectilinear obstacles. *IEEE Trans. Comput.*, 1987, pp.321-331.

Computing minimum length paths of a given homotopy class
(Extended Abstract)*

John Hershberger
DEC Systems Research Center

Jack Snoeyink†
Department of Computer Science
Utrecht University

Abstract

In this abstract, we use the universal covering space of a surface to generalize previous results on computing paths in a simple polygon. We look at optimizing paths among obstacles in the plane under the Euclidean and link metrics and polygonal convex distance functions. The universal cover is a unifying framework that reveals connections between minimum paths under these three distance functions, as well as yielding simpler linear-time algorithms for shortest path trees and minimum link paths in simple polygons.

1 Introduction

If a wire, a pipe, or a robot must traverse a path among obstacles in the plane, then one might ask what is the best route to take. For the wire, perhaps the shortest distance is best; for the pipe, perhaps the fewest straight-line segments. For the robot, either might be best depending on the relative costs of turning and moving.

In this abstract, we find shortest paths and closed curves that wind around the obstacles in a prescribed fashion—that have a certain homotopy type. We omit all pseudocode, most proofs, and some figures due to space constraints. We consider the Euclidean and link metrics for general paths, and convex and link distance functions for paths restricted to use c given orientations, such as rectilinear paths. Our work presents these distance functions in a unified framework and generalizes previous results for paths in simple polygons. It also results in simplified linear-time algorithms for Euclidean shortest path trees and minimum link paths and new algorithms for c-oriented paths in simple polygons.

In the remainder of this section, we review previous results on finding minimum Euclidean and link paths, closed curves, and c-oriented paths. In section 2, we define triangulated manifolds, homotopy classes, and the universal covering space—important topological tools for our algorithms. Section 3 investigates general paths under the Euclidean metric. It gives algorithms for Euclidean shortest paths of a certain homotopy class, shortest path trees, and shortest closed curves. Section 4 deals with the link metric, in which the length of a path is the number of its line segments. It gives algorithms for minimum link paths and discusses closed curves. Finally, section 5 develops similar algorithms for paths restricted to c previously chosen orientations, such as rectilinear paths.

---

*This research was supported by Digital Equipment Corporation and the ESPRIT Basic Research Action No. 3075 (project ALCOM).

†On leave from the Department of Computer Science of the University of British Columbia. Portions of this research was performed while at the Computer Science Department of Stanford University.

1.1 Euclidean shortest paths

Many researchers have investigated the problem of finding minimum Euclidean paths in simple polygons. Chazelle [6] and Lee and Preparata [24] gave algorithms that, in triangulated polygons, compute the shortest Euclidean path between two points in linear time. We extend their ideas to universal covers and also to maintaining the shortest path homotopic to a path given on-line. Guibas et al. [17] showed how to compute the tree of all shortest paths from one polygon vertex to all other vertices in linear time; they use this as a preprocessing step to solve several shortest path and visibility query problems. One application of our on-line path algorithm is to compute the shortest path tree using simpler data structures.

Shortest Euclidean paths have been studied as one of the tractable cases of river routing in VLSI. Leiserson and Maley [25] and Gao et al. [13] use an algorithm much like our Euclidean minimum path algorithm to compute the *rubber-band equivalent* of each wire as a basic preprocessing step. Finding minimum paths among obstacles when the homotopy class is not given is a more difficult problem, and one that we will not discuss; see Ghosh and Mount [15] and Kapoor and Maheshwari [22] for efficient algorithms.

There are special closed curves of interest to computational geometers that fit within the framework of this research. Toussaint and others have studied *relative convex hulls*, also called geodesic hulls, in connection with the separability of polygons under translation [4, 11, 34, 32, 33]. Czyzowicz et al. [9] have recently solved the "Aquarium Keeper's Problem;" essentially, they use the reflection principle to convert this problem to one of computing the shortest loop around a triangulated annulus or Möbius strip. Our results on closed curves, reported in the full version of the paper, generalize these solutions.

1.2 Link shortest paths

Researchers have also looked at finding minimum paths in simple polygons under the link metric, in which the length of a path is the number of its line segments. Suri [30] developed a linear time algorithm for computing the minimum path between two points in a simple polygon. Ghosh [14] recently gave a linear time algorithm as a consequence of his work on computing the visibility polygon from a convex set. Both algorithms are based on a triangulation and the shortest path tree algorithm of Guibas et al. [17]. We show how to extend our Euclidean minimum path algorithm to compute the link minimum path in time proportional to the number of triangles the path intersects. This gives yet another linear-time algorithm in a simple polygon, but one that is more direct and also has application to paths of given homotopy class among obstacles. In the full paper, we also discuss minimum link closed curves.

1.3 Paths restricted to c orientations

In some applications, most notably VLSI, the orientations of paths are restricted. Rectilinear paths are the most important and, thus, the most studied.

For computing rectilinear shortest paths among rectilinear barriers under the Manhattan, or L_1, metric, researchers have developed algorithms that work in simple polygons (e.g. [29]) and in the presence of obstacles [8, 12, 23]. Mark de Berg [10] has given an algorithm that finds a path that is both a minimum link and an L_1 shortest path in a simple polygon.

Güting [20] defined c-oriented polygons as a generalization of rectangles; he and others [21, 28, 31] have looked at various geometry problems with restricted orientations. We, however, seem to be the first to present algorithms for c-oriented paths in a simple polygon. We show that the shortest Euclidean path, measured under a convex distance function, has the length of the minimum c-oriented path. We also give algorithms to compute the minimum length and link c-oriented paths.

2 Preliminaries

We begin this abstract by sketching the definitions of some important topological objects. The full paper gives more complete definitions.

2.1 Manifolds, complexes, and metrics

Our results apply to *boundary-triangulated 2-manifolds* (BTMs), simplicial complexes in which all vertices are boundary vertices. Simplicial complexes can be represented by Guibas and Stolfi's quad-edge structure [18], by Baumgart's winged-edge structure [3], or by their dual graph. We require that each triangle be able to access its edges and each edge its incident triangles in constant time. If a polygonal region is given, we can triangulate it in $O(n \log n)$ time by a sweepline algorithm [27] or, if there are a constant number of boundary components, in linear time [7].

We assume that a path α is given in some form that can be traced through the data structure that represents a BTM M in time proportional to α's complexity, C_α, plus the number of times α crosses a triangulation edge, Δ_α. If α is piecewise linear, then C_α is the number of α's line segments and, using any of the previous data structures, we can compute a constant number of segment/segment intersection points and determine the next triangulation edge that would be hit if the path leaves the current triangle. Thus we traverse α in $O(C_\alpha + \Delta_\alpha)$ time.

We consider two metrics for unrestricted paths in a BTM: the Euclidean and link metrics. Under both metrics, the minimum length paths are composed of line segments.

A useful example of a BTM is a triangulated polygonal region R in the Euclidean plane: a set bounded by n line segments with disjoint interiors. If one considers the line segments as obstacles and looks at paths avoiding the obstacles, then one can form equivalence classes of paths by relating paths that can be deformed to each other within R.

2.2 Homotopy classes

The topological concept of *homotopy* formally captures the notion of deforming paths and curves. See a basic topology book for the formal definitions [2, 26]. A homotopy relation partitions paths or closed curves into equivalence classes. Thus, we can describe a homotopy class by giving a representative path, or closed curve, α. Given α, we seek to compute a minimum length representative of α's class under the Euclidean and link metrics.

Let's look at one concrete example of a path homotopy. In a BTM, a path gives a sequence of triangulation edges; we can form a *canonical path* that visits the midpoints of triangulation edges in the same sequence. It is easy to see that a path is homotopic to its corresponding canonical path—at times we will find it convenient to use the canonical path as the representative of a homotopy class.

A theorem of elementary topology says that, in a simply connected manifold, any two paths with the same starting and ending point are homotopic. This theorem becomes useful because we can always find a simply-connected covering space, which we now define.

2.3 Covering spaces

Informally, a topological space U is a covering space of a space X if, at each point $u \in U$, there is a corresponding point $x \in X$ such that things around u and x look the same in their respective spaces, but there may be many points of U mapping to the same point x. A space is always a covering space of itself under the identity map. For a more useful example, consider the covering space of a BTM M formed by the following procedure: Choose a base triangle of M, copy it, and make its edges active. Now, any triangle t with an active edge e is a copy of some triangle $t' \in M$ and of an edge e' of t'. There is another triangle $u' \in M$ incident to e'—copy it, forming u, and attach u to t along edge e. Make edge e inactive and the other two edges of u active. One can see that the function that sends the copy of a point to its original is a covering map. The covering space thus formed is the *universal covering space* of M.

Lemma 2.1 *The universal covering space U of a BTM is simply connected.*

Any path that begins in the base triangle has a unique lifting to the covering space, as indicated in figure 1. Formally, let $p: U \to M$ be a covering map. If a function f, from a space W to the BTM M is one-to-one and continuous, then a lifting of f is a map $\hat{f}: W \to U$ such

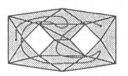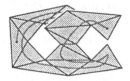

Figure 1: The lift of α

that the composition $p\hat{f} = f$. When we lift a path α, we use $U_\alpha \subset U$ to denote the BTM composed of the triangles of the universal cover U that intersect the path $\hat{\alpha}$.

One last lemma pulls all of the constructs in this section together:

Lemma 2.2 *If α is a path in a BTM, M, then we can construct U_α, the portion of the universal covering space of M that contains the lift of α, in $O(C_\alpha + \Delta_\alpha)$ time.*

Proof: The construction algorithm is simple: Begin with U_α equal to a copy of the triangle of M that contains the starting point of the path α. Then trace α through M and, simultaneously, the lift of α through the covering space—when α crosses a triangulation edge into a triangle of M, add a copy of the triangle to U_α if the lift has never crossed the corresponding edge before. (Otherwise, the triangle is already present.) We can trace the path α through the triangles of M in the stated time bound. ∎

3 The Euclidean metric

We investigate Euclidean shortest paths and closed curves in three subsections. In section 3.1, we remark how the algorithm of Chazelle [6] and Lee and Preparata [24] can be used to find shortest paths between two points of a given homotopy type. Section 3.2 extends this algorithm to maintain the shortest path on-line. As a by-product we can find shortest path trees in linear time without using finger search trees. This simplifies an important algorithm of Guibas et al. [17]. Section 3.3 shows how to compute shortest closed curves.

3.1 Funnels and the shortest path between two points

First we review funnels, a data structure used by many researchers to find shortest paths [6, 13, 17, 19, 24, 25]. Let p be a point and $\overline{uv}$ be a line segment in a simply connected BTM. The shortest paths from p to v and from p to u may travel together for a while. At some point a they diverge and are concave until they reach u and v. The region bounded by $\overline{uv}$ and the concave chains to a is called the *funnel*; a is the *apex* of the funnel.

We store the vertices of a funnel in a double-ended queue, a *deque*. Figure 2 shows that the extensions of funnel edges define wedges. When a point w is added to one end of the funnel, we pop points from the deque until we reach b, the apex of the wedge that contains w, then we push w. If the apex of the funnel is popped during the process, then b becomes the new funnel apex. Notice that the edge $\overline{bw}$ is on the shortest path to w.

The shortest path algorithms of Chazelle [6] and Lee and Preparata [24] both look for a path in a *sleeve polygon*—a triangulated simple polygon whose dual tree is simple path. We shall look for a *sleeve* BTM.

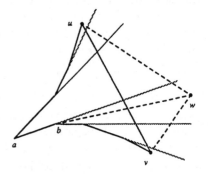

Lemma 3.1 *Let α be a path from p to q. One can compute, in $O(C_\alpha + \Delta_\alpha)$ time and space, a sleeve BTM that contains the Euclidean shortest path homotopic to α.*

By maintaining the funnel as we trace the canonical path through the sleeve, we compute the shortest path.

Figure 2: Splitting a funnel about w

Theorem 3.1 *The Euclidean shortest path that is homotopic to a given path α can be computed in $O(C_\alpha + \Delta_\alpha)$ time and $|U_\alpha|$ space.*

3.2 On-line shortest paths and shortest path trees

In this section, we show how to trace the lift of the path α from p to q and maintain the funnel used in shortest path computations. This procedure optimizes a path that is given on-line and is useful for interactive applications.

We can apply this procedure to compute shortest path trees in a simply connected BTM; the *shortest path tree* from a point p is the union of all shortest paths from p to vertices of the BTM. Guibas et al. [17] compute the shortest path tree of any triangulated simple polygon by splitting funnels—they use finger search trees to find the splitting vertices efficiently. We find the shortest path tree by tracing the boundary and maintaining the funnel; the edges added to the funnel compose the tree. Our algorithm runs in the same time and uses arrays in place of finger search trees.

First we describe a data structure that supports five operations on ordered lists:

Length(l)   Return the number of items in the list.

Index(l, i)   Return the ith item in the list.

Add(f, l, x)   Add the item x to the f=front (or b=back) of the list l.

Split(f, l, i)   Return the sublist of l in f=front (or b=back) of and including item i and discard the other half of the list.

Undo()   Undo the most recent **Add**() or **Split**() operation.

We store the list in the entries of an array with indices from *first* through *last*. When we perform an **Add()** or **Split()**, we record the changed array entries or indices in a history stack so that the **Undo()** operation can return the array to the previous state. Pseudocode has been omitted from this abstract. If we begin with an empty list, denoted by indices $first = n$ and $last = n - 1$, and perform at most n **Add()** operations, then an array of size $2n$ is sufficient.

Using this data structure, we can trace the lift of the path α from p to q and maintain a list that represents the funnel defined by the shortest paths from p to the current BTM edge. Again, pseudocode is omitted.

We begin in the base triangle with a funnel list containing only the point p. (Whenever the lift of α is in the base triangle, the funnel is the point p.) When α crosses an edge out of the base triangle, we add the edge's endpoints to the funnel.

Suppose the path α crosses an edge $\overline{uv}$ into a new triangle $\triangle uvw$ of U_α. If α leaves through one of the other edges of $\triangle uvw$, then the current funnel list is split into the funnel defined by $\overline{uw}$ or by $\overline{vw}$ as illustrated in figure 2. We compute the index i where the list splits when the point w is added and store i with the triangle $\triangle uvw$. We use an increasing increment search from the front of the list to find the index i in $2 \log i$ probes: check the extension of the 1st, 2nd, 4th, 8th, etc, edge of the funnel until we pass the point w, then perform binary search to find the wedge containing w. By searching from the front and back simultaneously, we find the splitting index in $O(\log d)$ steps, where $d = \min\{i, \mathbf{Length}(l) - i\}$. Finger search trees were used in [17] to implement the simultaneous increasing-increment search, but arrays avoid the extra pointer complexity.

Consider the dual graph of U_α—the triangles of the universal cover that intersect the lift of α—as a tree rooted at and directed toward the base triangle. When the path α encounters a triangulation edge, α is heading either away from or toward the base triangle. If α is heading away, then we perform a **Split()** indicated by the index stored in the current triangle and **Add()** the new triangle vertex to obtain the next funnel. If α is heading toward the base, then we undo the last split/add pair. Except for finding the splitting index—which one does once for each triangle of the universal cover U_α—one does a constant amount of work when visiting a triangle. The recurrence relation analysis of [17] shows that the time to find the splitting indices is also linear in $\triangle_\alpha$. Thus, we have established the following theorem.

Theorem 3.2 *One can trace a path α through the universal cover of a BTM and maintain the funnel in $O(C_\alpha + \triangle_\alpha)$ time and space.*

If P is a triangulated simple polygon and α is the path from a vertex p around the boundary of P and back to p, then the algorithm computes the edges of the shortest paths from p to vertices of the polygon—that is, the shortest path tree of P.

3.3 Shortest closed curves

If we know a point on the Euclidean minimum closed curve homotopic to a given curve, then we can use the funnel algorithm of section 3.1 to compute it. In the full paper, we reduce the problem to computing the shortest closed curve in a *band*—a BTM whose dual is a cycle. Then we extend the funnel algorithm to compute this curve by walking around the band twice, if it is orientable, or four times, if it is non-orientable. The running time, once the space has been triangulated, is linear in the number of triangles explored. This general algorithm matches the time bounds of algorithms that compute special cases of the minimum closed curve problem: relative convex hulls [11, 32, 34] and minimum perimeter inpolygons [9].

Theorem 3.3 *In a BTM M with a closed curve α, we can compute a band whose shortest closed curve is the lift of the shortest closed curve homomorphic to α in M. Computation time and space is $O(\Delta_\alpha)$.*

Theorem 3.4 *Given an orientable or non-orientable band M composed of n triangles, one can compute the shortest closed curve around M in linear time.*

4 The link metric

In the link metric, the length of a path or closed curve is the number of its line segments. Section 4.1 gives an algorithm to compute a minimum link path of a homotopy class in time proportional to the number of triangles that it intersects and Section 4.2 discusses minimum link closed curves.

4.1 Computing the minimum link path

In this section we show how to compute the minimum link path, α_{min}, homotopic to a given path α in time proportional to $C_\alpha + \Delta_\alpha + \Delta_{\alpha_{min}}$, the complexity of the path α plus the number of triangles intersected by both paths. Our approach is inspired by Ghosh's [14] observations about the relationship between minimum Euclidean and link paths in simple polygons. We compute the Euclidean shortest path and then use a greedy approach to minimize the number of line segments. Our algorithm is output-sensitive and is simpler than that of Ghosh in the simple polygon case; it avoids his middle step of computing a visibility polygon.

First, some definitions: Since we have enough time to compute the Euclidean shortest path in the homotopy class of α, we may assume that α is the shortest path from p to q. As before, U is the universal covering space and U_α consists of the triangles of U that intersect α.

Traversing α from p to q, we can label each vertex as a *left* or *right* turn. We call an edge $\overline{tu}$ of α an *inflection edge* if the labels of t and u differ; edges incident to p and to q can also be called inflection edges. (Ghosh calls such edges *eaves*.) The extension of a line segment $\overline{tu}$ in U, denoted $ext(\overline{tu})$, is the line segment, ray, or infinite line formed by extending $\overline{tu}$ until it hits boundary points of U. In a simple polygon, Ghosh observed that there is always a minimum link path including one line segment from the extension of each inflection edge. This is also true in a simply connected BTM:

Lemma 4.1 *If $\overline{tu}$ is an inflection edge of a Euclidean shortest path α, then a minimum link path homotopic to α can be assumed to use a subsegment of $ext(\overline{tu})$.*

Thus we can assume that any inflection edges are included in the minimum link path. We have reduced our problem to one of finding the minimal link path from $\overline{uv}$, a segment extending one inflection edge, to $\overline{u'v'}$, a segment extending another, where the shortest path from u to u' is concave.

If the extension segments $\overline{uv}$ and $\overline{u'v'}$ intersect in U_α, then no additional segments are needed. Otherwise, consider the Euclidean shortest path γ from v to v' in U_α; the path from u to u' and γ form what has been called the *hourglass* of $\overline{uv}$ and $\overline{u'v'}$. The path γ helps find a segment of the minimum link path.

Lemma 4.2 *The minimum link path joining $\overline{uv}$ and $\overline{u'v'}$ either has one segment or can be chosen to include an inflection edge of γ, the shortest path from v to v'.*

Proof: If γ is concave, then the concave chains can be separated by a line; one segment can join $\overline{uv}$ to $\overline{u'v'}$.

Otherwise, γ has an inflection edge. Let $\overline{bc}$ be the inflection edge closest to v as shown in figure 3. (We consider v to be labeled opposite u so that b may be v.) Because the paths from u to c and v to c are both concave, the extension of $\overline{bc}$ intersects $\overline{uv}$ at some point a. Let $\overline{cd}$ be the extension of $\overline{bc}$ through c in U_α. Any path from $\overline{uv}$ to $\overline{u'v'}$ must intersect both $\overline{bc}$ and $\overline{cd}$, so shortcutting the path with a subsegment of $\overline{ad}$ does not increase the number of its line segments. ∎

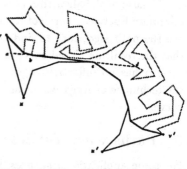

Figure 3: The shortest path γ has an inflection edge $\overline{bc}$

Finally, we discover the inflection edge $\overline{bc}$, if it exists, in time proportional to the number of triangles that $\overline{ac}$ intersects by the procedure outlined in the rest of this section.

Notice that $\overline{ad}$ is tangent to the concave chain. We find $\overline{ad}$ by moving the point a up the edge $\overline{uv}$ and maintaining the point c tangent to the chain. We stop the motion when one of three cases occurs:

1. The tangent $\overline{ac}$ becomes the extension of $\overline{u'v'}$: no extra segments are needed.

2. The moving point a reaches v: the segment $\overline{vc}$ is the inflection edge.

3. The tangent $\overline{ac}$ encounters a point b between a and c: the segment $\overline{bc}$ is the inflection edge.

The third case is the most difficult to detect; we use the following technical lemma:

Lemma 4.3 *The point b first encountered by the sweeping tangent is the endpoint of a triangulation edge that crosses the segment $\overline{ua}$ or the chain from u to c.*

Lemma 4.3 implies that we need look only at the convex hull of the endpoints of triangulation edges that we encounter during the sweep. These endpoints appear on the hull above $\overline{ac}$ in the same order as their edges appear along $\overline{ac}$. Points are added only at the ends of the segment $\overline{ac}$, so we can maintain the convex hull by a Graham scan [16] in a deque. Furthermore, the slope of $\overline{ac}$ changes monotonically, so we can also maintain b_0, the point of the hull having a tangent with this slope. When $\overline{ac}$ hits b_0, then $\overline{b_0 c}$ is an inflection edge that lemma 4.2 says can be used in a minimum link path.

Theorem 4.1 *A minimum link path α', homotopic to α, can be computed in space and time proportional to $O(C_\alpha + \Delta_\alpha + \Delta_{\alpha'})$.*

In a simply connected BTM, a minimum link path can cross any triangulation edge at most three times: any path that crosses a triangulation edge e four or more times can be shortcut by a portion of e, decreasing its length without changing its homotopy class since all paths with the same starting and ending point have the same homotopy class. Thus, the total time to compute minimum link paths in simple polygons is linear. Among many obstacles, a minimum link k-gon can intersect $\Theta(kn)$ triangulation edges.

4.2 Minimum link closed curves

As in section 3.3, if we know a vertex or edge of the minimum link closed curve, we can use the path algorithm to compute it. When the minimum link curve is convex, however, it seems difficult to find such a vertex or edge.

Because of the algorithm of section 3.3, we can assume that our closed curve α is the minimum Euclidean curve of its homotopy class. If α has an inflection edge, then we can use the path algorithm to find the paths between inflection edges. Lemma 4.1 implies that the resulting closed curve is a minimum link curve.

If α has no inflection edges, then the minimum link curve is convex. One can use the technique of Aggarwal et al. [1] as extended by Ghosh [14] to find the minimum link curve.

5 Paths with restricted orientations

For some applications, such as VLSI, the paths are restricted to c fixed directions or orientations; we call such paths *c-oriented*. Rectilinear paths with the four orientations of north, south, east and west are the most common. We show that the universal cover is also a good tool for finding minimal c-oriented paths of a given homotopy class. Due to space constraints, however, we have omitted many details from this abstract.

First, in section 5.1, we define convex polygonal distance functions appropriate to a given set of orientations. Then we show in section 5.2 that the length, under a convex distance function, of the path computed in section 3.1 equals the length of the shortest c-oriented path. Section 5.3 shows how to modify the minimum link algorithm of section 4.1 to compute minimum link c-oriented paths. Finally, section 5.4 shows that for paths restricted to three directions and for rectilinear paths, each homotopy class has a shortest path that is also a minimum link path. Mark de Berg [10] has independently noted this fact for rectilinear paths in simple polygons.

In all of the following sections, when we wish to construct paths restricted to c orientations explicitly, then we also restrict the boundary of the obstacles to the same set of c orientations. With such a restriction, there is always a path with at most $O(n)$ segments that follows obstacle boundaries. Without such a restriction, one can construct examples where any c-oriented path joining a given pair of points has infinitely many line segments.

5.1 Metrics versus distance functions

When paths are restricted, the link metric remains the number of line segment of a path. The Euclidean metric, however, should be replaced by a distance function that gives the length of the shortest restricted path between two points. One example is the Manhattan or L_1 metric for rectilinear paths, in which the length of a vector v is the sum of the lengths of the projections of v on the horizontal and vertical axes.

More generally, we can use Minkowski's convex distance functions [5]. Let A be a convex set whose interior contains the origin. The length of a vector v with respect to A is the amount that A must be scaled to include v; that is, $\|v\|_A = \liminf\{\lambda \geq 0 : v \in \lambda A\}$. The distance from point r to s is $\|s - r\|_A$. We do not call this a metric because it need not be symmetric: $\|v\|_A$ may not equal $\| - v\|_A$. It does, however, satisfy the triangle inequality [5]: if $u + v = w$ then $\|u\|_A + \|v\|_A \geq \|w\|_A$.

The points of the boundary of A are precisely the unit vectors of the distance function $\| \cdot \|_A$ [5]. Choosing A to be the unit circle gives the Euclidean metric; choosing A to be the diamond defined by the four unit vectors in the axial directions gives the L_1 metric. For a c-oriented path, a path restricted to follow the orientations of c unit length *basis vectors*, we choose A to be the convex hull of the basis vectors. Since the convex hull is a c-gon that contains the origin in its interior, any vector v can be expressed as a linear combination of basis vectors with positive coefficients. The next lemma proves that the length $\|v\|_A$ is the minimum sum of coefficients over all positive linear combinations of basis vectors that equal v and that this length is realized by using the one or two basis vectors adjacent to v in circular order.

Lemma 5.1 *Let U be a circularly-ordered set of basis vectors defining a convex distance function. Let v be a vector in the cone defined by adjacent basis vectors u and u'. If $v = au + bu'$, then $\|v\|_U = a + b$.*

5.2 Shortest paths under a convex distance function

For restricted paths, as for minimum link paths in section 4.1, we first compute the Euclidean shortest path α from p to q and use it as the representative of the homotopy class. This takes $O(C_\alpha + \Delta_\alpha)$ time.

The proofs leading to theorem 3.1 use only the triangle inequality to show that the path computed in section 3.1 is minimum under the Euclidean metric. But this implies

Theorem 5.1 *The Euclidean shortest path computed in section 3.1 is a shortest path under any convex distance function.*

To compute a c-oriented path of minimum length, it is enough to cut the Euclidean shortest path at all points with tangent vectors that are among the c basis vectors and compute c-oriented paths for the resulting pieces:

Lemma 5.2 *Let t be a point of the Euclidean shortest path α having a basis vector v as a tangent vector. Any minimum length path under the convex distance function goes through t.*

5.3 Minimum link c-oriented paths

We use a greedy method quite similar to that of section 4.1 to construct c-oriented paths. The time required will be proportional to the number of triangles explored, which, in this case, may be more than the number of triangles intersected by the path. However, the worst-case bounds are similar to those of section 4.1: If c is a constant, $O(nk)$ time is sufficient to construct a k link path of a given homotopy class and $O(n + k)$ time is sufficient in a simple polygon. If c is not a constant, but the directions are initially sorted, the time bounds increase to $O(nk \log c)$ and $O(n \log c + k)$.

5.4 Simultaneous minimization of length and links

We have seen how to compute the length of the shortest c-oriented path under a convex distance function and how to compute minimum link c-oriented paths. In this section we remark that a simplified version of the minimum link path algorithm can compute the path with fewest links of all shortest c-oriented paths. For rectilinear paths and paths restricted to three directions, we prove that this path is also a minimum link path—that length and links are minimized simultaneously. We omit the details here.

6   Conclusions

We have shown that the universal covering space of a triangulated region gives a useful framework for optimizing paths in the region under the Euclidean and link metrics. We have given simple direct algorithms for Euclidean shortest path trees and minimum link paths that use arrays in place of finger search trees. We have also given the first algorithms for computing minimum length and link c-oriented paths.

Acknowledgements

We would like to thank Joseph Mitchell and Leonidas Guibas for discussions on these and related problems. Both authors thank Digital Equipment Corporation and ALCOM for supporting this research.

References

[1] A. Aggarwal, H. Booth, J. O'Rourke, S. Suri, and C. K. Yap. Finding minimal convex nested polygons. *Information and Computation*, 83(1):98–110, Oct. 1989.

[2] M. A. Armstrong. *Basic Topology*. McGraw-Hill, London, 1979.

[3] B. Baumgart. A polyhedral representation for computer vision. In *Proceedings of the AFIPS National Computer Conference*, pages 589–596, 1975.

[4] B. Bhattacharya and G. T. Toussaint. A linear algorithm for determining translation separability of two simple polygons. Technical Report SOCS-86.1, School of Computer Science, McGill University, Montreal, 1986.

[5] J. W. S. Cassels. *An Introduction to the Geometry of Numbers*. Springer-Verlag, Berlin, 1959.

[6] B. Chazelle. A theorem on polygon cutting with applications. In *Proceedings of the 23rd IEEE Symposium on Foundations of Computer Science*, pages 339–349, 1982.

[7] B. Chazelle. Triangulating a simple polygon in linear time. In *Proceedings of the 31st IEEE Symposium on Foundations of Computer Science*, pages 220–230, 1990.

[8] K. L. Clarkson, S. Kapoor, and P. M. Vaidya. Rectilinear shortest paths through polygonal obstacles in $O(n \log^2 n)$ time. In *Proceedings of the Third Annual ACM Symposium on Computational Geometry*, pages 251–257, 1987.

[9] J. Czyzowicz, P. Egyed, H. Everett, D. Rappaport, T. Shermer, D. Souvaine, G. Toussaint, and J. Urrutia. The aquarium keeper's problem. In *Proceedings of the Second Annual ACM-SIAM Symposium on Discrete Algorithms*, pages 459–464, Jan. 1991.

[10] M. de Berg. On rectilinear link distance. Technical Report RUU–CS–89-13, Vakgroep Informatica, Rijksuniversiteit Utrecht, May 1989.

[11] M. de Berg. Translating polygons with applications to hidden surface removal. In *SWAT 90: Second Scandinavian Workshop on Algorithm Theory*, number 447 in Lecture Notes in Computer Science, pages 60–70. Springer-Verlag, 1990.

[12] P. J. de Rezende, D. T. Lee, and Y. F. Wu. Rectilinear shortest paths with rectangular barriers. *Discrete and Computational Geometry*, 4:41–53, 1989.

[13] S. Gao, M. Jerrum, M. Kaufmann, K. Mehlhorn, W. Rülling, and C. Storb. On continuous homotopic one layer routing. In *Proceedings of the Third Annual ACM Symposium on Computational Geometry*, pages 392–402, 1987.

[14] S. K. Ghosh. Computing the visibility polygon from a convex set and related problems. *Journal of Algorithms*, 12:75–95, 1991.

[15] S. K. Ghosh and D. M. Mount. An output sensitive algorithm for computing visibility graphs. In *Proceedings of the 28th IEEE Symposium on Foundations of Computer Science*, pages 11–19, 1987.

[16] R. Graham. An efficient algorithm for determining the convex hull of a finite pla-

nar set. *Information Processing Letters*, 1:132–133, 1972.

[17] L. Guibas, J. Hershberger, D. Leven, M. Sharir, and R. Tarjan. Linear time algorithms for visibility and shortest path problems inside triangulated simple polygons. *Algorithmica*, 2:209–233, 1987.

[18] L. Guibas and J. Stolfi. Primitives for the manipulation of general subdivisions and the computation of Voronoi diagrams. *ACM Transactions on Graphics*, 4(2):74–123, 1985.

[19] L. J. Guibas and J. Hershberger. Optimal shortest path queries in a simple polygon. *Journal of Computer and System Sciences*, 39(2):126–152, Oct. 1989.

[20] R. H. Güting. *Conquering Contours: Efficient Algorithms for Computational Geometry*. PhD thesis, Dortmund, 1983.

[21] R. H. Gütting and T. Ottmann. New algorithms for special cases of hidden line elimination problem. *Computer Vision, Graphics, and Image Processing*, 40:188–204, 1987.

[22] S. Kapoor and S. N. Maheshwari. Efficient algorithms for Euclidean shortest path and visibility problems with polygonal obstacles. In *Proceedings of the Fourth Annual ACM Symposium on Computational Geometry*, pages 172–182, 1988.

[23] R. C. Larson and V. O. Li. Finding minimum rectilinear paths in the presence of barriers. *Networks*, 11:285–304, 1981.

[24] D. T. Lee and F. P. Preparata. Euclidean shortest paths in the presence of rectilinear barriers. *Networks*, 14(3):393–410, 1984.

[25] C. E. Leiserson and F. M. Maley. Algorithms for routing and testing routability of planar VLSI layouts. In *Proceedings of the 17th Annual ACM Symposium on Theory of Computing*, pages 69–78, 1985.

[26] J. R. Munkres. *Topology: A First Course*. Prentice-Hall, Englewood Cliffs, N.J., 1975.

[27] F. P. Preparata and M. I. Shamos. *Computational Geometry—An Introduction*. Springer-Verlag, New York, 1985.

[28] G. J. E. Rawlins and D. Wood. Optimal computation of finitely oriented convex hulls. *Information and Computation*, 72:150–166, 1987.

[29] J.-R. Sack. *Rectilinear Computational Geometry*. PhD thesis, Carleton University, 1984.

[30] S. Suri. A linear time algorithm for minimum link paths inside a simple polygon. *Computer Vision, Graphics, and Image Processing*, 35:99–110, 1986.

[31] X.-H. Tan, T. Hirata, and Y. Inagaki. The intersection searching problem for c-oriented polygons. *Information Processing Letters*, 37:201–204, 1991.

[32] G. Toussaint. On separating two simple polygons by a single translation. *Discrete and Computational Geometry*, 4:265–278, 1989.

[33] G. T. Toussaint. Movable separability of sets. In G. T. Toussaint, editor, *Computational Geometry*, volume 2 of *Machine Intelligence and Pattern Recognition*, pages 335–376. North Holland, Amsterdam, 1985.

[34] G. T. Toussaint. Computing geodesic properties inside a simple polygon. *Revue D'Intelligence Artificielle*, 3(2):9–42, 1989. Also available as technical report SOCS 88.20, School of Computer Science, McGill University.

Approximation Algorithms for Selecting Network Centers

(Preliminary Version)

Judit Bar-Ilan *   David Peleg [†]

Abstract

This abstract concerns the issue of allocating and utilizing centers in a distributed network, in its various forms. The abstract discusses the significant parameters of center allocation, defines the resulting optimization problems, and proposes several approximation algorithms for selecting centers and for distributing the users among them. We concentrate mainly on *balanced* versions of the problem, i.e., in which it is required that the assignment of clients to centers be as balanced as possible. The main results are constant ratio approximation algorithms for the balanced κ-centers and balanced κ-weighted centers problems, and logarithmic ratio approximation algorithms for the ρ-dominating set and the k-tolerant set problems.

1 Introduction

The problem of allocating and utilizing centers in a communication network is a major issue in distributed network design. Among the various applications requiring the use of centers are distributed databases, routing, distributed data structures, etc. ([HR88,ML77,MK83,BG87,Pel90]). Using a collection of centers offers a convenient intermediate approach between the fully centralized and the fully distributed solutions, and provides a reasonable balance between the need for fault-tolerance and economical considerations. Unfortunately, all but the simplest center allocation problems are NP-hard, and therefore are considered unlikely to be tractable. In this paper, we discuss the significant parameters of center allocation, define the resulting optimization problems, and propose several approximation algorithms for selecting centers and for distributing the users among them.

In all variations of the center allocation problem considered in this abstract, our goal is composed of two parts. The first is to select a collection of centers $C = \{\theta_1, \ldots, \theta_\kappa\}$, and the second is to assign each of the remaining nodes in the network to one of the centers. We denote by $\varphi(v)$, the center that serves the node v. The pair (C, φ) determining the center assignment is referred to as the assignment pair.

---

*Department of Mathematics and Computer Science, University of Haifa, Haifa 31999, Israel.

[†]Department of Applied Mathematics and Computer Science, The Weizmann Institute, Rehovot 76100, Israel. Supported in part by an Allon Fellowship, by a Walter and Elise Haas Career Development Award and by a Bantrell Fellowship.

Let S_i denote the set of vertices assigned to the center θ_i, i.e., $S_i = \{v \mid \theta_i \in \varphi(v)\}$. A center assignment can thus be characterized alternatively by the collection of pairs

$$S = \{(\theta_1, S_1), \ldots, (\theta_\kappa, S_\kappa)\}$$

such that $\theta_i \in C$, $S_i \subseteq V$ for every $1 \leq i \leq \kappa$ and $\bigcup_i S_i = V$. We refer to the sets S_i as clusters, and to the collection S as an *assignment tuple*. We sometimes refer to S as a *partition*, when we wish to ignore the centers and consider only the collection of subsets S_i. We shall use the two representations, (C, φ) and S, interchangeably.

As accessibility is a major concern in distributed systems, the distance between clients and their respective centers is an important design parameter in center allocation problems. Consequently, given a collection of centers, one obvious assignment choice would be to select, for every client v, the center nearest to it. A potential problem with this choice is that it may conflict with another significant concern, namely, workload balancing. The *nearest center* assignment might result in all (or most) sites using a single center (or a small subset of centers). This is undesirable, since it overloads the chosen centers and may create bottleneck problems. A possible way to overcome this situation is to bound the maximum number of sites assigned to any particular center θ by $|S_i| \leq L$, for some bound L. (Naturally L has to be large enough, i.e., $L \geq n/\kappa$). This paper concentrates on center problems that insist on this maximal load requirement, referred to as *balanced* center problems.

It turns out that the two parts of the center allocation problems are not of equal difficulty. Once the centers are selected, it is possible to assign the clients optimally using standard flow techniques, as indicated in Section 2. We should therefore focus mainly on the harder part of the problem, involving the selection of the centers. Our aim is to select the "best" set of centers and the best assignment of nodes to the centers in a given network. There are two natural ways to approach this problem:

1. Fixing the number of centers, κ, and trying to minimize $\max\{(dist(v, \varphi(v))\}$.

2. Fixing a bound, ρ, on the maximal distance between a node and its center, and minimizing the number of centers.

These problems are sometimes referred to as the κ-*centers* and the ρ-*dominating set* problems. In both cases we consider the balanced versions of these problems, i.e., solutions in which no center is assigned more than L clients. Both these problems are NP-hard (for unbounded κ), even without the load constraint. Therefore, we direct our efforts toward attempting to approximate the optimum solution.

In the unbalanced case, i.e., with no constraint on the load, the two dual forms of the problem were given approximation algorithms before. The κ-centers problem was treated in [HS84], and given a polynomial time algorithm with approximation ratio 2. The ρ-dominating set problem can be formulated as a special case of the cover problem of [Lov75], for which the greedy algorithm described therein provides an approximation ratio of $\log n + 1$.

It turns out that the introduction of the balancing constraint affects the ρ-dominating set problem only little, and in the "positive" direction. In the full paper (see also

[BP90]), we extend the approximation technique of [Lov75] in a natural way to the balanced version of the ρ-dominating set problem, and get an approximation algorithm with ratio $\log L + 1$ (i.e., the ratio improves as L becomes smaller).

In contrast the κ-centers problem becomes harder with the introduction of the balancing requirement, and it is necessary to develop stronger techniques in order to overcome the difficulties resulting from the need to take the load constraint into account. In Section 3, we present our first main result, which is an approximation algorithm $Approx1(G, L, \kappa)$ for the balanced κ-center problem that achieves a constant approximation ratio. Let us now outline the strategy on which our solution is based. Our starting point is the elegant approximation technique of Hochbaum and Shmoys [HS84] for the unbalanced κ center problem. We start by choosing an initial set of centers using the algorithm of [HS84]. After the initial centers are chosen, we assign the clients to these centers in two phases using flow techniques. This initial assignment does not necessarily obey the load constraints. Now the centers are partitioned into two sets, namely, the "light" centers (those that have fewer than L clients), and the "heavy" ones (those that have more). For the light centers, this assignment is final, and we prove that the specific choice of the initial assignment guarantees that this does not harm the solution. For the heavy centers, however, some rebalancing is necessary. Each connected component of the heavy centers with their clients is treated separately. In each component, we construct a spanning tree, and apply a "tree-contraction" algorithm whose task is to balance the client loads on the centers. This algorithm works on the spanning tree from the leaves up, and moves clients along the edges of the tree, spreading them as needed among the selected centers and other nodes in their neighborhood, as necessary.

In Section 4 we consider the weighted variant of the problem, in which each node v has a nonnegative weight $\omega(v)$, and the feasibility constraint is that the selected collection of centers satisfies $\sum_{v \in C} \omega(v) \leq \kappa$. The algorithm proposed for this problem is based on a technique for converting solutions with approximation ratio α for the unweighted problem into solutions with approximation ratio $2\alpha + 1$ for the weighted problem. (The technique applies only for a specific type of solutions for the unweighted problem, referred to as *minimum cardinality* solutions, but fortunately the solutions generated by our unweighted algorithm $Approx1(G, L, \kappa)$ fall under this category.)

Finally, we also consider the issue of fault tolerance. Assigning multiple centers handles the problem of center crashes, but does not attempt to handle the problem of communication faults. We would like to exploit redundancy in order to enhance data *availability* in the face of communication failures, including possible network partitions. Towards this goal, we propose the concept of *k-tolerant* sets: Let $A \subseteq V$ be the set of potential servers, and $B \subseteq V$ the set of potential customers. A *k-tolerant A-set for B* (or simply a *k-tolerant set*) is a subset $C \subseteq A$, such that for every $v \in B$, either $v \in C$ (this is possible only if $v \in A \cap B$) or there are k vertex-disjoint paths from v to (k distinct vertices in) C. A solution to the *k-tolerant set* problem is such a set $C \subseteq A$ of minimal size. Note that when $A = V$, such a set exists in every graph, regardless of its connectivity. For example, Figure 1 depicts a 3-tolerant center set in a 1-connected graph, where $A = B = V$. An approximation algorithm for the k-tolerant set problem is described in the full paper (see also [BP90]), yielding an approximation

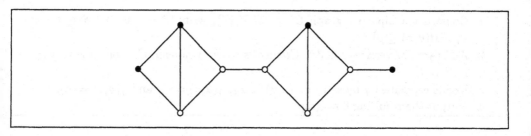

Figure 1: A 1-connected graph and a 3-tolerant center set for it (darkened vertices denote centers).

ratio of $k(\log n + 1)$.

2 Preliminaries

The network is described by a connected undirected graph $G = (V, E)$, $|V| = n$, with a weight $c_{(u,v)}$ for every edge $(u, v) \in E$, representing the length of the edge. The vertices represent the sites of the network (or the processors located at these sites) and the edges represent bidirectional communication channels between these sites. All the processors have distinct identities.

Let us define some concepts concerning graphs. For two vertices u, w in G, let $dist_G(u, w)$ denote the length of a shortest path in G between those vertices, where the length of a path is the sum of its edge weights. Let $D(G)$ denote the *diameter* of the network, i.e., $\max_{u,v \in V}(dist(u, v))$. (In the above notations, we sometimes omit the reference to G where no confusion arises.) Let (C, φ) be a given center assignment, and let $S = \{(S_1, \theta_1), ..., (S_\kappa, \theta_\kappa)\}$ be the induced assignment tuple. The following definitions formalize our measures for the quality of this assignment. We first define the appropriate *radius* measure. Let $\mathcal{R}(S_i)$ denote $\max_{v \in S_i}\{dist(v, \theta_i)\}$, and let $\mathcal{R}(S) = \max_i \mathcal{R}(S_i)$. Next, the *load* of a center θ_i is denoted $\mathcal{L}(\theta_i) = |S_i|$, and the maximal load of the assignment tuple S is $\mathcal{L}(S) = \max_i\{\mathcal{L}(\theta_i)\}$. We say that a pair (C, φ) is ρ-*dominating* if the induced assignment tuple S satisfies $\mathcal{R}(S) \le \rho$, and L-*balanced* if S satisfies $\mathcal{L}(S) \le L$.

Let us informally define the basic notion of an approximation algorithm. This is a polynomial time algorithm for an optimization problem with some performance guarantee on the quality of the produced solutions. The *approximation ratio* of an approximation algorithm for a minimization problem is the maximal ratio between the solution obtained by the algorithm and the optimal solution, where the maximum is taken over all input instances to the problem. A similar definition applies to maximization problems.

Next, we describe how to handle the situation in which the centers are already fixed, but we are given control over the assignment, φ, of servers to sites. Producing a balanced

---

1. Construct a bipartite graph $G' = (C, V, E')$, with $E' = \{(\theta, v) \mid \theta \in C,\ v \in V,\ dist(\theta, v) \leq \rho\}$.
2. Add two new vertices s and t. Connect s to every node in C. Connect every node in V to t.
3. Define capacities γ by setting $\gamma(s, \theta) = m_1$, $\gamma(\theta, v) = m_2$ and $\gamma(v, t) = m_3$.
4. Output the resulting flow-graph.

Figure 2: Procedure $Flow\_graph(C, V, \rho, m_1, m_2, m_3)$

assignment is a variant of a partitioning problem described in [BG87], Sect. 5.4.3, as the *concentrator location* problem, and solved via linear programming or the flow methods of [Ber85,BG89]. In that problem the minimized function is $\sum_{v \in V} dist(v, \varphi(v))$. Our solution for assigning centers to clients a slightly more involved variation of that solution.

The procedure for for finding a minimal radius, balanced assignment is as follows: the distances between the set of centers, C, and the set of clients, $V \setminus C$ are sorted. The procedure runs a binary search on the distances and checks whether a feasible solution exists with this distance as the radius. For each radius it builds an appropriate flow graph using procedure $Flow\_graph$. In this graph, each center has capacity L and a node is connected to a center iff its distance from that center is not greater than the current radius. If the maximum flow is $|V \setminus C|$, then each client got assigned to a center, and no center serves more than L clients, thus a feasible solution exists for this radius. The binary search on the distances finds the minimal feasible radius. For this radius, an assignment is constructed using the integer valued maximum flow in the corresponding flow-graph, where each node gets assigned to the center from which it gets the flow.

Procedure $Flow\_graph(C, V, \rho, m_1, m_2, m_3)$ is given in Figure 2, since it is to be used as a component in our later algorithms. From the above description we get:

Proposition 2.1 Given a graph G, a collection of centers $C \subseteq V$ and a bound L, such that $L \geq \frac{n}{\kappa}$, there is a polynomial time algorithm for computing an assignment $\varphi : V \mapsto C$ with an induced assignment tuple S satisfying $\mathcal{L}(S) \leq L$ and minimal radius $\mathcal{R}(S)$. ∎

3 The Balanced κ-center Selection Problem

In this section we turn to the selection problem, and present an approximate solution to the problem in which the number of centers, κ, and a load constraint L, are given, and we want to optimize the radius.

Definition 3.1 The *L-balanced κ-centers* problem is defined as follows:
Input: Graph $G = (V, E)$, integers L, $\kappa \geq 1$.
Goal: Select an L-balanced center assignment (C, φ), such that $|C| = \kappa$, minimizing the radius $\mathcal{R}(S)$ of the induced assignment tuple S.
Let us denote the optimal radius by $\mathcal{R}_{opt}(G, L, \kappa)$.

For $\iota = 1$ **to** $|E|$ **do**

   1. Choose a maximal independent set in G_ι^2. Let this set be $C = \{\theta_1, \theta_2, \dots, \theta_\ell\}$.

   2. If $\ell > \kappa$ or $|V| > L \cdot \kappa$ then return "failure".

   3. Let $\bar{G}_1, \dots, \bar{G}_m$ be the connected components of G_ι^2, let $\kappa_i = \lceil \frac{|\bar{V}_i|}{L} \rceil$, for $1 \le i \le m$, and let $C_i \subseteq C$ be the centers in connected component $\bar{G}_i$. If $\sum_i \kappa_i > \kappa$ then return "failure".

   4. For $i = 1, \dots, m$, call $\varphi \leftarrow Allocate(\bar{G}_i, L, C_i)$.

Figure 3: Algorithm $Approx1(G, L, \kappa)$

We assume that the input graph is a complete weighted graph $G = (V, E)$, with weights $c_{(u,v)}$ on the edges, where the edge weights satisfy the triangle inequality. If the graph does not obey these assumptions, we can modify it into a proper graph by setting $c_{(u,v)} = dist_G(u, v)$ for each edge (u, v). Without loss of generality, we label the edges so that $c_{e_1} \le c_{e_2} \le \dots \le c_{e_{\frac{n(n-1)}{2}}}$, and denote $c_j = c_{e_j}$ for all $j \ge 1$.

Following [HS84], for an integer ι, we define the bottleneck graph, $G_\iota = (V, E_\iota)$ to be an edge subgraph of G, where $E_\iota = \{e_j \in E \mid j \le \iota\}$. We also define the t-*closure* graph $(G')^t$, for any edge subgraph G' of G, as the unweighted graph in which two nodes are connected iff there is a path of at most t edges between them in the original graph G'.

We give an approximation algorithm to the problem, achieving a constant approximation ratio. Let us start with an overview of the algorithm. The algorithm considers the bottleneck graphs, G_ι, in increasing order of ι. For each such graph, it chooses a maximal independent set, C, in G_ι^2. As explained in [HS84] (see Claim 3.2 below), this set C indicates whether there exists a feasible solution in G_ι. If there is no feasible solution in G_ι, it considers the next bottleneck graph. Otherwise, it attempts to assign up to $L - 1$ clients from $V \setminus C$ to each center, using flow techniques. This process may leave some nodes of $V \setminus C$ unassigned. Next, each of these remaining nodes gets assigned, again using flow techniques, but this time ignoring the balance constraints. The goal of the first flow phase is to ensure that the assignment obtained by the two flow phases is "as balanced as possible", in the sense that if a center $\theta \in C$ has fewer than $L - 1$ clients, while some nodes are left unassigned, there is no way to improve the situation while still maintaining the balance constraint. We refer to the assignment obtained after both flow phases as the *initial* assignment, denoted φ_I.

Next, the set of centers, C, is partitioned into two sets: the set $\mathcal{E}$ consisting essentially of the light (or *empty*) centers, i.e., those having fewer than $L - 1$ clients (apart of themselves), and the set $\mathcal{F}$ of heavy (or *full*) centers, i.e., those having more than $L - 1$ clients. (A more precise definition of $\mathcal{E}$ and $\mathcal{F}$, detailing the classification of those centers with precisely $L - 1$ clients, appears in the description of Procedure *Partition* below.) For the nodes being assigned to centers in $\mathcal{E}$, this assignment is final. For the remaining nodes the algorithm appoints a minimal number of additional centers (if $|C| < \kappa$, and the load constraint is not satisfied). This is done by creating an auxiliary

1. Let $\theta_1, \ldots, \theta_m$ be the nodes of $\tilde{G}$.
2. For every j, $1 \leq j \leq m$, let $Carry\_in(\theta_j) \leftarrow \emptyset$
3. Construct a spanning tree in $\tilde{G}$.
4. **Repeat**
    (a) Pick an arbitrary leaf θ of the tree. Let $|Bin(\theta)| = iL + \epsilon$, where $0 \leq \epsilon < L$.
    (b) Pick $(i - 1)$ additional centers from the nodes in $Bin(\theta)$.
    (c) Distribute $i(L - 1)$ nodes among the i centers, assigning L nodes to each center (including itself), and assigning first the nodes in $Carry\_in(\theta)$.
    (d) Let $Carry\_out(\theta)$ be the set of unassigned nodes of $Bin(\theta)$, $|Carry\_out(\theta)| = \epsilon$.
    (e) Let θ' be the parent of θ in the tree.
    (f) Set $Carry\_in(\theta') \leftarrow Carry\_in(\theta') \cup Carry\_out(\theta)$,
    and set $Bin(\theta') \leftarrow Bin(\theta') \cup Carry\_out(\theta)$.
    (g) Remove θ from the tree.
5. **until** the tree consists of a single node.
6. For the last remaining node θ, execute steps (a) to (d). If $|Carry\_out(\theta)| > 0$, pick an additional center from this set and assign all the other members of $Carry\_out(\theta)$ to it.
7. Denote the resulting assignment for the nodes of $Bin(\tilde{G})$ by φ_F^j.

Figure 4: Procedure $TC(\tilde{G}_j)$

"neighborhood graph" connecting the centers, constructing a spanning forest of the centers of $\mathcal{F}$ in this graph, and applying a tree-contraction algorithm, TC, on each tree. After the algorithm is applied to all graphs G_i, the solution with the minimal radius (among all the solutions produced by the algorithm) is taken as our final solution for the balanced κ-center problem.

The main algorithm is presented in Figure 3. The algorithm $Approx1(G, L, \kappa)$ calls three other procedures: $Flow\_graph$, $Partition$ and $Allocate$. Procedure $Allocate(G, L, C$ is given in Figure 5, which in turn calls Procedure $TC(\tilde{G}_j)$ described in Figure 4. Procedure $Flow\_graph$ was presented in Figure 2. We now give a description of procedure $Partition$.

Procedure $Partition(C)$ partitions the centers into two sets $\mathcal{E}$ and $\mathcal{F}$, according to the initial assignment φ_I. We define $\mathcal{E}$ as follows: let $\mathcal{E}_0$ be the set of centers having at most $L - 2$ nodes currently assigned to them. Now set

$$\mathcal{E}_{j+1} = \mathcal{E}_j \cup \{\theta \in C \mid \exists v \in V \backslash C, \exists \theta' \in \mathcal{E}_j, \; dist(\theta, v) \leq 2 \cdot c_i, \; dist(\theta', v) \leq 2 \cdot c_i, \; \varphi_I(v) = \theta\}.$$

Let $\mathcal{E}$ be the largest set $\mathcal{E}_j$ obtained in this process, and let $\mathcal{F} = C \backslash \mathcal{E}$. Notice, that $\mathcal{E}$ consists of the set $\mathcal{E}_0$ of centers having at most $L - 2$ clients, plus all the centers that

1. Call $G_{F_1} \leftarrow Flow\_graph(C, V \setminus C, 2c_\iota, L-1, 1, 1)$. Compute the maximum flow in the graph. Define the partial assignment, φ_1 as follows: θ_j serves itself and those nodes in $V \setminus C$ that get flow from θ_j in the maximum flow. Let V' be the set of nodes assigned to centers.

2. Call $G_{F_2} \leftarrow Flow\_graph(C, V \setminus (V' \cup C), 2c_\iota, \infty, 1, 1)$. Compute the maximum flow in the graph. Assign each node a center according to the maximum flow. Denote this assignment by φ_2, and let $\varphi_I = \varphi_1 \cup \varphi_2$.

3. Call $Partition(C)$. Let $C = \mathcal{E} \cup \mathcal{F}$. For every center $\theta \in C$, let $Bin(\theta) = \{v \mid \varphi_I(v) = \theta\} \cup \{\theta\}$. Define $\tilde{G} = (\mathcal{F}, \tilde{E})$, where $\tilde{E} = \{(\theta, \theta') \mid \theta, \theta' \in \mathcal{F}, \exists u, v \in V$ s.t. $v \in Bin(\theta)$ and $u \in Bin(\theta')$ and there is an edge between u and v in $G_\iota^2\}$.

4. Let $\tilde{G}_1, \ldots \tilde{G}_l$ be the connected components of $\tilde{G}$. Let $Bin(\tilde{G}_j) = \bigcup_{\theta \in \tilde{G}_j} Bin(\theta)$. Call $\varphi_F^j \leftarrow TC(\tilde{G}_j)$ for every connected component.

5. Let $\varphi(v) = \varphi_I(v)$ if $\varphi_I(v) \in \mathcal{E}$, and let $\varphi(v) = \varphi_F^j(v)$ if $\varphi_I(v) \in \mathcal{F} \cap \tilde{G}_j$.

6. Output φ.

Figure 5: Algorithm $Allocate(G, L, C)$

can potentially transfer clients to the ones in $\mathcal{E}_0$. Also observe that the first flow phase guarantees that $\mathcal{E}$ contains no center with *more* than $L-1$ clients, since otherwise we could transfer clients and increase the flow in the flow graph $G_{F_1}(C)$ of that phase.

Let b be the radius of the optimal solution, let $X_{opt} = \{x_1, \ldots x_\kappa\}$ be the set of centers chosen in an optimal solution, and let φ_{opt} be the optimal assignment associated with these centers. Let $S_j^{opt} = \{v \mid \varphi_{opt}(v) = x_j\}$. It is easy to see that there exists an ι, such that $b = c_\iota$. Let us analyze the solution produced by our algorithm for this ι.

In step 2 of the algorithm $Approx1(G, L, \kappa)$, we return failure, when the size of the maximal independent set is greater than κ. This action is justified by the following claim.

Claim 3.2 [HS84] If the maximal independent set, C, selected in step 1 of algorithm $Approx1(G, L, \kappa)$ has size $|C| > \kappa$, then there is no solution to the κ-center problem (even without the balance constraint) with radius $\leq c_\iota$. ∎

Having an independent set of size κ is sufficient for producing a feasible solution in the κ-center problem with no balance constraint. In the balanced problem, some additional constraints must be satisfied, as can be seen from the next claim:

Claim 3.3 Let $\bar{G}_1, \ldots, \bar{G}_m$ be the connected components of G_ι^2, where $\bar{G}_j = (\bar{V}_j, \bar{E}_j)$. If $\sum_{j=1}^m \lceil \frac{|\bar{V}_j|}{L} \rceil > \kappa$, then there is no feasible solution for the problem with radius $\leq c_\iota$. ∎

Next we show that a feasible assignment exists, even if the assignment of the nodes assigned to centers in the set $\mathcal{E}$ by φ_I remains unchanged. Recall that each center in $\mathcal{E}$ has at most $L-1$ clients.

Claim 3.4 Let $\bar{V} = \{v \mid \varphi_I(v) = \theta_j \text{ where } \theta_j \in \mathcal{E}\}$. Then the assignment $\varphi_I \mid_{\bar{V}}$ can be extended to a total feasible assignment.

Outline of proof: Let $\theta \in \mathcal{E}$, and let $x \in X_{opt}$ be the center that serves θ in the optimal solution. If $\theta' \in \mathcal{E}$, $\theta \neq \theta'$, then x cannot serve θ' in the optimal solution, since the clients of x form a clique in G_i^2, and $\mathcal{E}$ is an independent set in G_i^2. Let $X_{\mathcal{E}} = \{x \in X_{opt} \mid \varphi_{opt}(\theta) = x, \theta \in \mathcal{E}\}$. We have shown that $|\mathcal{E}| = |X_{\mathcal{E}}|$. Let $U = \{v \mid \varphi_I(v) \in \mathcal{F}\}$, and let $W = \{v \mid \varphi_{opt}(v) \in X_{\mathcal{E}}\}$. Now we prove that $U \cap W = \emptyset$. Otherwise, let $v \in U \cap W$, and let $\varphi_{opt}(v) = x$. By the definition of $X_{\mathcal{E}}$, there exists a θ, such that $\varphi_{opt}(\theta) = x$, therefore $dist(v, \theta) \leq 2c_i$. But v is served by a center from $\mathcal{F}$, and could have been served by θ. Had it been served by θ, the maximum flow in the first flow graph could have been increased, and this is a contradiction.

Therefore, $W \subseteq V \setminus U = \bar{V}$, where $\bar{V}$ are the clients of centers in $\mathcal{E}$. Thus $\varphi_I \mid_{\bar{V}}$ can be extended to a total feasible assignment. ∎

We took care of the nodes assigned to centers in $\mathcal{E}$. The centers in $\mathcal{F}$ might be overloaded, so we have to appoint new centers. Next we show, that a feasible solution can be produced, even if we consider each connected component $\tilde{G}_j$ separately. That is, there exists a feasible solution where the nodes of $Bin(\tilde{G}_j)$ are served only by nodes from $Bin(\tilde{G}_j)$. Let l be the number of connected components of $\tilde{G}$:

Claim 3.5 Let $X_{\mathcal{E}} = \{x_1 \ldots, x_k\}$ be the set of optimal centers that are corresponding to the set of centers in $\mathcal{E}$. There exists a partition of $X_{opt} \setminus X_{\mathcal{E}}$ into l subsets $X^1, \ldots, X^l$, corresponding to $\tilde{G}_1, \ldots, \tilde{G}_l$ respectively, such that an $x \in X^j$ does not serve nodes in $Bin(\tilde{G}_i)$ for $i \neq j$.

Outline of proof: We show that for every $x \in X_{opt} \setminus X_{\mathcal{E}}$, one of the following holds:

1. $x \in \tilde{G}_m$, and $\{v \mid \varphi_{opt}(v) = x\} \subseteq Bin(\tilde{G}_m)$.

2. $\varphi_I(x) \in \mathcal{E}$, but still in the optimal solution, x serves nodes only from a *single* connected component $\tilde{G}_m$.

Therefore, each optimal center, that was not associated with a center in $\mathcal{E}$, is "dedicated" to one and only one connected component of $\tilde{G}$. ∎

During the tree contraction algorithm, new centers are chosen. Let $\mathcal{H}$ be the set of the new centers, and let $\mathcal{T} = \mathcal{F} \cup \mathcal{H}$. The set $\mathcal{T}$ can also be partitioned into l subsets $T^1, \ldots, T^l$, corresponding to $\tilde{G}_1, \ldots, \tilde{G}_l$ respectively, such that a $\theta \in T^j$ serves only nodes in $Bin(\tilde{G}_j)$, since every connected component was treated separately.

Lemma 3.6 For every $1 \leq j \leq l$, $|T^j| \leq |X^j|$.

Proof: The tree contraction algorithm, TC, assigns $\lceil \frac{|Bin(\tilde{G}_j)|}{L} \rceil$ centers to serve $Bin(\tilde{G}_j)$. The centers in X^j are dedicated to $Bin(\tilde{G}_j)$, but they might serve additional nodes, that in our solution are served by centers in $\mathcal{E}$. Therefore, the centers in X^j must serve at least $|Bin(\tilde{G}_j)|$ clients. These centers obey the balance constraint, thus $|X^j| \geq \lceil \frac{|Bin(\tilde{G}_j)|}{L} \rceil$. ∎

It remains to analyze the quality of the approximation. For the nodes in U', we assigned centers at distance $\leq 2b$. Now consider the nodes v, such that $\varphi_I(v) \in \mathcal{F}$. The final assignment is always determined at the leaves of the tree. Let $Bin_{init}(\theta)$ be the initial value of $Bin(\theta)$, and let $Bin_{final}(\theta)$ be the value of $Bin(\theta)$ when it becomes a leaf of the current tree. Let $v \in Bin_{final}(\theta)$. Recall that $Bin_{final}(\theta) = Bin_{init}(\theta) \cup Carry\_in(\theta)$. If $v \in Carry\_in(\theta)$, then v was moved to $Bin_{final}(\theta)$ from a descendant of θ in the original spanning tree. In the assignment process, we first assign centers to the nodes in the set $Carry\_in(\theta)$.

Lemma 3.7 All nodes in $Carry\_in(\theta)$ will be assigned centers from $Bin_{final}(\theta)$.

Proof: Let $|Carry\_in(\theta)| = kL + \epsilon$, then $|Bin_{final}(\theta)| \geq (k+1)L + \epsilon$, since $|Bin(\theta)| \geq L$, therefore we choose at least $(k+1)$ centers from the nodes of $Bin_{final}(\theta)$. ∎

From this it follows, that a node v is moved at most once. If $\varphi_I(v) = \theta$, then $\varphi_F(v)$ belongs either to $Bin_{final}(\theta)$ or to $Bin_{final}(\theta')$, where θ' is the parent of θ in the spanning tree. In $Bin_{final}(\theta')$ there are nodes from $Bin_{init}(\theta')$, or nodes v', such that $\varphi_I(v')$ is a child of θ'. Therefore, the worst case occurs, when v is being assigned to a center, c, such that $\varphi_I(v)$ is a sibling of $\varphi_I(c) = \theta''$. In this case:

$$dist(v,c) \leq dist(v,\theta) + dist(\theta,\theta') + dist(\theta',\theta'') + dist(\theta'',c)$$

Therefore

$$dist(v,c) \leq 2b + 6b + 6b + 2b = 16b$$

All the computations are polynomial in $|V|$. Therefore we have shown:

Proposition 3.8 There exists a polynomial time approximation algorithm for the L-balanced κ-center problem with approximation ratio 16. ∎

Remarks: Further improvement in the approximation ratio can be achieved by modifying the algorithm (particularly, the choice of centers and sets "$Carry\_in$" during the tree contraction phase), reducing the approximation ratio to 10. Also, if $\kappa = O(\log n)$, then there exists a different approximation scheme achieving an approximation ratio of 4. Details are omitted.

4 The Balanced Weighted Centers Problem

In this section we consider the weighted version of the L-balanced κ-centers problem. It is assumed that every node v has a weight $\omega(v)$, and we look for a solution with minimal radius, in which the sum of the weights of the centers is at most κ. For $U \subseteq V$, define $\omega(U) = \sum_{u \in U} \omega(u)$.

Definition 4.1 The *L-balanced, κ-weighted centers problem* is defined as follows:
Input: Graph $G = (V,E)$ with weights on the nodes and the edges, an integer L, and a real $\kappa > 0$.
Goal: Select an L-balanced center assignment (C,φ), such that $\omega(C) \leq \kappa$, minimizing the radius $\mathcal{R}(S)$ of the induced assignment tuple S.
Let us denote the optimal radius by $\mathcal{R}_{wopt}(G, L, \omega, \kappa)$.

For $\iota = 1$ to $|E|$ **do**

1. Choose a maximal independent set in G_ι^2. Let this set be $C = \{\theta_1, \theta_2, \ldots, \theta_\ell\}$.
2. Call Procedure $Allocate(G_\iota, L, C)$.
3. Construct $\hat{G} = (\hat{V}, \hat{U}, \hat{E})$ as follows: let $\hat{V}$ be the centers $\{\theta_1, \ldots, \theta_m\}$, returned by the procedure $Allocate$, and let $\hat{U} = V$. Let $(\theta_i, v) \in \hat{E}$ iff $dist(\theta_i, v) \leq (t+1)c_\iota$. The weight of the edge (θ_i, v) is $\omega(v)$.
4. Compute a minimum weight perfect matching in $\hat{G}$. If there is no perfect matching in the graph or the weight of the matching is greater thatn κ, return "failure".
5. Otherwise let $C' = \{v_1, \ldots v_m\}$, where θ_i was matched to v_i in the minimum weight perfect matching.
6. Call $Flow\_graph(C', V, 21 \cdot c_\iota, L-1, 1, 1)$. Assign each node a center according to the maximum flow.

Figure 6: Algorithm $Approx2(G, L, \kappa)$

As a basis for the approximation we shall make use of an initial solution for the *unweighted* problem, that enjoys the *minimum cardinality* property defined below.

Definition 4.2 Consider an instance of the L-balanced κ-weighted centers problem. Let $P = (C, \varphi)$ be a solution to the problem without the weight restriction. Then P is a *minimum cardinality solution* for the problem if for every optimal solution $X_{opt} = (C_{opt}, \varphi_{opt})$ for the weighted problem, and for every $A \subseteq C$, the subset of centers of $X \subseteq X_{opt}$ serving the clients of A satisfies $|X| \geq |A|$.

The minimum cardinality solution P is said to have approximation ratio t if its radius is at most $t \cdot \mathcal{R}_{wopt}(G, L, \omega, \kappa)$.

The approximation algorithm, $Approx2(G, L, \kappa)$ is given in Figure 6. The idea is to start with applying Procedure $Allocate$ and generate a minimum cardinality solution, and then use Hall's theorem to derive the ratio bound. The analysis is given by the following lemmas, whose proofs are omitted.

Lemma 4.3 The call to Procedure $Allocate$ in line 2 of algorithm $Approx2(G, L, \kappa)$ returns a minimum cardinality solution with approximation ratio 10. ∎

Lemma 4.4 Algorithm $Approx2(G, L, \kappa)$ is a polynomial time algorithm for the L-balanced κ-weighted centers problem with approximation ratio $2t + 1$. ∎

Proposition 4.5 There exists a polynomial time approximation algorithm for the L-balanced κ-weighted centers problem with approximation ration 21. ∎

Acknowledgments

We would like to thank Madan Gopal for helpful discussions.

References

[Ber85] D. Bertsekas. A unified framework for primal-dual methods in minimum cost network flow problems. *Math. Prog.*, 32:125–145, 1985.

[BG87] D. Bertsekas and R. Gallager. *Data Networks*. Prentice-Hall, Inc., Englewood Cliffs, NJ, 1987.

[BG89] A. Bouloutas and P.M. Gopal. Some graph partitioning problems and algorithms related to routing in large computer networks. In *Proc. 9th IEEE Conf. on Distributed Computing Systems*, pages 362–370, Newport Beach, CA, 1989.

[BP90] J. Bar-Ilan and D. Peleg. How to allocate network centers. Technical Report CS90-20, The Weizmann Institute of Science, 1990.

[HR88] A.R. Hevner and A. Rao. Distributed data allocation strategies. In *Advances in Computers, Vol. 27*, pages 121–155. Academic Press, 1988.

[HS84] D.S. Hochbaum and D. Shmoys. Powers of graphs: A powerful technique for bottleneck problems. In *Proc. 16th ACM Symp. on Theory of Computing*, pages 324–333. ACM, April 1984.

[Lov75] L. Lovász. On the ratio of optimal integral and fractional covers. *Discrete Mathematics*, 13:383–390, 1975.

[MK83] K. Murthy and J. Kam. An approximation algorithm to the file allocation problem in computer networks. In *Proc. 2nd ACM Symp. on Principles of Database Systems*, pages 258–266. ACM, 1983.

[ML77] H.L. Morgan and K.D. Levin. Optimal program and data locations in computer networks. *Comm. of the ACM*, pages 315–322, 1977.

[Pel90] D. Peleg. Distributed data structures: A complexity oriented view. In *4th Int. Workshop on Distributed Algorithms*, September 1990.

Facility Dispersion Problems: Heuristics and Special Cases [1]

(Extended Abstract)

S.S. Ravi   D.J. Rosenkrantz   G.K. Tayi
Department of Computer Science   School of Business Administration
SUNY at Albany   SUNY at Albany
Albany, NY 12222   Albany, NY 12222

Abstract

Facility dispersion problem deals with the location of facilities on a network so as to *maximize* some function of the distances between facilities. We consider the problem under two different optimality criteria, namely maximizing the minimum distance (MAX-MIN) between any pair of facilities and maximizing the average distance (MAX-AVG) between any pair of facilities. Under either criterion, the problem is known to be NP-hard, even when the distances satisfy the triangle inequality. We consider the question of obtaining near-optimal solutions. For the MAX-MIN criterion, we show that if the distances do not satisfy the triangle inequality, there is no polynomial time relative approximation algorithm unless $\mathbf{P} = \mathbf{NP}$. When the distances do satisfy the triangle inequality, we present an efficient heuristic which provides a performance guarantee of 2, thus improving the performance guarantee of 3 proven in [Wh91]. We also prove that obtaining a performance guarantee of less than 2 is NP-hard. For the MAX-AVG criterion, we present a heuristic which provides a performance guarantee of 4, provided that the distances satisfy the triangle inequality. For the 1-dimensional dispersion problem, we provide polynomial time algorithms for obtaining optimal solutions under both MAX-MIN and MAX-AVG criteria. Using the latter algorithm, we obtain a heuristic which provides a performance guarantee of $4(\sqrt{2}-1) \approx 1.657$ for the 2-dimensional dispersion problem under the MAX-AVG criterion.

1   Introduction

Many problems in location theory deal with the placement of facilities on a network so as to minimize some function of the distances between facilities or between facilities and the nodes of the network [HM79]. Such problems model the placement of 'desirable' facilities such as warehouses, hospitals, fire stations, etc. However, there are situations in which facilities are to be located so as to *maximize* some function of the distances between pairs of nodes. Such location problems are referred to as **dispersion** problems [Er89, Ku87, CD81] since they model situations in which proximity of facilities is undesirable. One example of such a situation is the distribution of business franchises in a city [Er89]. Other examples of dispersion problems arise in the context of placing 'undesirable' (also called 'obnoxious') facilities such as nuclear power

[1] Research supported by NSF Grants CCR-88-03278, CCR-89-05296 and CCR-90-06396.

plants, oil storage tanks and ammunition dumps [EN89, Er89, Ku87]. Such facilities need to be spread out to the greatest possible extent so that an accident at one of the facilities will not damage any of the others. Other applications of facility dispersion are discussed in [EN89, Er89].

Analytical models for the dispersion problem assume that the given network is represented by a set $V = \{v_1, v_2, \ldots, v_n\}$ of n nodes with a non-negative distance between every pair of nodes. The distances are assumed to be symmetric and so the network can be thought of as a complete undirected graph on n nodes with a non-negative weight on each edge. The weight of the edge $\{v_i, v_j\}$ $(i \neq j)$ is denoted by $w(v_i, v_j)$. We assume that $w(v_i, v_i) = 0$ for $1 \leq i \leq n$. The objective of the dispersion problem is to locate p facilities $(p \leq n)$ among the n nodes of the network, with at most one facility per node, such that some function of the distances between facilities is maximized. Two of the optimality criteria considered in the literature [Er89, EN89, Ku87] are MAX-MIN (i.e., maximize the minimum distance between a pair of facilities) and MAX-AVG (i.e., maximize the average distance between a pair of facilities). Under either criterion, the problem is known to be NP-hard, even when the distances satisfy the triangle inequality [EB90, Er89]. Although many researchers have studied the dispersion problem (see [EN89] for a survey and an extensive bibliography), except for [Wh91], the question of whether there are efficient heuristics with provably good performance has not been addressed. The main results of this paper provide an affirmative answer to that question, provided that the distances satisfy the triangle inequality. Specifically, we present a heuristic for the MAX-MIN dispersion problem and prove that it provides a performance guarantee of 2. We also show that no polynomial time algorithm can provide a better performance guarantee unless $\mathbf{P} = \mathbf{NP}$. For the MAX-AVG dispersion problem, we present a heuristic which provides a performance guarantee of 4. If the distances are not required to satisfy the triangle inequality, we show that there is no polynomial time relative approximation algorithm for the MAX-MIN dispersion problem unless $\mathbf{P} = \mathbf{NP}$. We also address the 1-dimensional and 2-dimensional versions of the dispersion problem. We present polynomial algorithms for obtaining optimal solutions for the 1-dimensional dispersion problem under both MAX-MIN and MAX-AVG criteria and a heuristic for the 2-dimensional dispersion problem under the MAX-AVG criterion.

The remainder of this paper is organized as follows. Section 2 contains the formal definitions and a brief discussion of the previous work on the dispersion problem. Sections 3 and 4 address the dispersion problem under the MAX-MIN and MAX-AVG criteria respectively. Section 5 discusses our results for the 1- and 2-dimensional versions of the dispersion problems. Due to space limitations, only sketches of proofs are included in this extended abstract. Detailed proofs will appear in a complete version of this paper.

2 Definitions and Previous Work

We begin with the specifications of MAX-MIN and MAX-AVG dispersion problems in the format of [GJ79]. Since these dispersion problems are trivial for one facility, we assume that the number of facilities (p) is at least 2.

(a) MAX-MIN Facility Dispersion (MMFD)

Instance: A set $V = \{v_1, v_2, \ldots, v_n\}$ of n nodes, a non-negative distance $w(v_i, v_j)$ for each pair v_i, v_j of nodes, and an integer $p \leq n$.

Requirement: Find a subset $P = \{v_{i_1}, v_{i_2}, \ldots, v_{i_p}\}$ of V, with $|P| = p$, such that the objective function $f(P) = \min_{x,y \in P} \{w(x,y)\}$ is maximized.

(b) MAX-AVG Facility Dispersion (MAFD)

Instance: As in MMFD above.

Requirement: Find a subset $P = \{v_{i_1}, v_{i_2}, \ldots, v_{i_p}\}$ of V, with $|P| = p$, such that the objective function $f(P) = \dfrac{2}{p(p-1)} \sum_{x,y \in P} w(x,y)$ is maximized.

The objective function for MAFD has the above form because the number of edges among the nodes in P is $p(p-1)/2$. Note that maximizing the average is equivalent to maximizing the sum.

Both MMFD and MAFD are known to be NP-hard, even when the edge weights satisfy the triangle inequality [EB90, Er89]. Much of the work on the dispersion problem reported in the literature (see the bibliography in [EN89]) falls into two categories. Papers in the first category deal with the design and implementation of branch-and-bound algorithms and heuristics (see [EB90, Er89, EN89, Ku87] and the references cited therein). However, except for [Wh91], only experimental studies of the performance of the heuristics have been reported. In [Wh91], White presents a heuristic for MMFD when the nodes are points in d-dimensional Euclidean space and the distance between a pair of points is their Euclidean distance. He shows that the heuristic provides a performance guarantee of 3. In the next section, we improve on that result by presenting a different heuristic which provides a performance guarantee of 2 for any instance of MMFD in which the distances satisfy the triangle inequality. We also show that unless **P = NP**, no polynomial time algorithm can provide a better performance guarantee.

Papers in the second category deal with restricted versions and variants of MMFD and MAFD. For example, [CG78] presents a polynomial algorithm for locating a facility on an edge of a connected (but not necessarily complete) network so as to maximize a weighted sum of the distances from the facility to the nodes of the network. Dasarathy and White [DW80] assume the nodes of the network to be a set S of n points in k-dimensional space and consider the problem of finding a point C within the convex hull of S so as to maximize the minimum Euclidean distance between C and points in S. They present polynomial algorithms for $k = 2$ and $k = 3$. For $k = 2$, this problem is referred to as the **largest empty circle** (LEC) problem in computational geometry literature (see Section 6.4 of [PS85]). The LEC problem can be solved in $O(n \log n)$ time, which is known to be optimal [PS85]. We note that the LEC problem is a variant of MMFD which allows a new facility to be placed at a point which is not necessarily one of the given points. In contrast, the dispersion problems addressed in this paper require that the facilities be placed at a subset of the given set of points. A problem similar to LEC but with a different distance function is studied in [MC86]. A weighted version of the problem for $k = 2$ is studied in [EO91]. In [CD81] a polynomial algorithm is presented for MMFD on tree networks under the assumption that the facilities can be placed on the edges of the tree. For a discussion of the other variants, we refer the reader to [EN89].

The main focus of this paper is on the analysis of heuristics for MMFD and MAFD. By a **heuristic** we mean a *polynomial time* approximation algorithm which produces feasible, but not necessarily optimal, solutions. Heuristics are commonly classified as *absolute* or *relative* depending on the types of performance guarantees that can be established for them (see for example, [FFL88]). An **absolute** approximation algorithm guarantees a solution which is within an *additive* constant of the optimal value for every instance of the problem. A **relative** approximation algorithm guarantees a solution which is within a *multiplicative* constant of the optimal value for every instance of the problem. It is easy to show, using the technique presented in [GJ79, pages 138-139], that there are no absolute approximation algorithms for MAFD or for MMFD, unless **P = NP**. So, we restrict our attention to the study of relative approximation algorithms.

3 Near-Optimal Solutions to MAX-MIN Dispersion

We first consider MMFD without requiring the distances to satisfy the triangle inequality. We have the following theorem.

Theorem 3.1 *If the distances are not required to satisfy the triangle inequality, then there is no polynomial time relative approximation algorithm for MMFD unless* $\mathbf{P} = \mathbf{NP}$.

Proof (sketch): We show that such an approximation algorithm (say A) can be used to devise a polynomial time algorithm for a known NP-complete problem, namely CLIQUE [GJ79], contradicting the assumption that $\mathbf{P} \neq \mathbf{NP}$.

Let $K \geq 1$ denote the performance guarantee provided by A and let $G(N, E)$ and integer J denote an instance of CLIQUE. We produce an instance of MMFD (without triangle inequality) as follows. The set V of nodes of the MMFD instance is in one-to-one correspondence with N. The number of facilities $p = J$ and the distances are defined as follows. We let $w(v_i, v_j) = K + 1$ if $\{n_i, n_j\}$ is in E; otherwise, $w(v_i, v_j) = 1$. It is possible to verify that the solution produced by A has a value greater than 1 iff G has a clique of size J. $\square$

Even though Theorem 3.1 provides a strong negative result, it is not applicable in many practical situations since distances often satisfy the triangle inequality. Therefore, it is of interest whether there is an efficient relative approximation algorithm for MMFD when the distances satisfy the triangle inequality (MMFD-TI). The remaining theorems in this section precisely characterize the performance guarantees obtainable for MMFD-TI.

A greedy heuristic (called GMM) for MMFD-TI is shown in Figure 3.1. We use P to denote the set of nodes at which GMM places the p facilities. As mentioned earlier, we assume that $p \geq 2$. The heuristic begins by initializing P to contain a pair of nodes in V which are joined by an edge of maximum weight. Subsequently, each iteration of GMM chooses a node v from $V - P$ such that the minimum distance from v to a node in P is the largest among all the nodes in $V - P$. Heuristic GMM terminates when $|P| = p$. The value of the solution produced by GMM is equal to $\min_{x,y \in P} \{w(x, y)\}$.

We first discuss an example to illustrate the GMM heuristic. Consider an MMFD-TI instance with 5 nodes, denoted by v_1, v_2, v_3, v_4, and v_5. Suppose $w(v_1, v_2) = 3$, $w(v_2, v_3) = w(v_2, v_4) = w(v_2, v_5) = 1$ and all other edge weights are equal to 2. Assume that the number of facilities (p) to be located is 3. To begin with, GMM will place two of the facilities at v_1 and v_2 since $w(v_1, v_2) = 3$ is the maximum edge weight. Now, no matter where the third facility is placed,

1. Let v_i and v_j be the endpoints of an edge of maximum weight.

2. $P \leftarrow \{v_i, v_j\}$.

3. **while** $(|P| \neq p)$ **do**

   begin

       (a) Find a node $v \in V - P$ such that $\min_{v' \in P}\{w(v, v')\}$ is maximum among the nodes in $V - P$.

       (b) $P \leftarrow P \cup \{v\}$.

   end

4. Output P.

Figure 3.1: Details of Heuristic GMM

---

the solution value of the placement is 1 since each of the remaining nodes (v_3, v_4 and v_5) has an edge of weight 1 to v_2. However, an optimal placement consists of the three nodes v_3, v_4 and v_5 and has a solution value of 2. Thus for this example, the solution value produced by GMM differs from the optimal value by a factor of 2. Our next theorem shows that the performance of GMM is never worse. Moreover, we will also show (Theorem 3.3) that unless $\mathbf{P} = \mathbf{NP}$, no polynomial time heuristic can provide a better performance guarantee.

Theorem 3.2 *Let I be an instance of MMFD-TI with $p \geq 2$. Let $OPT(I)$ and $GMM(I)$ denote respectively the values of an optimal placement and that produced by GMM for the instance I. Then $OPT(I)/GMM(I) \leq 2$.*

Proof (sketch): Consider the set-valued variable P in the description of GMM. Let $f(P) = \min_{x,y \in P}\{w(x, y)\}$. We will show that the condition

$$f(P) \geq OPT(I)/2 \tag{1}$$

holds after each addition to P. Since $GMM(I) = f(P)$ after the last addition to P, the theorem would then follow.

Since the first addition inserts two nodes joined by an edge of largest weight into P, Condition (1) clearly holds after the first addition. So, assume that the condition holds after k additions to P, where $k \geq 1$. We will prove by contradiction that the condition holds after the $(k + 1)^{st}$ addition to P as well.

To that end, let P^* denote the set of p nodes in an optimal placement. For convenience, we use l^* for $OPT(I)$. Let $P_k = \{x_1, x_2, \ldots, x_{k+1}\}$ denote the set P after k additions. (Note that $|P_k| = k + 1$, since the first addition inserts two nodes into P). Since we are assuming that Condition (1) does not hold after the $(k + 1)^{st}$ addition, it must be that for each $v \in V - P_k$, there is a node $x \in P_k$ such that $w(x, v) < l^*/2$. We describe this situation by saying that v is "blocked by" x.

Let $P' = P^* \cap (V - P_k)$. Note that $|P_k| = k + 1 < p$ since an additional node is to be added. Thus $|P'| \geq 1$. It is easy to verify that if $|P'| = 1$, then Condition (1) will hold after the $(k+1)^{st}$ addition. Therefore, assume that $|P'| = r \geq 2$. Furthermore, let $P' = \{y_1, y_2, \ldots, y_r\}$. Since $P' \subseteq P^*$, we must have $w(y_i, y_j) \geq l^*$ for $1 \leq i, j \leq r$, $(i \neq j)$. Now using the fact that the distances satisfy the triangle inequality, it is possible to show that no two distinct nodes in P' can be blocked by the same node $x \in P_k$. Using this result and the observation that none of the blocking nodes in P_k can be in P^*, it is possible to show that P_k must contain p nodes. However, this contradicts our initial assumption that $|P_k| < p$. Thus Condition (1) must hold after the $(k + 1)^{st}$ addition and the theorem follows. $\square$

Our next theorem shows that if $\mathbf{P} \neq \mathbf{NP}$, GMM provides the best possible performance guarantee obtainable in polynomial time for MMFD-TI.

Theorem 3.3 *If* $\mathbf{P} \neq \mathbf{NP}$, *no polynomial time relative approximation algorithm can provide a performance guarantee of* $(2 - \epsilon)$ *for any* $\epsilon > 0$ *for MMFD-TI.*

Proof (sketch): The proof is similar to that of Theorem 3.1 except that the edge weights are chosen as follows. Let $w(v_i, v_j) = 2 - \epsilon/2$ if $\{n_i, n_j\}$ is in E; otherwise, let $w(v_i, v_j) = 1$. Using the fact that $0 < \epsilon < 1$, it can be verified that the resulting distances satisfy the triangle inequality and that the value of the solution produced by A for the MMFD-TI instance is greater than 1 iff G has a clique of size J. $\square$

4 Near-Optimal Solutions to MAX-AVG Dispersion

In this section, we present a relative approximation algorithm for MAFD under the triangle inequality assumption (MAFD-TI). This heuristic which we call GMA, is shown in Figure 4.1. It is identical to the GMM heuristic shown in Figure 3.1, except that in Step 3(a), we choose a node $v \in V - P$ for which $\sum_{v' \in P} w(v, v')$ is maximum among all the nodes in $V - P$. Note that

1. Let v_i and v_j be the endpoints of an edge of maximum weight.

2. $P \leftarrow \{v_i, v_j\}$.

3. **while** $(|P| \neq p)$ **do**

   begin

       (a) Find a node $v \in V - P$ such that $\sum\limits_{v' \in P} w(v, v')$ is maximum
   among the nodes in $V - P$.

       (b) $P \leftarrow P \cup \{v\}$.

   end

4. Output P.

Figure 4.1: Details of Heuristic GMA

the solution value produced by GMA is equal to $\dfrac{2}{p(p-1)} \sum\limits_{x,y \in P} w(x, y)$. Our next theorem shows that GMA is indeed a relative approximation algorithm for MAFD-TI.

Theorem 4.1 *Let I be an instance of MAFD-TI with $p \geq 2$. Let $OPT(I)$ and $GMA(I)$ denote respectively the values of an optimal placement and that produced by GMA for the instance I. Then $OPT(I)/GMA(I) \leq 4$.*

Before sketching a proof of this theorem, we introduce some notation which will also be used in the next section. Suppose A and B are subsets of V. Let $W(A) = \sum\limits_{x,y \in A} w(x, y)$ and $W(A, B) = \sum\limits_{x \in A, y \in B} w(x, y)$. (Note that $W(A, B) = W(B, A)$.) Also, for $x \in A$, let $W(x, B) = \sum\limits_{y \in B} w(x, y)$. We can now state a lemma which is used in the proof of Theorem 4.1. The lemma is a consequence of the triangle inequality.

Lemma 4.2 *Let A and B be subsets of V with $|B| \geq 2$. Then $W(A, B) \geq |A|\, W(B)/(|B| - 1)$.*

Proof sketch for Theorem 4.1: We show by induction that after each addition, the average weight of an edge in P is at least $OPT(I)/4$.

The statement is clearly true after the first addition because an edge of maximum weight is added to P. So, assume that the statement holds after k additions, for some $k \geq 1$. For convenience, we use l^* for $OPT(I)$. Let P_k denote the set P after k additions ($|P_k| = k + 1$). We establish the following:

Claim: There is a node $x \in V - P_k$ such that $W(x, P_k) \geq (k + 1)\, l^*/4$.

This claim in conjunction with the induction hypothesis implies that the average weight is at least $l^*/4$ after the $(k+1)^{st}$ addition to P as well.

To outline a proof of the above claim, let P^* denote the set of p nodes in an optimal placement. We first show that if P_k and P^* are disjoint, then a node in P^* will satisfy the claim. So, we can assume that P^* and P_k have a non-empty intersection. Let $Y = P^* \cap P_k$ and $X = P^* - Y$. We note that

$$W(P^*) = \frac{p(p-1)}{2} l^* = W(X) + W(Y) + W(X,Y) \tag{2}$$

From (2) we can conclude that either $W(X)$ or $W(Y) + W(X,Y)$ must be at least $p(p-1) l^*/4$. In the former case, we can directly show that one of the nodes in X will satisfy the claim. In the latter case, we use Lemma 4.2 to show that $W(X, P_k) \geq |X| (k+1) l^*/4$ and as a consequence conclude that some node $x \in X$ will satisfy the claim. □

It is open whether GMA actually provides a performance guarantee of 2 for MAFD-TI. We now present an example which achieves the bound of 2 asymptotically. The MAFD instance I described below has a total of $2p$ nodes and p facilities are to be located. For convenience in description, we partition the set of $2p$ nodes into 3 sets called X, Y and Z. The set X has p nodes (denoted by $x_1, x_2, \ldots, x_p$), the set Y has $p - 2$ nodes (denoted by $y_1, y_2, \ldots, y_{p-2}$) and Z has 2 nodes (denoted by z_1 and z_2). The edge weights are chosen as follows. For any two distinct nodes x_i and $x_j \in X$, $w(x_i, x_j) = 2$. Also, $w(z_1, z_2) = 2$. All other edge weights are 1. It is straightforward to verify that the distances satisfy the triangle inequality. The set X is an optimal placement and its solution value $(OPT(I))$ is 2 (since every edge in X has a weight of 2). We can force GMA to place the first two facilities at z_1 and z_2 since $w(z_1, z_2) = 2$ is a maximum edge weight. It is easy to verify that in the subsequent $(p - 2)$ steps, GMA can be forced to choose all the $(p - 2)$ nodes from the set Y. Thus, we force GMA to return $Z \cup Y$ as the placement. The solution value $(GMA(I))$ corresponding to this placement is given by

$$GMA(I) = \frac{2}{p(p-1)} \left[2 + \left(\frac{p(p-1)}{2} - 1 \right) \right]$$

since there is only one edge (namely, that between z_1 and z_2) of weight 2 and the remaining edges are of weight 1. A bit of simplification shows that $GMA(I) = 1 + 2/(p(p-1))$. The ratio $OPT(I)/GMA(I)$ is given by

$$\frac{OPT(I)}{GMA(I)} = \frac{2}{1 + 2/(p(p-1))}$$

Clearly, the ratio can be made arbitrarily close to 2 by choosing p to be sufficiently large.

5 Dispersion Problems in One and Two Dimensions

The 1-dimensional dispersion problems are restricted versions of MMFD and MAFD where the vertex set V consists of a set of n points (denoted by $x_1, x_2, \ldots, x_n$) on a line. Thus $w(x_i, x_j) = |x_i - x_j|$. We denote these problems by 1D-MMFD and 1D-MAFD respectively and show that they can both be solved in polynomial time. We also show that the algorithm for 1D-MAFD can be used to obtain a heuristic with a performance guarantee of $4(\sqrt{2} - 1)$ for the 2-dimensional MAFD problem (2D-MAFD), where the nodes are points in $\Re^2$ and the distance between a pair of points is the Euclidean distance. Due to space limitations, we only discuss the main ideas behind these algorithms.

Consider 1D-MMFD. We first sort the points in $O(n \log n)$ time. Without loss of generality, let $x_1, x_2, \ldots, x_n$ denote the points in non-decreasing order. It is easy to see that there is always an optimal solution containing x_1. Now, in order to decide whether to include x_2 in the solution set, we test whether there is a placement of $(p-1)$ facilities on the $(n-2)$ point set $\{x_3, \ldots, x_n\}$ such that the solution value is *greater than* $(x_2 - x_1)$. This test needs only $O(n)$ time because the inclusion of a point in the solution set depends only on its nearest predecessor which has been included in the solution set. If there is a placement with solution value larger than $(x_2 - x_1)$, then x_2 is not included in the optimal solution being constructed. Otherwise, we can argue (as in the case of x_1) that there is an optimal solution which includes x_2. By repeating the above procedure with each of the remaining points, we obtain an $O(n^2)$ algorithm. By carrying out the tests in a careful manner, we can achieve a running time of $O(pn \log n)$. Thus the overall running time is $O(\min \{n^2, pn \log n\})$.

We can use dynamic programming to obtain a polynomial algorithm for 1D-MAFD. Again, sort the points in non-decreasing order and let $x_1, x_2, \ldots, x_n$ denote the points in sorted order. Consider a point x_j and an integer $k \le \max(j, p)$. Let $C = A \cup B$ be a set of p points consisting of a subset $A \subseteq \{x_1, \ldots, x_j\}$, with $|A| = k$, and a subset $B \subseteq \{x_{j+1}, \ldots, x_n\}$, with $|B| = p - k$. Using the notation of Section 4, we can show that

$$W(C) = W(A) + (p - k) W(x_j, A) + W(B) + k W(x_j, B) \tag{3}$$

Equation (3) shows that the choice of B has no effect on the choice of A. For a given subset $A \subseteq \{x_1, \ldots, x_j\}$, with $|A| = k$, define

$$f_{k,j}(A) = W(A) + (p - k) W(x_j, A) \tag{4}$$

1. Obtain the projections of the given set of points on each of the four axes defined by the equations $y = 0$, $y = x$, $x = 0$, and $y = -x$.

2. Find optimal solutions to each of the four resulting instances of 1D-MAFD.

3. Return the placement corresponding to the best of the four solutions found in Step 2.

Figure 5.1: Details of Heuristic PROJECT

Let $OPT_{k,j}$ be a subset of $\{x_1, \ldots, x_j\}$ that maximizes $f_{k,j}$. Note that $OPT_{k,j}$ either includes x_j or excludes it. If $x_j \notin OPT_{k,j}$, then the chosen subset must be $OPT_{k,j-1}$. Using (4) we can show that

$$f_{k,j}(OPT_{k,j-1}) = f_{k,j-1}(OPT_{k,j-1}) + k(p - k)(x_j - x_{j-1}) \tag{5}$$

If $x_j \in OPT_{k,j}$, then the chosen subset should be $OPT_{k-1,j-1} \cup \{x_j\}$ and again using (4) we can show that

$$f_{k,j}(OPT_{k-1,j-1} \cup \{x_j\}) = f_{k-1,j-1}(OPT_{k-1,j-1}) + (k - 1)(p - k + 1)(x_j - x_{j-1}) \tag{6}$$

Equations (5) and (6) show that given $OPT_{k,j-1}$, $OPT_{k-1,j-1}$, $f(OPT_{k,j-1})$, and $f(OPT_{k-1,j-1})$, we can compute $OPT_{k,j}$ and $f(OPT_{k,j})$. The boundary conditions are the following: $OPT_{0,j} = \emptyset$, $f(OPT_{0,j}) = 0$ $(1 \leq j \leq n)$, $OPT_{1,1} = \{x_1\}$, and $f(OPT_{1,1}) = 0$.

The running time is $O(n \log n)$ for sorting plus $O(pn)$ for to carry out the dynamic programming. Thus the overall running time is $O(\max(n \log n, pn))$.

It is open whether the 2-dimensional dispersion problems under Euclidean distance (denoted by 2D-MMFD and 2D-MAFD) are NP-hard. Interestingly, it is possible to use the algorithm for 1D-MAFD to obtain a heuristic for 2D-MAFD which provides a better performance guarantee than Heuristic GMA of Section 4. The input to the heuristic for 2D-MAFD is a set V of n points (x_1, y_1), (x_2, y_2), $\ldots$, (x_n, y_n), and the number of facilities p. The steps of this heuristic (called PROJECT) are shown in Figure 5.1. The performance guarantee provided by PROJECT is indicated in the following theorem.

Theorem 5.1 *Let I be an instance of 2D-MAFD. Let $OPT(I)$ and $PROJECT(I)$ denote respectively the solution values of an optimal placement and that produced by PROJECT for the instance I. Then $OPT(I)/PROJECT(I) \leq 4(\sqrt{2} - 1)$.*

The performance guarantee provided by PROJECT is approximately 1.657. By projecting the given set of points on more axes, it is possible to improve the performance guarantee further. As we increase the number of axes, the performance guarantee provided by this procedure approaches the value of $\pi/2 \approx 1.571$.

References

[CG78]   R.L. Church and R.S. Garfinkel, "Locating an Obnoxious Facility on a Network," Trans. Science, Vol. 12, No. 2, May 1978, pp 107-118.

[CD81]   R. Chandrasekharan and A. Daughety, "Location on Tree Networks: p-Centre and n-Dispersion Problems," Math. of OR, Vol. 6, No. 1, 1981, pp 50-57.

[DW80]   B. Dasarathy and L.J. White, "A Maxmin Location Problem," Oper. Res., Vol. 28, No. 6, 1980, pp 1385-1401.

[EB90]   E. Erkut, T. Baptie and B. von Hohenbalken, "The Discrete p-Maxian Location Problem," Computers in OR, Vol. 17, No. 1, 1990, pp 51-61.

[EN89]   E. Erkut and S. Neuman, "Analytical Models for Locating Undesirable Facilities,", Euro. J. of OR, Vol. 40, 1989, pp 275-291.

[EO91]   E. Erkut and T.S. Öncü, "A Parametric 1-Maximin Location Problem,", J. of Oper. Res. Soc., Vol. 42, No. 1, Jan 1991, pp 49-55.

[Er89]   E. Erkut, "The Discrete p-Dispersion Problem," Research Paper No. 87-5, Dept. of Finance and Mgt. Sci., Univ. of Alberta, Apr. 1989.

[FFL88] M.R. Fellows, D.K.Friesen and M.A.Langston, "On Finding Optimal and Near-Optimal Lineal Spanning Trees," Algorithmica, Vol. 3, No. 4, 1988, pp 549-560.

[GJ79]   M.R. Garey and D.S. Johnson, "Computers and Intractability: A Guide to the Theory of NP-Completeness," W.H. Freeman and Co., 1979.

[HM79]   G.Y. Handler and P.B. Mirchandani, "Location on Networks: Theory and Algorithms," MIT Press, Cambridge, MA, 1979.

[Ku87]   M.J. Kuby, "Programming Models for Facility Dispersion: The p-Dispersion and Maxisum Dispersion Problems," Geographical Analysis, Vol. 19, No. 4, Oct. 1987, pp 315-329.

[MC86]   E. Melachrinoudis and T.P.Cullinane, "Locating an Undesirable Facility with a Minimax Criterion," Euro. J. of OR, Vol. 24, 1986, pp 239-246.

[PS85]   F.P. Preparata and M.I. Shamos, "Computational Geometry: An Introduction", Springer-Verlag, Inc., New York, NY, 1985.

[Wh91]   D.J.White, "The Maximal Dispersion Problem and the 'First Point Outside the Neighborhood' Heuristic," Computers in Op. Res., Vol. 18, No. 1, 1991, pp 43-50.

Optimum Guard Covers and m-Watchmen Routes for Restricted Polygons[*]

Svante Carlsson[†]   Bengt J. Nilsson[††]   Simeon Ntafos[§]

Abstract

A watchman, in the terminology of art galleries, is a mobile guard. We consider several watchman and guard problems for different classes of polygons. We introduce the notion of vision spans along a path (route) which provide a natural connection between the (stationary) Art Gallery problem, the m-watchmen problem and the watchman route problem. We prove that finding the minimum number of vision points (i.e., static guards) along a shortest watchman route is NP-hard. We provide a linear time algorithm to compute the best set of static guards in a histogram (Manhattan skyline) polygon. The m-watchmen problem (minimize total length of routes for m watchmen) is NP-hard for simple polygons. We give a $\Theta(n^3 + n^2m^2)$-time algorithm to compute the best set of m (moving) watchmen in a histogram.

1 Introduction

The problem of placing guards in an art gallery so that every point in the gallery is visible to at least one guard has been considered by several researchers. If the gallery is represented by a polygon (having n vertices) and the guards are points in the polygon, then visibility problems can be equivalently stated as problems of covering the gallery with starshaped polygons. Chvàtal [Chv75] and Fisk [Fis78] proved that $\lfloor n/3 \rfloor$ guards are always sufficient and sometimes necessary. Many other results on art gallery problems and general visibility problems can be found in [ORo87].

Chin and Ntafos [CN87,CN88] consider the problem of finding shortest watchman routes for different classes of polygons. The route is the path traced by a moving guard who (in total) sees the whole polygon. They prove that the problem is NP-hard for polygons with holes and for 3-dimensional simple polyhedra. They also give a $\Theta(n)$-time algorithm that solves the problem for simple rectilinear polygons. In addition they provide a solution to the slightly different question of finding a shortest watchman route given a "door", a point on the boundary that the route must pass through. They give an $O(n^4)$-time algorithm that solves this problem for simple polygons without holes. (These bounds rely on the new linear time triangulation algorithm by Chazelle [Cha90].)

The stationary art gallery problem and the watchman route problem can be thought of as special cases of the m-watchmen problem. Suppose that we have m mobile watchmen and we want to design routes for them so that each point in a polygon is visible to some point along a route. The objective is to minimize the total length of the m routes. Clearly, the total length of an optimum solution increases (decreases) as the number of watchmen decreases (increases). At one limit, we have a situation where m is large enough so that the length of the optimum solution is zero (i.e., each watchman remains stationary) and we have the art gallery problem. At the other limit, if $m = 1$, we have the watchman route problem.

Another connection between static and mobile guard problems is obtained by considering a single mobile robot equipped with a vision system. We can think of the robot as performing two types of activity, locomotion and visual data acquisition. To minimize the distance traveled so that it

---

[*]The work of the second author was supported by the Deutsche Forschungsgemeinschaft under Grant No. Ot 64/5-4.
[†]Department of Computer Science, Lund University, Box 118, 221 00 Lund, Sweden.
[‡]Institut für Informatik, Universität Freiburg, Rheinstr. 10–12, D-7800 Freiburg, Fed. Rep. of Germany.
[§]Computer Science Program, University of Texas at Dallas, Richardson, TX 75083–0688, U.S.A.

sees all of the polygon, the robot traces a shortest watchman route. However, note that the robot's vision system does not need to be active throughout the route. Also, it is likely that the vision system performs better if the robot is not moving. We refer to points on the route at which the robot engages in vision activity as *vision points*, (a continuous portion of the route where the robot engages in vision activity is referred to as a *vision span*). Then we have the art gallery problem (minimize the number of vision points) with the restriction that the solution must lie along a path (e.g., the shortest watchman route).

In this paper we show that finding the best set of vision points along a path is NP-hard for simple polygons. Also, the m-watchmen problem for simple polygons is NP-hard. In view of these results, we consider the same problems for restricted classes of polygons. Nilsson and Wood [NW90] treat the class of spiral polygons. They give a linear time algorithm for the minimum stationary guard problem and a $\Theta(n^2)$ time algorithm for the multiple watchmen routes problem. In this paper we consider histograms (or Manhattan skyline polygons). A (vertical) *histogram* is an orthogonal polygon with a horizontal edge (the base) equal in length to the sum of the lengths of all the other horizontal edges. An interesting property of histograms is that the m-watchmen and the m-vision span (along a shortest watchman route) problems have a common optimum solution. Histograms are the most complex of the restricted polygons that have been studied in the literature to have this property.

In the next section we show that finding the best set of vision points on a path is NP-hard for simple polygons. In Section 2 we show that standard minimum guard covers for histograms can be found in linear time. In Section 4 we develop the multiple watchmen routes idea on histograms and prove that an optimum set of watchmen lies on the base of the histogram. This enables us to solve the m-watchmen problem with dynamic programming in $\Theta(n^3 + n^2m^2)$ time.

2 Minimum Guard Covers on Paths

In this section we study the problem of positioning stationary guards in simple polygons. We look at guards that are restricted to lie on a given curve, the *watchman route*. These guarding points are called the *vision points* of the route, and we prove that, in general, it is NP-hard to find the minimum number of vision points on a shortest watchman route.

DEFINITION 2.1 A *vision point* is a point along a given path in which the watchman will engage in visual data acquisition activity.

We take the *watchman route problem* to be the problem where a starting point on the boundary is specified. We refer to the watchman route problem when no starting point is specified as the *floating watchman route problem*.

THEOREM 1 *Finding the minimum number of vision points along a shortest floating watchman route in a simple polygon is NP-hard.*

PROOF: We modify the proof of Lee and Lin [LL86] and Aggarwal [Agg84], that finding a minimum guard cover for a simple polygon is NP-hard.

Their proof specifies a number of distinguished points so that a minimum guard cover uses a subset of these as guard locations if and only if the corresponding 3-SAT instance is satisfiable. We add a number of caves to the polygon in the construction so that the shortest watchman route will be forced to visit the distinguished points. Hence, the vision points for the caves together with the subset of distinguished points form a smallest set of vision points for the polygon. □

In the following we will look closer at a special class of polygons for which the vision point problem and the minimum guard cover problem are equivalent in the sense that the minimum guard cover lies on one (of many possible) shortest watchman routes.

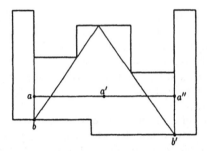

Figure 1: Optimum cover (guards b and b') is not on any shortest watchman route.
(e.g., guards a, a', and a'')

3   Minimum Guard Covers of Histograms

The problem of finding minimum guard covers (or equivalently minimum covers of starshaped polygons) for simple polygons has been proven NP-hard by Lee and Lin [LL86] and Aggarwal [Agg84]. The problem is still open for rectilinear polygons but is NP-hard when minimum covers with convex rectilinear polygons (rectangles) in rectilinear polygons are considered; see Culberson and Reckhow [CR88].

These results indicate that the problem of finding best guard covers is NP-hard even for rectilinear polygons. Therefore one may ask the following question: For what classes of polygons are there efficient solutions to the optimum guard covering problem? For example, for starshaped polygons there is a linear time algorithm to find the kernel of the polygon; see [PS85, pp 291–298]. For spiral polygons an algorithm exists which runs in linear time; see [NW90].

In this section we show that for histogram polygons there is a linear time algorithm to find the best guard set. We begin the section by defining this class of polygons.

DEFINITION 3.1 A *histogram* is a simple rectilinear polygon that is monotone with respect to the horizontal axis, and such that one of the monotone chains forming its boundary is a single straight line segment.

We will henceforth assume that the lower monotone chain is a straight line segment; see Figure 2, and call this segment the *base* of the histogram. Edelsbrunner *et al* [EOW84] define a vertical histogram as an orthogonal polygon with a horizontal edge (the base) equal in length to the sum of the lengths of the other horizontal edges.

Note that the monotone chain opposite of the base consists of a sequence of pairs of staircase patterns, one up staircase followed by a down staircase. We call the part of the polygon under such a pair a *pyramid*. Sack [Sac84] defines a pyramid to be horizontally convex vertical histogram. (Horizontal convexity is the same as monotonicity along a vertical line.)

The two staircases are connected by the *top edge*; see Figure 2. We will also use the term *west edge* to denote a vertical edge that has the interior to the right of the edge. An *east edge* is defined similarly as a vertical edge having the interior to the left. Note that a pyramid is a starshaped polygon with the kernel under the top edge.

We define the *supporting line* of a point s, lying on the base, and a pyramid Δ to be the line intersecting s, vertices of one staircase of Δ, and either the top edge or the other staircase. Note that the supporting line divides the pyramid into one section which is visible from s and another section where there may exist points that are not seen from s since the supporting line passes through a vertex. Should s lie below the top edge of the pyramid, we define the supporting line as the vertical line passing through s.

First, we prove some simple properties of histograms.

LEMMA 3.1 *For any histogram, a shortest floating watchman route and a minimum guard cover is positioned on the base of the histogram between the leftmost east edge and the rightmost west edge.*

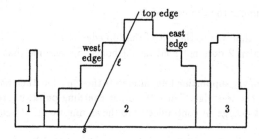

Figure 2: Histogram with three pyramids. ℓ is the supporting line of point s with respect to pyramid 2.

PROOF: To see that a shortest floating watchman route lies on the base between the two edges, note that in order for the watchman to see both the leftmost and the rightmost pyramid it must intersect the kernels of these pyramids, i.e., cross the extensions of the leftmost east edge and the rightmost west edge. Also, since the two extended edges are parallel, all horizontal segments between the two extended edges have the same length, thus we can pick the segment along the base to be our shortest route.

To see that a minimum guard cover lies on the base, take any guard cover and project the guards orthogonally onto the base. (This can be done since the polygon is monotone.) Take a point x visible from the original guard g. Let g' be the projection of the guard g. Project the point x onto the base giving point x'. The four points g', g, x, and x' define a convex area (trapezoidal) and, hence, x is seen by g'.

Any guard to the left of the leftmost east edge or to the right of the rightmost west edge can be moved, without affecting visibility, onto the intersection points of the extensions of the two edges and the base since they are still in the kernels of the "outermost" pyramids. □

Since an optimal set of guards lies on a shortest floating watchman route, the guarding points are also vision points on the route.

The following lemma provides the main basis for our algorithm and will be used extensively in the rest of the paper.

LEMMA 3.2 *Let s and t be two points on the base of a histogram, s being to the right of t, then s sees at least as much of the part of the polygon being to the right of s as t does.*

PROOF: Take a point x interior to the polygon and to the right of s, such that x is seen by t. We must prove that s also sees x. Project x orthogonally onto the base giving point x'. the triangle having corners at t, x, and x' contains s and is convex. Hence, s sees x. □

We state an algorithm to find a minimum guard cover of a histogram. The algorithm uses a greedy approach, walking from left to right. The first guard is placed on the base under the leftmost east edge. The algorithm then tries to place the next guard as far to the right of the previous one as possible.

We define the function $partner(s, \Delta)$ of a point s on the base and a pyramid Δ to be the furthest point from s, on the base, where s and the partner collectively see all of Δ. If s already sees all of Δ we let the partner be the right end point of the base.

The following lemma shows how to compute the partner of a given point with respect to some pyramid.

LEMMA 3.3 *Let s be a guarding point not under the top edge of a pyramid Δ. One of the following holds*

1. *The supporting line of $partner(s, \Delta)$ (w.r.t. Δ) intersects the supporting line of s on the top edge of Δ.*

2. $partner(s, \triangle)$ lies under the top edge of $\triangle$.

PROOF: Assume w.l.o.g. that s is to the left of the top edge of $\triangle$.

We need to prove that if 2 does not hold then 1 holds. So, assume that $partner(s, \triangle)$ does not lie under the top edge.

Suppose at most one of the supporting lines intersects the top edge, then one of s and $partner(s, \triangle)$ does not see any part of the top edge. Thus, the other guarding point, in order to see all of the top edge, must lie below the top edge. A contradiction to the assumption that neither s nor $partner(s, \triangle)$ lies below the top edge.

Suppose the supporting lines both intersect the top edge but at different points, and the two lines intersect outside the polygon, then the segment of the top edge between the intersection points is visible from both s and the partner so we can move the partner and increase the distance to s. A contradiction to the assumption that the partner is the furthest point from s retaining visibility.[1]

Suppose finally that the supporting lines intersect the top edge and each other but inside the histogram, then the segment of the top edge between the intersection points is not seen from any of the points. A contradiction to the visibility assumption. □

Given a point s on the base and a pyramid $\triangle$ the supporting line can be easily computed. First determine whether s lies to the left or right of the top edge of $\triangle$. Assume it lies to the left of $\triangle$, then follow the left staircase down, maintaining the interior line through s and the vertex while remembering the line having minimum slope.

The point $partner(s, \triangle)$ can be computed by taking the intersection point of the supporting line of s and the top edge of $\triangle$, then by following the right staircase down, maintaining the interior line through the intersection point and the vertex of the staircase. The line having maximum slope is the supporting line of the partner, and hence the partner is the intersection of this supporting line with the base. If the supporting line of s does not intersect the top edge of $\triangle$, the partner is the point on the base under leftmost east edge of $\triangle$.

If s lies to the right of $\triangle$, a similar scheme can be used to compute the supporting line and the partner.

We define $rlp(s)$, the *limit point to the right of* s to be the closest partner to the right of s. More formally

$$rlp(s) = \min_{\triangle_i \succeq s} (partner(s, \triangle_i)),$$

where the value of a point is its distance from the leftmost edge of the histogram; $\triangle_i$, $1 \leq i \leq k$, are the pyramids of the histogram; and $\succeq$ means "not to the left of". Similarly we define the limit point to the left of s, denoted $llp(s)$.

Next we show that the limit point to the right of a given point s is always the furthest point on the base which retains visibility between the two points.

LEMMA 3.4 *Let s be a point on the base of a histogram. There does not exist any point to the right of $rlp(s)$, on the base, which together with s sees all of the top monotone chain between the two points.*

PROOF: By the definition of limit point we know that $rlp(s)$ is the partner of s with respect to some pyramid $\triangle$. Thus, for any point v to the right of $rlp(s)$ the supporting lines of s and v, with respect to $\triangle$, will intersect the boundary of the polygon at distinct points and the portion of the boundary between the intersection points is neither visible from s nor from v. □

The right limit point of s can be computed by scanning the pyramids having their top edges to the right of s. This is done from left to right. For each pyramid the partner of s is computed and the leftmost partner found so far is remembered. The algorithm stops when the top edge of the next pyramid to be scanned is to the right of the leftmost partner found

This gives us the following algorithm for the minimum cover.

---

[1] By retaining visibility we mean that all of the histogram between a point and its limit point is visible from the two points.

Algorithm	*Histogram-Minimum-Cover(P)*
Step 1	Divide P into its pyramids, $\triangle_1, \triangle_2, \ldots, \triangle_k$
Step 2	Place the guard g_1 at the intersection of the base and
	the extension of the leftmost east edge of P
Step 3	**while** not all pyramids are guarded /* g_i was the last guard to be placed */ **do**
Step 3.1	$g_{i+1} := rlp(g_i)$
Step 3.2	$i := i + 1$
	endwhile
end	*Histogram-Minimum-Cover*

Suppose we have two sets of guards that each cover a histogram. Index the guards in the two sets from left to right as $\mathcal{G} = \{g_1, g_2, \ldots\}$ and $\mathcal{F} = \{f_1, f_2, \ldots\}$. Define the $<$ relation between two guards as

$$a < b \text{ if and only if } \begin{cases} \text{guard } a \text{ is to the left of guard } b \\ a \text{ and } b \text{ coincide and } a \text{ has smaller index} \end{cases}$$

THEOREM 2 *The algorithm Histogram-Minimum-Cover computes an optimum guard cover for a histogram polygon in linear time.*

PROOF: The subdivision into pyramids takes linear time since the histogram is monotone. Each pyramid is then scanned only once when the partner of the previously placed guard is computed. Thus, the algorithm above uses $\Theta(n)$ time.

To see that there does not exist a better guard cover, assume the contradiction. Suppose there is a guard cover having fewer guards. Let $\mathcal{G}$ be the solution obtained by our algorithm and $\mathcal{F}$ the solution, having less guards, that most closely matches $\mathcal{G}$ (going left to right). Find the first position where the solutions differ, i.e., $g_i = f_i$ but $g_{i+1} \neq f_{i+1}$. If $f_{i+1} < g_{i+1}$, then f_{i+1} can be moved to the right limit point of f_i, $rlp(f_i) = rlp(g_i) = g_{i+1}$, without affecting visibility. This contradicts our assumption that $\mathcal{F}$ matched $\mathcal{G}$ the best. On the other hand, $g_{i+1} < f_{i+1}$ is not possible because that implies that $\mathcal{F}$ is not a guard cover since $g_{i+1} = rlp(g_i)$, and by Lemma 3.4 this is the rightmost point which together with g_i sees all of the top monotone chain between the two guards.   □

4 Multiple Watchmen Routes

The problem of determining visibility properties using moving guards (watchmen) has received significantly less attention than the equivalent problems with static guards. However there are some results in this, harder, setting. If a watchman is allowed to "patrol" a line segment in the polygon it can be shown that $\lfloor n/4 \rfloor$ watchmen are necessary and sufficient to guard the polygon; see [ORo87].

In this section we show that the problem of computing the m shortest watchmen routes is NP-hard. For histogram polygons we are able to show that the problem is in P, and give a $\Theta(n^3 + n^2m^2)$ time algorithm based on dynamic programming. The algorithm computes $\Theta(n^2m)$ candidate routes and chooses the m routes that give the shortest total solution. It should be noted that, although histograms are polygons with a very simple structure, the watchmen routes problem becomes very hard even for this "simple" class of polygons.

DEFINITION 4.1 The m-*watchmen routes* are a set of m floating routes which together sees a polygon. We define the length of the m-watchmen routes as the sum of the lengths of the routes. The m-*watchmen routes problem* is the problem of finding the m-watchmen routes with shortest total length.

We prove that the m-watchmen routes problem is NP-hard.

THEOREM 3 *The m-watchmen routes problem in simple polygons is NP-hard.*

PROOF: It follows from the intractability of finding the minimum guard cover for a simple polygon; see [Agg84,LL86]. Suppose there exists an algorithm to compute the best m watchmen routes in polynomial time, then by running it for $m = 1, 2, \ldots, \lfloor n/3 \rfloor$ and, for each m, check whether the resulting routes have total length 0, we have a polynomial time algorithm for the minimum guard cover problem. A contradiction.   □

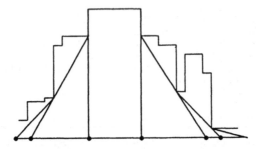

Figure 3: Pyramid with extended concavity chains and primary event points.

4.1 Multiple Watchmen Routes in Histograms

In this section we will look at how to compute the best set of routes, for any m moving watchmen in a histogram. It is interesting to note that this problem, although solvable in polynomial time, is much harder than finding the best set of static guards. To see this intuitively, consider the static guard problem of the previous section. Each guard is allowed to move only along the base, which gives each guard one degree of freedom. We will show that the moving watchmen also lie on the base but each watchman still has two degrees of freedom since both of the end points can move.

We sketch the algorithm we are going to apply. Instead of concerning ourselves with the watchmen routes (which lie on the base) we will treat the problem of computing the gaps between each pair of consecutive watchmen, i.e., compute the $m-1$ longest gaps such that the visibility requirement are not violated. The first thing to do is to define a bounded set of possible gaps between watchmen that provably always contain the solution. We then apply dynamic programming to choose choose the $m-1$ gaps with longest total length. The main problem lies in proving that a particular set of gaps always contains the solution.

We first prove that the optimum set of watchmen routes always lies on the base.

LEMMA 4.1 *There is an optimum solution in which all of the m floating routes are disjoint spans along the single watchman route on the base of the histogram.*

PROOF: Take any solution and project it orthogonally onto the base. Let point x be visible from some original watchman w. We need to show that the projected watchman w' also sees x. Since w sees x there is a point s_w on w for which the segment between x and s_w is interior to the polygon. By Lemma 3.1 the projected point s'_w also sees x and since w' is the projected watchman s'_w lies on w'.

The two end points of the single watchman route see the two extremal pyramids, so any watchman of the solution which lies outside the route can be moved to one of the end points without affecting visibility. Also, this can never increase the total length of the solution. □

We will henceforth refer to the single shortest watchman route on the base as the W-route.

It is easy to see that finding the $m-1$ gaps with the largest total length, such that the visibility requirement is fulfilled, is equivalent to the m-watchmen routes problem.

In order to make the problem tractable we wish to discretize it and find a bounded number of points on the base which can be end points of the watchmen routes. To find these points we begin by making the following definition.

DEFINITION 4.2 The *concavity chains* of the W-route and a pyramid are two reflex chains, one starting at the left end point of the W-route and ending at the left end point of the top edge of the pyramid and the other starting and ending at the two corresponding right end points, such that each chain is the shortest path between the end points; see Figure 3.

The *primary event points* of a pyramid $\triangle$ are the intersection points of the W-route and the extended links of the concavity chain that also intersect the top edge of $\triangle$; see Figure 3.

The *secondary event points* are all points s on the W-route such that $partner(s, \triangle_i) = partner(s, \triangle_j)$, for any two pyramids $\triangle_i$ and $\triangle_j$; see Figure 4.

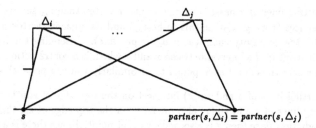

Figure 4: Secondary event points define local maxima of the distance function between guarding point and partner.

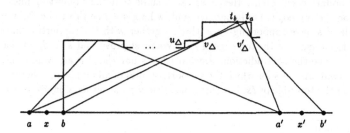

Figure 5: Gap $[b, b']$ is better than any gap starting between event points a and b. The points a', b', and x' are limit points to points a, b, and x, respectively.

In the following we denote the primary and secondary event points simply as event points when we don't care of what type they are.

The concavity chains are easily computed by scanning the top chain to right given the left end point of the $\mathcal{W}$-route, and we claim without proof that there can be $O(n^2)$ event points, that there exists examples of histograms with $\Omega(n^2)$ event points, and that the event points can all be computed in $O(n^2)$ time.

Each event point e has corresponding limit points $rlp(e)$ and $llp(e)$, as defined in Section 3, which define maximum visibility gaps from the event point, i.e., two watchmen positioned on the $\mathcal{W}$-route but stopping at the event point and one of the limit points cover the whole polygon and the limit point is also the furthest such point from the event point; see Lemma 3.4. The limit points can be computed using the technique given in the algorithm for the minimum stationary posting in Section 3.

The next Lemma shows that these gaps are better than any other set of gaps.

LEMMA 4.2 *Let $[x, rlp(x)]$ be a gap on the base of a histogram, such that x does not coincide with an event point or a limit point. There exists a gap, $[y, y']$, with y either slightly to the left of x or to the right of x, such that the gap $[y, y']$ is at least as long as the gap $[x, rlp(x)]$.*

PROOF: Let $x' = rlp(x)$. By the definition of right limit point we know that x' is the partner of x with respect to some pyramid Δ. Hence, by Lemma 3.3, the supporting lines of x and x' intersect on the top edge of Δ, and pass through the vertices v_Δ and v'_Δ of Δ. Let a be a point such that it is either an event point with $a' = rlp(a)$ or a left limit point with $a = llp(a')$ and a being the closest such point to the left of x. Similarly let b be the closest such point to the right of x.

Take a point z in the interval $[a, b]$. We need to ensure that z and $rlp(z)$ have supporting lines with respect to Δ which intersect on the top edge of Δ and also pass through the vertices v_Δ and v'_Δ of Δ.

Suppose this is not true. Either supporting lines of z and $rlp(z)$ intersect at the top of another pyramid or the lines pass through different vertices than v_Δ and v'_Δ of Δ.

- (If the supporting lines of z passes through an other vertex than v_Δ say u_Δ.) Take the upper of the two vertices (w.l.o.g. assume this to be v_Δ) and its adjacent vertex w lying in between v_Δ and u_Δ on the concavity chain ($w = u_\Delta$ is possible). Extending the link $[v_\Delta, w]$ gives a primary event point in the interval between x and z which is a contradiction to the assumption that a and b where the closest such points to x. Similarly you can prove this for $rlp(z)$.

- (If the supporting lines of z and $rlp(z)$ intersect on the top edge of another pyramid Δ_+). If the supporting lines of z and $rlp(z)$ also intersect on the top edge of Δ, then we immediately have a contradiction since then z is a secondary event point. Hence there is a point s between x and z where the intersection point of the supporting lines switches from pyramid Δ to pyramid Δ_+. (We can assume that there is no point between x and z for which the corresponding supporting lines intersect on the top edge of a third pyramid.) If the supporting lines of s and $rlp(s)$ intersect on the top edges of both Δ and Δ_+, we have again a contradiction since s is then a secondary event point. Hence we can assume that the supporting lines of s and $rlp(s)$ intersect on the top edge of only one pyramid, w.l.o.g. assume this to be Δ, whereas all points to one side of s have supporting lines which together with the supporting lines of their right limit points intersect on the top edge of Δ_+. We must prove that s or $rlp(s)$ is a primary event point, so assume the contradiction. Since neither s nor $rlp(s)$ is an event point there exists a neighbourhood around s for which the supporting lines of all points and their right limit points intersect on the top edge of Δ, but this is not true for s so either s or $rlp(s)$ is a primary event point.

This proves our claim.

Event points mark the points where supporting lines move from one vertex to another (see Figure 5, where supporting lines to the left of a intersect vertex u_Δ and supporting lines to the right of a intersect vertex v_Δ).

Let t_p be the point of intersection of the top edge of Δ and the supporting line of point p on the base. Now, since the triangles with corners at a, x, v_Δ and t_a, t_x, v_Δ, as well as triangles with corners at a', x', v'_Δ and t_a, t_x, v'_Δ, are similar, we have

$$\frac{|[a,x]|}{|[t_a,t_x]|} = \frac{|[a,v_\Delta]|}{|[t_a,v_\Delta]|} \quad \text{and} \quad \frac{|[a',x']|}{|[t_a,t_x]|} = \frac{|[a',v'_\Delta]|}{|[t_a,v'_\Delta]|}$$

and, hence,

$$\frac{|[a,x]|}{|[a',x']|} = \frac{|[a,v_\Delta]|}{|[a',v'_\Delta]|} \cdot \frac{|[t_a,v'_\Delta]|}{|[t_a,v_\Delta]|} = c, \text{ a constant.}$$

This gives us that the length of a gap $[y, rlp(y)]$, with $y \in [a,b]$ is

$$|[y, rlp(y)]| = |[x + \delta, rlp(x + \delta)]| = |[x, rlp(x)]| + (c-1)\delta.^2$$

When $c \le 1$ we can chose $\delta < 0$, i.e., y is to the left of x. Similarly, when $c > 1$ we chose $\delta > 0$ giving y to the right of x. In both cases $[y, rlp(y)]$ is at least as long as $[x, rlp(x)]$. $\qquad\square$

The constant c obtained in the proof we call the *sliding constant of gap* $g = [x, rlp(x)]$ or in short notation c_g.

One might think that the gaps defined by the event points and the limit points, in this way, would be a sufficient set to define the best m watchmen routes. However, this is not the case. The reason for this is that two gaps in the set might overlap, and by placing a static guard somewhere on the overlap the resulting gap (containing the static guard) could be larger than the combined original gaps.

We prove that it is always preferable to place the static guard at an end point of such an overlap, and thus for each gap it is only necessary to compute new gaps defined by $m - 2$ static guards to the left or right of any initially given gap.

We define the *propagation points* to be a sequence of points $p_0, p_1, \ldots, p_{m-1}$ such that p_0 is an event point and p_i is the limit point of p_{i-1}. Such a sequence of points define gaps with static guards

---

^{2}With a slight abuse of notation.

at the end points. For each event point we need to consider only $m - 1$ propagation points since we have at most m watchmen in total.

Each event point corresponds to two sequences of propagation points, one propagating to the left and one propagating to the right. This gives us $2(m-1) \cdot O(n^2) = O(n^2 m)$ gaps in total. We let the set of gaps defined in this way be G and prove that the $m - 1$ gaps giving an optimal solution are in this set, G.

LEMMA 4.3 *The set G contains $m - 1$ non-overlapping gaps that have total length no smaller than an optimum set of gaps.*

PROOF: We make a proof by contradiction and compare a solution contained in G with an allegedly optimal solution with at least one gap not in G.

Let $\mathcal{F} = (f_1, f_2, \ldots, f_{m-1})$ be an optimal solution where $f_1, f_2, \ldots, f_{m-1}$ are gaps ordered from left to right. We will show that there is a solution no smaller than the solution $\mathcal{F}$ having all its gaps in G. Let f_i be the leftmost gap of $\mathcal{F}$ not in G. Let g_j and g_{j+1} be the gaps in G have their end points closest to the end points of f_i, The left end point of g_j being to the left of the left end point of f_i and the left end point of g_{j+1} being to the right of the left end point of f_i. Assume also that f_i is chosen such that it matches g_{j+1} the closest.

The following cases arise.

f_i contains g_{j+1}.

By Lemma 3.4 g_{j+1} is the largest gap retaining visibility. Gap f_i therefore violates the visibility property.

f_i is contained in g_j.

The solution $(f_1, \ldots, f_{i-1}, g_j, f_{i+1}, \ldots, f_{m-1})$ improves on $\mathcal{F}$. (Should g_j and f_{i+1} overlap a static guard can be placed anywhere in the overlap.)

f_i and g_j overlap.

If f_i does not have an end point in common with neither f_{i-1} nor f_{i+1}, then by Lemma 4.2 there is a gap f_i' with $|f_i'| = |f_i| + (c_{f_i} - 1)\delta \geq |f_i|$.

If $c_{f_i} = 1$ the size of f_i' has not increased but we can then choose $f_i' = g_{j+1}$ or join f_i' to f_{i+1} with a static guard between them (whichever occurs first). This gives a contradiction since we assumed that f_i matched g_{j+1} the best already.

It is clear that f_i and f_{i-1} do not have a static guard between them since $f_{i-1} \in G$ and then the end points of f_i are propagation points of some event points. Hence, assume that the set $\{f_i, f_{i+1}, \ldots, f_{i+\ell-1}\}$ is a maximal set of consecutive gaps with only static guarding points between them. By repeated application of Lemma 4.2 we can choose $\{f'_i, f'_{i+1}, \ldots, f'_{i+\ell-1}\}$ to be a sequence of gaps such that $|f'_i| + |f'_{i+1}| + \cdots + |f'_{i+\ell-1}| = |f_i| + |f_{i+1}| + \cdots + |f_{i+\ell-1}| + (c_{f_i} c_{f_{i+1}} \cdots c_{f_{i+\ell-1}} - 1)\delta \geq |f_i| + |f_{i+1}| + \cdots + |f_{i+\ell-1}|$ and increase the total size of the solution. If $c_{f_i} c_{f_{i+1}} \cdots c_{f_{i+\ell-1}} = 1$, then again we can choose f'_i to be further to the right than f_i and thus match g_{j+1} better than f_i does.

f_i and g_j are disjoint.

This case can never occur since the union of the gaps in G span the complete $\mathcal{W}$-route. Therefore some gap in G will intersect f_i.

$\square$

We get the following algorithm

Algorithm	m-*Watchmen-Routes*(P)
Step 1	Divide P into its pyramids, $\Delta_1, \Delta_2, \ldots, \Delta_k$
Step 2	Compute the event points for pyramids Δ_1 to Δ_k
Step 3	for each event point do
Step 3.1	Compute the $m - 1$ gaps propagating to the right from the event point endfor
Step 4	Redo step 3 for the gaps propagating to the left
Step 5	Sort the gap end points on x-coordinate
Step 6	Apply dynamic programming to choose among the set of gaps
end m-*Watchmen-Routes*	

Let s be the right end point of some gap. If t is the left end point of the gap we call t the *mate* of s.

In Steps 3 and 4 we take care of the possibility of a sequence of propagation points. These points are computed using the same technique as in Section 3. The dynamic programming scheme will scan the gap end points from left to right which is why these points are sorted in Step 5. In Step 6 we assume a matrix M of size $m \times O(n^2 m)$. Item $M[i, j]$ contains the best solution of i watchmen from the left initial point (the left end point of the $\mathcal{W}$-route) to the jth gap end point. The best solution to item $M[i, j]$ is achieved by taking the best of two items. $M[i - 1, k]$, where $k < j$ and k is the index of the mate of gap end point having index j (Corresponds to placing a static guard at gap end point with index j). The other item is $M[i, j - 1]$ + length between gap end points having indices $j - 1$ and j (Corresponds to a watchman walking the distance between the $j - 1^{\text{st}}$ and the jth gap end points).

LEMMA 4.4 *The algorithm runs in* $\Theta(n^3 + n^2 m^2)$ *time and uses* $\Theta(n^2 m^2)$ *storage.*

PROOF: Steps 1 and 2 take $O(n^2)$ time by the discussion following Definition 4.2. Steps 3 and 4 can take up to $O(n^3)$ time since one scan of the polygon is performed to compute each sequence of propagation points and there are $O(n^2)$ of these. Sorting of the $O(n^2 m)$ gap end points takes $O(n^2 m \log n)$ time in Step 5. By the previous discussion Step 6 takes $O(n^2 m^2)$ time since the matrix has at most $n^2 m^2$ entries and each entry can be computed in constant time. This gives a time complexity of $O(n^2(n + m^2 + m \log n)) = O(n^2(n + m^2))$.

The storage bound follows from the fact that the matrix has $m \times O(n^2 m)$ entries.

Again, the example, having $\Omega(n^2)$ secondary event points, which we have already mentioned uses $\Omega(n^3 + n^2 m^2)$ time and $\Omega(n^2 m^2)$ storage which shows the lower bound for our algorithm. $\square$

Lemmas 4.3 and 4.4 give the following theorem.

THEOREM 4 *The* m-*Watchmen-Routes algorithm computes an optimum set of* m *watchmen routes in* $\Theta(n^3 + n^2 m^2)$ *time using* $\Theta(n^2 m^2)$ *storage.*

5 Conclusions

We have presented a number of results in this paper. The first is that to compute the restriction of a minimal guard set on a watchman route is still NP-hard. We have also presented a linear time algorithm to compute the minimal guard set in a histogram (Manhattan skyline) polygon, which is also a minimal set of vision points on the watchman route along the base of the histogram.

In the last section we have shown an algorithm to compute the best set of m watchmen in histogram polygons. Our algorithm runs in $\Theta(n^3 + n^2m^2)$ time and uses $\Theta(n^2m^2)$ storage.

Other problems related to the problem of finding the best watchmen for any value of watchmen (which is NP-hard in the general case) is that of finding best watchmen for a fixed number of watchmen, e.g., does there exist a polynomial time algorithm to compute the best 2-watchmen routes for simple polygons? Even for the single watchman route case the problem has not been completely answered, although, as stated in the introduction, a $\Theta(n)$ time algorithm exists for simple rectilinear polygons.

References

[Agg84]   A. AGGARWAL. *The Art Gallery Theorem: Its variations, applications and algorithmic aspects.* PhD thesis, Johns Hopkins University, 1984.

[CN87]   W. CHIN, S. NTAFOS. *Shortest Watchman Routes in Simple Polygons.* Technical Report, Dept. of Computer Science, Univ. of Texas at Dallas, 1987.

[CN88]   W. CHIN, S. NTAFOS. Optimum Watchman Routes. *Information Processing Letters*, 28:39–44, 1988.

[CR88]   J.C. CULBERSON, R.A. RECKHOW. Covering Polygons is Hard. In *Proc. 29th Symposium on Foundations of Computer Science*, pages 601–611, 1988.

[Cha90]   B. CHAZELLE. Triangulating a Simple Polygon in Linear Time. In *Proc. 31st Symposium on Foundations of Computer Science*, pages 220–230, 1990.

[Chv75]   V. CHVÀTAL. A Combinatorial Theorem in Plane Geometry. *Journal of Combinatorial Theory B*, 13(6):395–398, 1975.

[EOW84]   H. EDELSBRUNNER, J. O'ROURKE, E. WELZL. Stationing Guards in Rectilinear Art Galleries. *Comput. Vision, Graphics, Image Processing*, 27:167–176, 1984.

[Fis78]   S. FISK. A Short Proof of Chvàtal's Watchman Theorem. *Journal of Combinatorial Theory B*, 24:374, 1978.

[LL86]   D.T. LEE, A.K. LIN. Computational Complexity of Art Gallery Problems. *IEEE Transactions on Information Theory*, IT-32:276–282, 1986.

[NW90]   B.J. NILSSON, D. WOOD. *Watchmen Routes in Spiral Polygons.* Technical Report LU-CS-TR:90-55, Dept. of Computer Science, Lund University, 1990. An extended abstract of preliminary version was presented at the *2nd Canadian Conference on Computational Geometry*, pages 269–272.

[ORo87]   J. O'ROURKE. *Art Gallery Theorems and Algorithms.* Oxford University Press, 1987.

[PS85]   F.P. PREPARATA, M.I. SHAMOS. *Computational Geometry — an Introduction.* Springer Verlag, 1985.

[Sac84]   J.-R. SACK. *Rectilinear Computational Geometry.* PhD thesis, School of Computer Science, McGill University, 1984.

Applications of a New Space Partitioning Technique*

Pankaj K. Agarwal[†] and Micha Sharir[‡]

Abstract

We present several applications of a recent space partitioning technique of Chazelle, Sharir and Welzl [8]. Our results include efficient algorithms for output-sensitive hidden surface removal, for ray shooting in two and three dimensions, and for constructing spanning trees with low stabbing number.

1 Introduction

In a recent paper, Chazelle, Sharir and Welzl [8] have given a new quasi-optimal technique for simplex range searching in any dimension. The technique is based on a new hierarchical space partitioning scheme, which we briefly review below.

As it turns out, this partitioning scheme has several useful properties that make it applicable to a variety of other problems. Briefly, these properties are: (i) it yields quasi-optimal query time, which matches Chazelle's lower bound up to a factor of $O(n^\epsilon)$, for arbitrarily small $\epsilon > 0$; (ii) it applies in all dimensions; (iii) it admits trade-off between storage and query time; (iv) its preprocessing cost is low — close to the storage used; (v) it facilitates the use of multi-level data structures; and (vi) it can be efficiently dynamized. Properties (i)–(iv) are noted in [8]; property (v) is used there, but not to full generality; property (vi) is newly developed in this paper.

In this paper we study several of these applications and derive efficient solutions for them using the "CSW-scheme". The results that we obtain are:

1. **Efficient construction of a spanning tree with low stabbing number:** Given a set S of n points in $\mathbf{R}^d$, a (straight-edge) spanning tree T on S is said to have stabbing number κ if no hyperplane crosses more than κ edges of T. It has been shown in [9] that there always exists a spanning tree on S with stabbing number $\kappa = O(n^{1-1/d})$. Such spanning trees have been applied in simplex range searching [25, 9], computing a single face in arrangements of lines [13], and ray shooting in a

---

*Work on this paper by the second author has been supported by Office of Naval Research Grants N00014-89-J-3042 and N00014-90-J-1284, by National Science Foundation Grant CCR-89-01484, and by grants from the U.S.-Israeli Binational Science Foundation, the Fund for Basic Research administered by the Israeli Academy of Sciences, and the G.I.F., the German-Israeli Foundation for Scientific Research and Development. Both authors wish to acknowledge the support of DIMACS, an NSF Science and Technology Center, under grant STC-88-09684.

[†]Computer Science Department, Duke University, Durham, NC 27706

[‡]School of Mathematical Sciences, Tel Aviv University, Tel Aviv, Israel and Courant Institute of Mathematical Sciences, New York University, New York, NY 10012

collection of segments [2]. In the planar case the fastest (deterministic) algorithm for the construction of such a tree runs in time $O(n^{3/2} \log^2 n)$ [17] (see also [1, 9, 25, 15, 17]). No efficient algorithm is known for computing spanning trees of low stabbing number in higher dimensions. Using the CSW-scheme, we derive an improved (deterministic) algorithm that computes a spanning tree with stabbing number $O(n^{1-1/d+\epsilon})$ in time $O(n \log n)$ for $d = 2, 3$, and for any $\epsilon > 0$.[1] Thus, although the stabbing number that we get is slightly suboptimal, the time needed to construct the spanning tree is greatly reduced. Moreover, our algorithm allows to update the tree in amortized time $O(\log^2 n)$, as we insert or delete a point; the previously best known algorithm required roughly $\sqrt{n}$ amortized time to update the spanning tree [10].

2. **Ray shooting amidst segments in the plane:** In this basic problem we wish to preprocess such a collection into a data structure that supports fast *ray shooting* queries, in which we need to determine the first segment, if any, to be hit by a query ray. Even if the n given segments are intersecting, our solution requires $O(n \log^3 n)$ preprocessing time and storage, and answers queries in time $O(n^{1/2+\epsilon})$, for any $\epsilon > 0$. Solutions that match (and even slightly improve) the storage and query time are known (see for instance Agarwal [2], Bar Yehuda and Fogel [5], and Cheng and Janardan [10]), but a preprocessing as fast as ours appears to be new — previous bounds on the preprocessing cost are close to $O(n^{3/2})$. We can also obtain a whole range of resource bounds, in which we trade off storage (and preprocessing) for query time, as in [8]. Another advantage of our algorithm is that, unlike previous algorithms, we can efficiently maintain the structure dynamically as we insert or delete segments.

3. **Improved output-sensitive hidden surface removal:** Given n horizontal triangles in $\mathbf{R}^3$, we can compute their visibility map, as viewed from a point at $z = -\infty$, in time $O(n^{2/3-\epsilon} K^{1/3+\epsilon} \zeta^{1/3+\epsilon} + n^{1+\epsilon} + K\zeta^\epsilon)$, for any $\epsilon > 0$, where K is the number of vertices of that map and $\zeta \leq K$ is the maximum *contour* size (see Section 5 for details). It is based on our dynamic version of the ray shooting technique.

4. **Ray-shooting in 3-space:** Here we are given a collection of n (non-intersecting) triangles in $\mathbf{R}^3$ and wish to preprocess it into a data structure that supports fast ray shooting queries, defined in complete analogy with the planar case. This is a central problem in computer graphics, and, unfortunately, is considerably more difficult than its planar counterpart. Pellegrini has considered some related problems, but his approach fails to give an efficient algorithm for the ray shooting problem [20]. We obtain a solution that requires $O(n^{4/5+\epsilon})$ query time with $O(n^{1+\epsilon})$ storage, for any $\epsilon > 0$. Using the space/query-time tradeoff property of CSW-scheme and a recent result of Aronov and Sharir [4], we can answer a query in time $O(n^{16/15+\epsilon}/n^{4/15+\epsilon})$, for any $\epsilon > 0$. Recently and independently, de Berg et al. [6] have obtained similar results using a somewhat different technique. We also note that other solutions have been given earlier for several special cases of the general problem. For example, Schmitt et al. [21] discuss the case of axis parallel rectangles. Agarwal [1] (see also [18]) discusses the case of vertical ray shooting.

---

[1] In these complexity bounds the constants of proportionality depend on ϵ and generally increase to infinity as $\epsilon \downarrow 0$. This convention applies throughout the paper.

The paper is organized as follows. In Section 2 we will briefly review the CSW partitioning scheme and explain how to dynamize it. Section 3 describes an efficient construction of spanning trees with low stabbing number, based on the CSW-scheme. Section 4 presents the application of the CSW-scheme to planar ray shooting amidst a collection of line segments. This in turn is applied in Section 5 to obtain an improved output-sensitive hidden surface removal algorithm. Section 6 presents applications of the scheme to ray shooting in three dimensions and related problems.

2 The CSW Partitioning Scheme — An Overview

In this section we briefly review the space partitioning scheme of Chazelle, Sharir and Welzl [8]. This scheme produces a multi-level data structure, but we will describe here only its primary structure, which is rather similar to a geometric partition tree. In the following sections we will exploit this multi-level characteristic of the data structure, by attaching to its nodes substructures that will generally be different from those used in [8]. Since we now restrict our description to the primary structure, we obtain a structure that, without further enhancements, supports only efficient halfspace queries.

The structure has two types of nodes, 'simplex nodes' and 'triangulation nodes'. Each simplex node v of the structure is associated with a set S_v of n_v points in (some simplex in) d-space. We fix some sufficiently large constant integer parameter r. We start at the root node of the structure with the entire set S. Using random sampling, or alternative deterministic techniques, we construct $O(\log r)$ different triangulations of d-space, each consisting of $O(r^d)$ simplices, so that the following property holds: For any hyperplane h there exists at least one triangulation such that h crosses only $O(r^{d-1})$ simplices of the triangulation and that only $O(\frac{n}{r} \log r)$ points of S lie in those simplices. We say that such a triangulation is *sparse* for h. We now create $O(\log r)$ children of the root; they are triangulation nodes, and each corresponds to one of these triangulations. For each such child t, and for each simplex τ in the triangulation that t represents, we create a child of t that corresponds to τ; this is a simplex node that has the subset $\tau \cap S$ associated with it.

We now recurse with this partitioning scheme in each simplex τ of each triangulation, unless either (a) τ contains more than $c\frac{n}{r} \log r$ points of S, for some appropriate constant c, or (b) the number of points within τ is less than some parameter σ, chosen as a function of the storage s that we allow for the structure. We will refer to the structure resulting from this recursive process as the *top* part of the whole structure.

In case (a) the simplex becomes a so-called "fat leaf", and the (current level of the) structure is not expanded there any further. In case (b) we expand the structure at τ using the following different scheme. The points of S within τ are dualized to hyperplanes, and we continue the construction in the dual space. Applying Matoušek's algorithm [16], we choose a sample of r hyperplanes in the dual space and triangulate their arrangement so that each simplex intersects only $O(\frac{\sigma}{r} \log r)$ dual hyperplanes. There are $O(r^d)$ simplices in the triangulated arrangement. We associate with each resulting simplex τ the set $H(\tau)$ of hyperplanes that lie strictly above τ (and a similar set of hyperplanes that lie strictly below it). We continue recursively to process, for each simplex τ, the subset of hyperplanes crossing τ, in the same manner. The total storage (and preprocessing cost) required for this procedure is easily seen to be $O(\sigma^{d+\epsilon})$, for any $\epsilon > 0$ (see [8]). We will refer to this structures, constructed for 'leaf-simplices' of the top part, having less than σ points each, as the *bottom* part of the structure.

As explained in [8], a halfspace query, involving the halfspace lying, say above a query hyperplane h, is processed as follows. We find a sparse triangulation for h at the root of the structure. We then find all simplices of this triangulation that lie fully above h and those that are crossed by h. Simplices of the former type are fully within the query range, whereas simplices of the second type are processed further recursively. (Note that, since the triangulation is sparse for h, no simplex of the second type is a fat leaf, so the query will never be stuck at such a leaf.) If the number of points within a simplex is less than σ, we switch to the second kind of data structure. We locate the simplex τ that contains the point $h^\star$, dual to the query hyperplane h, and retrieve the associated set $H(\tau)$. We next recurse in the substructure for c. In total, we obtain a compact representation of the set of points that lie in our range, as the disjoint union of 'canonical' pre-stored subsets.

There are several issues to observe about the scheme. The first is the storage s (and preprocessing cost) needed for the structure. Actually, this can be set to any desired value between roughly linear in n and $n^{d+\epsilon}$, for any $\epsilon > 0$. The larger s is, the more "efficient" does a query become, where query efficiency is measured as the number of canonical subsets whose union constitutes the query output. Roughly speaking, to achieve storage s, we need to choose σ to be about $\left(\frac{s}{n}\right)^{\frac{1}{d-1}}$; the precise analysis is given in [8].

The next issue is the potential of the resulting structure to support multi-level substructures. We note that the output to a halfspace query is given as the disjoint union of 'canonical' subsets, where each subset is either the set of points of S within a simplex in one of the triangulations used in the top part of the structure, or is the primal version of a subset of the dual hyperplanes that pass above (or below) a simplex in the bottom part of the structure. In either case, we can take each such subset, process it further to obtain a second-level data structure, and attach it to the corresponding primary node. We note that the technique in [8] does just that, where the secondary structures are of the same kind as the primary structure; in fact, the structure produced in [8] is $d + 1$ levels deep. In the applications given in this paper, we will first apply the CSW–partitioning scheme recursively for a constant number of levels and then, at each cell, we construct an entirely different structure tailored to the specific problem that we are trying to solve. In our applications, the secondary structure will usually be on a set of different objects that stand in 1-1 correspondence with the original points of S — see below for examples.

Another issue is the "efficiency" of a query, as defined above. As argued in [8], the total number of canonical subsets that a query collects is $O(n^{1+\epsilon}/s^{\frac{1}{d}})$. Moreover, the analysis in [8] also shows that even if we perform a second-level query (of the same type as the top level query) at each of these subsets (actually going down any constant number of levels) the total number of subsets, of any level, that arise is still $O(n^{1+\epsilon}/s^{\frac{1}{d}})$ (with a larger constant of proportionality). Our multi-level structures will have a similar property, as will be observed below.

Finally, we show that we can maintain the CSW-partitioning structure dynamically as we insert or delete points. This is a new feature of the partitioning scheme, but, due to lack of space, we only give a brief description; the detailed analysis can be found in the full version of the paper [3]. Suppose we want to insert a point p to S (deletions are completely symmetric). We visit the CSW-partition tree in a top down fashion. At each node v visited, we perform one of the following two operations:

(i) Reconstruct the subtree rooted at v from scratch, including the secondary structure stored at v. Suppose m_v is the number of points in S_v when v was reconstructed last time. We perform this step again if S_v has been updated m_v/r times since that last reconstruction.

(ii) Insert p to the secondary structure stored at v, and to some of the children of v if v is not a leaf. If v is a leaf and there are no secondary structures, simply add p to S_v.

If the secondary structure is a CSW-partitioning tree, p is inserted recursively. Otherwise we assume that there is an efficient procedure for inserting p to the secondary structure. As mentioned earlier, in some applications the secondary structure is constructed on a different set of objects, in which case we add the object corresponding to p to the secondary structure, instead of p itself. This will become clearer when we consider a concrete example (e.g. see Section 4.2).

Next, if v is not a leaf, p is inserted to some of the children of v. If v stores the bottom part of the structure, we dualize p to the hyperplane p^* and descend to the children of v corresponding to the simplices (of v) that intersect p^*; there are $O(r^{d-1})$ such cells. Otherwise, for each of $O(\log r)$ triangulations stored at v, we determine the simplex τ containing the point p and descend to the child corresponding to τ; in this case p is inserted to $O(\log r)$ 'simplex children' of v.

A point can be deleted from S using the same approach. Analyzing the time spent by the above procedure, we can show (see [3] for details):

Theorem 2.1 *Given a set of n points in $\mathbf{R}^d$ and a parameter $n^{1+\epsilon} \leq s \leq n^d$, one can insert or delete a point from the CSW-partitioning structure in $O(s/n^{1-\epsilon})$ amortized time, and can answer a query in time $O(n^{1+\epsilon}/s^{1/d})$, for any $\epsilon > 0$.*

Corollary 2.2 *Simplex range searching, in any dimension d, can be performed dynamically, using quai-optimal resources: Allowing storage $n^{1+\epsilon} \leq s \leq n^d$, preprocessing cost is $O(s^{1+\epsilon})$, query time is $O(n^{1+\epsilon}/s^{1/d})$, and insertion and deletion of points costs $O(s/n^{1-\epsilon})$ amortized time per update.*

Remark 2.3: In 2 and 3 dimensions, if $\sigma = O(1)$, that is, if only the top part of the structure is being created, and we are creating a k-level structure, the preprocessing time and storage become $O(n \log^k n)$ and $O(n \log^{k-1} n)$, respectively, assuming that the kth level structure can be constructed in $O(n \log n)$ time and requires only $O(n)$ space; see the full version for details.

3 Spanning Trees with Low Stabbing Number

Let S be a set of n points in $\mathbf{R}^d$, $d = 2, 3$. In this section we will present an $O(n^{1+\epsilon})$ algorithm to compute a spanning tree of S with stabbing number $O(n^{1-1/d+\epsilon})$. As in [1, 15], the tree returned by our algorithm is actually a spanning path. For the sake of simplicity we describe the algorithm for $d = 2$; the same approach also works for $d = 3$.

The algorithm follows the approach of the CSW scheme, but constructs the tree directly instead of the data structure produced in [8]. We begin as in [8] with σ taken to be some constant. We choose some appropriate constant parameter r and construct $O(\log r)$ triangulations, of $O(r^2)$ triangles each, with the property that for each line ℓ there is at least one sparse triangulation, that is, ℓ crosses only $O(r)$ triangles of the triangulation and only $O(n/r \log r)$ points of S lie in those simplices. We now take these $O(\log r)$ triangulations, and superimpose them on each other. This yields one common convex subdivision, $\mathcal{M}$, of the plane with the property that the number of points of S in the cells crossed by any line is $O(\frac{n}{r} \log r)$. Moreover, the arguments in [8] imply that the overall complexity of $\mathcal{M}$ is $O(r^2 \log r)$, and that the number of cells of $\mathcal{M}$ crossed by a line is $O(r \log r)$ (see [3] for details).

We now construct our desired spanning path T as follows. For each cell c of $\mathcal{M}$ let $n_c = |S \cap c|$. We recurse within each c to construct a spanning path T_c of $S \cap c$ with low stabbing number. (If this set is empty there is nothing to be done, and if n_c is less than some constant, we take T_c to be any spanning path of $S \cap c$.) Now we connect all the paths T_c into the full spanning path T in any convenient manner.

Let $K(n)$ denote the maximum stabbing number that can arise in our construction for a spanning path on a set of n points. Then we have $K(O(1)) = O(1)$, and

$$K(n) \leq \sum_{i=1}^{ar \log r} K(n_i) + br^2 \log r \, ,$$

so that $\sum_i n_i \leq \frac{cn}{r} \log r$, for appropriate constants a, b, c. It is easily verified that the solution of this recurrence is $K(n) \leq A_\epsilon n^{1/2+\epsilon}$ for any $\epsilon > 0$. It is also easy to verify, that the time needed to construct T is $O(n \log n)$, for any $\epsilon > 0$. Finally, using the technique of dynamizing the CSW-scheme, we can also update the spanning tree in $O(\log^2 n)$ amortized time as we insert or delete a point; see [3] for details.

A similar technique applies in three dimensions, using another result in [8] on superimposing triangulated arrangements of planes in 3-space. We have thus shown

Theorem 3.1 *Given a set S of n points in $\mathbf{R}^d$, for $d = 2, 3$, and $\epsilon > 0$, one can construct, in time $O(n \log n)$, a spanning path T on S with stabbing number $O(n^{1-1/d+\epsilon})$. Moreover, a point can be inserted to or deleted from T in $O(\log^2 n)$ amortized time.*

Remarks 3.2: (1) The recurrence on $K(n)$ also applies to the time needed to find all the edges of T crossed by a query line or plane, hence this time is also $O(n^{1-1/d+\epsilon})$.
(2) This algorithm does not extend to $d > 3$ because we do not know how to bound the complexity of the map resulted by superimposing $O(\log r)$ triangulated arrangements of hyperplanes. This is the only step of our algorithm that does not extend to higher dimensions.
(3) If we choose $r = n^\epsilon$, the stabbing number improves to $O(n^{1-1/d}(\log n)^{O(\log \log n)})$, but the running time increases to $O(n^{1+\epsilon})$.

4 Ray Shooting in the Plane

The problem studied in this section is one of the basic problems in computational geometry. Let $\mathcal{G} = \{e_1, \ldots, e_n\}$ be a collection of n line segments in the plane. We wish to preprocess $\mathcal{G}$ so that, given any query ray ρ we can efficiently compute the first intersection, if any, of ρ with the segments in $\mathcal{G}$. In this section we obtain a solution that is efficient both in terms of preprocessing cost and of query time — it uses $O(n^{1+\epsilon})$ preprocessing time, $O(n^{1+\epsilon})$ storage, and $O(n^{1/2+\epsilon})$ query time, for any $\epsilon > 0$. As in the previous section, we lose slightly in terms of storage and query time, but gain considerably in terms of preprocessing. In addition, our technique can be dynamized efficiently.

We first present an algorithm for ray shooting amidst a collection of lines. We will need it as a subroutine to the general algorithm which is given next. We also discuss the issues of space query-time tradeoff and of dynamization.

4.1 Ray shooting among lines

In this subsection we assume that the segments of $\mathcal{G}$ are full lines. Dualize each line of $\mathcal{G}$ to a point, resulting in a set $\mathcal{G}^*$ of n points. Construct a single level CSW-partitioning structure on $\mathcal{G}^*$ with $\sigma = O(1)$ as described in Section 2. For a node v of the tree, let $\mathcal{G}_v$ be the set of lines corresponding to the points of $\mathcal{G}_v^*$. We construct the lower and the upper envelopes of $\mathcal{G}_v$, that is, the faces in the arrangement of $\mathcal{G}_v$ that lie above and below all the lines of $\mathcal{G}_v$, respectively. If we have already compute the lower and upper envelopes at all children of a node v, then the lower and upper envelopes of $\mathcal{G}_v$ can be computed in $O(|\mathcal{G}_v|)$ time, so the overall storage and preprocessing time are easily seen to be $O(n \log n)$.

Given a query ray ρ, let a be its origin point and let ℓ be the line supporting ρ. To answer a ray shooting query, we query the primary structure with the halfplane lying above a^*, the line dual to a. Let $\mathcal{G}_v^*$ be a canonical subset, corresponding to the a node v, in the output of the query, then a lies below all the lines dual to the points of $\mathcal{G}_v^*$. As a result, the first intersection point of ρ and $\mathcal{G}_v$ lies on its lower envelope. Since the lower envelope is a convex polygon, the first intersection point can be computed by a binary search. By repeating this step for all canonical subsets, we can find the first intersection point of ρ and the lines of $\mathcal{G}$ that lie above a. We now repeat the same procedure but query with the halfplane lying below a^*, and choose the intersection point that lies nearest to a.

The correctness of the algorithm is obvious. As for the query time, we spend logarithmic time at each canonical subset, and since there are $O(n^{1/2+\epsilon})$ canonical subsets, the query time is $O(n^{1/2+\epsilon})$, for any $\epsilon > 0$.

Theorem 4.1 *Given a collection of n lines, we can preprocess it, for any $\epsilon > 0$, in time $O(n \log n)$, into a data structure of size $O(n \log n)$, so that a ray shooting query can be answered in $O(n^{1/2+\epsilon})$ time.*

4.2 Ray shooting amidst arbitrary collections of segments

Let L be the set of the left endpoints of the segments in $\mathcal{G}$. Preprocess L for halfplane range queries, as described in Section 2, allowing quasi-linear storage. We next take each of the canonical subsets of L that are produced in the primary data structure and process it further as follows: Let L' be such a subset, $\mathcal{G}'$ the corresponding set of segments, and R' the set of right endpoints of the segments of $\mathcal{G}'$. We preprocess R' for halfplane range queries, and attach the resulting data structure to L' as a secondary structure. This is not the end yet: We next take each secondary canonical subset R'' and process the corresponding set of segments $\mathcal{G}''$ for ray shooting, as if they were full lines, using the technique in the preceding subsection. Note that, altogether, we have constructed a 4-level structure.

Given a ray ρ, let ℓ be the line supporting ρ, and let a be the origin of ρ. We first query the primary data structure with the halfplane lying above ℓ, then query each secondary data structure with the halfplane lying below ℓ. Let $\mathcal{G}''$ be a secondary canonical subset in the output of the query so far. We know that ℓ hits all segments in $\mathcal{G}''$, so extending them to full lines will not change the first one to be hit by ρ. We thus continue at this level with the ray shooting procedure described in the preceding subsection, applied to each of these $\mathcal{G}''$. We collect the outputs of all these 'sub-queries', and choose the one nearest to a as the final output.

The correctness of this query procedure is fairly obvious (see also [11], [14]). Concerning complexity, we pay logarithmic time at each 4th-level structure, and the number of such structures that are retrieved during the query processing is $O(n^{1/2+\epsilon})$, as follows from the discussion in Section 2. The resulting query time is thus $O(n^{1/2+\epsilon})$, for any $\epsilon > 0$. That our data structure uses quasi-linear preprocessing and storage also follows from the structure of the CSW scheme, as given in Section 2 (see Remark 2.3). We thus obtain

Theorem 4.2 *Given a set of n arbitrary segments in the plane and $\epsilon > 0$, one can preprocess it in time $O(n \log^3 n)$ into a data structure of size $O(n \log^3 n)$, so that, for any query ray ρ, the first segment that ρ hits can be found in time $O(n^{1/2+\epsilon})$.*

Since the CSW-partitioning scheme allows trade-off between storage (and preprocessing) and query time, we obtain the same tradeoff for our ray shooting algorithm; that is, when allowed s storage, it can answer a ray shooting query in time $O(n^{1+\epsilon}/s^{1/2})$; the preprocessing time is $O(s^{1+\epsilon})$, for any $\epsilon > 0$, as in [8].

Moreover, since each 4th-level structure is just a lower envelope of lines, we can insert or delete a segment into/from it in polylogarithmic time, which implies, as discussed in Section 2, that the structure can also be dynamized efficiently. Note that when we insert a new segment, we insert its left endpoint to the primary structure, its right endpoint to the secondary structure, and the line containing it to the tertiary and 4th-level structures. We thus conclude:

Theorem 4.3 *Given a collection $\mathcal{G}$ of n arbitrary segments in the plane, an $\epsilon > 0$, and a parameter $n^{1+\epsilon} \leq s \leq n^2$, we can preprocess $\mathcal{G}$ into a data structure of size s, in time $O(s^{1+\epsilon})$, which supports ray shooting queries in time is $O(n^{1+\epsilon}/s^{1/2})$, and which allows insertions and deletion of segments to be performed in amortized time $O(s/n^{1-\epsilon})$ per update.*

Finally, our ray shooting structure can be modified to report (or to count the number of) all segments intersected by a query segment e as follows. Recall that a segment of $\mathcal{G}''$, a secondary canonical subset in the output of the query, intersects e if and only if the line containing that segment intersects e. Therefore, we extend the segments of $\mathcal{G}''$ to full lines and dualize them to points. Let P'' be the resulting set of points, then a segment $e' \in \mathcal{G}''$ intersects e iff the corresponding point of P'' lies in the double-wedge dual to e. Hence, by preprocessing each secondary canonical subset for simplex range searching, we can report (or count the number of) all segments intersecting e; see [3] for more details. We thus obtain

Theorem 4.4 *Given a collection $\mathcal{G}$ of n segments in the plane, an $\epsilon > 0$, and a parameter $n^{1+\epsilon} \leq s \leq n^2$, we can preprocess $\mathcal{G}$ into a data structure of size s, in time $O(s^{1+\epsilon})$, so that, given any query segment e, we can report all k segments of $\mathcal{G}$ intersecting e in time $O(n^{1+\epsilon}/s^{1/2} + k)$, or can count the number of such segments in time $O(n^{1+\epsilon}/s^{1/2})$. As above, insertions and deletion of segments can be performed in amortized time $O(s/n^{1-\epsilon})$ per update.*

5   Efficient Output-Sensitive Hidden Surface Removal

In a recent series of papers, Overmars and Sharir [22, 18, 19] have studied the problem of output-sensitive hidden surface removal. For simplicity, assume that we are given a collection Δ of n horizontal triangles in 3-space, and we want to compute the portions of these triangles that are visible from $z = -\infty$. The xy-projections of these portions form a planar map $\mathcal{M}$, known as the *visibility map* of the triangles. If the combinatorial complexity (say, the number of vertices) of $\mathcal{M}$ is K, the goal is to compute $\mathcal{M}$ in time that depends on K (and on n) so that when K is small the algorithm runs faster. Overmars and Sharir give two output-sensitive solutions, the first [22] is very simple and runs in time $O(n\sqrt{K}\log n)$. An improved algorithm is given in [18]. It is a fairly complicated algorithm but its time complexity, $O(n^{4/3}\log^{2.89} n + K^{3/5}n^{4/5+\epsilon})$, is an improvement over the first algorithm. In this section we describe an $O(n^{2/3-\epsilon}K^{1/3+\epsilon}\zeta^{1/3+\epsilon} + n^{1+\epsilon} + K\zeta^{\epsilon})$ algorithm to compute the visibility map, where $\zeta \le K$ is the maximum contour size (see below). Our algorithm is significantly simpler than the second algorithm of Overmars and Sharir.

Our algorithm is based on the dynamic ray shooting technique developed in the preceding section. It constructs the visibility map incrementally by adding the triangles one by one in the nondecreasing order of their z-coordinates. Suppose we have computed $\mathcal{M}_i$, the visibility map of $\Delta_1, \ldots, \Delta_i$ and we are about to add Δ_{i+1}. Since we are adding the triangles in the nondecreasing order of their z-coordinates, Δ_{i+1} cannot obscure any portion of $\Delta_1, \ldots, \Delta_i$, so Δ_{i+1} can appear only in $\mathbf{R}^2 - \mathcal{M}_i$. We refer to the boundary of $\mathbf{R}^2 - \mathcal{M}_i$ as the *contour* of $\mathcal{M}_i$, and denote it by C_i. Let $\zeta = \max_i\{|C_i|\}$ be the maximum contour size. For the sake of convenience, we will use Δ_i to denote the triangle as well as its xy-projection.

The processing of Δ_{i+1} consists of computing $\mathcal{M}_{i+1}$ and C_{i+1} from $\mathcal{M}_i$ and C_i, respectively. The new vertices of $\mathcal{M}_{i+1}$, that is, the ones that were not in $\mathcal{M}_i$, are either the vertices of Δ_{i+1} or the intersection points of C_i and $\partial\Delta_{i+1}$. Moreover, every intersection point of C_i and $\partial\Delta_{i+1}$ is a vertex of $\mathcal{M}_{i+1}$. Once we know the new vertices of $\mathcal{M}_{i+1}$, its new edges can be computed in a straight forward manner. We maintain two data structures on C_i

- We preprocess C_i for efficient ray shooting queries as described in Section 4.2 so that the intersection points between C_i and a query segment can be reported quickly.

- We preprocess the left endpoints of the edges of C_i for triangle range searching queries so that the left endpoints lying in Δ_{i+1} can be computed quickly.

The updating of $\mathcal{M}_i$ and C_i consists of the following three steps:

(i) *Computing the vertices of $\mathcal{M}_i$:* For each edge e of Δ_{i+1} we compute the intersection points of e and C_i using the ray shooting data structure. To determine whether an endpoint a of e appears in $\mathcal{M}_{i+1}$, we shoot a ray from a along the line containing e but in the opposite direction. If the ray does not intersect C_i, a appears in $\mathcal{M}_{i+1}$, otherwise we can determine from the first intersection point whether a is a vertex of $\mathcal{M}_{i+1}$; see [3] for details.

(ii) *Deleting the edges of $C_i - C_{i+1}$:* For each edge $e \in C_i$, $e \cap \Delta_{i+1}$ does not appear in C_{i+1}, so if e intersects C_i, we replace e by $e - \Delta_{i+1}$; $e - \Delta_{i+1}$ may consist of two connected components. We delete e from C_i, insert the connected components of $e - \Delta_{i+1}$, and update both structures accordingly.

As for finding the edges e that intersect $\triangle_{i+1}$, e satisfies at least one of the following two conditions: (i) e intersects $\partial\triangle_{i+1}$, or (ii) the left endpoint of e lies in $\triangle_{i+1}$. The first type of edges can be determined in Step (i) while the second type of edges can be obtained using the range searching data structure.

(iii) *Adding the edges of $C_{i+1} - C_i$:* Let ab be an edge of $\triangle_{i+1}$. If ab does not intersect C_i but a appears in $\mathcal{M}_{i+1}$, then ab is a new edge of C_{i+1}. On the other hand if $\chi_1, \chi_2, \ldots, \chi_m$ are the intersection points of C_i and ab, then either $a\chi_1, \chi_2\chi_3, \chi_4\chi_5, \ldots,$ or $\chi_1\chi_2, \chi_3\chi_4, \ldots,$ are new edges of C_{i+1}. We add these edges to C_{i+1} and update the data structures.

Repeating this procedure for all triangles in order, we obtain the visibility map $\mathcal{M}$. Analyzing the running time of the algorithm and choosing the size of the ray shooting and range searching data structures appropriately, we can show

Theorem 5.1 *Given a set of n horizontal triangles in $\mathbf{R}^3$, their visibility map can be computed in time $O(n^{2/3-\epsilon}K^{1/3+\epsilon}\zeta^{1/3+\epsilon} + K\zeta^{\epsilon} + n^{1+\epsilon})$, where K is the output size and ζ is the maximum contour size.*

6 Ray Shooting in Three Dimensions

Let $\mathcal{T} = \{T_1, \ldots, T_n\}$ be a collection of n triangles in $\mathbf{R}^3$. We wish to preprocess them into a data structure that supports fast *ray shooting* queries, where each such query asks for the first triangle, if any, intersected by a query ray. Note that we do not put any restriction on the query ray — it can emanate from any point in any direction; nor do we require any special structure of the given collection of triangles — our technique can even handle intersecting triangles. We will show that, allowing s storage, we can preprocess $\mathcal{T}$ in time $O(s^{1+\epsilon})$, for any $\epsilon > 0$, into a data structure of size s, so that a ray shooting query can be answered in time $O(n^{16/15+\epsilon}/s^{4/15})$.

Let $e_1, \ldots, e_{3n}$ be the edges of the triangles in $\mathcal{T}$, and let ℓ_j denote the line containing e_j, for $j = 1, \ldots, 3n$. Let ρ be a query ray, and let λ be the line containing it. Let T be a triangle of $\mathcal{T}$, let e_1, e_2, e_3 denote its edges, and suppose that they lie respectively on the lines ℓ_1, ℓ_2, and ℓ_3. We orient the lines ℓ_i so that T lies to the right of each of them. The line λ intersects T if and only if it has the same *relative orientation* with respect to the three lines ℓ_i. The relative orientation of two oriented lines ℓ, λ in 3-space is defined to be the orientation of the simplex $abcd$, where $a, b \in \ell$, $c, d \in \lambda$, ℓ is oriented from a to b and λ is oriented from c to d. Equivalently, it is also the sign of the inner product between the two vectors in 6-space representing the *Plücker's coordinates* of the two lines. More details concerning Plücker's coordinates and relative orientations can be found in [7, 23].

These observations suggest the following approach. We describe a solution that uses roughly linear storage; the general case can be handled using the trade-off properties of the CSW scheme—see below. We first take one edge from each triangle $T \in \mathcal{T}$, and form the collection $\mathcal{L}_1$ of the lines containing these edges and oriented as above. We map each line ℓ in $\mathcal{L}_1$ to its Plücker point $\pi(\ell)$ in projective 5-space (see [7] for more details). We apply the CSW partitioning scheme to the resulting collection, $\mathcal{P}_1$, of points in 5-space. For each canonical subset of Plücker points, or equivalently of lines in 3-space, we take the corresponding subset of triangles of $\mathcal{T}$, pick a second edge in each, form the corresponding set of oriented lines containing these edges, transform them into

Plücker points, and apply the CSW scheme to these points. We attach the resulting data structure as a secondary substructure at the corresponding node of the primary structure. We then repeat the same process once more for each canonical subset in any secondary substructure, where now we use the third edge of each corresponding triangle of T.

We have thus obtained a 3-level structure, which uses $O(n^{1+\epsilon})$ space and takes $O(n^{1+\epsilon})$ time to construct. We use this structure to process our ray shooting query as follows. We map the line λ containing our query ray into its *Plücker's hyperplane* $\varpi(\lambda)$ in projective 5-space (see [7] for details). We query with this hyperplane in our structure to obtain all triples of lines in 3-space, each triple containing the edges of a single triangle in T, so that the three lines in a triple have the same relative orientation with respect to λ. In other words, the query output consists of all triangles stabbed by λ. Since we are in 5-space, the properties of the CSW scheme imply that the query time is $O(n^{4/5+\epsilon})$, and that the output of our query consists of $O(n^{4/5+\epsilon})$ pairwise disjoint canonical subsets of triangles.

We next have to shoot along ρ within each of these canonical subsets and report the hit point nearest to the origin of ρ. Let T' be one of these canonical subsets. Extend each triangle in T' to a full plane, and preprocess the resulting collection of planes for fast ray shooting. This can be done as follows. Dualize the collection of planes to obtain a set of points in 3-space. Apply the CSW partitioning scheme to this set. Take the origin point a of ρ, dualize it to a plane a^*, and query the resulting data structure with a^*. This gives us a collection of $O(n^{2/3+\epsilon})$ canonical subsets of dual points, each lying either fully above or fully below a^*. For each of these canonical subsets we go back to the primal space, and conclude that a lies either above all the corresponding planes, or below all of them. We can therefore precompute the upper and lower envelopes of this collection of planes and preprocess each of them for fast ray shooting, using the hierarchical representation of polytopes given by Dobkin and Kirkpatrick [12].

Using the properties of the CSW scheme, it is easy to verify that the overall time for processing a ray shooting query is $O(n^{4/5+\epsilon})$, for any $\epsilon > 0$. The preprocessing time and storage are both $O(n^{1+\epsilon})$.

The method just presented aims to achieve quasi-linear storage (and preprocessing). Using the trade-off property of the CSW scheme in conjunction with a recent result of Aronov and Sharir [4] on the complexity of the zone of an algebraic surface in a hyperplane arrangement, we can reduce query time at the cost of increasing storage: allowing s storage, the query time goes down to $O(n^{16/15+\epsilon}/s^{4/15})$, see [3] for details. We summarize our results in

Theorem 6.1 *Given a collection of n triangles in $\mathbf{R}^3$, and parameters $\epsilon > 0$ and storage size s that can vary between $n^{1+\epsilon}$ and n^5, we can preprocess the collection into a data structure of size s, in time $O(s^{1+\epsilon})$, which supports ray shooting queries in time $O(n^{16/15+\epsilon}/s^{4/15})$.*

The above scheme can be modified to report (or to count the number of) all k triangles that intersect a query segment e in time $O(n^{16/15+\epsilon}/s^{4/15} + k)$; see the full version for details [3].

References

[1] P.K. Agarwal, A deterministic algorithm for partitioning arrangements of lines and its applications, *Proc. 5th ACM Symp. on Computational Geometry*, 1989, pp. 11–22.

[2] P.K. Agarwal, Ray shooting and other applications of spanning trees with low stabbing number, *Proc. 5th ACM Symp. on Computational Geometry*, 1989, pp. 315–325.

[3] P. K. Agarwal and M. Sharir, Applications of a new space partitioning scheme, Tech. Rept. CS-1991-14, Dept. Computer Science, Duke University, 1991.

[4] B. Aronov and M. Sharir, On the zone of an algebraic surface in a hyperplane arrangement, *Proc. 2nd Workshop on Algorithms and Data Structures*, 1991.

[5] R. Bar Yehuda and S. Fogel, Good splitters with applications to ray shooting, *Proc. 2nd Canadian Conf. on Computational Geometry*, 1990, pp. 81–85.

[6] M. de Berg, D. Halperin, M. Overmars, J. Snoeyink, and M. van Kreveld, Efficient ray shooting and hidden surface removal, *Proc. 7th Symposium on Computational Geometry*, 1991.

[7] B. Chazelle, H. Edelsbrunner, L. Guibas, M. Sharir and J. Stolfi, Lines in space: Combinatorics and algorithms, Tech. Rept. 491, Dept. of Computer Science, New York University, February 1990.

[8] B. Chazelle, M. Sharir and E. Welzl, Quasi-optimal upper bounds for simplex range searching and new zone theorems, *Proc. 6th ACM Symp. on Computational Geometry*, 1990, pp. 23–33.

[9] B. Chazelle and E. Welzl, Quasi-optimal range searching in spaces of finite Vapnik-Chervonenkis dimensions, *Discrete Comput. Geom.* 4 (1989), pp. 467–489.

[10] S.W. Cheng and R. Janardan, Space efficient ray shooting and intersection searching: Algorithms, dynamization, and applications, in *Second SIAM-ACM Symposium on Discrete Algorithms*, 1991.

[11] D. Dobkin and H. Edelsbrunner, Space searching for intersecting objects, *J. Algorithms* 8 (1987), pp. 348–361.

[12] D. Dobkin and D. Kirkpatrick, A linear algorithm for determining the separation of convex polyhedra, *J. of Algorithms* 6 (1985), 381–392.

[13] H. Edelsbrunner, L. Guibas, J. Hershberger, R. Seidel, M. Sharir, J. Snoeyink and E. Welzl, Implicitly representing arrangements of lines and of segments, *Discrete Comput. Geom.* 4 (1989), pp. 433–466.

[14] L. Guibas, M. Overmars and M. Sharir, Ray shooting, implicit point location, and related queries in arrangements of segments, Tech. Report 433, Courant Institute, New York University, 1989.

[15] J. Matoušek, Spanning trees with low crossing numbers, to appear in *Informatique Theoretique et Applications*.

[16] J. Matoušek, Cutting hyperplane arrangements, *Proc. 6th ACM Symp. on Computational Geometry*, 1990, pp. 1–10.

[17] J. Matoušek, More on cutting hyperplanes and spanning trees with low crossing number, Tech. Rept., Freie Universität Berlin, 1990.

[18] M. Overmars and M. Sharir, An improved technique for output-sensitive hidden surface removal, Tech. rept. RUU-CS-89-32, Computer Science Department, University of Utrecht, December 1989.

[19] M. Overmars and M. Sharir, Merging visibility maps, *Proc. 6th ACM Symp. on Computational Geometry*, 1990, pp. 168–176.

[20] M. Pellegrini, Stabbing and ray shooting in three-dimensional space, *Proc. 6th ACM Symp. on Computational Geometry*, 1990, pp. 177–186.

[21] A. Schmitt, H. Müller and W. Leister, Ray tracing algorithms — Theory and algorithms, in *Theoretical Foundations of Computer Graphics and CAD* (ed. R. Earnshaw), NATO Series, Springer Verlag, 1988, pp. 997–1030.

[22] M. Sharir and M. Overmars, A simple output-sensitive hidden surface removal algorithm, Tech. rept. RUU-CS-89-26, Computer Science Department University of Utrecht, November 1989.

[23] D.M.H. Sommerville, *Analytical Geometry in Three Dimensions*, Cambridge, 1951.

[24] J. Stolfi, *Primitives for Computational Geometry*, Ph.D. Dissertation, Stanford University, 1989.

[25] E. Welzl, Partition trees for triangle counting and other range searching problems, *Proc. 4th ACM Symp. on Computational Geometry*, 1988, pp. 23–33.

Offline Algorithms for Dynamic Minimum Spanning Tree Problems

David Eppstein
Department of Information and Computer Science
University of California, Irvine, CA 92717

Abstract

We describe an efficient algorithm for maintaining a minimum spanning tree (MST) in a graph subject to a sequence of edge weight modifications. The sequence of minimum spanning trees is computed offline, after the sequence of modifications is known. The algorithm performs $O(\log n)$ work per modification, where n is the number of vertices in the graph. We use our techniques to solve the offline geometric MST problem for a planar point set subject to insertions and deletions; our algorithm for this problem performs $O(\log^2 n)$ work per modification. No previous dynamic geometric MST algorithm was known.

1. Introduction

For many years, algorithm researchers have studied problems of maintaining information about a dynamically changing graph. A classical problem in this field is maintaining the minimum spanning tree (MST) of a graph in which the weights of individual edges are subject to change [1, 4, 5, 9].

Each such update causes at most one edge to leave the MST and at most one other edge to take its place; one might expect algorithms for computing these changes to be quite efficient. Indeed, the best known algorithm takes $O(\sqrt{m})$ time per update, for a graph with n vertices and m edges [5]. However there is clearly room for improvement in this case. For planar graphs the situation is better; one can compute the changes to the MST in time $O(\log n)$ per update [4, 8].

All of the above algorithms are *online*: They accept as input one update at a time, and must output the corresponding changes to the MST before the next update is available to them. It is natural to consider *offline* algorithms for the same problem; such an algorithm would be given a single input consisting of a long sequence of updates, and only after the entire sequence is known would it output the corresponding sequence of MST changes. Offline algorithms are less general than online ones: an offline algorithm can not maintain the MST when updates depend on previous results of the algorithm. In exchange for this loss of flexibility one might expect improved time bounds. Surprisingly, there seems to be no previous work on offline dynamic MST algorithms.

This paper presents such an algorithm. The time per update is $O(\log n)$, greatly improving the previous $O(\sqrt{m})$ bound for the online problem and even matching the best known time bound for planar graphs.

We also examine the problem of maintaining a MST of a dynamic planar point set. It is well known that the MST is a subgraph of the Delaunay triangulation; therefore the static problem can be solved in time $O(n \log n)$. However no efficient algorithm was known for the dynamic geometric MST problem, in which updates consist of insertions and deletions of single points. We use techniques similar to those in our graph algorithm, and the graph algorithm itself as a subroutine, to achieve $O(\log^2 n)$ update time for the offline dynamic geometric MST problem. This time bound can be improved to $O(\log n \log \log n)$ for rectilinear MSTs.

2. Reduction and Contraction

Our offline graph MST algorithm works in a series of phases. In each phase we perform two steps: *reduction* and *contraction*. In a given phase we divide the sequence of edge weight updates into *blocks* and perform the reduction and contraction operations separately for each block.

Reduction consists of finding a set of edges in the graph which, given the update operations in the block, can not be used in any of the MSTs of any of the sequence of weighted graphs corresponding to the updates. Once these edges are found, they can be removed from the graph without changing the results of the computation in that block. Thus the graph can be made to have a smaller number of edges.

Contraction, similarly, consists of finding a set of edges which must be used in all the MSTs for the block. We then *contract* each of these edges, by merging sets of vertices that are connected by those edges. Thus the graph can be made to have a smaller number of vertices.

Reduction and contraction were used in a previous paper in which we described algorithms for finding a set of several different spanning trees having the minimum possible total weight [2]. However in that paper the actual process of reduction and contraction is performed differently. Curiously enough the problem solved in that paper had previously been attacked by using the (online) dynamic MST problem as a subroutine [5].

Both reduction and contraction are implemented in this paper using (static) MST computations. For this purpose we will use the recently discovered linear time MST algorithm of Fredman and Willard [6]. This algorithm uses a nonstandard model of computation, in which edge weights are binary integers; other MST algorithms do not specify the representation of weights, and only operate on them by comparisons. In our algorithm we allow $O(\log n)$ time per update, and we never operate on more than $O(n^2)$ updates at once. Therefore we have time to sort the weights using any $O(n \log n)$ time comparison based algorithm, and convert them to the integer representation needed by the MST algorithm.

Edge insertion and deletion operations can also be handled, by treating a deleted edge as having infinite cost. The obvious implementation of this would cause the algorithm to use $\Omega(n^2)$ space to keep track of the costs of all edges; we will show how to avoid this space penalty.

3. Reduction

The following fact is the basis of the reduction step.

Lemma 1. *Let G be a weighted graph, and S be a subset of its edges. Let T be the MST of $G - S$. Then no matter how the edge weights in S are changed, the MST of G will only contain edges in $T \cup S$.*

Proof: This follows immediately from the standard "dual greedy" MST algorithm in which the highest weight edge in any cycle is removed until the remaining graph has no cycles. If we break the cycles in $G - S$ first, all edges in $G - S - T$ will be removed. Therefore all edges in the MST will be in the remaining graph, $T \cup S$. □

Lemma 2. *Assume we are performing a reduction step in a block of k updates. Then we can reduce the number of edges in the graph from m to at most $n + k - 1$, in time $O(m)$.*

Proof: Let S be the set of edges updated in the block. Compute the MST of $G - S$ and apply Lemma 1. Then we need only keep the $n - 1$ MST edges, together with the edges in S; Lemma 1 shows that throwing away the other edges will not change the results of the computation in this block. □

4. Contraction

The following fact is the basis of the contraction step.

Lemma 3. *Let G be a weighted graph, and S be a subset of its edges. Let T be the MST of G, when the edges of S are given weights lower than those of any other edge in the graph. Then no matter how the edge weights in S are changed, the MST of G will always contain the edges in $T - S$.*

Proof: Consider changing the weights of the edges in S, one by one, from the weights in T to the new desired weight. At each such change, either the MST will not change, or the changed edge will leave the MST and some other edge will replace it. Therefore, the edges in $T - S$ will remain in the MST. □

Lemma 4. *Assume we are performing a contraction step in a block of k updates. Then we can reduce the number of vertices in the graph from n to at most $k + 1$, in time $O(m)$.*

Proof: Let S be the set of edges updated in the block. We assume that Lemma 2 has already been applied, so $G - S$ is a tree. Weight the edges in S lower than the minimum weight in $G - S$ and compute the MST T of G.

Construct a new graph G' as follows. Create a new vertex for each connected component of $T - S$, and replace each edge (x, y) in the graph with (x', y') where x' and y' are the vertices corresponding to the components containing x and y respectively.

Lemma 3 shows that each MST of G, for a given assignment of weights to the edges of S, can be found by computing the MST of G', and taking the union of the corresponding edges in G with the edges in $T - S$. In particular, each individual change to the MST of G (consisting of the removal of one edge from the MST, and the addition of one replacement edge) corresponds exactly to such a change in the MST of G', and vice versa. □

5. The Graph Algorithm

We now solve the offline dynamic graph MST problem. Recall that the algorithm is divided into *phases*, in each of which we perform reduce and contract steps within *blocks* of edge weight update operations. It remains to specify how many phases to use, and what block size to use within a phase.

We begin with all blocks of size m, the largest number of edges in the graph at any one time. Contraction and reduction are no help for such a large block, so we do not perform these steps in the first phase. Within a block we treat insertions and deletions as changes involving infinite edge weights, as discussed previously; the initial selection of block size m instead of $n^2/2$ means we only need $O(m)$ space instead of $O(n^2)$. At this point we can sort the weights used within the block (including the edge weights on entry to the block) in preparation for the MST algorithm of Fredman and Willard [6].

In each succeeding phase, we start with blocks of size b, and in each block the graph has been reduced, contracted, and reduced again. Therefore by Lemmas 2 and 4 it has at most $b+1$ vertices and b edges not involved in updates. There are of course at most b edges involved in updates. We then split each block into two smaller blocks of approximately equal length. These blocks will be used in the new phase, in which we again reduce, contract, and reduce the graph.

After $O(\log n)$ such phases, each block will consist of a single update and the reduced graph in each block will have two vertices and one non-updated edge, at which point we can easily tell whether the updated edge replaces or is replaced by the other edge.

This gives us a sequence of edge replacements, which we translate back into the unreduced form to produce the sequence of MST changes corresponding to the initial sequence of edge cost updates in the original graph.

Theorem 1. *Given a sequence of k edge weight modifications in a graph, starting from a state in which all weights are equal, we can compute the corresponding sequence of minimum spanning trees in time $O(k \log n)$ and space $O(m)$.*

Proof: The correctness of the algorithm follows from Lemmas 2 and 4. In each phase, for each block of size b, we perform $O(b)$ operations; therefore, for each update we perform $O(1)$ operations. There are $O(\log n)$ phases, so for each update we perform $O(\log n)$ total work. $\square$

6. The Geometric Algorithm

We now describe how to solve the offline MST problem for a planar point set, with updates consisting of point insertions and deletions. The geometric MST is simply the MST of the complete graph on the points, with edges weighted by distance. However each point update corresponds to the insertion or deletion of $O(n)$ edges in the complete graph, so a direct application of our graph algorithm would be no more efficient than recomputing the MST using the $O(n \log n)$ time static algorithm after each update.

Instead we use the following approach. We recursively break our update sequence into blocks, as in the graph algorithm. The processing at each level consists of identifying certain pairs of points as "interesting", adding the corresponding edges to a graph problem, and removing some of the points from lower levels of the recursion. After this decomposition, we will have identified $O(n \log n)$ interesting edges in each block of length n; all MST edges will occur in this list of interesting edges. Each such edge can be considered to be inserted in the graph when both of its two endpoints have been inserted in the plane, and deleted when one of its endpoints is deleted. Thus we reduce the problem to a graph problem in which there are $O(\log n)$ edge updates per point insertion or deletion, and which can therefore be solved in time $O(\log^2 n)$ per point update.

We assume throughout this section that point distances are measured using the Euclidean L_2 metric; similar techniques apply to other commonly used metrics. Define a *sextant* from point x to be an infinite closed wedge with 60° angle having x as its corner, with either one side of the wedge horizontal or both sides at 60° angles from horizontal. Each point corresponds to six possible sextants, and these six sextants together cover the plane. Our identification of interesting edges is based on the following well known facts.

Lemma 5. *Let (x, y) be an edge in the geometric MST of a planar point set S. Then y is the nearest point to x in the sextant from x containing y.*

Proof: Let z be closer. Then the edges xz and zy are both shorter than xy, so xy could not be a MST edge. $\square$

Lemma 6. *Given sets S and T, the nearest points in S in each sextant from each point of T can be found in time $O((|S| + |T|) \log |S|)$.*

Proof: For each sextant direction, one can construct in time $O(|S| \log |S|)$ an appropriate Voronoi diagram of the points of S. Point location queries in such a diagram can be performed for each point of T, in time $O(\log |S|)$ per query. Six such diagrams need be computed, for the six possible sextant directions, and six such queries need be performed for each point of T. $\square$

We now describe our offline geometric MST algorithm. As before, we consider blocks of update operations, perform certain reduction steps on them, and recursively divide them into pairs of smaller blocks.

Our block reduction works as follows. Let S be the *static* points of the block; that is, those points that are not inserted or deleted by the updates in the block. Let T be the *dynamic* points, that are updated within the block. Using lemma 6 we compute, for each point in $S \cup T$, the nearest points in S in each sextant. We declare each pair of points found in this step to be an interesting edge. Finally, we remove S from the point set when we consider the two recursive subblocks created by splitting our block. Thus in the recursive processing, the set of static points in each subblock will be a subset of the dynamic points in the other subblock.

Theorem 2. *Given a sequence of k point insertions and deletions in the plane, starting from an empty plane, we can compute the corresponding sequence of minimum spanning trees in time $O(k \log^2 n)$ and space $O(n)$, where n is the maximum number of points in the plane at any one time.*

Proof: We start with blocks of size $O(n/\log n)$. In the first reduction, there are $O(n)$ static points and therefore the time taken is $O(n \log n)$ per block. In subsequent reductions, when the block size is m, the number of static points is at most $m + 1$, the maximum size of the other block into which the block's parent is split; therefore, the time taken is $O(m \log m)$. The total work within each initial block per level of the recursive decomposition is $O(n)$, and the total work in all the $O(\log n)$ levels is $O(n \log n)$. Once the decomposition is performed, we will have identified at most six interesting edges per point per level, or $O(n)$ interesting edges altogether. Solving the offline graph MST problem on the sequence of insertions and deletions of interesting edges takes time $O(n \log n)$. Therefore the total time per block is $O(n \log n)$, and the time per update is $O(\log^2 n)$.

It remains to show that all MST edges are included in the set of interesting edges, and therefore that the algorithm correctly computes the sequence of MSTs for the changing point set. Let (x, y) be an edge in the MST after change c, and consider the largest block b containing c in which at least one of x and y would be static (if it weren't thrown away in a larger block). Such a block must exist, because only one point can be dynamic in the block containing c alone. Then neither x nor y can have been thrown away in a larger block, because only static points are thrown away. Assume without loss of generality that x is static in b. By lemma 5, x is the closest point to y in its sextant at time c. Since the static points of b are a subset of the input points existing after change c, point x must be the static point in block b that is closest to y in its sextant. Therefore it is included in the set of interesting edges. $\square$

7. Rectilinear Spanning Trees

We can improve theorem 2 for the important case of *rectilinear* minimum spanning trees. These are MSTs for the L_1 (or equivalently L_∞) planar metric. The following analogues of lemmas 5 and 6 hold; lemma 8 below is proved in [2]. We define a *quadrant* to be a quarterplane bounded by one horizontal and one vertical line.

Lemma 7. *Let (x, y) be an edge in the rectlinear MST of a planar point set S. Then y is the L_1-nearest point to x in the quadrant from x containing y.* $\square$

Lemma 8. *Given sets S and T, in two sorted orders, one for each coordinate, the L_1-nearest points in S in each quadrant from each point of T can be found in time $O((|S| + |T|) \log \log |S|)$.* $\square$

Lemma 8 speeds up the time spent identifying interesting edges. However there is another bottleneck in our algorithm, which is the $O(\log n)$ time spent processing each interesting edge. We deal with this by interleaving the graph edge reduction of theorem 2 with our geometric point reduction. To do this we maintain a *mixed* problem, consisting of a graph in which some of the vertices are geometric points, some of the vertices are just vertices, pairs of geometric points can be connected by their L_1-distances, and all vertices can be connected by the edges in the graph.

Lemma 9. *Given a mixed problem consisting of $O(n)$ points, non-point vertices, and edges, with the points sorted by each coordinate, the mixed rectilinear MST can be computed in time $O(n \log \log n)$.*

Proof: We compute a set of $O(n)$ interesting point-point distances using lemma 8, in time $O(n \log \log n)$. This gives us a pure graph problem with $O(n)$ vertices and edges, which can be solved using any $O(n \log \log n)$-time MST algorithm. $\square$

We cannot use the linear time MST algorithm of Fredman and Willard [6] because the coordinates of the input points are not assumed to be integers. However the best known non-integer MST algorithm takes time $O(n \log \log^* n)$ for our problem [7], easily meeting the $O(n \log \log n)$ requirement.

Theorem 3. *Given a sequence of k point insertions and deletions in the plane, starting from an empty plane, we can compute the corresponding sequence of rectilinear minimum spanning trees in time $O(k \log n \log \log n)$ and space $O(n)$, where n is the maximum number of points in the plane at any one time.*

Proof: We begin with blocks of size n. Within each initial block, we sort the points by each coordinate; these sorted orders will be maintained in the smaller blocks recursively split from the initial blocks. In each block, we maintain a mixed problem such that the following conditions hold.

- The MST of the mixed problem is a contraction of the MST of all static points in the block.

- Each edge in the MST of all static points that is removed by some change in the block corresponds either to a pair of geometric points in the mixed problem, or to an edge in the mixed problem.

- For each edge in an MST in the block that connects a static and dynamic point, the static point is one of the geometric points in the mixed problem.

- The size of the mixed problem is proportional to the size of the block.

In blocks consisting of a single change, we have a mixed problem of constant size in which one geometric point is inserted or deleted; the appropriate MST update can be found in constant time.

In larger blocks, we split each block into two smaller blocks and reduce the mixed problem appropriately. This is done as in the graph algorithm by computing two MSTs, one of the mixed problem without any dynamic points, and one of the mixed problem with all dynamic points. We also compute all interesting edges consisting of the closest static point in each quadrant to each dynamic point.

Then as in the graph algorithm we can remove all graph edges not in the first MST, and contract the edges that are in both MSTs. Whenever a contraction involves a geometric point, that point must become a graph vertex instead; therefore we cannot contract all edges that are in both MSTs. Instead, we mark all static points that are the nearest in some quadrant to some dynamic point; edges touching marked points will not be contracted. Because of lemma 7, in a block of b changes at most $4b$ static points will be marked, and each marked point can protect at most 4 edges from contraction. At most $3b$ static MST edges can be replaced when we include the dynamic points. Therefore the contraction results in a mixed problem having at most $19b$ edges and vertices, so the problem size is reduced appropriately. It can be verified that all the conditions above are maintained. The reduction process takes time $O(b \log \log b)$, for a total time of $O(\log n \log \log n)$ per change. $\square$

8. Conclusion

We described efficient algorithms for offline computation of MSTs in a changing graph or point set. We use reduction and contraction, which we introduced in our paper on finding several small spanning trees [2], and have recently applied to a persistent query version of the dynamic MST problem [3]. Reduction and contraction have thus proven useful for a variety of MST problems; perhaps they can be used in other graph algorithms.

The most pressing open problem suggested by this work is maintenance of geometric MSTs. No algorithm was known that achieved sublinear time bounds for both point insertions and deletions. We achieve $O(\log^2 n)$ time, at the expense of requiring offline operation. Our results may lead the way to efficient online algorithms for this problem.

Acknowledgements

I would like to thank Adam Buchsbaum and an anonymous referee for their helpful comments on an earlier draft of this paper.

References

[1] F. Chin and D. Houck. Algorithms for updating minimum spanning trees. *J. Comput. Syst. Sci.* 16 (1978) 333–344.

[2] D. Eppstein. Finding the k smallest spanning trees. *Proc. 2nd Scand. Worksh. Algorithm Theory*, Springer-Verlag LNCS 447 (1990) 38–47.

[3] D. Eppstein. Persistence, offline algorithms, and space compaction. Manuscript, 1991.

[4] D. Eppstein, G.F. Italiano, R. Tamassia, R.E. Tarjan, J. Westbrook, and M. Yung. Maintenance of a minimum spanning forest in a dynamic planar graph. *Proc. 1st ACM/SIAM Symp. Discrete Algorithms* (1990) 1–11.

[5] G.N. Frederickson. Data structures for on-line updating of minimum spanning trees, with applications. *SIAM J. Comput.* 14 (1985) 781–798.

[6] M.L. Fredman and D.E. Willard. Trans-dichotomous algorithms for minimum spanning trees and shortest paths. *Proc. 31st IEEE Symp. Found. Computer Science* (1990) 719–725.

[7] H.N. Gabow, Z. Galil, T. Spencer, and R. Tarjan. Efficient algorithsms for finding minimum spanning trees in undirected and directed graphs. *Combinatorica* 6 (1986) 109–122.

[8] H.N. Gabow and M. Stallman. Efficient algorithms for graphic matroid intersection and parity. *Proc. 12th Int. Conf. Automata, Languages, and Programming*, Springer-Verlag LNCS 194 (1985) 210–220.

[9] P.M. Spira and A. Pan. On finding and updating spanning trees and shortest paths. *SIAM J. Comput.* 4 (1975) 375–380.

An empirical analysis of algorithms
for constructing a minimum spanning tree

Bernard M.E. Moret and Henry D. Shapiro[1]

Department of Computer Science
University of New Mexico
Albuquerque, NM 87131

Abstract

We compare algorithms for the construction of a minimum spanning tree through large-scale experimentation on randomly generated graphs of different structures and different densities. In order to extrapolate with confidence, we use graphs with up to 130,000 nodes (sparse) or 750,000 edges (dense). Algorithms included in our experiments are Prim's algorithm (implemented with a variety of priority queues), Kruskal's algorithm (using pre-sorting or demand sorting), Cheriton and Tarjan's algorithm, and Fredman and Tarjan's algorithm. We also ran a large variety of tests to investigate low-level implementation decisions for the data structures, as well as to enable us to eliminate the effect of compilers and architectures.

Within the range of sizes used, Prim's algorithm, using pairing heaps or sometimes binary heaps, is clearly preferable. While versions of Prim's algorithm using efficient implementations of Fibonacci heaps or rank-relaxed heaps often approach and (on the densest graphs) sometimes exceed the speed of the simpler implementations, the code for binary or pairing heaps is much simpler, so that these two heaps appear to be the implementation of choice.

Some conclusions regarding implementation of priority queues also emerge from our study: in the context of a greedy algorithm, pairing heaps appear faster than other implementations, closely followed by binary, rank-relaxed and Fibonacci heaps, the latter two implemented with sacks, while splay trees finish a decided last.

1 Introduction

Finding spanning trees of minimum weight (often simply called "minimum spanning trees" or MSTs) is one of the best known graph problems; algorithms for this problem have a long history, for which see the article of Graham and Hell [7]. The best comparison-based algorithm to date, due to Gabow *et al.* [6], runs in almost linear time—its running time is $O(|E| \log \beta(|E|, |V|))$, where $\beta(m, n) = \min\{ i \mid \log^{(i)} n \leq m/n \}$; very recently, Fredman and Willard [5] have described a linear-time algorithm that uses address arithmetic and bit manipulations. Classical algorithms have slower asymptotic running times, but also tend to be simpler. Kruskal's algorithm runs in $O(|E| \log |V|)$ time; Prim's algorithm runs in $O(|E| \log |V|)$ time with simple implementations and in $O(|E| + |V| \log |V|)$ with more sophisticated priority queues; Cheriton and Tarjan's algorithm [1] runs in

---

[1]This author did his work while a visiting professor at the Institut für Grundlagen der Informationsverarbeitung und Computergestützte neue Medien, Technische Universität Graz, A-8010 Graz, Austria.

$O(|E| \log \log |V|)$ time. Because both Kruskal's and Prim's algorithms rarely attain their worst-case bounds and because their data structures can be maintained with very low overhead, one cannot determine the best algorithm solely on the basis of asymptotic analysis and so must resort to experimentation.

We studied four comparison-based algorithms in numerous implementations: Prim's algorithm, implemented with each of binary heaps, d-heaps, pairing heaps (for a description of which see [3]), Fibonacci heaps (see [4]), rank-relaxed heaps (see [2]), and splay trees (see [11]); Kruskal's algorithm, implemented with a first sorting pass and with demand sorting; Cheriton and Tarjan's algorithm (see [13]); and Fredman and Tarjan's algorithm (see [4]), implemented with Fibonacci heaps—we chose this last algorithm, which runs in $O(|E| \cdot \beta(|E|, |V|))$ time, rather than the very slightly faster algorithm of Gabow *et al.*, because the additional overhead of the latter seems certain to overwhelm its slight asymptotic gain. (We did not include the linear-time algorithm of Fredman and Willard in our study, for two reasons. First, it is so recent that little is yet known about implementation alternatives—the implementation described by its authors appears to suffer from significant overhead; and secondly, as it is closely related to the algorithm of Fredman and Tarjan, which we found not to be competitive, we believe that it would not be competitive either.)

Moreover, we experimented with both low- and intermediate-level implementations of the various priority queues used with Prim's algorithm, so that our study provides insights into the behavior of priority queue implementations when used within the setting of a typical greedy algorithm—thereby complementing the studies of Jones [8], who experimented with various implementations of priority queues under random distributions of operations (not including DECREASEKEY), and of Stasko and Vitter [12], who experimented with a variety of pairing heap designs under a variety of distributions of operations (including DECREASEKEY).

In order to avoid biassing our study because of a particular language, compiler, or architecture, we ran our tests on five different machines using two different languages. We also ran tests to evaluate the effect of paging and caching.

2 Methodology

As the actual performance of these implementations depends in various ways on the structure and density of the graphs (for instance, MSTs can be constructed in linear-time with a comparison-based algorithm for, among others, planar graphs), we decided to run experiments with graphs from four different families: (i) randomly generated graphs of specified densities; (ii) graphs of specified densities that represent worst cases for Prim's algorithm; (iii) sparse graphs induced by points placed at random in the Euclidean plane; and (iv) randomly generated trees. For the first two families, we chose densities corresponding to $|E| = |V|^{3/2}$ (dense graphs) and to $|E| = |V| \log_2 |V|$ (a density that represents a breakpoint, at least asymptotically, for Prim's algorithm). For each family and density of graphs, we generated random graphs in a range of sizes (up to 2^{17} nodes on sparse graphs and up to 2^{13} nodes and about 750,000 edges on dense graphs) and gathered running time data on 40 samples of each size.

We generate our random graphs of specified density by proceeding vertex by vertex. Let d be the density, $d = 2|E|/(|V| \cdot (|V| - 1))$; for vertex i, we generate exactly $(|V| - i) \cdot d$ edges connecting it to randomly chosen vertices of higher index and assign edge weights chosen from a uniform distribution. The worst-case graphs for Prim's method are constructed by connecting the first vertex to all other vertices (to foil the dynamic insertion versions), then using the same distribution of edges as for the first family, but running Prim's algorithm in adversary mode to determine connections and edge weights in order to ensure that every potential DECREASEKEY operation actually takes place and that, for binary and d-heaps, each DECREASEKEY operation makes its argument bubble all the way to the root (from the very bottom in the case of binary heaps). To generate graphs in the third family, points are randomly placed in the unit square, and, for each vertex, arcs are drawn to its five nearest neighbors; finally the graph is made undirected and duplicate edges are removed. We experimented with trees in the form of star graphs and line graphs, and with randomly generated trees; the latter are constructed by generating a random degree distribution and building the corresponding tree, then assigning edge weights by running Prim's algorithm in adversary mode in order to force each DECREASEKEY or INSERT operation to move all the way to the root in binary and d-heaps. As the lower density graphs thus generated in the first three families may not be connected, we simply reiterate the generation algorithm until a connected graph is generated.

In order to allow extrapolation and comparison with the asymptotic analysis of each algorithm, we computed the ratio of the running time of each implementation to that of a program that represents the ideal running time—i.e., a program that runs in $O(|E|)$ time with low coefficients (our actual baseline is a program that simply counts the number of edges in the graph). Our timing comparisons exclude the overhead of reading in or generating the data and of writing out the solution.

We ran our programs on five different machines: two CISC architectures (Vax-Station 2000 and SUN 3/60) and three RISC architectures (DEC 5000/200, SUN SparcStation II, and Silicon Graphics Iris 4D/340VGX), all running Unix (or Ultrix). Our programs are all coded extremely carefully and by the same author, so as to minimize any variations across implementations; in order to ascertain the effect of compilers on the running times, our programs were written in both Pascal and C—although we did all of our final testing in C. The code is written to be as efficient as possible, in terms of both time and space. Finally, we studied the effect of low-level implementation decisions on all architectures used in our study before settling on a standard implementation of each data structure: such considerations include the use of parallel arrays rather than records, the effect (in Pascal) of `with` statements on efficiency, the tradeoff between indirection and data moves, and so forth. In our final experiments, we used the Gnu compiler (`gcc`) on all machines except the MIPS-based machines (DEC 5000 and SG Iris), where we used the MIPS compiler (`cc`).

3 The Algorithms

We assume that the reader is already familiar with MST algorithms. Briefly stated, all algorithms operate greedily, choosing the next shortest edge that does not create a cycle and that may have to obey certain algorithmic constraints.

Kruskal's algorithm simply selects the next shortest edge that does not cause a cycle; it requires sorting the edges and a Union-Find data structure to detect cycles. Amortized over the execution of the algorithm, the latter data structure contributes a total cost of $O(|E| \cdot \alpha(|E| + |V|, |V|))$, where $\alpha(m, n)$ is the inverse Ackermann function. In the worst case, sorting takes $\Omega(|E| \log |V|)$ time and dominates the running time; demand sorting, either by using heaps or by using quicksort, has the same asymptotic worst case, but may be faster whenever Kruskal's algorithm produces a spanning tree without having to test each edge.

Prim's algorithm selects the the next shortest edge that touches the current subtree (effectively selecting a vertex to add to the tree) and requires a priority queue of vertices, with operations DELETEMIN and DECREASEKEY—the second operation being more common (especially in dense graphs) than the first. A binary heap or a d-heap both offer straightforward implementations, with logarithmic behavior for both operations; so do other heap structures, such as binomial queues, skew heaps, and leftist heaps. (For an overview of various implementations of priority queues, see [10], Section 3.1.) A splay tree or some other search tree, ordered by keys, also offers logarithmic behavior. The best data structure, in asymptotic terms, is a rank-relaxed or a Fibonacci heap, as both offer logarithmic DELETEMIN and amortized constant-time DECREASEKEY (Fredman *et al.* conjecture that pairing heaps offer the same asymptotic performance). Among data structures that we did not implement, finger search trees (balanced search trees with parent pointers and direct access to the leftmost leaf) offer amortized constant-time DELETEMIN and logarithmic DECREASEKEY, but, as the second operation is more common than the first, they give us a losing tradeoff over rank-relaxed and Fibonacci heaps. Regardless of the choice of priority queue, one is faced with a fundamental decision in the implementation of Prim's algorithm: whether to build the heap by preinserting all vertices with a key value of "infinity" or to build it dynamically, by inserting vertices when they become adjacent to the subtree under construction. The former immediately creates a priority queue of maximal size, but it can also take advantage of the special structure of the initial queue to build it in linear time; the latter spends additional time to ascertain whether an item is already in the heap, but generally keeps the priority queue much smaller. We experimented with both versions, although we used only dynamic insertion with pairing heaps, because preinsertion raises the unresolved issue of the ideal structure of a pairing heap when all keys are the same.

Fredman and Tarjan's algorithm may be viewed as a selective improvement of Prim's algorithm that eliminates the $|V| \log |V|$ expense of DELETEMIN operations by limiting the size of the priority queue, moving to a new subtree whenever the size of the current priority queue reaches a certain threshold; as such an approach eventually produce a spanning forest, each tree in the forest is then contracted (turned into a supervertex) and a new pass is started on the resulting graph. While any efficient heap structure can be adapted to work within this algorithm, we chose Fibonacci heaps without sacks; the algorithm also requires a Union-Find structure to maintain the supervertices.

Boruvka's algorithm, which we did not implement, proceeds in parallel by building a forest and adding to each tree of the forest the shortest edge with one endpoint in that tree; Cheriton and Tarjan's algorithm may be viewed as a selective serialization of that algorithm: it adds an edge to each subtree in the forest, but does so serially, using a queue-based scheduling and merging the pairs of subtrees as it goes. Like Kruskal's algorithm, Cheriton and Tarjan's algorithm selects edges and needs a Union-Find data structure to detect cycles; edges that are found to have both endpoints in the same subtree are not physically deleted (as the physical deletion would be too costly), deferring the actual deletion to further operations. A suitable implementation for meldable priority queues with this lazy deletion is the lazy leftist heap (see Tarjan [13]).

4 Results and Discussion

We begin with some observations concerning low-level implementation details and the effect of architectures. On the basis of these observations, we selected a style of implementation for each data structure and then ran tests (using Prim's algorithm) to compare various data structures used for implementing priority queues; we report the results of these comparisons in a second subsection. On the basis of these results, we chose one or two standard implementations for each of the four algorithms and ran tests to compare them; we present the results of these comparisons in the last subsection. (We collected far more data than are summarized here; further details will be found in a forthcoming technical report.)

4.1 Results on low-level implementation details

Apart from obvious differences in speed (in our experiments, the VaxStation 2000 is slowest; the SUN 3/60 runs about five to ten times faster; and the three RISC machines run yet faster, by a factor of five), we noted that the RISC machines were pretty much insensitive to low-level implementation choices, whereas such choices were very important for the CISC machines. In particular, the VAX implementations run faster when actually moving the data, while the SUN 3 implementations run faster when using indirection to avoid data moves (the SUN 3 is so efficient at pointer work that it executes our baseline routine almost as fast as the Sparc-Station II); in contrast, the RISC machines are essentially insensitive to these differences. Using C rather than Pascal attenuated many of the differences, both between machines (the code generated by the gcc compiler on the CISC machines was much faster than that generated by pc, whereas the code generated by either gcc or cc on the RISC machines was comparable to that generated by pc) and between choices of priority queues (for instance, rank-relaxed heaps consistently run faster than Fibonacci heaps when coded in Pascal, but not when coded in C, mostly as a result of the pointer arithmetic available in C.)

We also examined the influence of paging and caching. Our main conclusion regarding paging is the obvious one: if the various data structures occupy too much memory, the (wallclock) running times suffer enormously, because none of the algorithms exhibits strong locality of reference. The effectiveness of caching is strongly affected by the method used for storing graphs. Our baseline ran 4 times more slowly on graphs stored in the order in which they were generated than on the same graphs read in from a file where they had been stored in adjacency list order; the effect is far smaller (and more variable) on the MST algorithms, since they cause a lot of nonsequential memory references anyway. We used generation order for storage in our final experiments, since the effect of caching then tends to be cancelled when taking execution ratios.

Since our experimental results show that speed ratios as large as two can arise from seemingly arbitrary implementation choices on the VAX and SUN 3, it is clear that the RISC machines are preferable for the purpose of our experiments; since the Iris machine had the largest memory (65 Mbytes) of our various RISC platforms, we show below the results obtained on it.

4.2 Results on priority queues implementations

Our findings here are of two types: first, given a specific data structure, what improvements can be made that will make it more efficient? And secondly, how do the best versions of various data structures compare?

Driscoll *et al.* suggested an improvement for rank-relaxed heaps:[2] as the normal pointer implementation suffers from high coefficients, they suggested storing in each node an array of child pointers of approximately $\log_2 |V|$ in length and, in order to keep the storage down to $O(|V|)$, grouping approximately $\log_2 |V|$ vertices together into one heap node—which we call a "sack." Such an implementation should speed up the DECREASEKEY operation on average, since the change in key affects one element of the sack, but need not affect the sack's position within the heap. We realized that such an improvement could also be wrought on any tree-based implementations of priority queues and so experimented with sacks in Fibonacci and binary heaps as well as in rank-relaxed heaps. (Sacks give rise to further questions: should they be kept packed? and should storage within them be reclaimed? We found that packing the children as tightly as possible within each sack is effective for sparse graphs, becoming increasingly less useful with increasing density—but it seems to make little difference in any case.) Sacks proved very important for Fibonacci and rank-relaxed heaps: for all graph families except trees (where, with dynamic insertion, DECREASEKEY operations never occur), the versions using sacks always run faster than the standard implementations. The speed-up is most noticeable when dynamic insertion is not used, but it remains

---

[2]They also suggested two strategies for deletion, which they call "reclaiming" and "nonreclaiming." We found that reclaiming versions are slightly better for dense graphs, becoming less useful with decreasing edge density; however, the overall effect is minimal. We used reclaiming versions in final testing.

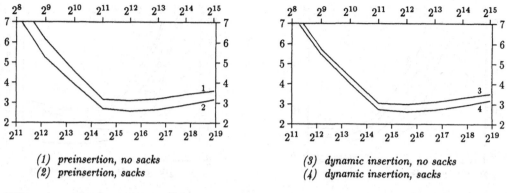

(1) *preinsertion, no sacks*
(2) *preinsertion, sacks*

(3) *dynamic insertion, no sacks*
(4) *dynamic insertion, sacks*

Figure 1: The effect of sacks on Prim's algorithm with Fibonacci heaps, using random graphs of density $|V|\log|V|$

significant in all cases. Figure 1 illustrates the behavior (as ratios to the baseline) of Prim's algorithm using Fibonacci heaps, in linked implementations, with and without sacks, and with static or dynamic insertion. (In this figure, as in all following ones, the size of the graph is shown on the logarithmic horizontal scale, with the number of vertices at the top and the number of edges at the bottom of the frame.) Another effect of sacks is a drastic reduction in storage requirements. In the case of Fibonacci heaps, and within the range of sizes we used. implementations with sacks used only half as much storage as implementations without; they oftne use less storage than conventional binary heaps.

We implemented binary heaps with the "bounce" heuristic of Floyd (see [10], Section 8.6); since this heuristic decreases the number of comparisons and does not significantly alter the number of moves, it routinely improves the behavior of the binary heaps. We tried various choices for storing the keys: in the heap; outside the heap with a pointer reference; both inside and outside. The last version is fastest, but requires additional $|V|$ storage; we chose to store the key outside the heap, a version that is barely slower and saves on storage. Although d-heaps offer better asymptotic performance than binary heaps for Prim's algorithm, their performance was uniformly worse in practice. The problem is due to the manner in which d was determined. Johnson [9] chose d so as to balance the work of the DELETEMIN operations ($|V|\cdot d\log_d|V|$) against the DECREASEKEY operations ($|E|\log_d|E|$), a choice that can be made when starting the algorithm. However, these two terms should be balanced based on the *actual* number of DECREASEKEY operations performed, call it m, not on the upper bound of $|E|$; but m cannot be computed ahead of time. Since m is much less than $|E|$ in most of our experiments, binary heaps actually yield a better balance than d-heaps; even for worst-case graphs, where m is maximized, d-heaps remain slower, due to larger overhead.

We implemented splay trees with top-down splaying—which offers a clearly better tradeoff between speed and storage than bottom-up splaying—and ran two versions: one with, and one without, parent pointers. The storage cost of the extra parent pointer is well justified, as the version with parent pointers consistently

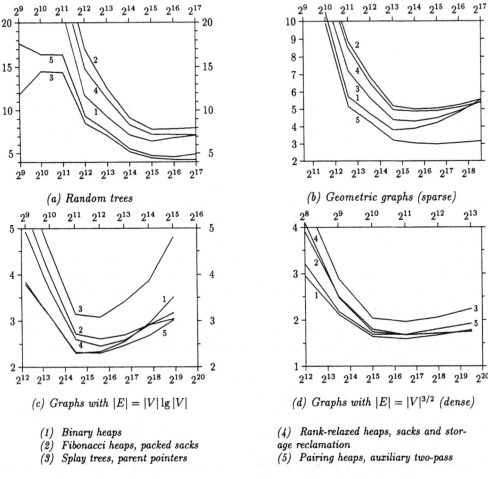

(a) Random trees

(b) Geometric graphs (sparse)

(c) Graphs with $|E| = |V| \lg |V|$

(d) Graphs with $|E| = |V|^{3/2}$ (dense)

(1) Binary heaps
(2) Fibonacci heaps, packed sacks
(3) Splay trees, parent pointers
(4) Rank-relaxed heaps, sacks and storage reclamation
(5) Pairing heaps, auxiliary two-pass

Figure 2: Experimental results with priority queues for Prim's algorithm

runs twice as fast as the version without. (In a splay tree, a DECREASEKEY is a DELETE followed by an INSERT and parent pointers allow most DELETE operations to remain local.)

Finally, we implemented four versions of pairing heaps: two-pass and multi-pass, with or without the auxiliary strategy proposed by Stasko and Vitter. We did not use sacks and used only insertion on demand. Unlike Stasko and Vitter, who reported that the auxiliary two-pass method consistently (if not always clearly) outperformed the others, we found it impossible to rank the four strategies. On the trees and the worst-case graphs for Prim's algorithm, the auxiliary methods ran slightly more slowly, while no difference could be discerned on the dense graphs.

Turning now to a comparison of the various priority queues, we conclude that pairing heaps are preferable. In our implementation (without sacks), they use a little more storage than binary heaps (25% more in the context of Prim's algorithm). As to speed, they are consistently the fastest or second fastest implementation, as evidenced by the data of Figure 2 (all methods illustrated in the figure use insertion on demand). Note that, even in the case of random trees, where (due to dynamic insertion) DECREASEKEY operations do not occur, pairing heaps remain very close to the best structure—which, in agreement with Jones' results, is the splay tree. An intriguing result is that pairing heaps do not seem to offer asymptotic performance equal to that of Fibonacci heaps, as the curves for the dense graphs, where Prim's algorithm runs in linear time, show; more testing on this issue is clearly necessary.

4.3 Results on the four algorithms

The remaining issue with Prim's algorithm is the desirability of dynamic insertion. We observed that the main advantage of dynamic insertion is that it keeps the priority queue small; this characteristic may not translate into smaller storage requirements, because our implementations allocate storage in large chunks (e.g., a whole array) in order to gain efficiency, but it may translate into smaller execution times. As a result, dynamic insertion should be more beneficial with sparse graphs, since these graphs generally give rise to small priority queues during execution. However, the tradeoff also depends on the cost of single insertions within the priority queue and on the cost and number of DECREASEKEY operations. We found that dynamic insertion is a clear winner on all sparse graphs (our geometric and tree families), for all priority queues; for the trees, in particular, the running times with preinsertion are often several times larger than those with dynamic insertion. On denser graphs, dynamic insertion remains the better choice for splay trees and rank-relaxed heaps (although the margin of gain decreases with increasing density), while the choice between it and preinsertion becomes indifferent for classical and Fibonacci heaps.

Turning to Kruskal's algorithm, we found, as expected, that sorting on demand is preferable to presorting in all graphs of moderate to high density; the difference is so marked that even the inefficient heaps yield better running times than the highly optimized quicksort presorting pass. In contrast, even the interruptible quicksort is slower than a presorting pass when run on our geometric and tree graphs. Overall, Kruskal's algorithm always runs significantly more slowly than Prim's and also uses considerably more storage (especially on dense graphs, because its storage requirements are proportional to $|E|$, whereas Prim's are proportional to $|V|$). On the other hand, using adjacency lists as our graph representation imposes a "preparation" penalty on Kruskal's algorithm, which must begin by scanning all edges and writing them into some auxiliary storage area; using an array of edges instead of adjacency lists alters the situation, assuming that the array can be overwritten during sorting. In order to get an estimate of the various costs, we recorded running times with and without the preparation time.

Cheriton and Tarjan's algorithm requires the most storage of all our implementations, closely followed by Fredman and Tarjan's algorithm. Note that the latter reduces to a standard Prim's algorithm with Fibonacci heaps whenever the density of the graph is at least $|E| = |V| \log |V|$ and also on line graphs; thus it cannot

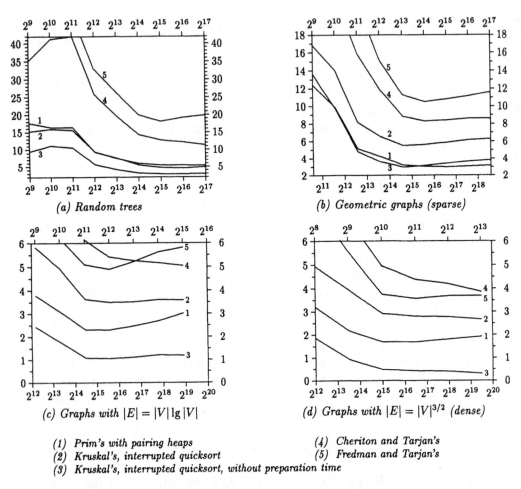

(a) Random trees

(b) Geometric graphs (sparse)

(c) Graphs with $|E| = |V| \lg |V|$

(d) Graphs with $|E| = |V|^{3/2}$ (dense)

(1) Prim's with pairing heaps
(2) Kruskal's, interrupted quicksort
(3) Kruskal's, interrupted quicksort, without preparation time
(4) Cheriton and Tarjan's
(5) Fredman and Tarjan's

Figure 3: Experimental results with the four comparison-based algorithms

be competitive on these graphs, since it suffers from significant overhead, in both time and space, when compared to Prim's algorithm. On random trees, Fredman and Tarjan's algorithm used two passes; but it switched to new subtrees so often (on a tree of 2^{17} vertices, it built a total of over 10^5 new subtrees!) that the overhead involved in such switches overwhelmed any advantage inherent in the smaller heaps. Neither Cheriton and Tarjan's nor Fredman and Tarjan's algorithm ever approached the speed of Prim's algorithm on any of our graph families.

Figure 3 shows the comparative running times of the four algorithms on our three main graph families: Prim's algorithm with pairing heaps, Kruskal's algorithm with interrupted quicksort, with and without the time needed to set up the edge array, Cheriton and Tarjan's algorithm, and Fredman and Tarjan's algorithm. Note that Kruskal's algorithm, if given an (unsorted, modifiable) array of edges, actually runs very quickly—often much faster than the best version of Prim's algorithm: on the dense graphs, it need only sort a fraction of the input before the spanning tree is complete, while on the sparse graphs, the very small overhead of

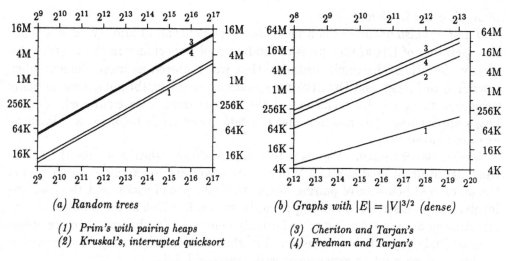

(a) Random trees

(b) Graphs with $|E| = |V|^{3/2}$ (dense)

(1) Prim's with pairing heaps
(2) Kruskal's, interrupted quicksort

(3) Cheriton and Tarjan's
(4) Fredman and Tarjan's

Figure 4: Storage requirements of the four algorithms

quicksort allows the complete sorting to proceed very quickly. However, not only must the input be in this special array form, but the worst-case behavior remains as poor as ever: we ran experiments with worst-case graphs—where we force demand sorting to process all edges by generating graphs made of two distinct components connected by a single, very long bridge—and found that Kruskal's algorithm runs slowly with or without the preparation time. Versions of Kruskal's algorithm based on adjacency lists use much more additional storage, as shown in Figure 4—showing the relative storage used (in addition to the graph itself) by the four methods on trees and on graphs of density $|E| = |V|^{3/2}$.

We have not shown any actual times, but let us simply note that, on the Iris, the source for the figures shown here (but also the slowest of the three RISC machines we used), the baseline for a dense graph with 8,192 vertices and about 750,000 edges takes about 3.5s and Prim's algorithm takes about 7.1s, while the baseline for a geometric graph with about 130,000 vertices and 390,000 edges takes 2.0s and Prim's algorithm takes about 6.0s. Thus very large graphs can be processed very quickly; indeed, the major limiting factor with these efficient algorithms is not time, but space: processing a graph with about 750,000 edges calls for 16 to 50 Mbytes of memory, depending on the algorithm chosen.

5 Conclusion and Future Work

We have presented the results of an exhaustive experimental investigation of comparison-based algorithms for the construction of a minimum spanning tree for a weighted graph. We conclude that the best algorithm for graphs given by their adjacency lists remains the simplest: Prim's algorithm implemented with dynamic insertion and with pairing or binary heaps. These implementations are least demanding of all in terms of storage and ran as fast as, or often much faster than, all other implementations.

Our study also uncovered new facts about the behavior of priority queues: since the operation of DECREASEKEY is of fundamental importance in the context of a greedy algorithm, the implementation that was reported as most efficient when restricted to INSERT and DELETEMIN operations—namely, the splay tree, actually turns out to be the least efficient of the various choices we examined. Pairing heaps appear best, but may not ofer the same asymptotic behavior as Fibonacci or rank-relaxed heaps.

Several issues remain. We want to compare these algorithms with the linear-time algorithm of Fredman and Willard. We also want to examine in more detail the amortized behavior of pairing heaps, the influence of sacks, and the tradeoffs involved in preinsertion. Most importantly, we want to look at the problem of constructing minimum spanning trees for truly enormous graphs (one or more orders of magnitude beyond those considered in this study), presumably in a geometric setting; and we want to experiment with real-world data.

References

[1] Cheriton, D., and R.E. Tarjan, "Finding minimum spanning trees," *SIAM J. Comput.* **5** (1976), pp. 724–742.

[2] Driscoll, J.R., H.N. Gabow, R. Shrairman, and R.E. Tarjan, "Relaxed heaps: an alternative to Fibonacci heaps with applications to parallel computation," *Comm. ACM* **11** (1988), pp. 1343–1354.

[3] Fredman, M.L., R. Sedgewick, D.D. Sleator, and R.E. Tarjan, "The pairing heap: a new form of self-adjusting heap," *Algorithmica* **1** (1986), pp. 111—129.

[4] Fredman, M.L., and R.E. Tarjan, "Fibonacci heaps and their use in improved network optimization algorithms," Proc. 25th Ann. IEEE Symp. Foundations Comput. Sci. FOCS-84, pp. 338–346; also in final form in *J. ACM* **34** (1987), pp. 596–615.

[5] Fredman, M.L., and D.E. Willard, "Trans-dichotomous algorithms for minimum spanning trees and shortest paths," Proc. 31st Ann. IEEE Symp. Foundations Comput. Sci. FOCS-90, pp. 719–725.

[6] Gabow, H.N., Z. Galil, T.H. Spencer, and R.E. Tarjan, "Efficient algorithms for minimum spanning trees on directed and undirected graphs," *Combinatorica* **6** (1986), pp. 109–122.

[7] Graham, R.L., and O. Hell, "On the history of the minimum spanning tree problem," *Ann. Hist. Comput.* **7** (1985), pp. 43–57.

[8] Jones, D.W. "An empirical comparison of priority-queue and event-set implementations," *Commun. ACM* **29** (1986), pp. 300–311.

[9] Johnson, D.B., "Priority queues with update and finding minimum spanning trees," *Inf. Process. Lett.* **4** (1975), pp. 53–57.

[10] Moret, B.M.E., and H.D. Shapiro. *Algorithms from P to NP. Volume I: Design and Efficiency.* Benjamin-Cummings, Menlo Park, CA, 1991.

[11] Sleator, D.D., and R.E. Tarjan, "Self-adjusting binary trees," Proc. 15th Ann. ACM Symp. Theory Comput. STOC-83, pp. 235–245; also in final form in *J. ACM* **32** (1985), pp. 652–686.

[12] Stasko, J.T., and J.S. Vitter, "Pairing heaps: experiments and analysis," *Commun. ACM* **30** (1987), pp. 234–249.

[13] Tarjan, R.E. *Data Structures and Network Algorithms.* SIAM, Philadelphia, 1983.

A linear time algorithm for computing the shortest line segment from which a polygon is weakly externally visible

by

Binay K. Bhattacharya
School of Computing Science
Simon Fraser University,
Burnaby, B.C., Canada, V5A 1S6.

Asish Mukhopadhyay
Department of Computer Science
Indian Institute of Technology,
Kanpur, India.

and

Godfried T. Toussaint
School of Computer Science
McGill University
Montreal, Canada.

Abstract

A simple polygon P is said to be weakly externally visible from a line segment if the line segment is outside P and if for every point x on the boundary of P there is a point y on the line segment such that the interior of the line segment xy does not intersect the interior of P. In this paper a linear time algorithm is proposed for computing the shortest line segment from which a simple polygon is weakly externally visible. This is done by a suitable generalization of a linear time algorithm which solves the same problem for a convex polygon.

1. Introduction

We assume that we have a simple polygon P which is weakly externally visible. This means that for every point p on the boundary of the polygon we can draw a ray from p such that it does not intersect the interior of P. P is weakly externally visible from a line L lying outside it if for any point p on the boundary of P there is a point y on L such that the line segment py does not intersect the interior of P. It is shown in [AT81] that this is equivalent to every vertex of P being weakly externally visible from L.

Associated with every vertex p of the polygon P, we have what is called an external cone of support, K(p), whose vertex is at p and whose bounding lines are the rays drawn from p to lie outside of P and touching it (see Fig. 1). We label the bounding rays of K(p) as clockwise (CW) and counter-clockwise (CCW) as follows. When we traverse the boundary of the cone in clockwise order starting from infinity (i.e. the interior of the cone lies to the right), the first bounding ray traversed is called the CW ray. Similarly, when we traverse the boundary of the cone in counter-clockwise order starting from infinity, then the first boundary ray traversed is called the CCW ray. These labels are shown in Fig.1. It is easy to see that a common transversal L of these cones is a line from which the polygon P is weakly externally visible. The classical problem of finding a common transversal of a finite set of simple geometric objects (usually convex) has been studied extensively [Gr58], [Le80], [AW87], [Ed85], etc. The best method to determine a common transversal for an arbitrary set of cones is due to Atallah and Bajaj[AB87]. Their algorithm takes $O(n\alpha(n)logn)$ in the worst case where $\alpha(n)$ is the inverse Ackermann function. However, if the cones are the external cones of support of the vertices of a weakly externally visible polygon then a common transversal can be computed in $O(n)$ time [BKT89].

Yan Ke [K88] considered the problem of detecting the weak visibility of a polygon from an internal line segment. He showed that the shortest such segment can be found in $O(nlogn)$ time. The problem of computing the shortest line segment from which P is weakly externally visible can be transformed to the problem of computing the shortest transversal for a set of external cones [BKT89]. If the cones are arbitrary then the shortest transversal can be computed in $O(nlog^*n)$ time [BCESTU91]. In [BT90], a linear time algorithm is given for computing the shortest line segment from which a convex polygon is weakly externally visible. In this paper we show that a linear time solution is possible even when the polygon is simple.

As a practical application of the problem, imagine a scenario in which we need to work out the shortest linear stretch a patrol would need to cover to keep watch on an enemy fort from outside. The above problem is an abstraction of this guarding problem.

The paper is organized as follows. In the next section, we discuss some of the preliminaries needed to understand the algorithm. The third section describes the algorithm and the following section contains an implementation of the algorithm. We round up in the fifth and final section.

2. Geometric properties

We first show that

Lemma 1: The shortest line segment must touch the convex hull of the polygon.

Proof: Clearly the line containing uv can not intersect the polygon; otherwise, the polygon can not be externally visible from uv. Therefore, the convex hull of the polygon must lie on one side of the line. Suppose the shortest line segment uv does not touch the convex hull of the polygon. We then determine two supporting lines of the polygon through u and v as shown in Fig.2. These two supporting lines must intersect at an extreme vertex p of the polygon, otherwise the polygon is not weakly visible from uv. Thus the polygon is contained in the triangle uvp thus formed.

Let $l_i = uv \cap K(p_i)$ for some vertex p_i of P. Let x_i be a point belonging to l_i. Then a ray joining p_i to x_i is either collinear with pu or pv or it intersects the interior of the line segment uv. Therefore, we can determine another line segment parallel to uv with end points lying on pu and pv which will be shorter than uv and from which the given polygon is weakly externally visible. This contradicts our earlier assumption. **Q.E.D.**

As mentioned earlier, it has been shown in [AT81] that if the vertices of P are weakly externally visible from a line L, the entire polygon is weakly externally visible from L. However, this is not generally true for a line segment s. In Fig. 3 all the vertices are visible from s. But the entire polygon is not visible from s since the boundary point x is not visible. Let p be a vertex of the convex hull of the polygon. Suppose s is the line segment from which the polygon is weakly externally visible and is constrained to touch the antipodal chain of p. We will refer s as the **line segment antipodal to p**. In Lemma 1 it has been observed that s must intersect both the bounding rays of p.

Our algorithm depends crucially on the algorithm developed in [BT90] for the case when P is a convex polygon. Key to this latter algorithm were the following two optimization results proved in [BT90].

Lemma 2: If p is a point lying inside a wedge bounded by the semi-infinite rays OQ and OR then the length of a line segment HK through p, with its end points H and K resting on OQ and OR respectively, is a unimodal function of θ for $\phi < \theta < \pi$, where ϕ is the angle of the wedge and θ is the angle HK makes with OR.

Lemma 3: Let P be a convex polygon contained in OQR, and HK a line segment tangent to P, with its end points H and K resting on OQ and OR respectively. The length of HK is a unimodal function of θ for $\phi < \theta < \pi$, where ϕ is the angle of the wedge and θ is the angle HK makes with OR.

First of all we show that the statement of Lemma 2 remains true even when we replace OQ and OR by convex polygonal chains as shown in Fig.4.

Lemma 4: Let OQ = $[\, Oq_1q_2q_3 \cdots q_m\,]$ and OR = $[Or_1r_2r_3 \cdots r_n]$ be two convex polygonal chain forming a concave wedge and p a point inside the wedge. If HK is line segment through p, with its end points H and K resting on OQ and OR respectively, then the length of HK is a unimodal function of θ, where $\alpha < \theta < \beta$; θ is the angle HK makes with Op, α and β are respectively the angles that rays parallel to $q_{m-1}q_m$ and $r_n r_{n-1}$ through p make with Op.

Proof: In Fig.4, HK is a line segment through p with H resting on $q_{i-1}q_i$ and K resting on $r_{j-1}r_j$. Let $r_{j-1}r_j$ and $q_{i-1}q_i$ be extended to meet at O'. Now we can appeal to Lemma 2 to determine if the length of HK, |HK|, increases or decreases if rotated counter-clockwise. We prove the unimodality of |HK| by showing that once HK starts increasing, it continues to do so.

Suppose |HK| increases as we rotate HK counter-clockwise. Sooner or later H reaches q_{i-1} or K reaches r_j. Suppose K reaches r_j earlier than H reaches q_{i-1}. Suppose H''K' is another placement which intersects the concave wedge at K''. Then |HK| < |H'r_j| < |H''K'| < |H''K''|. Now we switch to the cone formed by $r_j r_{j+1}$ and $q_{i-1}q_i$ intersecting at O''. We already know that |H'r_j| < |H''K''|. In this case the length of HK keeps increasing when rotated counter-clockwise due to the unimodality property established in Lemma 2. Exactly the same argument holds if the end point H reaches q_{i-1} earlier. In this case we switch to the cone formed by $r_{j-1}r_j$ and $q_{i-2}q_{i-1}$. If the corners are reached simultaneously we switch to the cone formed by $r_j r_{j+1}$ and $q_{i-1}q_{i-2}$ and similar arguments show that |HK| increases when further rotated counter-clockwise.

This proves that |HK| is a unimodal function of θ. **Q.E.D.**

3. The algorithm

We now take a brief look at how an O(n) time algorithm for finding the shortest line segment from which a convex polygon is weakly externally visible was designed in [BT90] based on Lemmas 2 and 3.

Let c_i be the vertex of a convex polygon P and suppose c_{i-1}, c_{i+1} are the vertices adjacent to c_i (see Fig.5). If the tangents to P parallel to $c_{i-1}c_i$ and $c_i c_{i+1}$ touch P at s_i to t_i respectively then the convex chain from s_i and t_i in counter-clockwise order is said to be the antipodal chain of p. We know from Lemma 1 that the shortest line segment from which P is weakly externally visible must touch P and if the line segment is antipodal to c_i, the line segment must touch P between s_i and t_i, and the end points must rest on the bounding rays (i.e. $c_i c_{i-1}$ and $c_i c_{i+1}$) of the cone of external support $K(c_i)$ of c_i.

The algorithm is now clear. For each vertex c_i, we make use of Lemmas 2 and 3 to find out the shortest line segment touching the antipodal chain of c_i with its end points on the bounding rays of $K(c_i)$. The shortest of all such line segments taken over all the vertices of P is the required line segment. The linear time solution of the method results from the fact that the antipodal chains of two adjacent vertices of P have only one vertex in common.

To extend the algorithm to the case of a simple polygon P, we need a suitable generalization of Lemma 3. What if the point p in Lemma 4 is replaced by a convex polygon lying inside the wedge, with one of its vertices at the apex of the wedge? This brings us to our next Lemma.

Lemma 5: Let C be a convex polygon lying inside a concave wedge bounded by two convex polygonal chains $OQ = [Oq_1q_2...q_m]$ and $OR = [Or_1r_2...r_m]$, with O also being one of the vertices of C. If HK is a line segment lying inside the wedge, touching C, with the end points H and K resting on OQ and OR respectively, then |HK| is a unimodal function of θ, where $\alpha < \theta < \beta$; θ, α and β are the angles that HK, $q_{m-1}q_m$ and r_nr_{n-1} make with a fixed direction through O respectively.

Proof: We will show that |HK| increases monotonically once it starts increasing as HK is rotated around C. Suppose |HK| is increasing at p when HK is rotated counter-clockwise. (Fig 6) Suppose H_1K_1 is the line segment through pq. Let H_2K_2 be the line through q parallel to HK. Since |HK| is increasing when rotated counter-clockwise, $|H_2K_2| < |HK| < |H_1K_1|$ [Fig.6]. Hence if we apply the result of Lemma 4 to point q, we observe that HK in the position H_2K_2 either decreases to a minimum and then increases or just keeps increasing so that in either case it is increasing as HK is rotated counter-clockwise around q. Therefore, HK in the position H_1K_1 is increasing in counter-clockwise direction. We can apply the same argument to the succeeding vertices of C.

This shows that |HK| is a unimodal function of θ. **Q.E.D.**

When the given polygon is simple, the generalization of the algorithm for a convex polygon becomes non-trivial in several ways. However a couple of pertinent observations gives us the required insights. We noted earlier that the shortest line segment must intersect all the external visibility cones of support.

Let uv be the shortest line segment from which the polygon is weakly externally visible. Suppose uv touches P at z. Suppose uv is antipodal to p. p and z are called antipodal vertices with respect to the direction of the line segment. p and z split the boundary of P into two chains, which will be known as CW and CCW chains from p to z. The CW chain from p to z is the polygonal chain of P from p to z traversed clockwise. Similarly, the counter-clockwise chain from p to z is the polygonal chain of P from p to z traversed counter-clockwise. We now show that

Lemma 6: u must lie on the CCW bounding ray of the external cone of support of a vertex in the CCW chain and v must lie on the CW bounding ray of the external cone of support of a vertex in the CW chain.

Proof: It is easy to see that uv must be contained in K(z). It is also true that uv must intersect both the bounding rays of K(p). We also notice that the CCW bounding ray of K(q) for each q, q ≠ z, p, in the CCW chain intersects the line segment zu, say at a_q. The intersection of the other bounding ray of K(q) with zu, if it exists, must be closer to u than a_q. Since uv is the shortest line segment, u must be determined by a CCW bounding ray of some vertex in the CCW chain. Similarly, we can show that v must be determined by a CW bounding ray of some vertex in the CW chain. **Q.E.D.**

The above Lemma suggests that in order to find the shortest line segment externally visible to P with its slope constrained to lie in a range with respect to which z and p are antipodal vertices, we group all the CCW bounding rays of the cones of external support of the vertices in the CCW chain, determined by p and z, into one. We then compute the envelope of these CCW bounding rays, such that for any point x on the envelope, a supporting ray, starting from x and away from the interior of the cones of external support of the vertex on whose CCW bounding ray x lies, intersects each of the cone of external support of the vertices of the CCW chain from p to z. This envelope will be referred to as the CCW envelope for the vertices of the polygonal chain from p to z in counter-clockwise direction, denoted as CCW_envelope [p,z].

Fig.7 shows the CCW_envelope [p,z] of a polygon P. We can similarly define the clockwise envelope of a polygon P for a given z and p, denoted as CW_envelope [z,p].

We can define the CW and CCW envelopes for the vertices in each pocket of P analogously. If C = $[c_1 c_2...c_k]$ is the convex hull of P, then the portion of the polygon P between successive vertices c_i and c_{i+1} of C constitute a pocket of P, denoted by $[c_i, c_{i+1}]$. Suppose the vertices of C are given in CCW order. Then the CCW envelope of the pocket $[c_i, c_{i+1}]$ is the envelope determined by the counter-clockwise bounding rays of the vertices in the pocket including the vertices c_i and c_{i+1}. This envelope is denoted by CCW_envelope$[c_i,c_{i+1}]$. The CW envelope of the pocket $[c_i, c_{i+1}]$ is similarly defined and is denoted by CW_envelope$[c_i,c_{i+1}]$.

We will use the term envelope in a broader sense. Suppose we are interested in merging two CW or two CCW envelopes. The merged picture should be such that for any x on the merged picture, the supporting ray r_x starting from x away from the interior of the external cone of support of the vertex on whose bounding ray x lies, intersect each of the bounding rays of the involved envelopes. We call this merged picture as envelope also. It should be noted that r_x may not intersect all the bounding rays which are represented by the envelopes. This generalization allows one to define envelopes of the form CW_envelope $[c_i,c_i]$ or CCW_envelope $[c_i, c_i]$ for an arbitrary c_i.

Let s_i and t_i be the points of contact of tangents to P drawn parallel to $c_{i-1}c_i$ and $c_i c_{i+1}$ respectively. From the above discussion it follows that the shortest line segment antipodal to c_i, touches P at some vertex u of the antipodal chain of c_i and its end points rest on CW_envelope [u, c_i] and CCW_envelope $[c_i,$u].

Suppose we want to find the shortest line segment which is externally visible to P and is antipodal to c_i. Suppose we start the search from s_i. The initial direction of the line segment tangent at s_i is parallel to $c_{i-1}c_i$ and the terminating position is given by the tangent at t_i parallel to $c_i c_{i+1}$. These are determined by the fact that in the least the shortest line segment antipodal to c_i intersects both the bounding $K(c_i)$.

At this stage an important difference shows up as compared to the convex polygon case. While trying to find the shortest line segment antipodal to the vertex c_i, the bounding rays on which the end points must lie change dynamically. Suppose we have determined the shortest line segment antipodal to c_i and supporting P at s_i such that its endpoints lie on the CW_envelope $[s_i,c_i]$ and the CCW_envelope $[c_i,s_i]$. If u is the next convex hull vertex which will be the new point of contact, the CCW_envelope $[c_i,s_i]$ needs to be merged with the CCW_envelope $[s_i,$u] to obtain the CCW_envelope $[c_i,$ u]. Similarly, the CW_envelope $[s_i,c_i]$ needs to be updated by removing from it the CW_envelope $[s_i,$u] to obtain the CW_envelope [u, c_i]. This process will be referred to as 'unmerging' of the CW_envelope $[c_i,s_i]$ by the CW_envelope $[s_i,$ u].

Having found the shortest line segment antipodal to the vertex c_i, we find the shortest line segment for the next vertex c_{i+1} and repeat the process. Here again the CW_envelope $[t_i, c_i]$ and the CCW_envelope $[c_i,t_i]$ need to be updated. Let the convex chain from s_{i+1} to t_{i+1} be the antipodal chain of the vertex c_{i+1}. Clearly, $s_{i+1} = t_i$. The CCW_envelope $[c_{i+1},s_{i+1}]$ can be obtained from the CCW_envelope $[c_i,t_i]$ by 'unmerging' the CCW_envelope $[c_i,c_{i+1}]$ from it. Similarly, the CW_envelope$[s_{i+1}, c_{i+1}]$ can be obtained from the CW_envelope $[t_i,c_i]$ by merging the CW_envelope $[c_i,c_{i+1}]$ with it. When all the hull vertices are examined this way, we determine the shortest line segment from which the simple polygon is weakly externally visible.

We can now state the algorithm more formally as follows:

Algorithm shortest-line-segment

Input: An n-vertex weakly externally visible simple polygon.
Output: The shortest line segment from which P is weakly externally visible.

Step 1: Find the convex hull $C = [c_1 c_2 .... c_k]$ of P;
Step 2: Compute the external cone of visibility for each vertex of P.
Step 3: Compute the CW and CCW envelopes of each pocket $[c_j, c_{j+1}]$, j= 1,2, ...,k, where $c_{k+1} = c_1$.
   Set $i \leftarrow 1$; $s \leftarrow \infty$;
Step 4: Determine the antipodal chain for the vertex c_i {Let the chain be from s_i and t_i in counter-clockwise direction.}
Step 5: Determine the envelopes CW_envelope $[s_i, c_i]$ and CCW_envelope $[c_i, s_i]$.
Step 6: Determine the shortest line segment antipodal to c_i, dynamically updating the envelopes.
Step 7: If the length of the shortest line segment < s then reset s to the length of the current shortest line segment.
Step 8: Determine the envelopes CW_envelope $[t_i, c_{i+1}]$ and CCW_envelope $[c_{i+1}, t_i]$ from the respective CW_envelope $[t_i, c_i]$ and CCW_envelope $[c_i, t_i]$.
Step 9: Determine the antipodal chain for the vertex c_{i+1}. {Let the chain be from s_{i+1} and t_{i+1} in CCW direction.}
Step 10: $i \leftarrow i+1$; if i > k stop; else repeat steps 6 through 9.

4.   A linear time implementation of the algorithm.

We now prove the following Lemma.

Lemma 7: The envelopes of a pocket $[c_j, c_{j+1}]$ can be computed in time proportional to the number of vertices in the pocket.

Proof: We describe the computation of the CCW envelope. The CW envelope is similarly computed. We compute the envelope as we go counter-clockwise from c_j to c_{j+1} along the polygon boundary. The basic step is to merge the current envelope with the next (counter-clockwise) bounding ray.

We first show that a bounding ray a can intersect the current envelope at most once. If possible suppose it intersects twice [Fig.8(a)]. Let b and c be the bounding rays that contain the edges of the current envelope which intersect a. Suppose a, b and c are the bounding rays of the vertices p_a, p_b and p_c respectively. Let the rays b and c intersect at x. The region determined by the polygonal chain from p_b to p_c and the line segments $p_b x$ and $p_c x$ can not contain any vertex of P in its interior. p_a also can not lie on the polygonal chain from p_b to p_c. Therefore, the bounding ray of $K(p_a)$ can not intersect the current envelope more than once without violating the simplicity property of P or the weak external visibility property of P. The merging of the current envelope with the bounding ray can be implemented by following the current envelope starting from the semi-infinite edge [Fig.8(b)]. The time taken is proportional to the portion of the current envelope not present in the merged picture. We are implicitly assuming that the current envelope is available in a doubly linked list. A simple accounting scheme shows that the procedure is linear in the number of vertices considered. Q.E.D

Using the arguments given in Lemma 7 we can also show that

Lemma 8: Two CW or CCW envelopes can intersect at most once and can be merged in time proportional to the portion of the envelopes not present in the merged envelope.

Suppose we know CW_envelope $[s_i, c_i]$ and CCW_envelope $[c_i, t_i]$. These two envelopes must meet at c_i thus forming a wedge W_i. The convex hull of P is the convex polygon inside the wedge W_i with c_i forming the apex. We now show that

Lemma 9: The shortest line segment with end points resting on CW_envelope $[s_i, c_i]$ and CCW_envelope $[c_i, t_i]$ and touching the convex chain from s_i to t_i is the shortest line segment antipodal to c_i.

Proof: We prove the Lemma by showing that during the searching process for each vertex u in the antipodal chain from s_i to t_i, the relevant parts of the CW_envelope $[u, c_i]$ for the searching vertex u and the CCW_envelope $[c_i, u]$ are exactly the same as the relevant parts of the CW_envelope $[s_i, c_i]$ and the CCW_envelope $[c_i, t_i]$ for the searching vertex u.

Consider any line segment xy touching P at u with x and y resting on the CW_envelope $[s_i, c_i]$ and the CCW_envelope $[c_i, t_i]$ respectively. Consider the envelopes CW_envelope $[u, t_i]$ and CCW_envelope $[s_i, u]$. The CCW_envelope $[u, t_i]$ can not interesect uy. Therefore, y lies on the envelope determined by the CCW bounding rays of the vertices from c_i to u only. Hence the relevant parts of the CCW_envelope $[c_i, t_i]$ and the CCW_envelope $[c_i, u]$ are the same when the examining line segments touch u. Similarly, we can show that the parts of the CW_envelope $[s_i, c_i]$ and the CW_envelope $[u, c_i]$ relevant for the vertex u are also the same. **Q.E.D.**

We therefore notice that step 6 can be considered as a static problem of finding the shortest line segment in a concave wedge touching a convex polygon. Once the wedge, the convex chain and the valid placement of a line segment are known, we can compute the shortest line segment, using Lemma 5, in $O(n_i)$ time where n_i is the number of vertices in the antipodal chain. The last placement of the line segment during the processing of vertex c_i becomes the first placement of the line segment during the processing of the vertex c_{i+1}.

Suppose c_j is a convex hull vertex. Let the convex chain from s_j to t_j be the antipodal chain of c_j. Suppose s is any vertex in the CCW polygonal chain c_j to s_j. Also suppose t is any vertex in the CW polygonal chain c_j to t_j. Using arguments similar to those put forward in Lemma 9 we can show that

Lemma 10: The shortest line segment touching the convex chain from s_j to t_j with end points resting on CW_envelope $[s, c_j]$ and CCW_envelope $[c_j, t]$ is the shortest line segment antipodal to c_j.

Suppose the convex chain from s_1 to t_1 forms the antipodal chain of c_1. Suppose c_m is the vertex such that its antipodal chain from s_m to t_m either contains c_1 or $s_m = c_1$ or $t_m = c_1$. It is easy to show that $c_m = s_1$.

We divide our problem into two similar subproblems. One subproblem is to find the shortest line segment antipodal to each of the vertices c_j, j= 1,2, ..., m-1. In this case we start with the CW_envelope $[s_1, c_1]$ and the CCW_envelope $[c_1, c_1]$. In order to determine the shortest line segment antipodal to some c_j between c_1 and c_m, we work with the envelopes CW_envelope $[s_1, c_j]$ and CCW_envelope $[c_j, c_1]$. The shortest line segment antipodal to c_j thus obtained is the required solution, according to Lemma 10. The other subproblem is to find the shortest line segment antipodal to each of the vertices c_j, j= m,m+1, ..., k. Here the initial envelopes are CW_envelope $[s_m, c_m]$ and CCW_envelope $[c_m, c_m]$. We discuss the first problem below. The second problem is solved similarly.

Suppose we are interested in computing the shortest line segment antipodal to c_1. We compute the CCW_envelope $[c_1, c_1]$ by merging the CCW envelopes of the pockets $[c_k, c_1]$, $[c_{k-1}, c_k]$, ..., $[c_1, c_2]$ in order. The basic step is to merge the current envelope, obtained so far, with the CCW envelope of the next pocket. We know from Lemma 8 that two CCW envelopes can intersect at most once. Therefore we can easily perform the merging step in time proportional to the portion of the envelopes being merged which are not present in the merged picture. A simple accounting scheme shows that the above procedure to compute the CCW_envelope $[c_1, c_1]$ is linear in the number of vertices considered.

Once the history of the computation of the CCW_envelope $[c_1, c_1]$ as described above is maintained, we can easily 'unmerge' the CCW_envelope $[c_1, c_2]$ from the CCW_envelope $[c_1, c_1]$ to obtain the CCW_envelope $[c_2, c_1]$. The time required is proportional to the size of the CCW_envelope $[c_1, c_2]$. This is possible because the CCW_envelope $[c_1, c_2]$ was the last envelope of the pocket $[c_1, c_2]$ that was merged while forming the CCW_envelope $[c_1, c_1]$. Again once the CCW_envelope $[c_2, c_1]$ is known, we can obtain the CCW_envelope $[c_3, c_1]$ by 'unmerge' the CCW_envelope $[c_2, c_3]$ from the CCW_envelope $[c_2, c_1]$. Therefore, we can 'unmerge' the CCW envelopes of pockets $[c_1, c_2], [c_2, c_3], ..., [c_k, c_1]$ in order from the CCW_envelope $[c_1, c_1]$. The total time required to implement this sequence of 'unmerge' is linear in the number of vertices involved.

We compute the CW_envelopes $[s_1, c_1]$ by merging one pocket at a time in order. The total time needed is linear in the number of vertices.

We now describe a detailed version of the algorithm **shortest-line-segment** to find the shortest line segment.

Algorithm shortest-line-segment_detailed

Input: An n-vertex weakly externally visible simple polygon.
Output: The shortest line segment from which P is weakly externally visible.

Step 1: Find the convex hull $C = [c_1 c_2 .... c_k]$ of P;
Step 2: Compute the external cone of visibility for each vertex of P.
Step 3: Compute the CW and CCW envelopes of each pocket$[c_j, c_{j+1}]$, j= 1,2, ...,k, where $c_{k+1} = c_1$.
Step 4: Determine the antipodal chain for the vertex c_1. {Let the chain be from s_1 to t_1.}
Step 5: Determine the vertex c_m whose antipodal chain from s_m to t_m contains c_1 or $s_m = c_1$ or $t_m = c_1$.
Step 6: Set index $\leftarrow$ 1; final_index $\leftarrow$ m; s $\leftarrow \infty$.
Step 7: Compute CW_envelope $[s_{index}, c_{index}]$ and CCW_envelope $[c_{index}, c_{index}]$.
Step 8: Set i $\leftarrow$ index.
Step 9: Determine the antipodal chain for the vertex c_i {Let the chain be from s_i and t_i in counter-clockwise direction.}
Step 10: Search the antipodal chain of c_i for the shortest line segment with its end points on CW_envelope $[s_{index}, c_i]$ and CCW_envelope $[c_{index}, c_i]$.
Step 11: If the length of the shortest line segment < s then reset s to the length of the current shortest line segment.
Step 12: Merge CW_envelope $[c_i, c_{i+1}]$ to CW_envelope $[s_{index}, c_i]$ to obtain CW_envelope $[s_{index}, c_{i+1}]$.
Step 13: 'Unmerge' CCW_envelope $[c_i, c_{i+1}]$ from CCW_envelope $[c_i, c_{index}]$ to obtain CCW_envelope $[c_{i+1}, c_{index}]$.
Step 14: i $\leftarrow$ i+1; if i < final_index then go to step 9.
Step 15: If final_index < k then set index $\leftarrow$ m; final_index $\leftarrow$ k+1; go to step 7.

From the discussion detailed above we can conclude that

Theorem 1: The algorithm shortest_line_segment_detailed is correct and can be implemented in O(n) time.

5.   Conclusions

In this paper we have described an O(n) algorithm for computing the shortest line segment from which a polygon P is weakly externally visible. The method uses a generalization of the minimization problem described in [BT90]. The algorithm finds the shortest line segment intersecting the cones of visibility of the vertices of P. The algorithm works in a dynamic environment where insertions and deletions of the visibility cones are performed. It would be interesting to see whether the approach presented here could be used to find the shortest internal line segment from which the polygon is internally visible. What happens when we try to generalize this result to 3 dimension would be a topic of further interesting research.

References

[AB87] Atallah, M. and Bajaj, C., "Efficient algorithms for common transversals," Information Processing Letters, Vol. 25, pp.87-91, 1987.

[AT81] Avis, D. and Toussaint, G.T., "An optimal algorithm for determining the visibility of a polygon from an edge," IEEE Transaction on Computers, Vol. C-30, No. 12, 1981, pp. 910-914.

[AW87] Avis, D. and Wenger, R., "Algorithms for line stabbers in space," Proc. 3rd ACM Symposium on Computational Geometry, pp.300-307, 1987.

[BCETSU91] Bhattacharya, B., Czysowicz, J., Egyed, P., Toussaint, G., Stojmenovic, I. and Urrutia, J., "Computing shortest transversals of sets," Forthcoming Proc. 7th ACM Symposium on Computational Geometry, 1991.

[BKT89] Bhattacharya, B.K., Kirkpatrick, D. and Toussaint, G.T., "Determining sector visibility of a polygon," Proc. 5th ACM Symposium on Computational Geometry, pp.247-254, 1989.

[BT90] Bhattacharyya, B.K. and Toussaint, G.T., "Computing shortest transversals," Tech. Report SOCS 90.6, McGill University, April 1990.

[Ed85] Edelsbrunner, H., "Finding transversals for sets of simple geometric figures," Theoretical Computer Science, Vol.35, pp.55-69, 1985.

[Gr58] Grunbaum, B., "On common transversals,"Arch. Math., Vol.9, pp. 465-469, 1958.

[Ke88] Ke, Yan, "Detecting the weak visibility of a simple polygon and related problems," John Hopkins University, manuscript, 1988.

[Le80] Lewis, T., "Two counter-examples concerning transversals for convex subsets of the plane," Geometriae Dedicata, Vol.9, pp. 461-465, 1980.

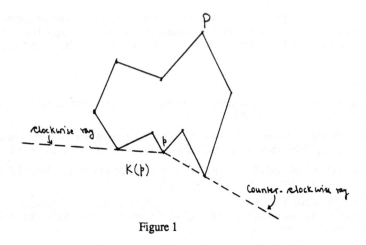

Figure 1

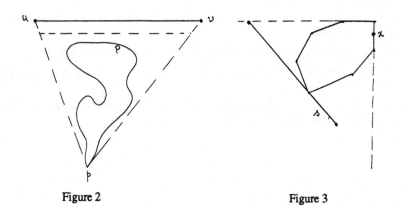

Figure 2

Figure 3

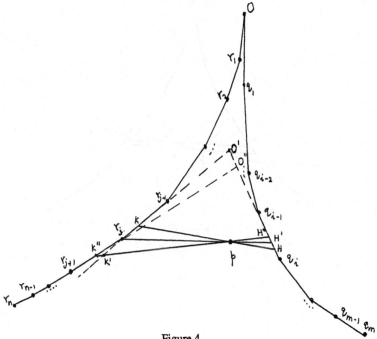

Figure 4

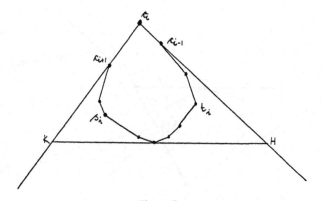

Figure 5

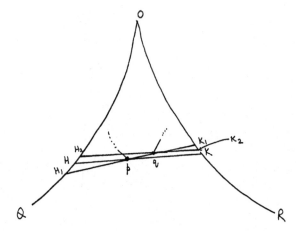

Figure 6

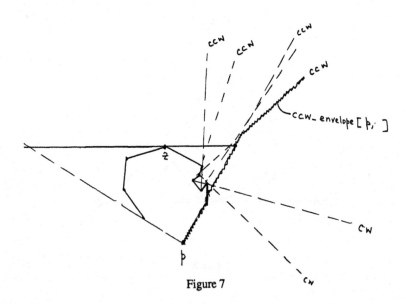

Figure 7

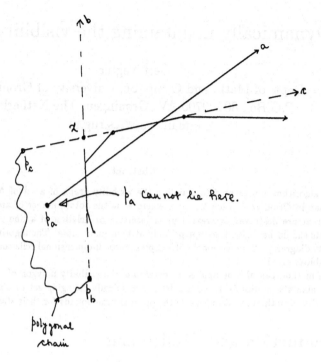

Figure 8 (a)

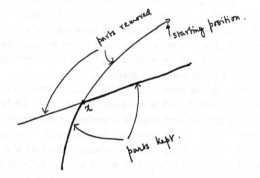

Figure 8 (b)

Dynamically maintaining the visibility graph

Gert Vegter

Dept. of Math. and Comp. Sc, University of Groningen,
P.O.Box 800, 9700 AV Groningen, The Netherlands
e-mail: gert@cs.rug.nl

Abstract

An algorithm is presented to maintain the visibility graph of a set of N line segments in the plane in $O(\log^2 N + K \log N)$ time, where K is the total number of arcs of the visibility graph that are destroyed or created upon insertion or deletion of a line segment. The line segments should be disjoint, except possibly at their end-points. The algorithm maintains the visibility diagram, a 2-dimensional cell complex whose 0-dimensional cells correspond to arcs of the visibility graph.

The method can also be applied to determine the visibility polygon of a query point, and also to plan the motion of a rod amidst a dynamically changing set of obstacles. The time complexity of both applications meets the optimal time bounds for their static counterparts.

1 Introduction and applications

In recent years several visibility problems have been considered in computational geometry. The algorithms for solving these problems have important applications in hidden-line elimination, finding shortest paths and planning the motion of simple robots amidst polygonal obstacles.

Consider a set S of N line-segments (obstacles) in the plane which don't intersect except possibly at their boundary points. In particular S may consist of the edges of mutually disjoint simple polygons. The set $Bd(S)$ of (relative) boundary points of segments in S is the set of *nodes* of the visibility graph $G_S = (V_S, E_S)$ of S. For two boundary points p and q the pair (p, q) is an *arc* of G_S if and only if the points p and q can see each other, i.e. the line segment (p, q) is disjoint from all segments in S. In this paper G_S is a *directed* graph.

Several *static* algorithms for computing the visibility graph have been developed: in [10] the first algorithm with a non-trivial time bound is developed, [4] contains the first optimal algorithm, and in [7] a sub-optimal, but in practice very feasible algorithm is developed.

In this paper a *dynamic* algorithm is developed, that maintains the visibility graph when the set of obstacle segments is allowed to change upon insertion or deletion of a line segment. To this end we maintain the *visibility diagram* $VIS(S)$ of S, which is a subdivision (cell complex) of a certain two-dimensional space whose points correspond to lines in the plane, see also [9]. This geometric object may be considered as a *generalization of the visibility graph* of the set S of obstacle edges. $VIS(S)$ implicitly represents the visibility graph: its 0-cells correspond to arcs of G_S. Due to the additional topological information, viz. the set of 1- and 2-dimensional cells, this subdivision can be maintained efficiently when the set S changes dynamically. We stress that cells of $VIS(S)$ of dimension 0, 1 and 2 will be referred to as vertices, edges and faces, respectively. To avoid confusion elements of V_S and E_S will be called *nodes* and *arcs* of the visibility graph, respectively. Our main result is:

Theorem 1.1 *The visibility diagram of S is stored in $O(|E_S|)$ space. It can be maintained in $O(\log^2 N + K \log N)$ time, where K is the total number of arcs of the visibility graph of S that are destroyed or created upon insertion or deletion of a line segment.*

The method for maintaining the visibility diagram dynamically has some important applications. The problems treated in [9] in a static setting can now also be solved in the dynamic context. In particular the visibility polygon of a query point in the plane with respect to a dynamically changing set of obstacles S can be determined in time $O(\log^2 N + k \log(N/k))$, where k is the size of the visibility polygon. Moreover we can plan a feasible motion of a rod amidst the set S of obstacle segments in $O(N^2)$ time—which is optimal in a certain sense, cf. [6], [5]—even if the set S is allowed to change dynamically. A motion planning query involves not only the initial and final positions of the rod, but also the length of the rod is a query parameter, see [9].

To make the paper largely self-contained we introduce the visibility diagram in section 2. In section 3 we show how to compute the visibility diagram of the set S_θ obtained from S by adding two degenerate segments, each consisting of a single end-point of the obstacle segment s_0 to be inserted. The vertices of $VIS(S_\theta)$ that are created correspond to new arcs of the visibility graph. In section 4 $VIS(S')$ is constructed from $VIS(S_\theta)$, where $S' = S \cup \{s_0\}$. We briefly indicate how to update the visibility diagram upon *deletion* of an obstacle segment. In section 5 possible extensions are mentioned.

The author is grateful to Mark de Berg for pointing out some omissions in an earlier version of this paper.

2 Review of the visibility diagram

The duality transform

As observed e.g. in [1] and [9], when dealing with visibility problems natural objects are *directed lines* rather than points. Since conceptually and algorithmically points are easier to deal with it is appropriate to endow the set of lines, which is a 2-dimensional manifold, with (local) coordinates so that each line is represented by a point.

The set of directed lines in the plane is denoted by $\mathcal{L}$. For a line segment s the set of directed lines intersecting s is denoted by $\mathcal{L}_s$. Similarly for a point p in the plane $\mathcal{L}_p$ is the set of directed lines containing p.

To introduce coordinates on $\mathcal{L}$ we exploit a so called *duality transform*. This transform maps the set of non-vertical lines in the plane to points in the plane. More precizely, if x_1, x_2 are Cartesian coordinates in the plane, the duality transform $\mathcal{D}$ maps a non-vertical line with equation $x_2 = h_1 x_1 + h_2$ to the point (h_1, h_2). To avoid confusion the plane with Cartesian coordinates x_1, x_2 is called the *primal* plane, and the image of $\mathcal{D}$ is called the *dual plane*. The dual plane has coordinates h_1, h_2. It is easy to check that the set $\mathcal{L}_p$ of lines containing a point $p = (p_1, p_2)$ in the primal plane is mapped onto the set $\{(h_1, h_2) \mid h_2 = -p_1 h_1 + p_2\}$, which is a line in the dual plane.

To cover the set of non-vertical *directed* lines we need two coordinate-systems: one to represent the set of lines $\mathcal{L}^+$ crossing the vertical axis from left to right, the other one to represent the set $\mathcal{L}^-$ of lines crossing the vertical axis from right to left. A natural representation for the set of directed lines is the so-called *two-sided* plane, introduced in [1]. Also see [9] for a description of the topology of the sets $\mathcal{L}$ and $\mathcal{L}_s$.

Next consider the set $\mathcal{L}_s^+ \stackrel{\text{def}}{=} \mathcal{L}^+ \cap \mathcal{L}_s$, $s \in S$. The duality transform maps $\mathcal{L}_s^+$ onto a wedge-shaped region in the dual plane, bounded by the lines that are the dual images of the end-points of the segment s. This set can be turned into a 2-dimensional cell-complex. Its 2-cells (faces) correspond to the maximal connected subsets consisting of lines whose visibilities are points on the (relative) interior of the same obstacle segment. Here the *visibility* of a directed line $l \in \mathcal{L}_s$ is the point on an obstacle segment that is seen by an observer standing at the point of intersection of l and s and watching in the direction of l. A line $l_0 \in \mathcal{L}_s^+$ belongs to an edge or a vertex if l_0 has *critical visibility*, i.e. if its visibility belongs to $Bd(S)$.

We describe this cell complex (subdivision) in more detail. Consider the subset of $\mathcal{L}_s^+$ consisting of lines with critical visibility. Its dual image is a set of line segments in the dual plane. To see this observe that the sets of lines in $\mathcal{L}_s^+$ containing some boundary point give rise to a straight-line subdivision (arrangement) of the wedge-shaped region $\mathcal{D}(\mathcal{L}_s^+)$. In Figure 1b, where $s = [q_1, q_2]$, this subdivision consists of both the solid and dashed parts. Vertices of this subdivision correspond

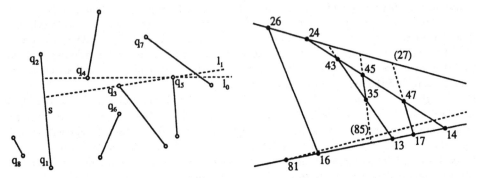

Figure 1: a. The primal plane   b. The subdivision of $\mathcal{L}_s^+$

to lines containing at least two end-points of obstacle segments. In this paper we assume that the obstacle segments are in *general position*, i.e. no three boundary points lie on the same line. Therefore interior vertices have degree 4 and boundary vertices have degree 3. In Figure 1a vertex marked 'ij' is the dual of the directed line $l(q_i, q_j)$.

Next consider an edge of this subdivision, lying in the dual image of the set $\mathcal{L}_q$ for $q \in Bd(S)$. Point q will be called the *source* of the edge. Suppose the vertices incident upon this edge are the dual image of lines l_0 and l_1. Then the edge is the dual image of the set of lines obtained by rotating l_0 around q until it coincides with l_1. See Figure 1, where l_0 and l_1 correspond to vertices '45' and '35'. During this rotation no other obstacle-endpoint is hit.

An edge of this subdivision is the dual image of a set of lines in the primal plane with critical visibility iff. the visibility of each of these lines is the source of the edge. Therefore we erase any edge whose source is not visible along the lines whose dual images constitute this edge. The remaining subdivision is shown solid in Figure 1. The set $\mathcal{L}_s^-$ contains a similar subdivision. Together these subdivisions form a subdivision of $\mathcal{L}_s$, which is denoted by $\mathcal{V}is(s, S)$.

Vertices of $\mathcal{V}is(s, S)$; relation with the visibility graph

Consider a vertex v of the dual arrangement of $\mathcal{L}_s^+$, corresponding to line $l = l(q, q')$, $q, q' \in Bd(S)$. We distinguish the following cases:

1. *Both q and q' are not visible from s along l.* Then v is not a vertex of $\mathcal{V}is(s, S)$, and none of the edges incident upon v is (part of) an edge of $\mathcal{V}is(s, S)$; an example is vertex '85' in Figure 1.

2. *q is visible from s along l.* If q and q' can't see each other v is not a vertex of $\mathcal{V}is(s, S)$: both edges incident upon v with source q' are erased, and the other pair of edges are joined. See vertex '27' in Figure 1. If q and q' can see each other v is a vertex of $\mathcal{V}is(s, S)$ of degree 3, since q' is visible from s when we rotate $l(q, q')$ slightly about q' in one direction and invisible if we rotate in the opposite direction. An example is vertex '35' in Figure 1.

3. *q' is a boundary point of s.* If q and q' can't see each other v is not a vertex of $\mathcal{V}is(s, S)$. Otherwise for technical reasons (see below) we consider v as a vertex of degree 2. An example is vertex '81'.

This classification of vertices shows that there is a 3:1–correspondence between vertices of the visibility diagram and arcs of the visibility graph. Since $\mathcal{V}is(s, S)$ is a planar subdivision, the sets of faces and edges have asymptotically the same size as the set of vertices. Therefore the size of the collection $\mathcal{V}is(s, S)$, $s \in S$, is asymptotically the same as the size of the visibility graph.

For an end-point q of $s \in S$ the sequence of vertices lying in the dual image of $\mathcal{L}_q$ contained in the boundary of $\mathcal{V}is(s, S)$ corresponds to the clockwise sequence of arcs of the visibility graph incident upon q. In section 3 we use that this part of the boundary of $\mathcal{V}is(s, S)$ also corresponds to the subdivision $\mathcal{V}is([q, q], S)$ of the degenerate point-segment $[q, q]$.

Faces of $Vis(s, S)$

The previous analysis also shows that at a vertex of $Vis(s, S)$ none of the angles formed by successive edges is concave. Therefore each face of $Vis(s, S)$ is a *convex polygon*. When the dual image of $l \in \mathcal{L}_s^+$ ranges over a face f of $Vis(s, S)$, in the primal plane the line segment connecting $l \cap s$ with the visibility of l sweeps out a region bounded by two concave chains, the segment s and the obstacle segment $\bar{s}$ containing the visibility of any line whose dual image lies in f, see Figure 2 (also cf. [1]). This polygonal region, called the *basin* of f, will be denoted by $\Pi(f, S)$.

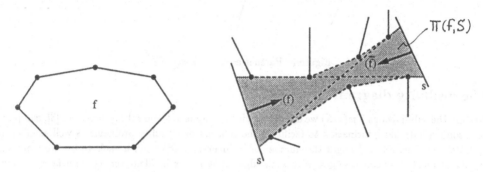

Figure 2: A face of $Vis(s, S)$ and its basin

In Figure 2 we have drawn a small arrow labeled (f), whose base lies on the segment s and whose direction is the direction of l. This convention is adopted to stress that l is considered as an element of $\mathcal{L}_s$, whose dual image lies on face f. A similar convention is adopted with regard to vertices and edges of $Vis(s, S)$.

The boundary of face f consists of two convex chains bounded by the vertices with minimal and maximal h_1-coordinate. These chains are called the *left* and *right* boundary of f. Vertices on the boundary of f correspond to certain arcs of the visibility graph contained in the basin of f. These arcs—drawn dashed in Figure 2—form a non-simple polygon, called the *skull* of f. The vertex with minimal (maximal) h_1-coordinate and the corresponding edge of the skull will be called counterclockwise (ccw)-minimal (ccw-maximal).

If f is a face of $Vis(s, S)$ we call s the *base-segment* of f, denoted by $base(f)$. The *partner* of f is the face $\bar{f}$ having the same basin as f, whose base-segment $\bar{s}$ is the segment containing the visibilities of any $l \in f$. The elements of $\bar{f}$ are obtained from elements of f by reversing their direction, see Figure 2.

Edges of $Vis(s, S)$

Finally we note that edges also come in pairs, although the correspondence is different from the way faces are paired up. Consider an edge e of $Vis(s)$, and suppose its source p does not belong to s. Then a line l whose dual image lies on e has visibility p. Let $\bar{s}$ be the segment containing p, then l may also be considered as an element on the boundary of $\mathcal{L}_{\bar{s}}$. In the subdivision $Vis(\bar{s}, S)$ the edge $\bar{e}$ containing the dual image of l is called the *partner* of e, see Figure 3. Similarly e is called the partner of $\bar{e}$. One of the technical reasons for introducing boundary vertices of degree 2 (like the dual of l_0 in $Vis(\bar{s}, S)$, see above) is to achieve that each edge has a partner corresponding to the same set of lines. To this end we also have to introduce an *ideal segment* s_∞, which is an infinite segment at infinity. We refer to [9] for a precize setting. For this paper it is sufficient to assume that when moving backward along any line $l \in \mathcal{L}_p$, $p \in Bd(S)$, we hit some segment of S (possibly s_∞), as indicated in Figure 3.

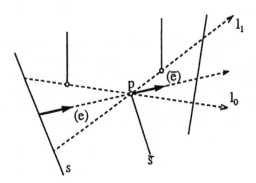

Figure 3: Partner-edges of $\mathcal{V}is(s,S)$

The visibility diagram

To store the subdivisions $\mathcal{V}is(s,S)$ we use the incidence graphs of these subdivisions, see [3], equipped with some additional information to facilitate the solution of visibility problems as well as efficient updates. For vertices and edges the number of incidences is $O(1)$: with each vertex we store the set of—at most 4—incident edges, with each edge we store the incident vertices and faces. A face is represented by a balanced ordered tree, whose sequence of leaves corresponds to the sequence of edges incident upon this face. This tree is both accessed at the root in a search for certain edges or vertices—e.g. when searching for vertices corresponding to destroyed arcs of the visibility graph—and at the leaves during a split or concat operation on the boundary of the associated face—e.g. when updating the visibility graph, in which case faces incident upon a certain edge have to be joined. Therefore we equip this tree with references from father to sons and vice versa. In this way both searching (top-down) and splitting and concatenating edge-sequences (bottom up) can be performed in logarithmic time—with respect to the size of the trees involved.

With each edge and face we store its partner. In section 4 we introduce a dynamic data structure that enables access of the base segment of a face in $O(\log N)$ time.

The collection of subdivisions $\mathcal{V}is(s,S)$, $s \in S$, augmented with the secondary information for faces and edges described above, is called the *visibility diagram* for the set of obstacles S, denoted by $VIS(S)$. As indicated above it can be be stored in $O(|E_S|)$ space. We refer to [9] for a static construction in $O(N^2)$ time.

3   Arcs created due to insertion of a line segment

When a new segment is inserted in the set of obstacle segments the visibility graph changes due to both the creation and destruction of arcs. Since there is a 3:1–correspondence between the set of arcs of the visibility graph and the set of vertices of the visibility diagram the creation of a new arc of the visibility graph gives rise to the creation of 3 new vertices of the visibility diagram. In this section we consider this creation in more detail.

The set of arcs of the visibility graph that are created corresponds to new pairs of endpoints of obstacle segments that can see each other. Therefore one of these endpoints lies on the inserted line segment s_0. In other words, the new arcs are also found if we insert two *degenerate segments* $[p_0,p_0]$ and $[p_1,p_1]$, corresponding to the boundary points of s_0. The *intermediate set* of obstacles $S \cup \{[p_0,p_0],[p_1,p_1]\}$ is denoted by S_∂. Note that $E_{S'} \setminus E_S = E_{S_\partial} \setminus E_S$ is the set of created arcs of the visibility graph.

Proposition 3.1 $VIS(S_\partial)$ can be constructed from $VIS(S)$ in $O(\log^2 N + K_c \log N)$ time, where K_c is the number of created arcs of the visibility graph.

We shall describe the construction of $VIS(S')$ in the simple case where $S' = S \cup \{[p,p]\}$ for a single free point p. This construction generalizes the computation of the visibility polygon of p with respect to S, since obviously $\mathcal{V}is([p,p],S)$ is a representation of this visibility polygon, see section 2.

The algorithm determines the visibility polygon of p by rotating a line l about p, starting from a vertically downward position. During this rotational sweep it maintains the obstacle segment s from which the query point is seen along l. A transition occurs when the sweep–line passes a boundary point that either lies on s or is visible from s along l, see Figure 4. The pair consisting of p and this boundary point determines a new edge of the visibility graph.

At such a transition the dual image of l lies on an edge of the subdivision $\mathcal{V}is(s,S)$. Between two transitions the dual of l describes a line–segment connecting two edges on the boundary of a face. Each endpoint of this line–segment is a new vertex of the visibility diagram, and the line–segment connecting two successive new vertices is a new edge.

Disregarding initialization we assume that the first of these edges, corresponding to the counterclockwise earlier transition, has been determined. The second edge can then be determined by walking along the boundary of the face, meanwhile checking whether the edge that is passed contains the dual of a line through p.

If at a transition p hides the segment s and the other end–point from view along l, we don't have to update s: just after the transition p is still visible from s along the sweep–line. If however p is on s, or p does not hide the other end-point from view, s has to be updated. It is not hard to see that the new value s' of s can be determined in $O(1)$ time using the information stored with edge e of $\mathcal{V}is(s,S)$ that contains the dual of the sweep-line at the current transition: we continue our walk at the partner of edge e, which is an edge of the subdivision $\mathcal{V}is(s',S)$, see Figure 4. (In this figure the dual image of $\mathcal{L}_{q_i}$ has label i.)

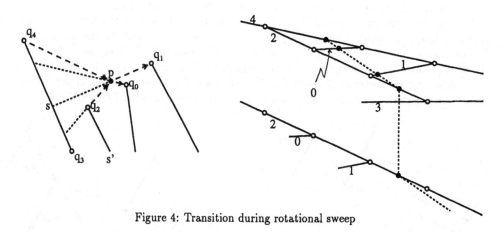

Figure 4: Transition during rotational sweep

The initialization step of the algorithm uses a point-location based data structure. Applying a result of [8] (see also [2]) it can be shown that this data structure can be maintained dynamically in $O(\log^2 N + K_c \log N)$ time, without increasing the space complexity. It can also be shown that the total number of edges of $VIS(S)$ passed is $O(N)$. This yields the time bound as stated above. We again refer to [9] for details.

4 Arcs destroyed due to insertion of a line segment

In this section we discuss the destruction of arcs of the visibility graph, or equivalently, of vertices of the visibility diagram upon insertion of an obstacle segment. To this end we show how $VIS(S')$, with $S' = S \cup \{s_0\}$, is constructed from $VIS(S_\theta)$. We shall prove the following result, which together with proposition 3.1 yields theorem 1.1.

Proposition 4.1 *The subdivision $VIS(S')$ can be constructed from $VIS(S_\partial)$ in $O(\log^2 N + K_d \log N)$ time, where $K_d \stackrel{\text{def}}{=} |E_S \setminus E_{S'}|$ is the number of destroyed arcs of the visibility graph.*

In the previous section created arcs were determined in the order in which they appear in the visibility polygons of the endpoints of the new segment. The set of destroyed arcs is also determined in a certain order, corresponding to the *visibility polygon* of the new segment. We first introduce this generalization of the visibility polygon for a free segment s_0. This geometric object corresponds to a linear order on the set of faces of $\mathcal{V}is(s_0, S')$. We then show that for each old segment $s \in S$ any face of the new subdivision $\mathcal{V}is(s, S')$ is obtained by *joining* a collection of faces of the subdivision $\mathcal{V}is(s, S_\partial)$. This collection turns out to be a contiguous subsequence of the collection of faces of $\mathcal{V}is(s, S_\partial)$ with respect to the order referred to above. The latter property is crucial for efficient updates of the base-segment stored with each face of the visibility diagram. Finally we describe how to construct $\mathcal{V}is(s_0, S')$ efficiently by reusing edges and vertices of the subdivisions $\mathcal{V}is(s, S_\partial)$, $s \neq s_0$, that are removed from these subdivisions upon insertion of the segment s_0. For convenience we describe separately the update of the subdivisions $\mathcal{V}is(s, S')$, $s \neq s_0$, the construction of $\mathcal{V}is(s_0, S')$ and the method for updating secondary information stored with faces of these subdivisions. These operations should be merged into a one-pass update of the visibility diagram. This trivial modification is left to the reader.

The visibility polygon of a line segment

For a face f of $\mathcal{V}is(s, S)$ let $vis(f)$ be the subsegment described by the visibility of a line l when its dual image ranges over f. In particular $vis(f)$ is an edge of the boundary of the basin $\Pi(f, S)$ of face f. We give $vis(f)$ a direction so that this basin lies in its left half-plane.

Let $\mathcal{F}(s, S)$ be the set of faces of $\mathcal{V}is(s, S)$. The collection of subsegments $vis(f)$, $f \in \mathcal{F}(s, S)$, is assembled into the *visibility polygon* of segment s with respect to the set of obstacles S. This is done by establishing a linear order on $\mathcal{F}(s, S)$, and hence on this collection of subsegments. The endpoint of a subsegment will then be connected to the beginpoint of its successor by a line segment.

To determine the successor of a face $f \in \mathcal{F}(s, S)$ let $q \in Bd(S)$ be the vertex of the *left* convex chain bounding the basin $\Pi(f, S)$ of f, that is nearest the end-point of $vis(f)$, see Figure 5.

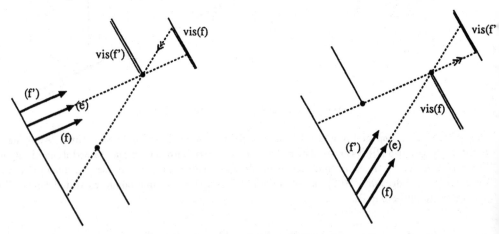

Figure 5: Successive edges of the visibility polygon of s

Then q is the source of exactly one edge e contained in the *left* boundary of face f. It is not hard to check that q is not an end-point of s, so e is not contained in the boundary of the dual image of $\mathcal{L}_s$. Therefore e is also contained in the (right) boundary of a second face $f' \in \mathcal{F}(s, S))$, which by definition is the *successor* of f. The corresponding order on $\mathcal{F}(s, S)$ is denoted by $\lhd_s$. Connecting the endpoint of $vis(f)$ and the beginpoint of $vis(f')$ we obtain a—not necessarily simple—polygon, which is called the *visibility polygon* of s. See also Figure 7, where successive faces—labeled $1, \cdots, 7$—of $Vis(s, S_\theta)$ are depicted. The corresponding edges of the visibility polygon are labeled accordingly.

Construction of $Vis(s, S')$ for $s \neq s_0$

We first focus on the collection of subdivisions $Vis(s, S')$, $s \neq s_0$. It turns out that these subdivisions are obtained by joining—taking the union of—faces of $VIS(S_\theta)$. In the sequel faces of $VIS(S_\theta)$ will be denoted by lower case symbols, faces of $VIS(S')$ by upper case symbols.

A face f of the subdivision $VIS(S_\theta)$ is of exactly one of the following types.

Active: s_0 intersects the skull of f. Each of the—2 or 4—edges of the skull of f that intersect s_0 corresponds to a destroyed arc of the visibility graph, and hence to a destroyed vertex of the visibility diagram, see Figure 6a.

Passive: s_0 is disjoint from the skull of f, but it does intersect the basin of f. In this case f is also a face of $VIS(S')$, since its skull is not affected by the insertion of s_0. However, if s_0 separates the skull of f from the base–segment of f, as in Figure 6b, the base–segment of f in $VIS(S')$ is s_0. Otherwise the base–segment of the partner $\overline{f}$ of f in $VIS(S')$ is s_0.

Neutral: the basin of f is disjoint from the segment s_0. Now f is also a face of $VIS(S')$, and the base–segment of neither f nor $\overline{f}$ changes.

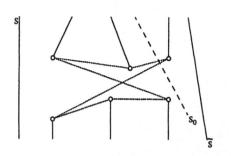 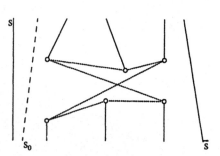

Figure 6: a. Skull of active face   b. Skull of passive face

Note that a face and its partner are of the same type.

Lemma 4.2 *The number of active faces of $VIS(S_\theta)$ is proportional to the number of destroyed arcs of the visibility graph.*

Proof. The skull of an active face with respect to S_θ has at least two and at most four edges that intersect s_0. Each of these edges corresponds to a destroyed arc of the visibility graph. Each arc of the directed visibility graph corresponds to exactly three vertices of $VIS(S_\theta)$, each lying in the boundary of an active face. From these observations the claim easily follows. $\qquad\Box$

Our goal is to update the visibility diagram by merely processing *active* faces. In view of lemma 4.2 we can charge $O(\log N)$ time to each active face to achieve the update time stated in proposition 4.1. The following result concerns the relation between faces of $VIS(S_\theta)$ and faces of $VIS(S')$.

Lemma 4.3 *Let s be an obstacle segment, different from s_0.*

  (i) *Each face F of $Vis(s, S')$ is obtained by joining a collection L_F of faces of $Vis(s, S_\theta)$. The collection L_F consists of successive faces of the subdivision $Vis(s, S_\theta)$.*

  (ii) *For every active face f of $Vis(s, S_\theta)$ there is a unique face F of $Vis(s, S')$ such that $f \in L_F$.*

  (iii) *If the sequence L_F consists of one element, this element is either a neutral or a passive face of $Vis(s, S_\theta)$.*

(iv) If L_F contains at least two elements, its first and last elements are active faces of $Vis(s, S_\theta)$. The sequence of faces in L_F that are incident upon the boundary of F coincides with the subseqence consisting of all active faces in L_F.

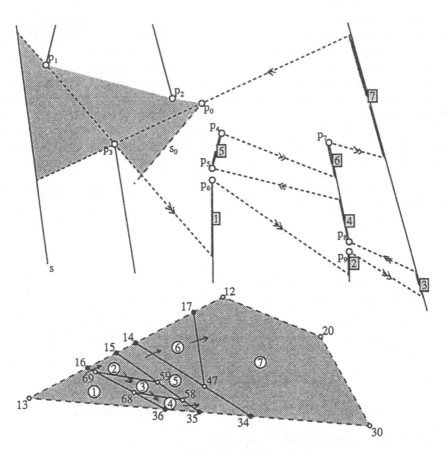

Figure 7: Joining faces into a new face

We refer to Figure 7 where a face F (dashed boundary) of $Vis(s, S')$ and its basin are depicted. The new segment s_0 hides a sequence of successive edges of the visibility polygon of s with respect to S_θ from view along lines $l \in F$. Faces $f \in L_F$ and their visiblities $vis(f)$ are labeled $1, \cdots, 7$. The edge shared by a face and its successor in the subdivision $Vis(s, S_\theta)$ is marked by an arrow pointing into the successor.

Face $f \in L_F$ is active iff. it is incident upon the boundary of F. In Figure 7 face '3' is passive, the remaining faces are active. Destroyed faces in the boundary of F are indicated by fat dots.

Face F is called an *inherited* face if L_F has one element. Otherwise F is called an *assembled* face. Lemma 4.3.iv shows that an assembled face F can be constructed by merely considering active faces in L_F. Below we show how this subsequence can be determined efficiently. We note here that joining faces belonging to L_F amounts to 'erasing' part of the subdivision $Vis(s, S_\theta)$, viz. the collection of edges and vertices that belong to the interior of the region F. This part of the subdivision will be reused in the construction of $Vis(s_0, S')$.

The sequence L_F to which an active face f belongs is also denoted by L_f. In particular for all active faces f belonging to L_F we have $L_f = L_F$.

Lemma 4.4 *If an active face f is the first or last element of the sequence L_f, its partner $\overline{f}$ in the*

Proof. Let F and F' be the faces obtained by joining the faces in L_f and $L_{\overline{f}}$, respectively. Suppose f is the first element of L_f. Then its ccw-minimal vertex is also a vertex of F. In particular s_0 does not separate $s = base(f)$ from the ccw-minimal edge of the skull of f, see Figure 6a. But then s_0 does separate $\overline{s} = base(\overline{f})$ from this edge, which coincides with the ccw-minimal edge of the skull of $\overline{f}$. Therefore the ccw-minimal vertex of $\overline{f}$ is not a vertex of F', so $\overline{f}$ is not the first element of $L_{\overline{f}}$.

The segment s_0 does intersect the ccw-maximal edge of the skull of f, which coincides with the ccw-maximal edge of the skull of $\overline{f}$. Therefore the ccw-maximal vertex of $\overline{f}$ is destroyed, so in particular $\overline{f}$ is not the last element of $L_{\overline{f}}$. $\qquad\square$

Active faces are processed in a certain order, which we shall describe now. First consider the sequence $\mathcal{F}(s_0, S')$ of faces of $\mathcal{V}is(s_0, S')$, ordered according to $\lessdot_{s_0}$. By definition the partner $\overline{F}$ of an element F of this sequence is a face of $\mathcal{V}is(s, S')$, for some $s \neq s_0$, so it is obtained by joining the faces belonging to the sequence $L_{\overline{F}}$, cf. lemma 4.3. Appending the sequences $L_{\overline{F}}$ as F ranges over $\mathcal{F}(s_0, S')$ we obtain a sequence of faces of $VIS(S_\vartheta)$, containing all active faces. Therefore this sequence defines a linear order on the set of active faces of $VIS(S_\vartheta)$. The sequence of active faces will be denoted by $\mathcal{F}_{act}$.

Lemma 4.5 *The successor of an active face in the sequence $\mathcal{F}_{act}$ can be determined in $O(\log N)$ time.*

Proof. First we note that the partner of the successor of a face f in $\mathcal{F}_{act}$ is equal to the successor of $\overline{f}$ in $\mathcal{F}_{act}$. Therefore we maintain both f and $\overline{f}$ during the computation of $\mathcal{F}_{act}$. In view of lemma 4.4 it then suffices to show that, if f is not the last element of L_f, the successor of f in $L_f \cap \mathcal{F}_{act}$ can be determined in $O(\log N)$ time.

So suppose f is not the last element of L_f. Let F be obtained by joining all faces in L_f. The sequence of active faces in L_f is obtained by joining two—not disjoint—sequences L_{left} and L_{right}, consisting of faces of L_f incident upon the left and right boundary of F, respectively. See Figure 7, where L_{left} is the sequence of faces $1, 2, 5, 6, 7$. Suppose $f \in L_{left}$. It obviously suffices to show that the successor of f in L_{left} can be determined in $O(\log N)$ time.

To this end we first determine the ccw-maximal vertex v of $\mathcal{V}is(s, S_\vartheta)$ that lies on the boundaries of both f and and F. Since v is one of the—1 or 2—destroyed vertices in the *left* boundary of f, it can be determined by a binary search on the sequence of vertices in this left boundary. We omit the—trivial—details, but conclude that v can be determined in $O(\log N)$ time. Since the successor of f in L_{left} is also incident upon v, it can be determined in $O(1)$ additional time. $\qquad\square$

Corollary 4.6 *The collection of subdivisions $\mathcal{V}is(s, S_\vartheta)$, $s \neq s_0$, can be constructed from $VIS(S_\vartheta)$ in $O(\log^2 N + K_d \log N)$ time.*

Proof. An assembled face F can be constructed by splitting the boundaries of active faces in L_F at destroyed vertices, and successively concatenating appropriate parts of these split-boundaries (those drawn dashed in Figure 7). This takes $O(\log N)$ time per destroyed vertex. Lemma 4.5—and its proof—shows that the sequence of destroyed vertices can be determined in $O(\log^2 N + K_d \log N)$ time. (Without proof we mention that initialization takes $O(\log^2 N)$ time, cf. [8] or [2].) $\qquad\square$

Construction of $\mathcal{V}is(s_0, S')$

To construct the subdivision $\mathcal{V}is(s_0, S')$ we shall cut the set $\mathcal{L}_{s_0}$ more or less naturally into pieces, which together form a jig-saw puzzle. On a piece of this puzzle a set of vertices and edges is depicted, forming the restriction of $\mathcal{V}is(s_0, S')$ to this particular piece. The subdivision $\mathcal{V}is(s_0, S')$ will therefore emerge upon completion of the jig-saw puzzle. Pieces that completely fall within a face of this subdivision are called *trivial*. A piece that is not trivial is called *non-trivial*. By merely assembling the non-trivial pieces we obtain an efficient construction method for the subdivision $\mathcal{V}is(s_0, S')$.

First we introduce the *collection of pieces* of the jig-saw puzzle. To this end consider the set $\mathcal{F}_0$ of faces of $VIS(S')$ whose visibility is s_0. By definition $\mathcal{F}_0$ is the set of partners of faces of $\mathcal{V}is(s_0, S')$

With each $F \in \mathcal{F}_0$ corresponds a region P_F in $\mathcal{L}_{s_0}$: for $l \in \mathcal{L}_{s_0}$ we define $l \in P_F$ iff. l is a line intersecting the base-segment s of F and, considered as an element of $\mathcal{L}_s$, l belongs to F. Note that in this way there is a natural *identification* of sets P_F and F. The collection of closures of sets P_F, $F \in \mathcal{F}_0$, covers $\mathcal{L}_{s_0}$. We consider these sets as the pieces of a jig-saw puzzle.

Next we consider the *subdivision* of each of these pieces. To this end note that for $F \in \mathcal{F}_0$ the sequence L_F of faces of $\mathcal{V}is(s, S_\partial)$ that are merged into F (see lemma 4.3) defines a subdivision of F, and hence of P_F.

Lemma 4.7 *Consider a face $F \in \mathcal{F}_0$ with base-segment s. The restriction of the subdivision $\mathcal{V}is(s_0, S')$ to P_F is isomorphic to the restriction of the subdivision $\mathcal{V}is(s, S_\partial)$ to F (the isomorphism being realized by the identification of F and P_F described above).*

Proof. The proof is a straightforward consequence of the definition of faces, edges and vertices of the subdivisions $\mathcal{V}is(s_0, S')$ and $\mathcal{V}is(s, S_\partial)$. □

The subdivision of P_F is non-trivial iff. the sequence L_F contains at least two elements, or equivalently, iff. F is an *assembled* face. To obtain the subdivision on a non-trivial piece P_F we *need not create* every edge and vertex, but it suffices to *reuse* the part of subdivision $\mathcal{V}is(s, S_\partial)$ lying in $F \subset \mathcal{L}_s$, where s is the base-segment of F. This part is deleted from $\mathcal{V}is(s, S_\partial)$ by 'breaking' the incidences of edges at *destroyed* vertices.

We refer to Figure 7 for a non-trivial piece of the jig-saw puzzle. Identifying the region P_F with face F, whose boundary is drawn dashed, we see that the solid lines correspond to the the part of $\mathcal{V}is(s, S_\partial)$ that is reused. Along the edge of piece P_F bounded by vertices '15' and '12' we glue a piece $P_{F'}$ corresponding to a face F' of $\mathcal{V}is(s_1, S')$, where s_1 is the obstacle segment containing p_1. We refer to the full version for details.

Lemma 4.8 *The subdivision $\mathcal{V}is(s_0, S')$ can be constructed in $O(\log^2 N + K_d \log N)$ time from $VIS(S_\partial)$.*

Proof. Constructing $\mathcal{V}is(s_0, S')$ amounts to glueing the parts of $VIS(S_\partial)$ to be reused at *destroyed* vertices, drawn fat in Figures 7. The sequence of destroyed vertices is found during the construction of the sequence $\mathcal{F}_{act}$ described above. The glue-operations are similar to those described in the proof of Corollary 4.6. □

Updating face-information

To determine the base-segment of a face we store the sequence of faces of $\mathcal{V}is(s, S)$ in the leaves of a balanced tree. The segment s is stored in the root. Therefore the base segment can be determined in $O(\log N)$ by walking from the leaf, storing the face, to the root. Since during updates of $\mathcal{V}is(s, S)$ *contiguous subsequences* of faces are merged, see lemma 4.3, these trees can be maintained in $O(\log N)$ time per destroyed vertex. To construct the tree storing the faces of $\mathcal{V}is(s_0, S')$ we reuse parts of the old trees, as we did with part of the old subdivisions, cf. lemmas 4.7 and 4.8, charging $O(\log N)$ time to each destroyed vertex. Again we omit further details.

Deleting an obstacle segment

Updating the visibility diagram upon deletion of a line segment s_0 is essentially the inverse of the process described above. First we disassemble the jig-saw puzzle by cutting $\mathcal{L}_{s_0}$ into non-trivial pieces. The subdivisions on these pieces are then projected onto corresponding faces of subdivisions $\mathcal{V}is(s, S_\partial)$, $s \neq s_0$. In this case S_∂ is the set of obstacles obtained from S by merely removing the open segment which is the (relative) interior of s_0.

Finally we erase from $VIS(S_\partial)$ the edges whose source is one of the boundary points of s_0. This operation has time complexity $O(\log^2 N + K \log N)$. We omit further details from this version of the paper.

5 Conclusion and further research

We have introduced the visibility diagram of a set of disjoint line segments in the plane. This object is a 2-dimensional cell complex which may be considered as a generalization of the visibility graph. A dynamic algorithm is developed that maintains the visibility diagram upon insertion or deletion of a line segment. This work has applications to visibility and motion planning problems in the plane.

Future research will be concerned with adapting the method to obstacles of special shape, e.g. convex curves or convex polygons. We also intend to extend the work to 3-space.

References

[1] B. Chazelle and L. Guibas. Visibility and shortest paths in plane geometry. *Proceedings of the ACM Conference on Computational Geometry, Baltimore*, pages 135–146, 1985.

[2] Cheng and Javardan. *Proceedings FOCS'90*, pages 96–105, 1990.

[3] H. Edelsbrunner. *Algorithms in Combinatorial Geometry.* Springer–Verlag, New York, Heidelberg, Berlin, 1987.

[4] S.K. Ghosh and D. Mount. An output sensitive algorithm for computing visibility graphs. *Proc. 23th Annual Symp. on Found. of Computer Science*, pages 11–19, 1987.

[5] Y. Ke and J. O'Rourke. Lower bounds on moving a ladder in two and three dimensions. *Journal of Discrete and Computational Geometry*, 3:197–217, 1988.

[6] D. Leven and M. Sharir. An efficient and simple motion-planning algorithm for a ladder moving in 2-dimensional space amidst polygonal barriers. *Journal of Algorithms*, 8:192–215, 1987.

[7] M. Overmars and E. Welzl. New methods for computing visibility graphs. *Proceedings of the fourth ACM Symposium on Computational Geometry*, pages 164–171, 1988.

[8] R. Tamassia. Dynamic data structures for two-dimensional searching. *Technical Report, Coord. Sc. Lab, Univ. of Illinois at Urbana-Champaign*, 1988.

[9] G. Vegter. The visibility diagram, a data structure for visibility and motion planning. *Proceedings 2nd Scandinavian Workshop on Algorithm Theory, Springer Lecture Notes in Computer Science*, 447:97–110, 1990.

[10] E. Welzl. Constructing the visibility graph for n line segments in the plane. *Information Processing Letters*, 20:167–171, 1985.

An Optimal Algorithm for Computing Visibility in the Plane

Paul J. Heffernan*
Joseph S. B. Mitchell[†]
School of Operations Research and Industrial Engineering
Engineering & Theory Center
Cornell University
Ithaca, NY 14853

Abstract

We give an algorithm to compute the visibility polygon from a point among a set of h pairwise-disjoint polygonal obstacles with a total of n vertices. Our algorithm uses $O(n)$ space and runs in optimal time $\Theta(n + h \log h)$, improving the previous upper bound of $O(n \log h)$.

1 Introduction

Let $\mathcal{D}$ be a planar polygonal domain with h holes and n vertices: $\mathcal{D}$ is a connected closed subset of the plane whose boundary consists of a set of n line segments. If $h = 0$, then $\mathcal{D}$ is simply connected and is called a *simple polygon*. If $h > 0$, then $\mathcal{D}$ is multiply connected, and its holes form a set $\mathcal{P} = \{P_1, P_2, \ldots, P_h\}$ of pairwise-disjoint simple polygons in the plane. The *visibility polygon* with respect to a point $s \in \mathcal{D}$ is the locus of all points $q \in \mathcal{D}$ such that $\overline{sq} \subset \mathcal{D}$. The problem of computing the visibility polygon with respect to a given point s is known as the "hidden line removal" problem and is fundamental in computational geometry.

Algorithms to compute the visibility polygon have been known for some time; a clear summary of the many known visibility algorithms is given in chapter 8 of O'Rourke's book [O'R]. For the case of a simple polygon P, optimal $O(n)$ algorithms have been given by [EA,Le2] and by [JS], who correct a minor error in [EA,Le2] while simplifying the algorithm of [Le2]. For the case of a polygon P with holes, straightforward $O(n \log n)$ time algorithms can be based on plane (rotational) sweep about s (as in [Le1,SO]) or based on divide-and-conquer (as in [AM]). In fact, by using the linear-time algorithms

---

*Email: heff@orie.cornell.edu. Supported by an NSF graduate fellowship.

[†]Email: jsbm@cs.cornell.edu. Partially supported by a grant from Hughes Research Laboratories, Malibu, CA, and by NSF Grant ECSE-8857642.

for simple polygons to compute the visible portion of the boundary of each hole, and then merging these "profiles", one can obtain a simple $O(n \log h)$ algorithm for computing the visibility polygon (see [AM,AAGHI]).

There is an $\Omega(n + h \log h)$ lower bound (from sorting) for computing a visibility polygon [AAGHI,O'R,SO]. Optimal algorithms that achieve this time bound were known for the special case in which the holes P_i are convex ([AM,AAGHI]) or star-shaped ([AM]).

Until now, it has been an open problem to determine an algorithm for the general visibility polygon problem that is linear in n, with *any* dependence on h. ([AAGHI] pose the problem of determining an $O(n + h^2)$ bound.) We resolve this problem by giving a simple algorithm (not requiring linear-time triangulation) whose time bound is $O(n + h^2)$.

We also give an optimal-time algorithm, requiring time $O(\tau(n) + h \log h)$, where $\tau(n)$ is the time required to triangulate a simple polygon. By Chazelle's recent breakthrough [Ch], we know that $\tau(n) = O(n)$, and there are several deterministic and randomized algorithms giving bounds of $O(n \log \log n)$ ([KKT,TV]) or $O(n \log^* n)$ ([CTV,Se]).

Below, we list the specific results of this paper:

(1) We give a very simple $O(n + h^2)$ algorithm, that does not require triangulation.

(2) The algorithm of (1) can be transformed into a dynamic insertion procedure with time complexity $O(\tau(n) + h^2 \log \bar{n})$. Here, $\bar{n} \leq n$ is the number of sides of the most complex hole. This algorithm is given in the full version of this paper [HM].

(3) We have an $O(\tau(n) + h \log^2 h)$ algorithm that is relatively simple, but relies on triangulation. This algorithm is not discussed here, since it does not directly lead to our strongest theoretical result (see [HM]).

(4) We have an $O(\tau(n) + h \log(\bar{n} + h))$ algorithm for the special case in which all of the holes are *stabbed* by a line. This algorithm, which is somewhat involved, is omitted here, due to space limitations; see [HM].

(5) We give an $O(\tau(n) + h \log \log h \log^2 \bar{n})$ algorithm based on applying the result (4) to $O(\log h)$ classes of obstacles, and then merging the resulting set of visibility polygons.

(6) Finally, we show how the algorithm of (5) can be modified to yield an optimal time bound of $\Theta(n + h \log h)$ for the general problem.

Concurrent with our work, [BG] have given an $O(\tau(n) + h^{1+\epsilon})$ algorithm for "merging" the h holes (thereby permitting linear-time visibility computation). Most recently, [BC] have tightened the bound of [BG] to $O(\tau(n) + h \log h)$, allowing a different method of achieving our optimal-time result.

2 Notation and Basic Properties

The *visibility profile* of a set of pairwise-disjoint polygonal obstacles $P_1, \ldots, P_h$ is the "lower envelope" of the obstacles — it is what is seen by an observer at $y = -\infty$.

Converting to a profile problem. In order to solve our problem of computing the visibility polygon from a point $s \in \mathcal{D}$, we claim that it suffices to consider the problem of computing the visibility profile of the set of holes of $\mathcal{D}$. We reduce the first problem to the second by centering a polar coordinate system (θ, r) at s. The basic idea of the reduction is to center a polar coordinate system, (θ, r), at s, and map the holes into an orthogonal coordinate system, (x, y), where, for a point p of a polygonal hole, $y(p) = r(p)$ and $x(p) = \theta(p) + 2\pi k$ (for some integer k). A line of sight in the polar system, a ray with terminus s, corresponds to a vertical line directed upwards in the orthogonal system. If a hole does not intersect the ray $\theta = 0$, then we let the integer k equal 0 for all points on the hole. The situation becomes more complicated when a hole intersects $\theta = 0$; we will need to cut such a hole into two pieces. If P is such a hole, let t and t' be the points of $P \cap \{(\theta, r) | \theta = 0\}$ with minimum and maximum r-coordinates, respectively. For a point $p \in P$, we define a value $x^*(p)$ to be the amount of winding (*not* taken mod 2π) on $P_{CW}(t, p)$, the clockwise subchain from t to p (note that $x^*(p) = \theta(p) + 2\pi k$, for some k). Lemma 1 (below) tells us that if we set $x(p) = x^*(p)$ for points $p \in P_{CW}(t, t')$, and $x(p) = x^*(p) + 2\pi$ for $p \in P_{CW}(t', t)$, then the visibility profile of P in the orthogonal system over the x-range $[0, 2\pi]$ corresponds to the visible portion of P around s. We have replaced P with two polygonal chains; these chains can be made into polygons if we "double" each vertex. While we have not mapped segments into straight-line segments, we have preserved the basic properties necessary to compute visibility. The visibility polygon problem in the polar system has now been reduced to computing the visibility profile from below and outputting the portion over the x-range $[0, 2\pi]$.

Lemma 1 *On subchain $P_{CW}(t, t')$, only points p with $x^*(p) \in [0, 2\pi)$ can be visible from s in the polar system, and on $P_{CW}(t', t)$, only points p with $x^*(p) \in (-2\pi, 0]$ can be visible.*

We let $x(p)$ and $y(p)$ denote the x- and y-coordinates of a point p in the plane. Let $\rho_d(p)$ and $\rho_u(p)$ represent the rays with root p in the direction straight down and straight up, respectively. If a value x represents an x-coordinate, then let x^- and x^+ denote the x-coordinate values infinitesimally left and right of x, respectively.

A polygon P_i contains two vertices, l_i and r_i, of minimum and maximum x-coordinate, respectively. Since we assume that the observer is at $y = -\infty$, the chain obtained by traversing P_i clockwise from l_i to r_i is completely blocked from the view of the observer by the counterclockwise chain of P_i from l_i to r_i. We therefore use only this lower chain of P_i when computing the visibility profile. In fact, in the remainder of this paper, we assume that $P_1, \ldots, P_h$ are polygonal chains joining their left endpoints (l_i) to their right endpoints (r_i).

For any set $S \subseteq \mathcal{P} = \{P_1, \ldots, P_h\}$, we let $VP(S)$ denote the visibility profile of the chains in S, and we let $VP(S; x)$ denote the point of $VP(S)$ with x-coordinate x. We abuse notation slightly and write $VP(i, \ldots, j)$ and $VP(i, \ldots, j; x)$, instead of $VP(\{P_i, \ldots, P_j\})$ and $VP(\{P_i, \ldots, P_j\}; x)$.

We can think of the profile $VP(S)$ as a piecewise-continuous function over the domain $[x(l_{min}), x(r_{max})]$, where $x(l_{min}) = \min_{i \in S} x(l_i)$ and $x(r_{max}) = \max_{i \in S} x(r_i)$. The points x of discontinuity of $VP(S)$ are of two types:

- x is a *jump* if $VP(S)$ coincides with the same chain P_i at both x^- and x^+; and

- x is a *leap* if $VP(S)$ coincides with distinct chains P_i and P_j ($i \neq j$) at x^- and x^+.

A maximal connected subdomain of $[x(l_{min}), x(r_{max})]$ not containing a leap in its interior is called a *piece* (the corresponding section of $VP(S)$ over this domain is also called a piece). Since a piece of $VP(S)$ corresponds to a section of a specific chain P_i, we say that P_i *appears* in $VP(S)$ with this piece. These definitions are illustrated in Figure 1

While we have thought of $VP(S)$ as a function in order to define jumps and leaps, our algorithms will store a visibility profile $VP(S)$ as a polygonal chain. A vertical edge of the chain $VP(S)$ corresponds to a jump or leap. We call the vertical edge of a jump of a profile $VP(i)$ a *lid*, since its interior is disjoint from P_i. We will usually represent a lid by $\overline{ab}$, where $y(a) < y(b)$; therefore, if $\overline{ab}$ is a lid of $VP(i)$, then $VP(i; x(a)) = a$.

A leap x between chains P_i and P_j can occur in one of two manners. The leap x may be caused by the left or right endpoint of one of the two chains, or it may occur where one profile intersects a lid of the other. These cases are illustrated in Figure 1: Coordinates x_1 and x_3 correspond to leaps at a left endpoint, x_5 and x_6 correspond to leaps at a right endpoint, and x_2 and x_4 correspond to leaps at lids.

We will often use the expression "p is below q" to indicate that $y(p) < y(q)$. Similar use is made of the terms "above", "left", and "right". We say that profile $VP(S)$ is below profile $VP(S')$ at x-coordinate x if $y(VP(S; x^-)) < y(VP(S'; x^-))$ or if $y(VP(S; x^+)) < y(VP(S'; x^+))$. It is possible for one but not both of these conditions to hold if one of the profiles has a jump or leap at x.

We will assume without loss of generality that all chains $P_1, \ldots, P_h$ lie completely above the x-axis, so that the point $p = (x, 0)$ is below $VP(i)$ for any x and any profile $VP(i)$.

Lemma 2 *If $x(l_i) < x(l_j)$, then P_j appears at most once in $VP(i,j)$; that is, at most one piece of $VP(i,j)$ is contributed by $VP(j)$.*

Proof. Suppose that $VP(j)$ contributes two pieces to $VP(i,j)$. Let p and q be points of $VP(j)$ on each of the two pieces, with p on the left piece and q on the right piece. Let r be a point of $VP(i)$ that lies on a piece between the two pieces contributed by $VP(j)$, so that $x(p) < x(r) < x(q)$. Refer to Figure 2.

Now consider the closed Jordan curve given by starting at point $(x(p), 0)$, going up to p, following chain P_j to q, going down to $(x(q), 0)$, and then returning to $(x(p), 0)$ along the x-axis. Since point r is on the profile $VP(i,j)$, it must lie in the bounded component defined by this Jordan curve. On the other hand, the left-most point l_i must lie in the unbounded component defined by the closed curve, since l_i is to the left of l_j. This implies that P_i must cross the Jordan curve, which is a contradiction, since p and q are on the profile, and P_i and P_j do not cross. □

Lemma 3 *If $x(r_j) < x(r_i)$, then P_j appears at most once in $VP(i,j)$.*

We can now give a full characterization of $VP(i,j)$, for chains P_i and P_j whose x-coordinate domains overlap. Assume without loss of generality that $x(l_i) < x(l_j)$. Refer to Figure 3.

(1). If $x(r_i) < x(r_j)$, then clearly P_i and P_j each appear at least once in $VP(i,j)$, and by the previous two lemmas, each appears at most once. The profile $VP(i,j)$ therefore consists of a piece from $VP(i)$ lying left of a piece from $VP(j)$.

(2). If $x(r_j) < x(r_i)$, there are two possibilities:

(a). Profile $VP(j)$ may lie completely above $VP(i)$, in which case $VP(i,j)$ is the single piece $VP(i)$.

(b). If P_j appears once in $VP(i,j)$, then P_i must appear exactly twice, since pieces alternate, and the left- and right-most pieces are from P_i.

We now state a combinatorial lemma of fundamental importance:

Lemma 4 *The profile $VP(S)$ has $O(|S|)$ pieces.*

Proof. Consider the (ordered) sequence σ of indices of chains P_i that contribute pieces to $VP(S)$. There are $|S|$ different indices, and by the definition of pieces, no index i can appear twice consecutively in σ. By Lemma 2 (or Lemma 3), it is not possible to have a subsequence of the form $\ldots, i, \ldots, j, \ldots, i, \ldots, j, \ldots$. This implies that σ is a Davenport-Schinzel sequence of order 2, so its maximum length is given by $\lambda_2(|S|) = 2|S| - 1$. (See [Sh] for background on the theory of Davenport-Schinzel sequences.) $\square$

3 An $O(n + h^2)$ Algorithm

We describe now a simple $O(n+h^2)$-time algorithm for computing the visibility profile of a collection of disjoint polygons. Not only is the algorithm relatively easy to implement, but it resolves the previously open theoretical question of whether or not an algorithm *linear* in n exists.

Assume that we have indexed the chains $\mathcal{P} = \{P_1, \ldots, P_h\}$ so that their left endpoints $l_1, \ldots, l_h$ are sorted by decreasing x-coordinate. The algorithm simply considers the profiles one-by-one according to this order: Step i consists of adding $VP(i)$ to $VP(1, \ldots, i-1)$ to obtain $VP(1, \ldots, i)$. The time to update the profile when we insert $VP(i)$ is linear in h and the size of P_i, implying the claimed overall time bound.

The algorithm maintains a sorted list of the leaps, $x_1, \ldots, x_K$, of the current profile $VP(1, \ldots, i-1)$. Each leap x_k in the list stores a pointer to the point $VP(1, \ldots, i-1; x_k)$. To add $VP(i)$ to the profile, we traverse $VP(i)$ to place pointers on the points $VP(i; x_1), \ldots, VP(i; x_K)$. Now, for each leap x_k, we compare the points $VP(1, \ldots, i-1; x_k)$ and $VP(i; x_k)$, to determine whether $VP(i)$ is below $VP(1, \ldots, i-1)$ at this x-coordinate. If so, we have identified a piece of $VP(i)$ in $VP(1, \ldots, i)$; we simultaneously traverse $VP(i)$ and $VP(1, \ldots, i-1)$ to the left, maintaining our pointers at the same approximate x-coordinate, until we reach the x-coordinate, x_l, where $VP(i)$ is no longer below $VP(1, \ldots, i-1)$; x_l is the left endpoint of this piece of $VP(i)$, and consequently is a leap in $VP(1, \ldots, i)$. Similarly, we simultaneously traverse $VP(i)$ and $VP(1, \ldots, i-1)$ to the right to obtain the right endpoint, x_r. The portion of $VP(1, \ldots, i-1)$ between x_l and x_r is replaced by the corresponding portion of $VP(i)$, and the leaps at x_l and x_r are incorporated into the list, along with pointers to $VP(1, \ldots, i; x_l)$ and $VP(1, \ldots, i; x_r)$. Of course, the interval $[x_l, x_r]$ may contain leaps of $VP(1, \ldots, i-1)$ other than x_k, but this poses no difficulty to the algorithm. The following lemma establishes that the new profile obtained in this manner is in fact $VP(1, \ldots, i)$, and that the updated list of leaps is the list for $VP(1, \ldots, i)$.

Lemma 5 *Each piece of $VP(i)$ in $VP(1,\ldots,i)$ must cover a leap of $VP(1,\ldots,i-1)$; that is, for each piece contributed by $VP(i)$, there exists a leap x of $VP(1,\ldots,i-1)$ such that $VP(1,\ldots,i;x) \in VP(i)$.*

Proof. Suppose we have a piece $[x_l, x_r]$ of $VP(i)$ in $VP(1,\ldots,i)$ that lies between consecutive leaps x_k and x_{k+1} of $VP(1,\ldots,i-1)$. Since $[x_k, x_{k+1}]$ is a single piece of $VP(1,\ldots,i-1)$, the points $VP(1,\ldots,i-1;x_k^+)$ and $VP(1,\ldots,i-1;x_{k+1}^-)$ lie on the same profile $VP(j)$. We have that $VP(j)$ lies below $VP(i)$ at x_k^+ and x_{k+1}^-, and that $VP(i)$ lies below $VP(j)$ at x_l^+, where $x_k < x_l < x_{k+1}$. But this contradicts Lemma 2, since $x(l_i) < x(l_j)$, by our ordering of the polygonal chains. $\square$

We now analyze the time complexity of the algorithm. The initial indexing of the polygonal chains requires time $O(h \log h)$, to sort the left endpoints of the chains. Adding $VP(i)$ to $VP(1,\ldots,i-1)$ requires that we traverse $VP(i)$ twice — once to place pointers to $VP(i;x_1),\ldots,VP(i;x_K)$, and once during the simultaneous traversals of $VP(i)$ and $VP(1,\ldots,i-1)$. The time spent in all steps except the traversing of $VP(1,\ldots,i-1)$ is $O(|P_i|+h)$, implying a total of $O(n+h^2)$ over all steps. The simultaneous traversals of the updating step require that we traverse sections of $VP(1,\ldots,i-1)$, which consists of profiles that have already been processed. However, the portions we traverse are deleted from the current profile $VP(1,\ldots,i)$, and are never traversed again. Thus, the entire algorithm runs in time $O(n+h^2)$.

4 An Optimal Algorithm

We turn our attention now to a different algorithm, one which attains the optimal $\Theta(n + h \log h)$ time bound. We will describe first an algorithm that runs in time $O(n + h \log \log h \log^2 n)$, and will then modify it to perform in optimal time.

We begin by sorting the x-coordinates of the endpoints of $P_1,\ldots,P_h$, thereby obtaining a list $x_1,\ldots,x_{2h}$. If $\mathcal{P} = \{P_1,\ldots,P_h\}$, define S_1 to be the chains of $\mathcal{P}$ stabbed by the vertical line $x = x_h$. Define S_2 to be the chains of $\mathcal{P} \setminus S_1$ stabbed by $x = x_{\lfloor h/2 \rfloor}$ or $x = x_{\lfloor 3h/2 \rfloor}$. Continuing in this way, we obtain a partitioning of $\mathcal{P}$ into a class of subsets $S_1,\ldots,S_{\lceil \log 2h \rceil}$. Below we will show that the visibility profile of a set S' of h' polygonal chains stabbed by a vertical line can be computed in time $O(n' + h' \log \bar{n}')$, where n' is the total number of vertices in S', and $\bar{n}'$ is the number of vertices on the largest chain in S'. Therefore $VP(S_1),\ldots,VP(S_{\lceil \log 2h \rceil})$ can be computed in time $O(n + h \log \bar{n})$, where $\bar{n}$ is the size of the largest chain in $\mathcal{P} = \{P_1,\ldots,P_h\}$. We will also show how to merge $VP(S')$ with $VP(S'')$ in $O(\bar{h} \log^2 \bar{n})$ time, for two subsets S' and S'' of $\{S_1,\ldots,S_{\lceil \log 2h \rceil}\}$, with $max\{i|S_i \in S'\} < min\{j|S_j \in S''\}$, where $\bar{h}$ is the total number of polygonal chains in the sets comprising S' and S''. This allows one to compute $VP(\mathcal{P})$ by recursively computing $VP(S_1 \cup \ldots \cup S_{\lceil (\log 2h)/2 \rceil})$ and $VP(S_{\lceil (\log 2h)/2 \rceil + 1} \cup \ldots \cup S_{\lceil \log 2h \rceil})$ and then merging them. Each step of the recursion requires time $O(h \log^2 \bar{n})$, and the recursion has depth $O(\log \log h)$, giving a total algorithm run-time of $O(n + h \log \log h \log^2 \bar{n})$.

Throughout our algorithm, we will wish to make logarithmic-time queries that we call *lid queries*. Lid queries are bases on planar point location, a familiar notion in computational geometry. Basically, given a chain P_i and a point in the plane p, a lid query of p on P_i asks where p is with respect to P_i. The result of such a query will give us important information about $VP(\{P_i, c\})$, for any chain c containing p such that $c \cap P_i = \emptyset$. We now describe lid queries in detail.

Consider a lid $\overline{ab}$ of $VP(i)$, the visibility profile of chain P_i. The lid $\overline{ab}$ and the subchain of P_i between a and b form a simple polygon, which we call a *pocket*, such that all points in the interior of the pocket are non-visible from below. If a point p in the pocket lies on a chain c that does not intersect P_i, then c can be below $VP(i)$ only if it crosses $\overline{ab}$. Since $\overline{ab}$ is on $VP(i)$, crossing $\overline{ab}$ is also a sufficient condition for c to be below $VP(i)$ somewhere. We now state formally the information that we want from a *lid query*.

Definition 1 (Lid Query) *For a point p not on P_i, one of the following is true:*

  1. p is below $VP(i)$,

  2. p is in a pocket of $VP(i)$,

  3. p lies in the region above the simple, infinite chain $c = \rho_u(l_i) \cup P_i \cup \rho_u(r_i)$.

If (1) is true, a lid query of p on P_i returns $VP(i; x(p))$ (in Figure 4, for example, a query on p_1 returns $VP(i; x(p_1))$). If (2) is true, the lid query returns the lid $\overline{ab}$ that defines the pocket (e.g. a query on p_2 in the figure returns lid $\overline{a_2 b_2}$). If (3) is true (as it is for p_3 in the figure), then the rays $\rho_u(l_i)$ and $\rho_u(r_i)$ together have the property that we desire in the lids, so the lid query returns them.

Lid queries will be used often in our algorithm to determine quickly if and where a chain c containing a point p lies below the current profile $VP(S)$. Often, we will try to add a chain c to $VP(S)$ to form $VP(S \cup \{c\})$, knowing only that c contains a certain point p and that c and S are disjoint. We formalize this notion by defining the *lid property*.

Definition 2 (Lid Property) *An x-coordinate x has the lid property for point p and profile $VP(S)$ if, for any chain c that contains p and is disjoint from S, $VP(\{c\})$ is below $VP(S)$ somewhere only if $VP(\{c\})$ is below $VP(S)$ at x.*

When a lid query of a point p on a chain P_i returns a lid $\overline{ab}$, the x-coordinate $x(a)$ satisfies the lid property for p and $VP(i)$. Our algorithm will produce, through the use of lid queries, x-coordinates that satisfy the lid property for the current profile $VP(S)$; typically these will be the x-coordinates of either leaps or jumps of $VP(S)$.

Lid queries are basically planar point location. We first obtain the vertical visibility map of P_i (in $O(n)$ time for all chains, by [Ch]). Then P_i can be preprocessed to return the trapezoid of the map containing a query point in logarithmic time (by [Ki], for example); this is sufficient to answer the query for case (1). For cases (2) and (3), we need to do some more work. Consider the dual graph of the trapezoidal decomposition, with the edge corresponding to the decomposition ray $\rho_u(r_i)$ deleted. This graph is a tree, and the trapezoids that comprise any given pocket correspond to a subtree. Each lid of $VP(i)$ corresponds to an edge in the tree, that separates the subtree of the adjacent pocket from the rest of the tree. For each lid (including $\rho_u(l_i)$), we begin at the corresponding edge in the dual graph, and traverse through the subtree of the pocket, assigning to each node a pointer to the lid. Therefore, if the planar point location query encounters a trapezoid with a pointer to a lid, the lid query returns that lid. We see that the chains $P_1, \ldots, P_h$ can be preprocessed in $O(n)$ total time to handle lid queries in time $O(\log \bar{n})$, where $\bar{n}$ is the number of vertices on the largest of $P_1, \ldots, P_h$.

At this point we mention that our algorithm requires a careful representation of a current profile $VP(S)$. We said earlier that $VP(S)$ is stored as a polygonal chain. It is important that this be done by adding pointers (along with necessary added vertices b at the top of leaps) to the original profiles $VP(i)$, $P_i \in S$. Therefore, to traverse a section of $VP(S)$, we begin by traversing a section of the appropriate profile $VP(i)$, and, upon encountering a leap at x-coordinate x, leap to the new appropriate profile $VP(j)$, by means of a pointer from $VP(i; x)$ to $VP(j; x)$. We take care to mention the necessity of this representation of $VP(S)$, because our procedures will sometimes ask for a point $VP(S; x)$ by performing a lid query of $p = (x, 0)$ on the appropriate profile $VP(i)$. Since the lid query preprocessing is done on each original chain P_i, and we cannot afford to do preprocessing on an intermediate, composite profile like $VP(S)$, it is imperative that we can use a lid query on an individual chain P_i to find a point $VP(S; x)$.

4.1 The Profile for a Set of Stabbed Polygons

In the full paper ([HM]), we describe a procedure for computing the visibility profile of a collection of polygonal chains $P_1, \ldots, P_h$, all of which are stabbed by a vertical line ℓ. The basic method is to construct the profile incrementally, adding the chains in order of increasing "bottom" point (the *bottom point* of chain P_i is the lowest crossing point of P_i with ℓ). With each addition, we must do a type of binary search on the current profile, in order to determine how the newly added chain affects it. The search is conducted efficiently by a careful analysis of the structure of the profile of stabbed chains, using lid queries.

4.2 Merging

We describe the merging of the visibility profiles of two subsets S' and S'' of $S = \{S_1, \ldots, S_{\lceil \log h \rceil}\}$, where $max\{i | S_i \in S'\} < min\{j | S_j \in S''\}$. We have a family of vertical lines such that every chain in S' is stabbed by at least one of the lines, but no chain of S'' is stabbed by a line. Since the vertical lines separate the elements of S'', we can individually consider the interval between each pair of consecutive lines. Therefore, we consider a vertical strip bordered by the lines π_l and π_r. All chains of S' that appear in $VP(S')$ in the strip are stabbed by either π_l or π_r, and no chain of S'' appearing in $VP(S'')$ is stabbed by either line.

Inductively, we assume that we have, for $VP(S')$ $(VP(S''))$ over the strip, a sorted list of all leaps, and for each leap x of $VP(S')$ $(VP(S''))$, a pointer to $VP(S'; x)$ $(VP(S''; x))$. We merge the lists to form a single sorted list $x_1, \ldots, x_K$ of all leaps in $VP(S')$ and $VP(S'')$. For a leap x_k from $VP(S')$ $(VP(S''))$, we must compute a pointer to $VP(S''; x_k)$ $(VP(S'; x_k))$. We do this through lid queries, as follows. We can assume that in forming the list $x_1, \ldots, x_K$, every leap x_k from $VP(S')$ knows which profile from S'' contributes $VP(S''; x_k)$ (this consists of knowing the leaps from $VP(S'')$ that are nearest to x_k to the left and right). Since we know the profile $VP(j)$ that contributes $VP(S''; x_k)$, we can compute $VP(j; x_k) = VP(S''; x_k)$ by querying the point $(x_i, 0)$ on P_j. Since $(x_k, 0)$

lies below $VP(j)$ (by our assumption that chains lie above the x-axis), the query returns $VP(j; x_k)$.

We define a *subpiece* of $VP(S')$ or $VP(S'')$ as the portion of the profile between two consecutive leaps in the merged list $x_1, \ldots, x_K$. Note that a subpiece is a subset of some piece. The following lemma motivates the merge procedure:

Lemma 6 *Suppose we have a subpiece over the interval $[x_k, x_{k+1}]$, such that $VP(i)$ and $VP(j)$ contribute this subpiece to $VP(S')$ and $VP(S'')$, respectively. Then $VP(S')$ is below $VP(S'')$ somewhere in $[x_k, x_{k+1}]$ only if $VP(i)$ is below $VP(j)$ at x_k or x_{k+1}.*

Proof. We know that $VP(j)$, as an element of $VP(S'')$, does not intersect π_l nor π_r, whereas $VP(i)$, as an element of $VP(S')$, does intersect at least one of the two. This implies that either $x(l_i) < x(l_j)$ or $x(r_i) > x(r_j)$, which, by Lemma 2 or Lemma 3, means that $VP(j)$ appears at most once in $VP(i, j)$. This means it is impossible for $VP(j)$ to be below $VP(i)$ at both x_k and x_{k+1} when $VP(i)$ is below $VP(j)$ somewhere in between x_k and x_{k+1}. $\square$

Now we combine the two profiles over the subpiece $[x_k, x_{k+1}]$. If $VP(j)$ is below $VP(i)$ at both x_k and x_{k+1}, then the entire subpiece is contributed by $VP(j)$. Below we describe a procedure for the case when $VP(i)$ is below at one of x_k and x_{k+1}, and $VP(j)$ is below at the other. We then show how this procedure can be modified to handle the case where $VP(i)$ is below at both x_k and x_{k+1}.

If $VP(i)$ is below at one of x_k and x_{k+1}, and $VP(j)$ is below at the other, then our task consists of finding the unique leap in $VP(i, j)$ between x_k and x_{k+1}, without traversing portions that are part of $VP(S' \cup S'')$. A naive scheme could take time linear in the size of the portions of $VP(S')$ and $VP(S'')$ between x_k and x_{k+1}, but our approach requires only polylog time. Assume that the vertices of each original profile have been numbered in left-to-right order, and placed in a data structure so that if we are given a pointer to a vertex of the profile, we can in constant time return the numbering of the vertex. Our procedure maintains two pointers to $VP(i)$, denoted p_l^i and p_r^i, which are initialized to $VP(i; x_k)$ and $VP(i; x_{k+1})$, and pointers p_l^j and p_r^j to $VP(j)$, initialized to $VP(j; x_k)$ and $VP(j; x_{k+1})$. The pointers p_l^i and p_l^j will be maintained at the same x-coordinate, as will p_r^i and p_r^j. Initially we know that there is exactly one leap of $VP(i, j)$ between $x(p_l^i) = x(p_l^j)$ and $x(p_r^i) = x(p_r^j)$; the procedure maintains this property while moving $x(p_l^i) = x(p_l^j)$ and $x(p_r^i) = x(p_r^j)$ closer together, eventually sandwiching the leap. The procedure alternates steps on the pointer pairs (p_l^i, p_r^i) and (p_l^j, p_r^j). We describe a step on the pair (p_l^i, p_r^i):

1. Query the numbering of the vertices of $VP(i)$ nearest p_l^i and p_r^i, and assign these numberings to p_l^i and p_r^i.

2. Find q, the vertex of $VP(i)$ whose numbering is halfway between the numberings of p_l^i and p_r^i.

3. Compute $VP(j; x(q))$.

4. Compare the y-coordinates of $VP(j; x(q))$ and $q = VP(i; x(q))$; this tells us whether the leap is left or right of $x(q)$; accordingly, update either p_l^i and p_l^j, or p_r^i and p_r^j.

Upon completion of this step on the pair (p_l^i, p_r^i), perform a symmetric step on (p_l^j, p_r^j), and continue to alternate the steps. Eventually the total number of vertices on $VP(i)$ between p_l^i and p_r^i and on $VP(j)$ between p_l^j and p_r^j is less than a small, pre-set constant, so in constant time we find the leap and complete this subpiece of $VP(S' \cup S'')$.

A slight modification of the above procedure handles the case where $VP(i)$ is below $VP(j)$ at both x_k and x_{k+1}. Query $VP(j; x_k)$ on $VP(i)$, to find a lid $\overline{ab}$ which $VP(j)$ must cross in order to contribute a piece to $VP(i, j)$. If $x(a) \notin [x_k, x_{k+1}]$, then $VP(j)$ is not below $VP(i)$ anywhere in the interval $[x_k, x_{k+1}]$. If $x(a) \in [x_k, x_{k+1}]$, then perform a lid query on the point $(x(a), 0)$ to compute $VP(j; x(a))$. If $VP(i)$ is below $VP(j)$ at $x(a)$, then the subpiece between x_k and x_{k+1} is contributed totally by $VP(i)$; otherwise, we break the subpiece $[x_k, x_{k+1}]$ into two subpieces, $[x_k, x(a)]$ and $[x(a), x_{k+1}]$, and process each subpiece with the above procedure.

Consider the total time of merging $VP(S')$ and $VP(S'')$. Let $\bar{h}$ represent the total number of chains in S' and S'', and $\bar{n}$ the number of vertices on the largest chain in the set $S' \cup S''$. Creating the combined sorted lists of leaps $x_1, \ldots, x_K$ for all strips takes time $O(\bar{h})$, because we already have the sorted lists of leaps for S' and S'' separately. Computing $VP(i; x_k)$ and $VP(j; x_k)$ for every leap x_k requires time $O(\bar{h} \log \bar{n})$, since it consists of performing one lid query per leap. We then process each of the $O(\bar{h})$ subpieces separately, perhaps breaking some subpieces into two subpieces with the help of a single lid query. Processing a subpiece consists of alternating steps on the pairs of pointers (p_l^i, p_r^i) and (p_l^j, p_r^j). Each step consists of one lid query plus some constant time work, and is therefore $O(\log \bar{n})$. Because every two steps eliminate at least half of the vertices of $VP(i)$ between p_l^i and p_r^i and of $VP(j)$ between p_l^j and p_r^j, the number of steps is $O(\log \bar{n})$. Therefore each subpiece requires $O(\log^2 \bar{n})$ time, for a total of $O(\bar{h} \log^2 \bar{n})$ time to merge $VP(S')$ and $VP(S'')$.

4.3 Putting It Together: An Optimal Algorithm

The above subsections describe how to compute $VP(\mathcal{P})$ for $\mathcal{P} = \{P_1, \ldots, P_h\}$ in time $O(n + h \log \log h \log^2 \bar{n})$, where n is the total number of vertices in $\mathcal{P}$, and $\bar{n}$ is the number of vertices on the largest chain of $\mathcal{P}$. A modification allows this algorithm to compute $VP(\mathcal{P})$ in optimal $\Theta(n + h \log h)$ time. The modification consists of breaking $\mathcal{P}$ into two groups, the "large" chains and the "small" ones, computing the visibility profile of each group separately, and then merging the profiles with a final linear-time merge.

The first observation to be made is that if $h = O(n / \log^3 n)$, then the algorithm's complexity is $O(n)$. Motivated by this observation, we break $\mathcal{P}$ into two groups as follows: all chains in $\mathcal{P}$ with more than $\log^3 n$ vertices are placed in the "large" group, and the rest in the "small" group. The large group can have no more than $n / \log^3 n$ members, so the algorithm can compute the visibility profile of this group in $O(n)$ time.

Assume that the small group has $h = \Omega(n / \log^3 n)$ members (if not, the algorithm is $O(n)$ on this group). No chain of the small group has more than $\log^3 n$ vertices, implying that $\bar{n} \leq \log^3 n$. The complexity of the algorithm is therefore $O(n + h \log \log h \log^2(\log^3 n)) = O(n + h \log \log h(\log \log n)^2)$. Since $h = \Omega(n / \log^3 n)$, we have that $\log \log n = O(\log \log h)$, giving a complexity of $O(n + h(\log \log h)^3) = O(n + h \log h)$. Therefore, the visibility profiles of both the small group and the large group can be computed in $O(n + h \log h)$, giving $VP(\mathcal{P})$ in the same time bound.

References

[AM] E. Arkin and J.S.B. Mitchell, "An Optimal Visibility Algorithm for a Simple Polygon With Star-Shaped Holes", Technical Report No. 746 School of Operations Research and Industrial Engineering, Cornell University, June, 1987.

[AAGHI] T. Asano, T. Asano, L. Guibas, J. Hershberger, and H. Imai, "Visibility of Disjoint Polygons", *Algorithmica*, Vol. 1, (1986), pp. 49-63.

[BC] R. Bar-Yehuda and B. Chazelle, private communication, 1991.

[BG] R. Bar-Yehuda and R. Grinwald, "An $O(N + H^{1+\epsilon})$-Time Algorithm to Merge H Simple Polygons", Technical Report 657, Computer Science Dept., Technion, 1990.

[Ch] B. Chazelle, "Triangulating a Simple Polygon in Linear Time", CS-TR-264-90, Princeton Univ., May 1990.

[CTV] K. Clarkson, R.E. Tarjan, C. Van Wyk, "A Fast Las Vegas Algorithm for Triangulating a Simple Polygon", *Proc. Fourth Annual ACM Symposium on Computational Geometry*, pp. 18-22, 1988. Also Princeton Technical Report CS-TR-157-88.

[EA] H.A. El Gindy and D. Avis, "A Linear Algorithm for Computing the Visibility Polygon From a Point", *Journal of Algorithms*, Vol. 2 (1981), pp. 186-197.

[HM] P.J. Heffernan and J.S.B. Mitchell, "An Optimal Algorithm for Computing Visibility in the Plane", Technical Report No. 953, School of Operations Research and Industrial Engineering, Cornell University, 1990.

[JS] B. Joe and R.B. Simpson, "Correction to Lee's Visibility Polygon Algorithm", *BIT*, **27** (1987), pp. 458-473.

[Ki] D.G. Kirkpatrick, "Optimal Search in Planar Subdivisions", *SIAM Journal on Computing*, **12** (1983), No. 1, pp. 28-35.

[KKT] D.G. Kirkpatrick, M.M. Klawe, and R.E. Tarjan, "Polygon Triangulation in $O(n \log \log n)$ Time with Simple Data-Structures", *Proc. Sixth Annual ACM Symposium on Computational Geometry*, Berkeley, CA, June 6-8, 1990, pp. 34-43.

[Le1] D.T. Lee, "Proximity and Reachability in the Plane", Ph.D. Thesis, Report R-831, Dept. of Electrical Engineering, University of Illinois at Urbana-Champaign, Nov. 1978.

[Le2] D.T. Lee, "Visibility of a Simple Polygon", *Computer Vision, Graphics, and Image Processing*, Vol. 22 (1983), pp. 207-221.

[O'R] J. O'Rourke, *Art Gallery Theorems and Algorithms*, Oxford University Press, 1987.

[Se] R. Seidel, "A Simple and Fast Incremental Randomized Algorithm for Computing Trapezoidal Decompositions and for Triangulating Polygons", Manuscript, October, 1990.

[Sh] M. Sharir, "Davenport-Schinzel Sequences and their Geometric Applications", pp 253-278, NATO ASI Series, Vol. F40, Theoretical Foundations of Computer Graphics and CAD, R.A. Earnshaw (Ed.), Springer-Verlag Berlin Heidelberg, 1988.

[SO] S. Suri and J. O'Rourke, "Worst-Case Optimal Algorithms For Constructing Visibility Polygons With Holes", *Proc. Second Annual ACM Symposium on Computational Geometry*, Yorktown Heights, NY, June 1986, pp. 14-23.

[TV] R.E. Tarjan and C. Van Wyk, "An $O(n \log \log n)$-Time Algorithm for Triangulating a Simple Polygon", *SIAM Journal on Computing*, **17** (1988), No. 1, pp. 143-178.

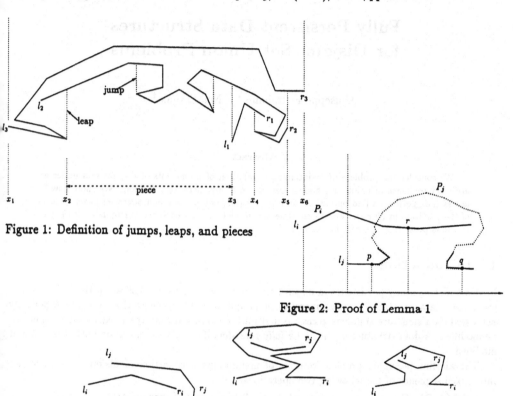

Figure 1: Definition of jumps, leaps, and pieces

Figure 2: Proof of Lemma 1

Figure 3: Structure of $VP(i,j)$

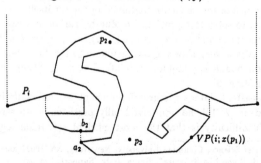

Figure 4: Results of lid queries of points p_1, p_2, p_3 on chain P_i

Fully Persistent Data Structures
for Disjoint Set Union Problems

Giuseppe F. Italiano [*]   Neil Sarnak [†]

Abstract

We consider the problem of maintaining a partition of n elements of disjoint sets under an arbitrary sequence of union and find operations in a fully persistent fashion. We show how to access (i.e., perform a find operation) and modify (i.e., perform a union operation) each version of the partition in $O(\log n)$ worst-case time and in $O(1)$ amortized space per update. No better bound is possible for pointer based algorithms.

1   Introduction

In this paper we study persistence in disjoint set union data structures. Following Driscoll et al. [8], we define a data structure to be *ephemeral* when an update destroys the previous version. A *partially persistent* data structure supports access to multiple versions, but only the most recent version can be modified. A data structure is said to be *fully persistent* if every version can be both accessed and modified.

The disjoint set union problem consists of maintaining a collection of disjoint sets under an intermixed sequence of the following two operations.

union(A, B) : Combine the sets named A and B into a new set named A;

find(x) : Return the name of the set containing element x.

Initially the collection consists of n singleton sets $\{1\}, \{2\}, \ldots, \{n\}$, and the name of set $\{i\}$ is i. Due to the definition of the union and find operations, two invariants hold at any time. First, the sets are always disjoint and define a partition of elements into equivalence classes. Second, the name of each set corresponds to one of the elements contained in the set itself.

The best algorithms to solve this problem are due to Tarjan and van Leeuwen [15, 18]. These algorithms require $O(n)$ space and $O(k\alpha(k + n, n) + n)$ time, where k is the number of find operations performed and α is a very slowly growing function, the functional inverse of Ackermann's function [18]. No better bound is possible for pointer based algorithms [4, 11, 16], and in the cell probe model of computation [9].

While partially persistent data structures for disjoint set union problems have been already proposed in the literature [2, 13], no non-trivial fully persistent data structures for this problem are known. The main difficulty is that the most efficient set union algorithms use linked data

---

[*]Department of Computer Science, Columbia University, New York, NY 10027 and Dipartimento di Informatica e Sistemistica, Università di Roma "La Sapienza", Roma, Italy. Supported in part by an IBM Graduate Fellowship and by the ESPRIT Basic Research Action Program of the European Communities under contract No. 3075 (Project ALCOM). Work done while visiting the IBM T. J. Watson Research Center.

[†]IBM Research Division, T.J. Watson Research Center, Yorktown Heights, NY 10598.

structures whose nodes have degrees not necessarily bounded by a constant (see for instance [18]); and unfortunately the techniques proposed by Driscoll et al. [8] to make data structures persistent are not applicable in this case.

The *fully persistent disjoint set union problem* can be defined as follows. Throughout any sequence of operations, multiple versions of a set union data structure are maintained (i.e., multiple versions of the partition are maintained). Union operations are updates, while find operations are accesses. If the j-th union operation applies to version v, $v < j$, then the result of the update is a new version j. The operations on the fully persistent data structure can be defined as follows (we use upper case initials to distinguish them from the corresponding operations on the ephemeral data structure).

Union(x, y, v) : Denote by X and Y the two sets in version v containing respectively x and y. If $X = Y$, then do nothing. Otherwise create a new version in which X and Y are combined into a new set. The new set gets the same name as X.

Find(x, v) : Return the name of the set containing element x in version v.

Initially the partition consists of n singleton sets $\{1\}, \{2\}, \ldots, \{n\}$ and the name of set $\{i\}$ is i. This is version 0 of the set union data structure. The restricted case in which Union operations are allowed to modify only the most recent version defines the *partially persistent disjoint set union problem*.

In the remainder of this paper, we denote by m the total number of Union operations performed. Before stating our bounds, let us review some trivial solutions to the fully persistent disjoint set union problem. If we store each version of the partition explicitly, then each Find can be performed in $O(1)$ time and each Union in $O(n)$ time, at the expense of $O(n)$ storage per update. If we do not store all the versions but just the sequence of Union operations, then each Union can be performed in $O(1)$ time and $O(1)$ space. However, a Find requires now $O(m)$ time. With a little more care, this time can be reduced to $O(n)$.

We show how to perform Union and Find operations in any version in $O(\log n)$ [1] worst-case time and $O(1)$ amortized space for update. No better pointer based algorithm is possible, since in this setting Mannila and Ukkonen [13] showed a lower bound of $\Omega(\log n)$ amortized time for the partially persistent disjoint set union problem.

We then consider a variant of the above problem, called the *fully persistent disjoint set union problem with deletions*, and in which besides Unions and Finds we allow deletions of an element from a set:

Delete(x, v) : Denote by X the set in version v containing element x. If $|X| = 1$ then do nothing. Otherwise, create a new version of the partition by deleting x from X. In the new version, element x is in a set by itself and the name of set $\{x\}$ is x. The name of set $X' = X - \{x\}$ in the new version is defined as follows. If the name of X was not x, then X' gets the same name as X. Otherwise X' gets as a new name arbitrarily one of its elements.

We show that each Union, Find and Delete operation can be supported in $O(\log n)$ worst-case time and $O(1)$ space per update.

Motivations for studying these problems arise in several application areas. For instance modern high level languages such as Hermes [3] and SETL [6] support sets as basic data types and need fast set manipulation primitives on multiple versions of the same set. Another application is logic programming memory management [12, 19]. In Prolog, for example, variables of clauses correspond to elements of a set, and unifications imply disjoint set union operations. In this case, the availability of multiple versions of disjoint sets allows one to support backtracking and branching techniques, and to trace different program executions at the same time.

---

[1] All the logarithms are assumed to be to the base 2 unless explicitly specified otherwise.

The remainder of this paper consists of three sections. Section 2 shows how to make set union data structures fully persistent in $O(\log^2 n)$ time and $O(1)$ amortized time per update. We improve the time bound to $O(\log n)$ and show how to deal with Delete operations in section 3. Section 4 contains some concluding remarks.

2 An $O(\log^2 n)$ Fully Persistent Data Structure

In this section we describe techniques to make set union data structures fully persistent.

To perform Union and Find operations, we use the basic set union tree data structures (see for instance [18]), which represent sets making use of rooted trees. Each tree corresponds to a set. Nodes of the tree correspond to elements of the set. The name of the set is stored in the root of the tree, which is referred to as the *canonical element* of the set. Each node of the tree points to its parent, except for the root that points to itself.

We recall that in such an ephemeral data structure, a find(x) can be performed by starting from the node x and by following the pointer to the parent until the tree root (i.e., the canonical element) is reached. The name of the set stored in the canonical element is then returned. Therefore, the time bound of a find operation is proportional to the longest path from a leaf to its root in a set union tree. To keep those paths short, one of the following two *union rules* can be applied while performing a union operation.

union by size : make the root of the smaller tree point to the root of the larger, arbitrarily breaking a tie. This requires maintaining the number of descendants for each node throughout all the sequence of operations.

union by rank : make the root of the shallower tree point to the root of the other, arbitrarily breaking a tie. This requires maintaining the height of the subtree rooted at each node, in the following referred to as the *rank* of a node, throughout all the sequences of operations.

Using either *union rule* yields trees of path length no greater than $O(\log n)$. In the following, we describe our data structure when *union by size* is used, but the same results can be easily extended to the case of *union by rank*.

As mentioned in the introduction, we cannot apply the techniques of Driscoll et al. [8] to make the forest of set union trees fully persistent. Indeed, as a consequence of how union operations are carried out, the in-degree of a node in a set union tree is not necessarily bounded by a constant. Our solution is as follows. Throughout the sequence of operations, we maintain a fully persistent balanced search tree T. Version v of T contains information about version v of the forest of set union trees, as follows. Each version of T contains n items $1, 2, \ldots, n$ corresponding to the n elements of the disjoint sets. For $1 \leq x \leq n$, item x in version v of T has associated three fields: *parent, size*, and *name*. We will refer to those three fields as $parent(x, v)$, $size(x, v)$ and $name(x, v)$ respectively. Field $parent(x, v)$ stores the parent of element x in version v of the forest of set union trees, if x is a non-root node; otherwise if x is a root, $parent(x, v) = x$. Field $size(x, v)$ is equal to the size of the subtree rooted at x in version v of the forest of set union trees. Finally, $name(x, v)$ is defined only if x is a root in version v of the forest of set union trees, and contains the name of the set; if x is not a root in version v, then $name(x, v)$ contains a special null value. Version 0 of T is initialized as follows. For each element x, $1 \leq x \leq n$, $parent(x, 0) = x$, $size(x, 0) = 1$ and $name(x, 0) = x$. This corresponds to version 0 of the set union data structure and can be done in $O(n)$ time. The operations defined on the fully persistent search tree T are: $Lookup(i, v)$ which returns the three fields associated with element i in version v (namely $parent(i, v)$, $size(i, v)$, and $name(i, v)$), and $Store(i, f_1, f_2, f_3)$, which creates a new version of the balanced search tree by storing values f_1, f_2 and f_3 in the three fields of element i.

We now describe how to perform Union and Find operations. To perform a Find(x, v), we use the same algorithm as in the ephemeral data structure; that is we follow the path from x to its root in version v of the forest of set union trees. However, now we have to be able to navigate in version v of the forest of set union trees, disregarding all Union operations with no influence on version v. We do this with the help of the fully persistent balanced search tree T as follows. We start by computing the parent of x by performing a Lookup(x, v) operation on T. If x is a root (i.e., if the *parent* field returned by the Lookup operation is equal to x) we stop and output the canonical information retrieved. Otherwise, we repeat the same operation for the parent of x, the grandparent of x and so forth until the root is reached. More formally, we can describe the algorithm using the following pseudo-code.

```
Procedure Find(x, v);
begin
    ⟨parent, size, name⟩ ← Lookup(x, v);
    y ← x;
    while y ≠ parent do begin
        y ← parent;
        ⟨parent, size, name⟩ ← Lookup(y, v)
    end;
    return(⟨parent, size, name⟩)
end;
```

The correctness of the Find operation hinges upon our ability to update the fully persistent balanced search tree T after a Union(x, y, v) operation. This can be done by first retrieving the size of the sets containing elements x and y in version v and then by merging those two sets by means of a union by size:

```
Procedure Union(x, y, v);
begin
    ⟨root_x, size_x, name_x⟩ ← Find(x, v);
    ⟨root_y, size_y, name_y⟩ ← Find(y, v);
    if root_x ≠ root_y then
        if (size_x ≥ size_y) then begin
            Store(root_x, root_x, size_x + size_y, name_x);
            Store(root_y, root_x, size_y, null)
        end
        else begin
            Store(root_y, root_y, size_x + size_y, name_x)
            Store(root_x, root_y, size_x, null);
        end
end;
```

Lemma 2.1 *Procedure Find(x, v) correctly returns the canonical element of the set containing x in version v by performing at most $O(\log n)$ Lookup operations on T.*

Proof: Let $\mathcal{F}_v$ be the version v of the forest of set union trees built through Union and Find operations. We remark that $\mathcal{F}_v$ is not maintained explicitly by our algorithm. It can be easily proved by induction on the number of operations that, for each element x and version v, the fields $parent(x,v)$, $size(x,v)$ and $name(x,v)$ correspond respectively in $\mathcal{F}_v$ to the parent of x, to the size of the subtree rooted at x, and to the name of the set containing x if x is a tree root. As a consequence, all the Union operations correctly follow the union by size rule. Thus, in any version of the set union data structure, the path from a leaf to its root is of length at most $\log n$. Therefore, each Find operation performs at most $O(\log n)$ Lookup operations on T. □

Theorem 2.1 *The fully persistent disjoint set union problem can be solved in $O(\log^2 n)$ worst-case time per operation, and in $O(1)$ amortized space per update.*

Proof: Denote by $A(n)$ and $S(n)$ the time required to perform respectively a Lookup and a Store operation on a fully persistent balanced search tree T. A Union operation requires $O(S(n))$ time plus the time required by two Find operations. To bound the time required by a Find, we notice that as a consequence of Lemma 2.1 Find(x,v) examines at most $O(\log n)$ different nodes in the path from x to the root. Therefore, a Find operation can be supported in $O(A(n)\log n)$ time. Choosing the fully persistent balanced search trees of Driscoll et al. [8] gives $A(n) = S(n) = O(\log n)$ in the worst case, giving the $O(\log^2 n)$ worst-case time per operation. With this implementation of T, the space is $O(1)$ amortized per update. □

By using the fully persistent array of Dietz [7] to support Lookup and Store operations, we can achieve a slightly better expected bound:

Corollary 2.1 *The fully persistent disjoint set union problem can be solved in $O(\log n \log \log m)$ expected amortized time per operation, where m is the total number of Union operations performed. The space required is $O(1)$ amortized per update.*

We recall that we do not create a new version of our fully persistent data structure each time a Union(x,y,v) with x and y being already in the same set in version v is executed (i.e., we do not count such a Union as an actual operation). Therefore $m \leq O(n!)$, and $\log \log m \leq O(\log n)$. This implies that the bounds in Corollary 2.1 are never worse than the bounds in Theorem 2.1.

The time bounds can still be improved by making use of a more efficient set union tree data structure proposed by Blum [5] and called a k-UF tree. This data structure is able to support each union and find operation in the ephemeral data structure in $O(\frac{\log n}{\log \log n})$ time in the worst case. We need to know the following facts about k-UF trees, and we refer the reader to [5] for all the other details of the method.

For any $k \geq 2$, a k-UF tree is either a singleton node or a rooted tree T such that the root has at least two children, each internal node has at least k children, and all leaves are at the same level. As a consequence of this definition, the height of a k-UF tree with n leaves is not greater than $\lceil \log_k n \rceil$. We refer to the root of a k-UF tree as *fat* if it is either a singleton node or it has more than k children, and as *slim* otherwise. A k-UF tree is said to be *fat* if its root is fat, otherwise it is referred to as *slim*. Disjoint sets can be represented by k-UF trees as follows. Each set is a k-UF tree. The elements of the set are stored in the leaves of the tree. There is a canonical element for each set storing the name of the set; once again the canonical element is the tree root. Furthermore, the root also contains the height of the tree and a bit specifying whether it is fat or slim.

A find(x) is performed as described before by starting from the leaf containing x and returning the name stored in the root. This can be accomplished in $O(\log_k n)$ worst-case time. A union(A,B) is performed by first accessing the roots r_A and r_B of the corresponding k-UF trees T_A and T_B. Blum assumed that his algorithm obtained r_A and r_B in constant time before performing a union(A,B). If this is not the case, r_A and r_B can be obtained by means of two finds (i.e., find(A) and find(B)), due to the property that the name of each set corresponds to one of the items contained in the set itself. We now show how to unite the two k-UF trees T_A and T_B. Assume without loss of generality that $height(T_B) \leq height(T_A)$. Let w be the node on the path from the leftmost leaf of T_A to r_A with the same height as T_B. Clearly, w can be located by following the leftmost path starting from the root r_A in exactly $height(T_A) - height(T_B)$ steps. When combining T_A and T_B, only three cases are possible, which give rise to three different types of unions.

Type 1 - Root r_B is fat and w is not the root of T_A. Then r_B is made a sibling of w.

Type 2 - Root r_B is fat and w is fat and equal to r_A (the root of T_A). A new (slim) root r is created and both r_A and r_B are made children of r.

Type 3 - This deals with the remaining cases, i.e., either root r_B is slim or w is equal to r_A and slim. If root r_B is slim, then all the children of r_B are made the leftmost children of w. Otherwise, w is equal to r_A and is slim. In this case, all the children of $w = r_A$ are made the rightmost children of r_B.

Figure 1 shows the three different types of unions in k-UF trees. Since each union involves traversing a path in a k-UF tree and re-directing at most k pointers (for type 3 unions), it can be supported in $O(\log_k n + k)$ time. Choosing $k = \left\lceil \frac{\log n}{\log \log n} \right\rceil$, gives a bound of $O(\frac{\log n}{\log \log n})$ time for each union and find. We can make Blum's data structure fully persistent with the following bounds.

Theorem 2.2 *The fully persistent disjoint set union problem can be solved either in $O(\frac{\log^2 n}{\log \log n})$ worst-case time per operation or in $O(\frac{\log n \log \log m}{\log \log n})$ expected amortized time per operation and in $O(\frac{\log n}{\log \log n})$ amortized space per update.*

Proof: We follow the same approach as in the proof of Theorem 2.1. Namely we maintain a fully persistent balanced search tree T, such that version v of T stores version v of the forest of k-UF trees. Once again, let us denote by $A(n)$ and $S(n)$ respectively the times required to perform a Lookup and a Store operation in a fully persistent balanced search tree of size n. Each time we make a structural change in the forest of k-UF trees, the change is recorded with a Store operation in T. Information about version v of the forest of k-UF trees can be retrieved by means of Lookup operations on T. However, now there are two differences with the basic set union tree data structure used in the proof of Theorem 2.1.

The first difference is that the size of T is no longer n. Indeed the forest of k-UF trees contains n leaves plus the internal nodes. However, the number of internal nodes is bounded above by n due to the properties of k-UF trees. As a result, the size of T will be $2n$ instead of n. There is a further complication, since we need to have entries in T also for those internal nodes. But this can be easily handled, since we can assign a unique integer in $[n, 2n]$ to an internal tree node as soon as it is created by simply incrementing a counter.

The second difference is that now a union operation in the ephemeral data structure can change as many as $O(\frac{\log n}{\log \log n})$ pointers, which causes $O(\frac{\log n}{\log \log n})$ Store operations in T. As a result, the time for a Union operation becomes $O(\frac{\log n}{\log \log n} S(n))$.

Due to the properties of k-UF trees, the bound for a Find operation becomes $O(\frac{\log n}{\log \log n} A(n))$. Choosing the fully persistent trees of Driscoll et al. [8] for the implementation of T gives the $O(\frac{\log^2 n}{\log \log n})$ time bound. If Store and Lookup operations are supported with a fully persistent array instead, we achieve an $O(\frac{\log n \log \log m}{\log \log n})$ expected amortized bound per operation. In either case the space required becomes $O(\frac{\log n}{\log \log n})$ per update, since there can be that many Store operations required in the worst case by a Union(x, y, v). $\square$

Type 1:

Type 2:

Type 3:

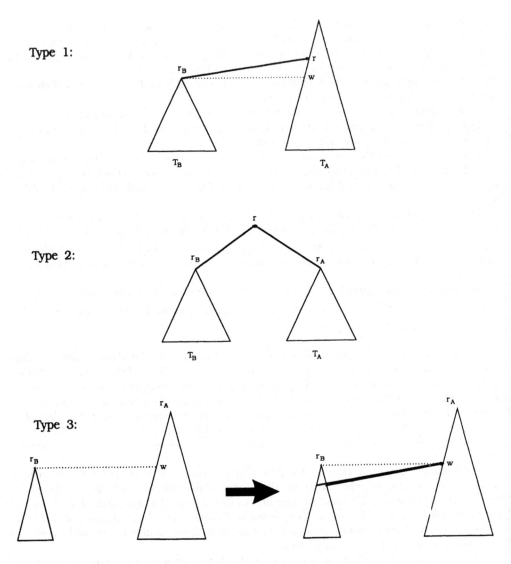

Figure 1: The three types of unions in a k-UF tree

3 An Improved Fully Persistent Data Structure

In this section we show how to improve to $O(\log n)$ the time bound for the fully persistent disjoint set union problem. In order to achieve this improvement, we do not use an extra persistent data structure such as the balanced search tree or the array used in the previous section. Rather, we store the disjoint sets in a different ephemeral data structure, and make this data structure persistent.

The ephemeral data structure we use can be described as follows. An (a, b)-tree, for $b \geq 2a$, is a tree with all the leaves at the same level and such that each node but the root has at least a and at most b children. The root has at least 2 and at most b children. In what follows, we denote by $\delta(v)$ the degree of a node v. We represent each disjoint set by an (a, b)-tree, with the elements of the set stored in the leaves. The canonical element storing the name of the set is again the root of the tree. As usual, a find(x) in this ephemeral data structure can be performed by following the path from the leaf corresponding to element x to the tree root, in order to return the name stored there. This requires $O(\log n)$ time. A union(A, B) can be performed by simply concatenating the two (a, b)-trees storing sets A and B, plus some extra bookkeeping to store the name of the new root. A concatenate(T_1, T_2, T) takes two (a, b)-trees T_1 and T_2 and produces a new (a, b)-tree T which has all the leaves of T_1 to the left of the leaves of T_2. The relative order of the leaves previously in T_1 (and T_2) stays unchanged. This can be done in $O(\log n)$ by using a well known algorithm (see for instance [1]).

Huddleston and Mehlhorn [10] showed how to support any sequence of k insertions and deletions in an (a, b)-tree starting from an empty tree in a total of $O(k)$ time. This result can be actually extended to concatenate operations. Before showing this, we need to describe the concatenate algorithm. Let h_1 and h_2 be respectively the heights of T_1 and T_2. Let ρ_1 be the rightmost leaf in T_1 and let σ_2 the leftmost leaf in T_2. With some additional bookkeeping in the root of (a, b)-trees we can access its leftmost and rightmost leaf in constant time. We climb up from ρ_1 and σ_2 towards their roots alternating among them one step at the time. We stop when the first root is entered. Without loss of generality assume that $h_1 \geq h_2$ (otherwise interchange the role of T_1 and T_2 in what follows). Let r_1 and r_2 be respectively the roots of T_1 and T_2, and let v_1 be the node of T_1 in the path from ρ_1 to r_1 at the same height as r_2. T_2 can be implanted in T_1 as follows.

If $\delta(v_1) + \delta(r_2) \leq b$, we combine v_1 and r_2 into one node and stop. Otherwise, we share children among v_1 and r_2 so as to satisfy the degree constraint; then we insert an edge from r_2 to the parent of v_1. We refer to this as a *sharing step*. Let x be the parent of v_1. Because of the sharing step, the degree of x is increased by 1. If it is still no greater than b, we stop. Otherwise, $\delta(x) = b + 1$. We create a new node x' and move to it the $\left\lceil \frac{b+1}{2} \right\rceil$ rightmost children of x. Then we insert an edge from x' to the parent of x. We call this a *splitting step*. We apply the splitting step recursively while going up to the root until either the degree constrained is satisfied or we reach the root r_1. If also the root has to be split, we create a new root r of degree 2. Since each sharing and splitting step can be implemented in constant time, the total cost of concatenate is $O(\min\{h_1, h_2\} + \ell)$, with ℓ being the number of splitting steps applied.

Lemma 3.1 *Each concatenate operation on (a, b)-trees can be performed in $O(1)$ amortized time.*

Proof: To prove the lemma, we show that any sequence of k concatenate operations starting from n singleton (a, b)-trees can be performed in a total of $O(k)$ time. We use the potential technique of Sleator and Tarjan (see for instance [17]). To each node in an (a, b)-tree, we assign a potential function φ defined as follows. Let $A = a - 1$. If v is a non-root node then

$$\varphi(v) = \frac{\delta(v) - a}{A}.$$

Otherwise, if r is a root of an (a, b)-tree of height h

$$\varphi(r) = h - 1 + \frac{\max\{0, \delta(r) - a\}}{A}.$$

We define the potential $\varphi(T)$ of an (a, b)-tree T to be $\varphi(T) = \sum_{x \in T} \varphi(x)$. Given a forest of (a, b)-trees, the potential Φ is the sum of the potential of all the trees in the forest. The amortized time of an operation is the actual time plus the change it causes in the potential function. Since the potential is always non-negative and it is initially 0, the total time of a sequence of operations is bounded above by the sum of the amortized times of the operations.

To compute the amortized cost of a concatenate operation, we bound the change in potential due to a sharing step, due to a splitting step, and due to the change of height in the resulting tree.

Consider a sharing step between nodes x and y, and let x borrow $p \le (b - a)$ children from y. Denote by x' and y' the nodes x and y after the sharing step. Then $\varphi(x') \le \varphi(x) + \frac{p}{A}$ and $\varphi(y') = \varphi(y) - \frac{p}{A}$. As a consequence, $\varphi(x') + \varphi(y') - \varphi(x) - \varphi(y) \le 0$. The other possible change in the potential is caused by the insertion of at most two edges leaving x' and y' (if x and y are both roots). This is at most $\frac{2}{A}$. Therefore $\Delta\Phi_{sharing} \le \frac{2}{A} = O(1)$.

As for a splitting step, assume that a non-root node x is split into x' and x''. Since $\delta(x) = b + 1$ and $\delta(x'') = \lceil \frac{b+1}{2} \rceil$, then $\delta(x') = \lfloor \frac{b+1}{2} \rfloor$. Therefore $\varphi(x') + \varphi(x'') - \varphi(x) = -\frac{a}{A}$. Furthermore the introduction of an edge from x'' to the parent of x causes an increase of potential of at most $\frac{1}{A}$ in the parent of x. Therefore for a non-root node $\Delta\Phi_{splitting} \le -1$. If x is a root, the only difference is that now there will be two edges leaving x' and x''. Thus in this case $\Delta\Phi_{splitting} \le -1 + \frac{1}{A}$.

We are left to bound the change of potential due to change in heights. Before we had two trees of height h_1 and h_2. This contributes $h_1 + h_2 - 2$ to the potential. After a concatenate operation we get a tree of height at most $\max\{h_1, h_2\} + 1$. Consequently, $\Delta\Phi_{height} \le -(\min\{h_1, h_2\} - 1)$.

We now compute the amortized complexity of a concatenate operation. Let ℓ be the number of splitting steps involved. As mentioned before, the actual time is $\ell + \min\{h_1, h_2\}$. To compute the change in potential, we notice that we can have at most one sharing step; furthermore, among the ℓ splitting steps, at most one can be a root splitting. The change in potential is therefore $\Delta\Phi \le O(1) + (-\ell + \frac{1}{A}) - (\min\{h_1, h_2\} - 1) \le O(1) - \ell - \min\{h_1, h_2\}$. This implies the $O(1)$ amortized bound for a concatenate operation. $\square$

Theorem 3.1 *The fully persistent disjoint set union problem can be solved in $O(\log n)$ worst-case time per operation, and in $O(1)$ amortized space per update.*

Proof: Lemma 3.1 implies that each union operation produces $O(1)$ amortized structural change in our ephemeral data structure based on (a, b)-trees. Furthermore, each node of an (a, b)-tree has bounded degree. Therefore, we can apply the technique of Driscoll et al. [8] to make (a, b)-trees fully persistent under any sequence of concatenate operations in $O(\log n)$ worst-case time per operation and with $O(1)$ amortized extra space per update. $\square$

We now turn to the fully persistent disjoint set union problem with deletions, which consists of supporting any sequence of Union, Find and Delete operations. Union and Find operations are performed as before. We support a Delete(x, v) operation as follows. The deletion of an item from the ephemeral (a, b)-tree is carried out using the classical algorithm (see for instance [1, pag. 151]), and we perform this operation in a fully persistent fashion by using the techniques of Driscoll et al. [8]. We need some more care in handling the names of the sets. Denote by X the set containing element x in version v. In the new version created by the Delete(x, v) operation, the name of set $X' = X - \{x\}$ is defined as follows: if the name of X was not x, then X' gets the same name as

X; otherwise X' gets as a new name arbitrarily one of its elements. The new name of X' can be computed as follows. The name of X in version v can be found in by a Find(x, v) operation. If this is different from x, then we stop. Otherwise, we have to find (in version v) an element different from x in the set X. This can be done by going in version v from the root of the (a, b)-tree T_X storing set X to the leftmost and to the rightmost item of T_X.

In order to show that Delete operations can be supported efficiently by using the techniques of Driscoll et al. [8], we have to make sure that each operation in the ephemeral data structure implies an $O(1)$ amortized structural change. We do this by generalizing Lemma 3.1 to a larger repertoire of operations, such as inserting an item into an (a, b)-tree, deleting an item from an (a, b)-tree, and concatenating two (a, b)-trees.

Lemma 3.2 *Any sequence of k concatenate operations, insertions and deletions on (a, b)-trees can be performed in a total of $O(k)$ time.*

Proof: We follow the same potential proof technique used in Lemma 3.1. Huddleston and Mehlhorn [10] showed that k insertions and deletions on an (a, b)-tree require a total of $O(k)$ time. We adapt their proof by defining the potential of a non-root node v to be

$$\varphi(v) = \left(\left| \delta(v) - \left\lceil \frac{b}{2} \right\rceil \right| + a - 1 \right).$$

For a root r

$$\varphi(r) = h - 1 + \max \left\{ 0, \delta(r) - \left\lceil \frac{b}{2} \right\rceil + a - 1 \right\}.$$

The potential of the forest of (a, b)-trees is defined as in Lemma 3.1.

Using this potential function, the amortized time of an insertion or of a deletion can be shown to be $O(1)$ by using techniques similar to the ones used in [10]. Furthermore, the same argument used in the proof of Lemma 3.1 shows that the amortized time of a concatenate operation is $O(1)$. □

Lemma 3.2 yields the following theorem.

Theorem 3.2 *There is fully persistent data structure that supports each Union, Find and Delete in $O(\log n)$ worst-case time each, and that requires $O(1)$ amortized space per update.*

Proof: Lemma 3.2 guarantees that the structural change in (a, b)-trees is $O(1)$ amortized per operation. Therefore, we can use the techniques in [8] to make (a, b)-trees fully persistent in $O(\log n)$ worst-case time per insertion, deletion and concatenate operation and $O(1)$ amortized space per update. Furthermore, to compute the new name of the set produced by a Delete operation we need at most two accesses in an (a, b)-tree of version v. Consequently, each Union, Find and Delete operation can be implemented in $O(\log n)$ by means of a constant number of operations on fully persistent (a, b)-trees. □

4 Conclusions

In this paper we have shown how to maintain efficiently multiple versions of a partition undergoing disjoint set union operations. In particular, we have described a fully persistent data structure that is able to support Union, Find and Delete operations in $O(\log n)$ worst-case time each and that requires $O(1)$ amortized extra space per update. This bound is tight since any pointer based algorithm requires $\Omega(\log n)$ amortized time to solve the partially persistent disjoint set union problem, as shown by Mannila and Ukkonen [13]. However, this lower bound does not hold if address arithmetic is allowed. Can we achieve better algorithms for our problems by using the extra power of a Random Access Machine?

From both a theoretical and practical point of view, it would be interesting to perform a more general repertoire of operations in a fully persistent fashion, such as set unions, set differences, set intersections and set equality tests on non-disjoint sets. However, we are not aware of any non-trivial data structure for this problem.

References

[1] A. V. Aho, J. E. Hopcroft, and J. D. Ullman, *The Design and Analysis of Computer Algorithms*, Addison-Wesley, Reading, MA, 1974.

[2] A. Apostolico, G. Gambosi, G. F. Italiano, M. Talamo, "The set union problem with unlimited backtracking", Tech. Rep. CS-TR-908, Department of Computer Science, Purdue University, 1989.

[3] D. Bacon, V. Nguyen, R. Strom, D. Yellin, "The Hermes language reference manual", IBM Tech. Rep., 1990.

[4] L. Banachowski, "A complement to Tarjan's result about the lower bound on the complexity of the set union problem", *Inform. Processing Lett.* 11 (1980), 59–65.

[5] N. Blum, "On the single operation worst-case time complexity of the disjoint set union problem", *SIAM J. Comput.* 15 (1986), 1021–1024.

[6] R. B. K. Dewar, E. Schonberg, J. T. Schwartz, "Introduction to the use of the Set-Theoretic Programming Language SETL", Courant Institute of Mathematical Sciences, Computer Science Department, New York University, 1981.

[7] P. Dietz, "Fully persistent arrays", *Proc. Workshop on Algorithms and Data Structures* (WADS 1989), *Lecture Notes in Computer Science* vol. 382, Springer-Verlag, Berlin, 1989, 67–74.

[8] J. R. Driscoll, N. Sarnak, D. D. Sleator, R. E. Tarjan, "Making data structures persistent", *J. Comput. Sys. Sci.* 38 (1989), 86–124.

[9] M. L. Fredman, M. E. Saks, "The cell probe complexity of dynamic data structures", *Proc. 21th Annual ACM Symp. on Theory of Computing*, 1989, 345–354.

[10] S. Huddleston, K. Mehlhorn, "A new data structure for representing sorted lists", *Acta Informatica* 17 (1982), 157–184.

[11] J. A. La Poutré, "Lower bounds for the union-find and the split-find problem on pointer machines", *Proc. 22nd Annual ACM Symposium on Theory of Computing*, 1990, 34–44.

[12] H. Mannila, E. Ukkonen, "On the complexity of unification sequences", *Proc. 3rd International Conference on Logic Programming, Lecture Notes in Computer Science* 225, Springer-Verlag, Berlin, 1986, 122–133.

[13] H. Mannila, E. Ukkonen, "Time parameter and arbitrary deunions in the set union problem", *Proc. 1st Scandinavian Workshop on Algorithm Theory* (SWAT 88), *Lecture Notes in Computer Science* vol. 318, Springer-Verlag, Berlin, 1988, 34–42.

[14] N. Sarnak, R. E. Tarjan, "Planar point location using persistent search trees", *Comm. ACM* 29 (1986), 669–679.

[15] R. E. Tarjan, "Efficiency of a good but not linear set union algorithm", *J. Assoc. Comput. Mach.* 22 (1975), 215–225.

[16] R. E. Tarjan, "A class of algorithms which require non linear time to maintain disjoint sets", *J. Comput. Syst. Sci.* 18 (1979), 110–127.

[17] R. E. Tarjan, "Amortized computational complexity", *SIAM J. Alg. Disc. Meth.* 6 (1985), 306–318.

[18] R. E. Tarjan, J. van Leeuwen, "Worst-case analysis of set union algorithms", *J. Assoc. Comput. Mach.* 31 (1984), 245–281.

[19] D. H. D. Warren, L. M. Pereira, "Prolog – the language and its implementation compared with LISP", *ACM SIGPLAN Notices* 12 (1977), 109–115.

Algorithms For Generating All Spanning Trees Of Undirected, Directed And Weighted Graphs

Sanjiv Kapoor   H. Ramesh *
Department of Computer Science
Indian Institute of Technology
Hauz Khas, New Delhi 110016

Abstract

We present algorithms for enumerating all spanning trees in directed, undirected and weighted graphs. The algorithms use a search tree technique to construct a computation tree. The computation tree is used to output all the spanning trees by listing the relative differences between spanning trees rather than the entire spanning trees themselves. For the undirected case, both the construction and the listing of the trees are shown to require $O(N + V + E)$ time for a graph with V vertices, E edges and N spanning trees. This algorithm is optimal and is based on exchanging non-tree edges with tree edges in a fundamental cycle. A similar exchange technique is used to construct the computation tree for a directed graph in $O(NV + V^3)$ time. The time for listing out the trees remains $O(N)$. For a weighted graph, we show how to sort the spanning trees by weight using the computation tree in $O(N \log V + VE)$ time.

1   Introduction

Spanning tree enumeration in undirected graphs is an important issue in many problems encountered in network and circuit analysis. Applications are given in [4]. Weighted spanning tree enumeration would find an application in the electrical wiring problem: n pins are to be connected together using the least amount of wire, with the connection satisfying some additional complicated constraints, e.g., a minimum separation between wires. A possible solution is to enumerate spanning trees in order of weight till a suitable connection is found. Further applications are envisaged in fault tolerant communication networks.

This problem has a long history. [5], [1], and [6] constitute the earliest work. Previous techniques used for solving the problem include depth first search [3], selective enumeration and testing [1], and edge exchanging [2].

---

*Present Address: Courant Institute of Mathematical Sciences, New York University, New York, NY 10012

For undirected graphs, Gabow and Myers' algorithm [3] seems to be the fastest. It runs in time $O(NV)$ on a graph with V vertices, E edges and N spanning trees. If all spanning trees are required to be explicitly output then this algorithms is optimal. For many practical applications, a computation tree giving relative differences between spanning trees suffices as output. Complete information about any spanning tree can be easily obtained from these differences. Since the size of the output now is $O(N)$, a faster generation scheme is desirable.

In this paper, spanning trees are enumerated by listing out differences between them, i.e., a tree is output in the form of differences from the previous tree output. Each node of the computation tree that describes this procedure represents a spanning tree of the graph. The spanning trees represented by a node and its parent in the computation tree differ in exactly one pair of edges. Note that an edge external to a spanning tree may be exchanged for any edge in its fundamental cycle to give a new spanning tree at distance one from the original spanning tree. By repeating this for all external edges, all spanning trees at distance one from the original spanning tree can be generated. The computation tree is generated in a depth-first manner by repeatedly applying this procedure. Repetitions of spanning trees are avoided using a search tree inclusion-exclusion technique. The algorithm for undirected graphs runs in $O(N + V + E)$ time and uses $O(VE)$ space.

An algorithm for sorting the nodes of the computation tree in increasing order of weight in $O(N \log V + VE)$ time is also presented. It is based on a branch and bound strategy applied to the computation tree. This sorting strategy betters the $O(N \log N)$ heapsort scheme in [2].

Finally, an algorithm for generating spanning trees in directed graphs is presented. The method of swapping non-tree edges with tree edges is used in this case also. The algorithm runs in $O(NV + V^3)$ time and uses $O(V^2)$ space. This improves the previously best known bound of $O(NE + V + E)$ in [3].

Sections 2, 3 and 4 describe, respectively, the enumeration of spanning trees in undirected, weighted, and directed graphs.

2 Undirected Graphs

Let G be a graph with V vertices, E edges and N spanning trees.

2.1 Algorithm Outline

The algorithm begins by generating one spanning tree T. It then generates all other spanning trees by replacing edges in T by edges outside T. A edge not in T can replace any of the tree edges in its fundamental cycle. Each such exchange gives a new spanning tree. This computation is represented by a computation tree with tree T at its root and and the exchange trees at its sons. To generate other trees, the sons of the root are expanded recursively in the same manner.

Repetitions of spanning trees are avoided by a search tree type computation which uses the inclusion/exclusion principle. This is achieved by implicitly maintaining 2 sets,

IN and OUT, at each node in the computation tree. For a node x in the computation tree, the set IN_x consists of edges which are necessarily included in each spanning tree generated in the subtree rooted at x. OUT_x consists of edges which are necessarily excluded from these spanning trees.

We now formally define the computation tree $C(G)$ for graph G. Each node x of $C(G)$ has a spanning tree S_x of G associated with it. For the root r of $C(G)$, S_r is any arbitrary tree. IN_r and OUT_r are both empty. Let x be any node of $C(G)$. Let f be an edge such that $f \notin S_x$ and $f \notin OUT_x$. Let the fundamental cycle of f w.r.t S_x contain the following edges of $S_x - IN_x : e_1, e_2, \ldots, e_k$. Then x has sons b_i, $1 \leq i \leq k+1$. Each S_{b_i}, $1 \leq i \leq k$, is obtained by replacing e_i by f in S_x. $S_{b_{k+1}}$ is defined to be the same as S_x. For all nodes y in the subtrees rooted at b_i, $1 \leq i \leq k$, S_y contains edge f. For all nodes y in the subtree rooted at b_{k+1}, S_y does not contain edge f. For the sons of x, the IN and OUT sets are defined as follows:

$IN_{b_j} = IN_x \cup \{e_1, e_2, \ldots, e_{j-1}\} \cup \{f\}$, for $1 \leq j \leq k$
$OUT_{b_j} = OUT_x \cup \{e_j\}$, for $1 \leq j \leq k$
$IN_{b_{k+1}} = IN_a$
$OUT_{b_{k+1}} = OUT_x \cup \{f\}$

For convenience, we define OD_x to be the set of edges not present in $OUT_x \cup S_x$. We also define $EXCH_{b_j}$ to be the pair (e_j, f). $EXCH_{b_{k+1}}$ and $EXCH_r$ are defined to be ϕ.

The next lemma follows from the inclusion/exclusion principle and shows that $C(G)$ suffices to generate all spanning trees of G.

Lemma 2.1: $C(G)$ has at its nodes all the spanning trees of G. Moreover, 2 nodes are associated with the same spanning tree if and only if one of them is a descendant of the other along a rightmost path.

$C(G)$ is generated recursively in a depth-first manner. To generate the sons of a node x in $C(G)$, we choose an edge $f \in OD_x$. We determine the edges in $S_x - IN_x$ occurring in the fundamental cycle of f w.r.t S_x. For each such edge, we construct a new son of x and recursively construct the subtree rooted at that son of x. Then we add f to OUT_x and recursively construct the subtree rooted at the last son of x.

At each son of x, the fundamental cycles with respect to the new spanning tree need to be computed. Two strategies can be adopted to do this. The first strategy scans the cycles which are modified and computes each new fundamental cycle in time proportional to its size. Since each fundamental cycle generates a number of new spanning trees equal to the number of tree edges in it, the running time of the algorithm in linear in N. Note that this strategy allows us to begin with any arbitrary tree at the root of $C(G)$ and also to choose edges from OD_x for replacement in any arbitrary fashion. This strategy, however, involves repeated scanning of edges which occur in more than one cycle. Here we give a more efficient strategy.

In order to compute and modify fundamental cycles efficiently, we begin with the DFS tree of G at the root r of $C(G)$. This ensures that all edges in OD_r are back edges w.r.t S_r. For any edge f in OD_r, its fundamental cycle w.r.t S_r is now simply found by going up S_r starting from the lower endpoint of f and ending at the upper endpoint. We ensure that this property is also maintained at each node x of $C(G)$. This is done as follows. We order the edges in OD_x by the postorder number of their upper endpoint. Here, the postorder number refers to postorder traversal of S_x. This order is used to select edges in OD_x for replacement. Also note that S_x may contain edges which being in IN_x are not candidates for replacement. At each node x, we maintain a form of S_x called T_x, in which edges in IN_x are effectively contracted.

Henceforth, whenever an edge is referred to by a pair of vertices (a, b), a refers to the lower endpoint and b refers to the upper endpoint w.r.t the tree in question.

The actual mechanics of edge exchange are as follows. At node x in $C(G)$, the first edge in OD_x is selected and removed from OD_x. Call this edge $f = (x_0, x_k)$. The fundamental cycle of f w.r.t T_x is computed. Let it be $(x_0, x_1,, x_k, x_0)$. Then each edge $e_i = (x_{i-1}, x_i)$, $1 \leq i \leq k$, is replaced in turn by f to generate the sons $b_1, b_2, \ldots, b_k$ of x. T_{b_i} $(2 \leq i \leq k)$ is computed from $T_{b_{i-1}}$ in the following manner. Edge e_i is removed. Then all children z of vertex x_{i-1} having the following property are determined: there exists an edge in OD_{b_i} whose lower endpoint is a descendant[1] of z in $T_{b_{i-1}}$ and whose upper endpoint is an ancestor of x_{i-1}. (Note that by virtue of our ordering on OD_x, the upper end point of such an edge must be an ancestor of x_k). These vertices are now made children of x_k. T_{b_1} is computed from T_x by doing the same for the children of x_0. Also, the lower endpoint of each edge in OD_{b_i} incident on x_{i-1} is now changed to x_k. This short circuits the edges added to IN_{b_i} at node x. However, it leads to the formation of multiple edges and self loops. Self loops are removed immediately from OD_{b_i}. Multiple edges are identified and bundled together so that they can be handled collectively. This is useful in showing the linear time bound of the algorithm.

Next, we describe the data structures required to perform the above operations efficiently.

2.2   Data Structures

The preceding algorithm outline describes various sets associated with each node of $C(G)$. However, the data structures we use are global. Changes made to these data structures are stored prior to recursion and restored on return. The data structure $CHANGES$ is used to store these changes.

The data structure G_OD is the global version of OD. It is implemented as an ordered list. The order is the same as the ordering on OD. Each list element is a sublist. Each sublist corresponds to a set of edges which together form a multiedge. Each multiedge has a field called $LOWER$ which stores the lower endpoint of that multiedge. Associated with G_OD is an array AOD. $AOD[u]$ contains an unordered list containing sublists of G_OD which have lower endpoint u. Insertions and deletions in G_OD are assumed to modify AOD accordingly.

The data structure GT is the global version of T. It is maintained as an array of parent pointers. Also, each node u has a list $L[u]$ consisting of those children z of u in GT which have the following property: there exists an edge in G_OD, one of whose endpoints is a descendant of z, and whose other endpoint is an ancestor of u.

---

[1] A descendant (ancestor) of node z includes all proper descendants (ancestors) of z plus z itself.

2.3 Algorithm Description

The algorithm consists of two main procedures, *Gen* and *Main*. *Main* constructs the data structures corresponding to the root of the $C(G)$ and then calls *Gen*. Given these data structures, *Gen* recursively constructs $C(G)$. It uses the procedures *Modify0*, *Restore0*, *Modify1*, *Restore1*, *Modify2* and *Restore2*. Each of the *Modify* procedures modifies the global data structures as described below and stores these modifications in *CHANGES*. These modifications are undone by the corresponding *Restore* procedures. As a result, the state of the global data structures before a call to *Gen* and after return from that call is the same.

Modify0 traverses G_OD collecting together multiedges and changing AOD accordingly. *Modify1* modifies GT to reflect the effect of the exchange performed. *Restore1* undoes all modifications done by the various calls made to *Modify1* in the *For* loop. *Modify2* makes the necessary modification to L due to the addition of an edge to the *OUT* set when the last son of the current node in $C(G)$ is generated.

The output of the entire algorithm consists of the first spanning tree computed by *Main*, followed by a sequence of edge pairs for each of the other spanning trees. For each spanning tree, the sequence of edge pairs associated with it is accumulated in *REP*. This sequence gives the replacements that need to be made to obtain this spanning tree from the previous one.

The procedures are given below. The *Restore* procedures and *Modify0* have been left out.

Algorithm *Main(G)*
    $GT \leftarrow$ DFS tree of G;
    Output GT;
    $G\_OD \leftarrow$ Ordered list of edges $\notin GT$;
    Construct the data structures AOD, L;
    $CHANGES \leftarrow \phi$;
    $REP \leftarrow \phi$;
    If $G\_OD \neq \phi$ then *Gen()*;
end *Main*

Algorithm *Gen()*
    $mf \leftarrow$ The first multiedge in G_OD;
    $f \leftarrow$ An edge in mf; {Let $f = (x_0, x_k)$}
    Determine the fundamental cycle of f w.r.t GT;{Let it be $(x_0, x_1, \ldots, x_k, x_0)$ }
    For $i := 1$ to k **do**
        $e \leftarrow (x_{i-1}, x_i)$;
        Add the edge pair (e, f) to *REP*;
        Output the edge pairs in *REP*;
        {A new spanning tree is output}
        {This corresponds to the creation of a new son of the

          current node in $C(G)$}
       $REP \leftarrow \phi$;
       $Modify1(i,k)$;
       $Modify0()$;
       If $G\_OD \neq \phi$ then $Gen()$;
          {Generate the subtree rooted at the new son created above}
       $Restore0()$;
       Add the edge pair (f, e) to REP;
    End for;
    $Restore1()$;
    Remove edge f from mf in G_OD;
       {If f is the only edge in mf, mf is also removed from G_OD }
    $Modify2(k)$;
    If $G\_OD \neq \phi$ then $Gen()$;
       {Generate the subtree rooted at the last son of the current
       node in $C(G)$}
    $Restore2()$;
    Restore edge f to mf in G_OD;
End Gen

Algorithm $Modify1(i,k)$
    If $x_{i-1} \in L[x_i]$ then remove x_{i-1} from $L[x_i]$;
    If $AOD[x_{i-1}] \neq \phi$ or $L[x_{i-1}] \neq \phi$ then
        $j \leftarrow i$;
       While $L[x_j] = \phi$ and $AOD[x_j] = \phi$ and $j < k$ **do**
          Remove x_j from $L[x_{j+1}]$;
          $j \leftarrow j+1$;
       End While;
    For each multiedge me in $AOD[x_{i-1}]$ **do**
       If the upper endpoint of me is x_k **then**
          Remove me from $AOD[x_{i-1}]$ and G_OD; {Remove self loops}
       else
          Remove me from $AOD[x_{i-1}]$ and add me to $AOD[x_k]$;
          Change $me.LOWER$ to x_k;
    End for;
    For each vertex v in $L[x_{i-1}]$ **do**
       Make x_k the parent of v and remove v from $L[x_{i-1}]$;
       Add v to $L[x_k]$;
    End for;
    Stack the above modifications on $CHANGES$;
End $Modify1$

Algorithm $Modify2(k)$
    For $i := 1$ to k **do**
       If $AOD[x_{i-1}] = \phi$ and $L[x_{i-1}] = \phi$ then

Remove x_{i-1} from $L[x_i]$;
  End for;
  Stack the above modifications on $CHANGES$;
End *Modify2*

2.4   Correctness and Complexity

Lemma 2.2: At the point when *Gen* is called to construct the subtree of $C(G)$ rooted at node x

a) G_OD consists of all edges in OD_x which are not self loops in the graph obtained by contracting edges in IN_x. These edges are ordered by the postorder number (w.r.t S_x) of their upper endpoints.

b) For any edge $f = (x_0, x_k)$ in G_OD, the edges in GT on the path from x_0 to x_k are precisely those edges in $S_x - IN_x$ which occur in the fundamental cycle of f w.r.t S_x.

c) AOD and L, which have been defined w.r.t G_OD and GT, are correctly computed.
Proof: We show this by induction on the levels of $C(G)$. For the base case, i.e., the root of $C(G)$, this is true by virtue of the the preprocessing done in *Main*. As the induction hypothesis, assume that the above is true for node x of $C(G)$. We prove that properties a, b and c hold for each son of x.

Let $f = (x_0, x_k)$ be the edge in G_OD chosen as the replacement edge. Let its fundamental cycle be $x_0, x_1, \ldots, x_k, x_0$. Then b_i, the ith son of x $(1 \leq i \leq k)$ is obtained by swapping the edge $e_i = (x_{i-1}, x_i)$ in S_x for edge f. Let X refer to the set of vertices $x_0, x_1, \ldots, x_{i-1}$. The edges $(x_j, x_{j+1}), 0 \leq j < i - 1$, are now in IN_{b_i}. By our ordering on G_OD, all edges in G_OD incident on descendants of X have upper endpoints which are ancestors of x_k. These are the only edges in G_OD whose fundamental cycles are modified due to the replacement of e_i by f. *Modify1* changes the lower endpoint of all edges in G_OD incident on a vertex in X to x_k. Note that this does not affect the order of edges in G_OD. For each vertex v in the L list of some vertex in X, *Modify1* makes v the son of x_k. This ensures that the edges (x_j, x_{j+1}), $0 \leq j < i - 1$, which are added to the IN set, are removed from these new fundamental cycles. This preserves property b. Any self loops formed above are immediately removed from G_OD. The only edge present in OD_x but absent from OD_{b_i} is f. *Modify1* removes edge f from G_OD in the process of removing self loops. This preserves property a. Whenever G_OD is modified or the lower endpoint of an edge in G_OD is changed, AOD is modified accordingly. *Modify1* also changes L to reflect these changes, thus preserving property c. Thus these three properties are true for the first k sons of x. For the last son of x, f is removed from G_OD. *Modify2* makes the resulting changes to L. Hence these properties are true for all sons of x. $\square$

Theorem 2.1: *Main* correctly outputs the first spanning tree and the sequences of differences between the remaining spanning trees of G, without repetition.
Proof: From lemma 2.2 it follows that $C(G)$ is generated correctly. The output of the algorithm corresponds to a depth first traversal of $C(G)$ with $EXCH_x = (e, f)$ added to REP when node x is first visited and (f, e) added to REP when backtracking up from x. REP is set to ϕ whenever a tree (i.e, the edge pairs in REP) is output. Thus at any

point in the algorithm, *REP* stores the differences between the tree at the current node of $C(G)$ and the last tree output. A spanning tree is output only at those nodes which are not the last of their parents' children. This theorem now follows from lemma 2.1. □

Before analysing the complexity of the algorithm, we define a compressed computation tree $C'(G)$. $C'(G)$ is the same as $C(G)$ except that all nodes along any rightmost path in $C(G)$ are condensed into one node. We define the work done at a node x in $C'(G)$ to sum of the work done at each of the nodes in $C(G)$ which were condensed to give x. Let $s(x)$ and $g(x)$ be, respectively, the number of sons and grandsons of x in $C'(G)$.

Lemma 2.3: At node x in $C'(G)$, the work done by *Gen* (except the work for outputting *REP*) is $O(s(x) + g(x))$.
Proof: Let u be any node of $C'(G)$. Let u be formed by condensing the nodes $u_1, u_2, \ldots, u_m$ of $C(G)$. Then just prior to the call to *Gen* which generates the subtree of $C(G)$ rooted at u_1, G_OD contains exactly $m - 1$ edges. Since G_OD has no self loops, each of these edges is used to generate at least one son of one of $u_1, u_2, \ldots, u_{m-1}$. This fact is used to account for the work done by *Gen*.
The total work done by the calls to *Modify0* at node x in $C'(G)$ is at most $O(g(x))$. The work done by *Modify2* at node x is at most $O(s(x))$. Only the work done by *Modify1* is left to be accounted for. The work done by the *while* loop over all calls to *Modify1* at x is at most $O(s(x))$. In each call to *Modify1*, at most one self loop is detected. This is because multiple edges have been collected together in G_OD. The time taken by the two *for* loops over all calls to *Modify1* is at most $O(g(x))$. The lemma follows. □

Lemma 2.4: The space required by the algorithm is $O(VE)$.
Proof: All data structures except *CHANGES* require $O(V + E)$ space. *CHANGES* requires $O(VE)$ space. □

Theorem 2.2: *Main* correctly generates all spanning trees of G in $O(N + V + E)$ time and $O(VE)$ space.
Proof: The preprocessing in *Main* takes $O(V + E)$ time. The theorem then follows from theorem 2.1, lemma 2.3, lemma 2.4, and the fact that the size of the output is at most $2N$. □

3 Weighted Graphs

The algorithm for sorting the nodes of $C'(G)$ by weight follows a branch and bound strategy. $C(G)$ is constructed with the minimum spanning tree (*MST*) of G at the root. Note that this involves using the first of the two strategies mentioned in section 2.1. This strategy allows us to generate $C(G)$ with *MST* at the root. At each node $x \in C(G)$, this strategy also allows us order the edges in OD_x by weight and choose the smallest weight edge in OD_x as the replacement edge. The time for generating $C(G)$

is $O(N + E \log E + V)$. The space required is $O(V^2E)$. We need the following lemmas before describing the sorting algorithm.

Lemma 3.1: For each node x of $C(G)$, if $EXCH_x = (e, f)$, then f has weight greater than or equal to e.
Proof: Let F be the set of edges of G not in MST and having weight less than or equal to f. Then e lies in the fundamental cycle (w.r.t MST) of an edge in F. This is because at every node v, the smallest weight edge in OD_v is chosen as the replacement edge. The lemma follows. $\square$

Corollory 3.1 For any pair of nodes $x, y \in C(G)$ such that y is an ancestor of x, $weight(S_x) \geq weight(S_y)$.

Lemma 3.2: The number of different edge pairs which can occur as $EXCH_x$ for some node x of $C(G)$ is at most $(V - 1)(E - V + 1)$.
Proof: The first entry of the edge pair has to be an edge in MST. The second has to be an edge not in the MST. $\square$

The algorithm follows. It uses $(V - 1)(E - V + 1)$ queues, one for each possible exchange edge pair. It is based on the above lemmas and the fact that the application of a particular exchange to each of a sorted list of trees results in a sorted list of new trees.

Algorithm *Genwt()*
        Initialize queues to ϕ;
        $N \leftarrow$ Root node of $C'(G)$;
        Repeat
            Output N;
            Put each son x of N in the queue indexed by $EXCH_x$;
            $N \leftarrow$ The minimum weight node among the heads of all queues;
            Remove N from its queue;
        Until all queues are ϕ
End *Genwt*

Theorem 3.1: *Genwt* sorts the nodes of $C'(G)$ by weight in $O(N \log V + VE)$ time and $O(N + V^2E)$ space.
Proof: The above algorithm performs N delete-mins from the heads of VE (approximately) queues. This takes $O(N \log VE)$ time. Initializing the queues takes $O(VE)$ time. The space required is $O(V^2E)$ for constructing $C'(G)$, $O(N)$ for storing $C'(G)$, and $O(VE)$ for the queues. $\square$

Note, however, that outputting spanning trees in the form of differences from the previous tree leads to a $O(NV)$ size output and takes $O(NV)$ time.

4 Directed Graphs

We describe an algorithm to generate the computation tree $C(G, v)$ for all spanning trees rooted at a vertex v of G. To generate all spanning trees of G, this procedure can be repeated for each vertex of G.

The algorithm is similar to the undirected case. However, exchanging edges in a fundamental cycle is no longer a valid strategy. Instead, for a spanning tree T, we classify edges not in T as *back* and *non-back* edges. *Back* edges lead from a vertex to its ancestor in T. The remaining edges are called *non-back* edges. The following properties are used to perform exchanges.

Property 1: Given a spanning tree T rooted at v, a *non-back* edge f can replace exactly one edge in T to give a new spanning tree rooted at v. This edge is the edge having the same destination as f.

Property 2: A *back* edge can not be exchanged with any edge of T to get a new spanning tree rooted at v.

$C(G, v)$ is defined in a manner analogous to the undirected case. However, note that each edge not in T can be exchanged for at most one edge in T. This implies that the $C(G, v)$ is a binary tree for the directed case.

As in the undirected case, we begin with the DFS tree T of G rooted at v, at the root r of $C(G, v)$. This ensures that all *non-back* edges w.r.t T lead from a vertex with higher postorder number to a vertex with lower postorder number in T. To maintain this property at each node x of $C(G, v)$, the set of edges in OD_x that are *non-back* w.r.t S_x is kept ordered by the increasing postorder number (w.r.t S_x) of the destination vertex. The first edge in this ordered set is chosen for replacement. Then, the following property describes the change in the status of *back* and *non-back* edges when an exchange is made.

Property 3: Let spanning tree T' be obtained from spanning tree T by replacing tree edge e by the *non-back* edge f. Let a be the source vertex of f and b be the destination vertex. Let c be the least common ancestor of a and b in T. All edges (except f) which are *non-back* w.r.t T remain *non-back* w.r.t T'. The only edges which are *back* w.r.t T and *non-back* w.r.t T', are those whose sources are descendants of b in T, and whose destinations lie on the path from b to c in T (b, c excluded).

Next, we describe the data structures required. Again, we use global data structures, storing modifications before recursive calls and undoing modifications on return. The data structure B stores, at each node x of $C(G, v)$, the set of edges in OD_x which are *back* edges w.r.t S_x. B is maintained as an array of size V. For each vertex u, $B[u]$ is a list containing edges with source u. This list is sorted by increasing postorder number (w.r.t S_x) of the destination vertices of these edges. The list $B[u]$ is also indexed by an array $B[u].A$. $B[u].A[v]$ points to the first edge in $B[u]$ whose destination is a proper ancestor of v in S_x. $B[u]$ has a field $B[u].base$. Any edge in $B[u]$ incident upon a descendant of $B[u].base$ in S_x is redundant for further computations. $B[u].base$ is initialized to u. The second data structure required in NB. At each node x of $C(G,v)$, NB stores the set of edges in OD_x which are *non-back* w.r.t S_x. Additionally, each edge f in NB satisfies the following condition: the edge in S_x having the same destination as f is not in IN_x. NB is organized as a list of sublists. Each sublist stores edges having a common destination. The various sublists are ordered by the postorder number of their destination vertices. NB can be indexed by an array ANB. $ANB[v]$ points to the first sublist in NB whose destination vertex has greater postorder number than v.

We give the algorithm now. As in the undirected case, it consists of a procedure *Maindir* which does the preprocessing and then calls *Gendir* to generate the computation tree.

Algorithm *Maindir(G,v)*
      $GT \leftarrow$ DFS tree of G rooted at v;
      Output GT;
      Construct the data structures NB and B;
      $CHANGES \leftarrow \phi$;
      $REP \leftarrow \phi$;
      If $NB \neq \phi$ then *Gendir()*;
end *Maindir*

Algorithm *Gendir()*
      $mf \leftarrow$ The first sublist in NB;
      $NB \leftarrow NB$ - mf;
      $f \leftarrow$ An edge in mf; {Let f be directed from a to b}
      $e \leftarrow$ The edge in GT incident on b; {Let e be directed from d to b}
      Make b the son of a in GT;
      Add the edge pair (e,f) to REP;
      Output the edge pairs in REP;
            {A new spanning tree is output}
            {This corresponds to the creation of a new son of the
            current node in $C(G)$}
      $REP \leftarrow \phi$;
      Back_to_Non-back(a,b);
      If $NB \neq \phi$ then *Gendir()*;
            {Generate the subtree rooted at the new son created above}
      Restore();
            {Undoes the effect of *Back_to_Non-back* on B,NB}
      Restore b to be the son of d in GT;
      Add the edge pair (f,e) to REP;
      $NB \leftarrow NB +(mf - f)$;
      If $NB \neq \phi$ then *Gendir()*;
            {Generate the subtree rooted at the last son of the current
            node in $C(G)$}
      Restore edge f to mf in NB;
End *Gen*

Algorithm $Back\_to\_Non\text{-}back(a,b)$

    $c \leftarrow$ Least common ancestor of a and b in GT;

    For each vertex w in the subtree of GT rooted at b **do**

        If $B[w].base < c$ **then**

            $B[w].base \leftarrow c$;

            Find all edges in $B[w]$ with destinations on the path from b to c;

                {With both endpoints excluded}

            Insert these edges into NB;

    End For;

    Stack the above modifications on $CHANGES$;

end $Back\_to\_Non\text{-}back$

Let $C'(G,v)$, $s(x)$ and $g(x)$ be defined in a manner analogous to the undirected case. We state the following lemmas.

Lemma 4.1: *Gendir* does $O(V * s(x) + g(x))$ work at node x of $C'(G,v)$.

Lemma 4.2: $C(G,v)$ is generated by *Maindir* in $O(NV + V^2)$ time and $O(V^2)$ space.

Theorem 4.1: All rooted directed spanning trees of G can be output in the form of inter-tree differences in $O(NV + V^3)$ time and $O(V^2)$ space.

5   Conclusions

Faster algorithms have been obtained for generating the computation tree for enumerating spanning trees in undirected and directed graphs, and for sorting the trees by weight in a weighted graph. The problems of generating the computation tree for directed graphs in linear time, and enumerating spanning trees in order in a weighted graph using less than $O(N)$ space, remain open.

References

([1]) J. P. Char: Generation of trees, 2 trees and storage of master forests, IEEE Trans. Circuit Theory, CT-15, 1968.

([2]) Harold N. Gabow: Two algorithms for generating weighted spanning trees in order, SIAM J. Comp., Vol 6, No. 1, March 1977.

([3]) H. N. Gabow and E. W. Myers: Finding all spanning trees of directed and undirected graphs, SIAM J. Comp., Vol 7, No. 3, Aug 1978.

([4]) W. Mayeda: Graph Theory, John Wiley, NY 1972.

([5]) G. J. Minty: A simple algorithm for listing all trees of a graph, IEEE Trans. Circuit Theory, CT-12, 1965.

([6]) R. C. Read and R. E. Tarjan: Bounds on backtrack algorithms for listing cycles, paths and spanning trees, Networks, 5, No. 3, 1975.

Sorting Multisets and Vectors In-Place*

J. Ian Munro   Venkatesh Raman
Department of Computer Science
University of Waterloo
Ontario, Canada N2L 3G1.

Abstract

When a list to be sorted consists of many repeated elements (i.e. when it is a *multiset*), the familiar $\Omega(n \lg n)$ lower bound on the number of comparisons is replaced by a bound that depends on the multiplicities of the elements in the list. We adapt heapsort for multisets and provide the first *in-place* algorithm for multisets that achieves the optimal bound up to lower order terms. We, then, obtain an optimal in-place algorithm to lexicographically sort an array of multidimensional vectors, by applying the multiset sorting algorithm in each coordinate.

We exploit the relationship between the two sorting problems to improve the lower order term of a known lower bound[2] for sorting multisets. It follows that the ("non in-place") algorithms developed by Munro and Spira [8] for multisets are within $O(n)$ time optimal. Our improvement to the lower bound for sorting multisets also improves a lower bound (in the lower order term) for determining the mode, the most frequently occurring element in the multiset.

1   Introduction

Let S be a list of size n containing m distinct elements where the i-th distinct element appears n_i times. Such a list (consisting of repeated elements) is called a *multiset*. Munro and Spira[8] proved using a ternary decision tree approach, that the number of three-way comparisons required to sort S on the average is $n \lg n - \sum_{i=1}^{m} n_i \lg n_i - (n - m) \lg \lg m - O(n)$. They also gave adaptations of mergesort and treesort whose running times match this bound up to lower order terms. Dobkin and Munro[2] later extended the lower bound to $n \lg n - \sum_{i=1}^{m} n_i \lg n_i - n \lg(\lg n - (\sum_{i=1}^{m} n_i \lg n_i)/n) - O(n)$.

Both the adaptations of Munro and Spira follow the same technique to save the unnecessary comparisons between equal valued keys: whenever two items are found to be equal, one is discarded and a record of the number of occurrences of the item is kept with the remaining copy. Full implementation of these techniques require substantial extra

---

*Research supported by Natural Sciences and Engineering Research Council of Canada grant No.A-8237 and the Information Technology Research Centre of Ontario.

space to identify the (non)discarded items and to keep count of the duplicates. We adapt heapsort to provide the first *in-place* algorithm (one that uses only a constant number of extra data locations and pointers besides the storage for the input) for multisets that is optimal up to lower order terms.

This algorithm is then applied to solve the following problem which finds application in computational geometry[3], databases[10], and text searching[7]: Given an array of k-dimensional vectors, sort them lexicographically in-place using optimal number of scalar comparisons. The obvious sorting algorithm that treats the vectors as one-dimensional elements and compares them using at most k comparisons each, requires $\Theta(kn \lg n)$ key comparisons in the worst case. However, the best known lower bound for this problem is $n \lg n + (k-1)n - O(n)$ [1]. This bound can be easily seen by considering the case in which all the n vectors are identical in the first $(k-1)$ coordinates and distinct under the last coordinate. Adaptations of quicksort and binary insertion sort have been reported for this problem ([1], [10], [3]). These adaptations are either not in-place, or not efficient for the worst case. By applying our method to sort a multiset in each coordinate, we obtain the first in-place algorithm that sorts lexicographically using the optimal number (up to lower order terms) of scalar comparisons in the worst case.

By exploiting the relationship between the problems of sorting vectors and multisets, we improve the lower bound of Dobkin and Munro for sorting multisets, to eliminate the $n \lg(\lg n - (\sum_{i=1}^{m} n_i \lg n_i)/n)$ term. This proves that the algorithms for multisets provided by Munro and Spira [8] are within $O(n)$ time optimal, and our in-place adaptation of heapsort is within $O(n \lg^* n)$ time optimal. The improvement in the lower bound for sorting multisets also eliminates the $n \lg(\lg n - (\sum_{i=1}^{m} n_i \lg n_i)/n)$ term from the best known lower bound [2] for determining the mode, the most frequently occurring element in a multiset.

The next section gives an in-place algorithm for sorting multisets optimally, and the subsequent section applies the algorithm to lexicographically sort a list of vectors. Section 4 presents the improvement to the lower bound for sorting multisets. Section 5 presents conclusions and open problems.

Throughout this paper, lg denotes logarithm to the base 2 and lg* denotes the number of times the logarithm has to be taken before the quantity becomes less than 1. Also in this paper, we do not attempt to perform stable sorts (those that keep the equal keys in their original order).

2 Sorting Multisets In-Place

Quicksort can be adapted to sort a multiset by doing a three way partition at each recursive step[9]. Bentley and Saxe[1], while building a data structure to efficiently search a list of multidimensional vectors, extend this quicksort type approach for multisets to achieve the bound (within a constant factor) for the worst case as well. However, the method is not in-place and is inefficient as it finds the exact median of the elements at each recursive step. We develop an efficient, heapsort based, in-place method to sort a multiset whose performance achieves the optimal bound up to lower order terms.

Procedure Multiset_Heapsort(S);
  heapify (quit when the key value of each child is $\leq$ that of the parent);
  while heap is not empty do *Delete_allmax(root, heap);*
Procedure Delete_allmax(node, heap);
  if value(left_child(node))=value(node)
        then Delete_allmax(left_child(node), heap);
  if value(right_child(node))=value(node),
        then Delete_allmax(right_child(node), heap);
  delete(node, heap);

Figure 1: *Heapsort – Adapted for Multisets*

2.1 Heapsort – Adapted for Multisets

Heapsort works by performing n deletemax operations after first building the (max) heap. Our adaptation modifies the deletemax operation to delete all the keys whose values are the same as the maximum value of the heap, in chunks. As all those keys will be closer to the root, they can be found by performing a post-order traversal of the tree, and the values are deleted starting from the leftmost node whose value is different from those of its children. Thus, when a node with key value x is actually deleted, no two elements with value x are compared. For, the value of the node's children are strictly smaller than its value when the node is being deleted.

Figure 1 explains the algorithm. In the figure, *heapify* is the standard ($2n$ comparisons) procedure that transforms an array in-place into one that satisfies the heap property. The subroutine *delete(node, heap)* deletes the node and heapifies the portion of the heap that has the *node* as root. As in the standard heapsort, the deleted elements form a part of the growing sorted list at the end of the array, and are no longer part of the heap.

The procedure *Delete_allmax* deletes all values in the heap that are equal to the value of the root. It first searches for a node whose value is equal to that of the root and is different from those of its children, and deletes that node. This procedure is repeatedly followed in a bottom-up fashion until the root gets deleted. For simplicity, the procedure is presented recursively. The recursion can be unraveled by following a post order method for deleting the nodes, once the leftmost node whose value is the same as that of the root, but different from that of its both children, is reached.

Standard implementation of the delete operation costs $2\lg f$ comparisons and $\lg f$ data movements, if f is the number of elements in the subheap rooted at the node to be deleted. If there are n_i elements having the same value as that of the root, then one step of *Delete_allmax(root, heap)* which deletes all the nodes having the same value as the root, takes at most $2(n_i \lg n - n_i \lg n_i) + O(n_i)$ three way comparisons and $n_i \lg n - n_i \lg n_i + O(n_i)$ data movements. This is based on the observation that a node deleted at an intermediate level of the tree needs to travel less than $\lg n$ distance to a leaf. The bound follows directly from the following lemma:

Lemma 1 *Let T be a complete binary tree with n nodes, and let B be a subtree rooted at the root of T consisting of n_i nodes. Then the sum of the path lengths in T, from each node of B to a leaf of T is at most $n_i \lg n - n_i \lg n_i + O(n_i)$.*

Proof: Let the nodes of B be labeled arbitrarily from 1 to n_i. The required sum is $\sum_{j=1}^{n_i}$ (length of the path from node j to a leaf of T)

$$\leq \sum_{j=1}^{n_i} (\lceil \lg n \rceil - (\text{length of the path from root of } T \text{ to node } j))$$

$$\leq n_i \lg n + n_i - \sum_{j=1}^{n_i} (\text{length of the path from root of } B \text{ to the node } j)$$

$$= n_i \lg n + n_i - \text{the internal path length of } B$$

As the internal path length of a binary tree having n nodes is at least $n \lg n - O(n)$ [5], the required sum is at most $n_i \lg n - n_i \lg n_i + O(n_i)$.

$\square$

So overall, the adaptive heapsort performs at most $2(n \lg n - \sum_{i=1}^{m} n_i \lg n_i) + O(n)$ comparisons and $n \lg n - \sum_{i=1}^{m} n_i \lg n_i + O(n)$ data movements. Thus we have proved the following theorem.

Theorem 1 *Let L be a multiset of n items consisting of m distinct keys, where the i-th distinct item appears n_i times. L can be sorted in-place using $2(n \lg n - \sum_{i=1}^{m} n_i \lg n_i) + O(n)$ comparisons and $n \lg n - \sum_{i=1}^{m} n_i \lg n_i + O(n)$ data movements, in the worst case.*

The number of comparisons performed in one *delete* operation can be reduced by using the technique of Gonnet and Munro[4] to $\lg f + \lg^* f$ comparisons for a heap of size f. If this approach is used for the *delete* procedure in Figure 1, then one step of *Delete_allmax(root, heap)* requires at most $n_i \lg n + O(n_i) + n_i \lg^* n - n_i \lg n_i$ comparisons, in the worst case. If we have an output buffer, then each *delete* operation can be modified to use at most $\lg f$ comparisons, by placing the deleted elements in the output buffer. Thus the adaptive algorithm can be improved to prove the following theorem.

Theorem 2 *Let L be a multiset of n items consisting of m distinct values, where the i-th distinct item appears n_i times. L can be sorted in-place using $n \lg n - \sum_{i=1}^{m} n_i \lg n_i + O(n)$ data movements and $n \lg n - \sum_{i=1}^{m} n_i \lg n_i + O(n) + n \lg^* n$ comparisons in the worst case. If we have an output buffer of size n, then L can be sorted using $n \lg n - \sum_{i=1}^{m} n_i \lg n_i + O(n)$ comparisons and $n \lg n - \sum_{i=1}^{m} n_i \lg n_i + O(n)$ data movements, in the worst case.*

This yields an information theoretically optimal (up to lower order terms) in-place algorithm for sorting a multiset. Without the output buffer, the time bound of the algorithm has an additive $n \lg^* n$ term.

```
Procedure In-Place_lexisort(L)
Multiset_Heapsort(L) under the first key.
for i = 2 to k do,
    j_start ← 1 ;
        while j_start ≤ n do
            j_end ← j_start + 1;
                while j_end ≤ n and L[j_end][i − 1] = L[j_start][i − 1] do j_end = j_end + 1;
                j_end := j_end − 1;
                Multiset_Heapsort L[j_start, j_end] under the i-th key.
                if j_end < n then j_start := j_end + 1;
        endwhile
endfor
```

Figure 2: *Optimal In-Place Algorithm to Lexicographically Sort Vectors*

---

3   Sorting Vectors In-Place

For sorting a list of k-dimensional vectors, Bentley and Saxe[1] obtained an optimal algorithm, by repeatedly inserting elements into their multidimensional data structure constructed for fast searches in lexicographically sorted list. This algorithm performs $n \lg n + (k − 1)n$ comparisons albeit using $O(n)$ space and $O((n \lg n + kn) \lg n)$ other "bookkeeping" operations. We obtain a near optimal in-place algorithm for the problem by using the adaptive method of Section 2.1 to sort under each key. It performs at most $n \lg n + O(kn) + O(kn \lg^* n)$ comparisons and $n \lg n + O(kn)$ data movements.

Let L be a k-dimensional array of vectors to be sorted. Element at the i-th cell of L is denoted by $L[i]$ and the j-th coordinate of an element x is denoted by $x[j]$. So $L[i][j]$ denotes the j-th coordinate of the i-th element of the array L. The pseudocode of Figure 2 explains the algorithm. The algorithm first sorts the vectors under the first coordinate using our multiset algorithm of previous section. Then it repeatedly, for $i = 2$ to k, identifies all the vectors having the same value under the previous $((i − 1)$-th) key, and for each such value, sorts those vectors using our multiset algorithm under the i-th key.

The claimed bounds follow from the fact that the number of comparisons and data movements saved between equal valued items by the adaptive heapsort for each coordinate, is precisely the number spent on those elements, for sorting under the next coordinates. Thus we have the following theorem from Theorem 2 and from the algorithm of Figure 2.

Theorem 3 *An array of n k-dimensional vectors can be lexicographically sorted in-place using $n \lg n + O(kn) + O(kn \lg^* n)$ comparisons and $n \lg n + O(kn)$ data (vector) movements in the worst case.*

4   Lower Bound For Sorting Multisets

Dobkin and Munro[2] have shown that the number of three-way comparisons required, on average, to sort a multiset S is $n \lg n − \sum_{i=1}^{m} n_i \lg n_i − n \lg(\lg n − (\sum_{i=1}^{m} n_i \lg n_i)/n) − O(n)$

even if the n_is are given. The proof is based on the ternary decision tree approach. We reduce the problem of lexicographically sorting vectors to the multiset sorting problem, use the lower bound for sorting vectors to improve the lower order term of the above lower bound, and prove the following theorem.

Theorem 4 *Let S be a multiset with multiplicities $n_1, n_2, ...n_m$ (where $n = \sum_{i=1}^{m} n_i$). Then at least $n \lg n - \sum_{i=1}^{m} n_i \lg n_i - n \lg e + O(\lg n)$ three-way comparisons are required on the average to sort S (where e is the base of the natural logarithm).*

Proof: Consider an algorithm A that sorts S using the optimal number T of ternary comparisons on the average. Pad each element in S with some arbitrary value which becomes its second coordinate, such that at the end, we obtain a list L of n 2-dimensional distinct vectors (i.e. no two vectors are same in both coordinates). Consider an algorithm that lexicographically sorts L using the optimal number T_l of ternary comparisons on the average. Then it is well known[6] that

$$T_l \geq n \lg n - n \lg e + O(\lg n) \tag{1}$$

as the vectors are all distinct, and there are $n!$ orders between them. Now consider the following approach as in Figure 2 to sort L lexicographically.

- Sort the first coordinate of L (i.e. the set S) using A. Mark the boundaries of each block (of size n_i, for $i = 1, 2...m$) of vectors having the same first coordinate.

- Sort each block (of size n_i, for $i = 1, 2, ..m$) of vectors having the same first coordinate, using the (standard) mergesort.

As mergesort performs at most $n \lg n$ comparisons on the average to sort a list of size n, the above approach performs $T + \sum_{i=1}^{m} n_i \lg n_i$ comparisons on the average. So from Equation 1, we get

$$T + \sum_{i=1}^{m} n_i \lg n_i \geq T_l \geq n \lg n - n \lg e + O(\lg n)$$

from which it follows that

$$T \geq n \lg n - \sum_{i=1}^{m} n_i \lg n_i - n \lg e + O(\lg n).$$

$\square$

Corollary 1 *The adaptations of mergesort and treesort for multisets developed by Munro and Spira [8] are within $O(n)$ of optimal.*

Corollary 2 *The adaptation of Heapsort described in Section 2.1 is within $O(n \lg^* n)$ of optimal.*

In [8] Munro and Spira proved a lower bound for finding the *mode*, the most frequently occurring element in a multiset, by relating it to the problem of sorting multisets. They show that once the mode (and r, its frequency of occurrence) is found, then we can "finish" sorting the multiset in $n \lg r - \sum_{i=1}^{m} n_i \lg n_i + O(n)$ comparisons. So, our improvement to the sorting lower bound also improves the lower order term of the known lower bound $(n \lg(n/r) - O(n-m) \lg \lg m - O(n))$ for finding the mode of a multiset.

Corollary 3 $n \lg(n/r) - O(n)$ *ternary comparisons are necessary on the average, to determine the mode of a multiset.*

The best known upper bound for determining the mode due to Dobkin and Munro [2], however, remains within $O(n \lg \lg(n/r))$ of the optimal.

5   Conclusion

We have adapted heapsort to obtain optimal (up to lower order terms) in-place algorithms for sorting a multiset and an array of vectors. We have also made an improvement to the lower order term of the known lower bound for sorting multisets. This improvement immediately improves the lower order term of a lower bound for determining the mode and makes the known "non in-place" algorithms for multisets optimal within an $O(n)$ term. It would be interesting to see whether the $O(n \lg^* n)$ term from the upper bound for in-place sorting multisets can be removed. In this context we note that, it is not even known (to the best of our knowledge) whether there is an in-place comparison based sort that performs $n \lg n + O(n)$ comparisons and $O(n \lg n)$ data movements in the worst case even if all elements are distinct.

Another important issue that arises when sorting multisets is stability (which requires the relative order of equal valued items to be maintained before and after the sort). Our heapsort adaptation is unstable due to the highly unstable nature of heapsort. It would be of interest to develop a stable in-place method to sort a multiset optimally.

References

[1] J. L. Bentley and J. B. Saxe, *Algorithms on Vector Sets*, SIGACT News, Fall (1979) 36-39.

[2] D. Dobkin and J. I . Munro, *Determining the Mode*, Theoretical Computer Science **12** (1980) 255-263.

[3] Th. M. Fischer, *Refined Bounds on the Complexity of Sorting and Selection in D-Dimensional Space*, Proceedings of MFCS, Lecture Notes in Computer Science, Springer Verlag 233 (1986) 309-314.

[4] G. H. Gonnet and J. I. Munro, *Heaps on Heaps*, SIAM Journal of Computing 15 (4) (1986), 964-971.

[5] D. E. Knuth, *The Art of Computer Programming. Volume I: Fundamental Algorithms*, Addison-Wesley (1973).

[6] D. E. Knuth, *The Art of Computer Programming. Volume III: Sorting and Searching*, Addison-Wesley (1973).

[7] U. Manber and G. Myers, *Suffix Arrays: A New Method for On-Line String Searches*, Proceedings of the first annual ACM-SIAM Symposium on Discrete Algorithms (1990) 319-327.

[8] I. Munro and P. M. Spira, *Sorting and Searching in Multisets*, SIAM Journal of Computing, 5 (1) (1976) 1-8.

[9] L. M. Wegner, *Quicksort for Equal Keys*, IEEE Transactions on Computers, C-34 (4) (1985) 362-367.

[10] J. Wiedermann, *The Complexity of Lexicographic Sorting and Searching*, Proceedings of MFCS, Lecture Notes in Computer Science, Springer Verlag 74 (1979) 517-522.

Probabilistic Leader Election on Rings of Known Size [*]

Karl Abrahamson [†] Andrew Adler [‡] Lisa Higham [§]
and David Kirkpatrick [¶]

Abstract

We are interested in the average case complexity of leader election (and related problems) on asynchronous processor rings, whose size, n, is known to all constituent processors.

Duris and Galil [10] prove an $\Omega(n \log n)$ lower bound for the average (over all assignments of identifiers) of the number of messages required by any *deterministic* leader election algorithm, when n is a power of 2 and the processor identifier space is sufficiently (exponentially) large. More recently Bodlaender [7], using substantially different techniques, has generalized this result to apply to arbitrary n and to identifier spaces as small as cn, for any $c > 1$.

Both Duris and Galil's and Bodlaender's lower bounds extend naturally (but significantly) to the expected message complexity of *Las Vegas* algorithms (randomized algorithms that terminate, always correctly, with probability 1) [7, 11, 5]. These bounds, in turn, can be shown to apply to the evaluation of several natural functions on rings, including AND, OR and XOR [7].

We show that the lower bound technique introduced by Duris and Galil can be modified to provide a direct proof of an $\Omega(n \log n)$ lower bound for the expected message complexity of Las Vegas leader election on *anonymous* (identifier free) rings, that is substantially simpler than the original. This simplicity not only serves to highlight the important structure of technique but also facilitates its extension to both arbitrary ring size and to *Monte Carlo* algorithms (randomized algorithms that err with probability at most ϵ). Specifically, we prove that the expected message complexity of any probabilistic algorithm that selects a leader with probability at least $1 - \epsilon$ on an anonymous ring of known size n, is $\Omega(n \min(\log n, \log \log(1/\epsilon)))$. A number of common function evaluation problems (including AND, OR, PARITY, and SUM) on rings of known size, are shown to inherit this complexity bound; furthermore these bounds are tight to within a constant factor.

[*]This research was supported in part by the Natural Sciences and Engineering Research Council of Canada, the British Columbia Advanced Systems Institute and the Killam Foundation.

[†]Computer Science Department, Washington State University, Pullman, WA 99164-1210, U.S.A.

[‡]Department of Mathematics, University of British Columbia, Vancouver, B.C. V6T-1W5, Canada

[§]Computer Science Department, University of Calgary, Calgary, Alberta, T2N-1N4, Canada

[¶]Department of Computer Science, University of British Columbia, Vancouver, B.C. V6T-1W5, Canada

1 Introduction and background

Leader election is the problem of distinguishing a single processor from among all the processors of a distributed network. This problem has been recognized as a fundamental problem in distributed computing because it forms a building block for many more involved algorithms including the evaluation of functions whose arguments are distributed over a network. Leader election has been extensively studied, with particular attention being paid to asynchronous processor rings. Within this model, many versions of the problem arise depending upon whether or not the processors have distinct identifiers and whether or not the ring size is known to processors in the network.

Even if the ring size is unknown (that is, the same algorithm must work over a range of ring sizes), unique identifiers make it possible to elect a leader by a deterministic algorithm using $O(n \log n)$ (unidirectional) messages [9, 15]. Furthermore, without knowledge of the ring size $\Omega(n \log n)$ messages are required (in the worst case [8], the average case [14] or even the expected case (for Las Vegas algorithms) [13]), for sufficiently large identifier spaces.

In the absence of identifiers, (i.e., on an *anonymous* ring), no deterministic algorithm can elect a leader. Furthermore, unless the ring size is known to within a multiple of two, even Las Vegas leader election is impossible [1]. However, when the ring size is known to within a factor of two, Las Vegas leader election can be achieved on anonymous rings in $O(n \log n)$ expected messages [12] or bits [1]. Furthermore, $\Omega(n \log n)$ bits are required, even in the best case, when ring size is known only to within a constant factor [1].

When the ring size is known exactly the lower bound techniques discussed so far do not seem to be sufficiently powerful to produce $\Omega(n \log n)$ message lower bounds. In particular, the Las Vegas leader election lower bounds of [1], which apply even to best case computations, can provide nothing better than an $\Omega(n\sqrt{\log n})$ bit lower bound when n is known. (In fact, this lower bound is tight, for best case computations [3, 4, 2].) Burns [8] presents an $\Omega(n \log n)$ lower bound for the worst case number of messages, assuming a sufficiently large identifier space. Bodlaender [6] extends this to an average case result for a restricted class of algorithms. The first $\Omega(n \log n)$ average case lower bound that applies to arbitrary algorithms on bidirectional rings was provided by Duris and Galil [10]. Their result, however, assumes that the identifier space is sufficiently (in fact, exponentially) large and that the ring size is a power of two. Using a substantially different approach, Bodlaender [7] was able to remove both of these restrictions. Specifically, he provides an $\Omega(n \log n)$ lower bound on the average (over all assignments of identifiers) of the number of messages required to elect a leader on a ring of arbitrary (known) size n, with an identifier space as small as cn, for any $c > 1$.

Both Duris and Galil's and Bodlaender's bounds extend without difficulty to include Las Vegas algorithms [7, 11, 5]. Indeed, Bodlaender [7] shows that, under fairly general conditions, randomization does not help in the presence of identifiers. As we show in this paper, a simpler and more direct proof of an $\Omega(n \log n)$ expected message complexity for Las Vegas leader election on anonymous rings can be formulated by modifying the techniques introduced by Duris and Galil. In their deterministic model, counting arguments are used to ensure that the characteristic of distinct identifiers is maintained. In the anonymous Las Vegas model these combinatorial techniques are not needed. The resulting simplification facilitates extension of the Duris and Galil's result in two directions, first for ring sizes that are not necessarily a power of two, and second for Monte Carlo algorithms (randomized algorithms that permit error with probability at most ϵ). Our lower bound for Monte Carlo algorithms is new; furthermore, it is not clear whether Bodlaender's approach lends itself to the analysis of Monte Carlo complexity.

Leader election can be decomposed into two problems called attrition and solitude verification [12, 1]. Attrition is the problem of reducing the original collection of contenders for leadership to exactly one contender. Solitude verification is the problem of confirming that only one contender remains. Our lower bounds for randomized leader election when the ring size is not known exactly all follow by reduction from solitude verification. However, when the ring size is known exactly, the complexity of solitude verification drops to $\Theta(n\sqrt{\log n})$ bits [3, 4, 2]. The randomized leader election lower bound presented in this paper follow by reduction from attrition. We refer to a Monte Carlo attrition procedure that deadlocks with probability at most ϵ by ϵ-attrition. We show that every ϵ-attrition procedure for rings of known size n has expected complexity $\Omega(n \min \{\log n, \log \log (1/\epsilon)\})$ messages.

Attrition is a subproblem of a number of common functions such as AND, OR, PAR-

ITY, and SUM as well as of leader election. For example, a Monte Carlo OR algorithm that errs with probability at most ϵ can be converted into a ϵ-attrition procedure. Section 4 contains such a reduction that permits the ϵ-attrition lower bound to extend to OR. Similar reductions extend the lower bound to several other functions. Section 4 also describes Monte Carlo algorithms for these problems, which demonstrate that the lower bounds are tight. A number of fundamental problems including AND, OR, PARITY, SUM and leader election are thus shown to have expected complexity $\Theta\left(n \min\left\{\log n, \log\log\left(1/\epsilon\right)\right\}\right)$ messages on rings of known size n.

Section 2 outlines the model of computation used for the lower bound result and reviews the definitions of relevant terms.

2   Model

An attrition procedure is required to reduce the original n contenders on a ring to exactly one while ensuring that all contenders are not eliminated. The definition of attrition can be generalized to permit deadlock with low probability. A Monte Carlo procedure is an *ϵ-attrition* procedure if it deadlocks with probability at most ϵ and all computations that do not deadlock eventually reduce the number of contenders on the ring to exactly one. Nondeadlocking computations of ϵ-attrition do not terminate. Therefore the complexity of ϵ-attrition is defined to be the expectation over all nondeadlocking computations of the number of messages sent until exactly one contender remains.

In the natural description of Monte Carlo ϵ-attrition for an anonymous ring, each processor runs the same randomized process. A processor's next state and next output message is determined by its current state, its last input message, and the result of a random experiment. Random choices occur throughout the run of the algorithm. However, these random choices can be simulated by a single random choice by each processor at the beginning of the algorithm. A processor randomly chooses a function from "internal state, input message" pairs to "internal state, output message" pairs. (Essentially, a processor preselects all its random coin tosses.) The resulting model pushes all the randomization to a single random experiment by each processor at the beginning of the computation. The rest of the computation proceeds deterministically. Hence, a randomized distributed procedure for ϵ-attrition on an anonymous ring is modelled as a probability space of deterministic processes available for assignment to processors on the ring.

It will suffice to assume that the behaviour of each deterministic process is entirely determined by its current state, where the current state records all information the process can know. Since this is all that is assumed of a process, the result applies even when the original randomized processes are not necessarily message-driven.

3   Lower Bounds for ϵ-attrition

This section bounds the expected message complexity of ϵ-attrition. It will be shown that any ϵ-attrition procedure has expected message complexity $\Omega\left(n \min\{\log n, \log\log(1/\epsilon)\}\right)$. The expected message complexity of Las Vegas (nondeadlocking) attrition, $\Omega(n \log n)$, follows from the general result by setting the allowable probability of deadlock, ϵ, to less than $1/2^n$.

The proof uses two techniques that are adapted from those introduced by Duris and Galil [10]. The first technique argues that expected computation for parts of the ϵ-attrition procedure cannot be too low because otherwise deadlock will occur, under a specified scheduler, with intolerably high probability. This is the essence of Lemma 3.1. The second technique, used here in Theorem 3.2, sums these expected message complexities for disjoint parts of the ϵ-attrition procedure to get a lower bound on total expected computation.

Definitions and Notation:

Let α be the probability space of deterministic processes associated with an ϵ-attrition procedure for anonymous rings of known size n. Denote by α^l the product space formed from l copies of α under the induced product probability measure. Let $\mathcal{R}$ be a ring of n processes each from α and let Π be any sequence of consecutive processes in $\mathcal{R}$. Let $len(\Pi)$ denote the number of processes in sequence Π. If $len(\Pi) = l$, then Π is called an l-process.

Imagine placing barriers on the links before and after sequence Π, and executing Π with the barriers remaining in place until the computation on the associated ring segment is *quiescent,* that is, all remaining messages are queued at barriers and computation cannot proceed until at least one barrier is removed. (There is no loss of generality in assuming that with probability 1, a random sequence of processes reaches quiescence for every scheduler, because otherwise a large amount of message traffic can be forced.) A *computation of* Π is any pattern of message traffic on the segment corresponding to Π that could occur from the beginning of the computation up to the point when it reaches quiescence. A *partition* of Π is a sequence of nonempty subsequences of processes whose concatenation is Π. A partition of Π into subsequences $\Pi_1, \ldots, \Pi_k$ is denoted $\Pi_1|\Pi_2|\ldots|\Pi_k$. A *decomposition* of an integer L is any sequence $l_1, \ldots, l_k$ of positive integers that sum to L and is denoted $l_1|l_2|\ldots|l_k$. The partition $\Pi_1|\Pi_2|\ldots|\Pi_k$ is said to be *consistent* with decomposition $l_1|l_2|\ldots|l_k$ if $l_i = len(\Pi_i)$ for $1 \leq i \leq k$.

Suppose that barriers are placed on the links between adjacent members of some partition $\Pi_1|\Pi_2|\ldots|\Pi_k$ of Π. It is intended to measure the number of *additional* messages sent on sequence Π after removal of all the barriers between adjacent segments of the partition while the links at either end of Π remain blocked. The scheduler, however, is only partially constrained by the barriers. Several different computations may still arise depending on the scheduling of messages within each of the segments in the partition and on the scheduling after the barriers are removed. A *scheduling function* is a function that assigns a fixed schedule to each sequence of processes. If S is a scheduling function, then $S(\Pi)$ is a *scheduled process sequence.* If Π is a sequence of deterministic processes, then there is exactly one computation that can arise from $S(\Pi)$. The notation is extended so that $S(\Pi_1|\Pi_2|\ldots|\Pi_k)$ denotes the schedule S applied separately to each subsequence of processes Π_i in $\Pi_1|\Pi_2|\ldots|\Pi_k$. To avoid the ambiguity caused by undetermined schedulers, the following will establish a scheduling function that assigns a fixed scheduled process $S^*(\Pi)$ to an arbitrary sequence Π.

Let S' be any scheduling function that schedules Π by first running $S(\Pi_1|\Pi_2|\ldots|\Pi_k)$ and then removing the barriers between the segments of the partition $\Pi_1|\Pi_2|\ldots|\Pi_k$ and continuing with the computation until Π is quiescent. S' is an *extension of S for* $\Pi_1|\Pi_2|\ldots|\Pi_k$. The computation that occurs under S' after removal of the barriers in $\Pi_1|\Pi_2|\ldots|\Pi_k$ and up to the point where Π is quiescent, is the *continuation of* $S(\Pi_1|\Pi_2|\ldots|\Pi_k)$ *under S'.* The extension, S', of S for $\Pi_1|\Pi_2|\ldots|\Pi_k$ that minimizes the number of messages sent in the continuation of $S(\Pi_1|\Pi_2|\ldots|\Pi_k)$ under S' is the *minimum extension of S for* $\Pi_1|\Pi_2|\ldots|\Pi_k$. The *cost of the continuation of* $S(\Pi_1|\Pi_2|\ldots|\Pi_k)$ is the number of messages sent during the continuation of $S(\Pi_1|\Pi_2|\ldots|\Pi_k)$ under S' where S' is the minimum extension of S for $\Pi_1|\Pi_2|\ldots|\Pi_k$.

As a first step toward defining S^*, a fixed partition is specified for each sequence of processes. Let $L \leq n$ and let k be the integer satisfying $\lfloor \frac{n}{8^k} \rfloor \leq L < \lfloor \frac{n}{8^{k-1}} \rfloor$. Let $l = \lfloor \frac{n}{8^{k+1}} \rfloor$. Define the decomposition $d(L) = l_1|l_2|l_3|l_4|l_5|l_6$ by $l_1 = l_2 = l_3 - 1 = l_4 - 1 = l_5 = l$ and $l_6 = L - \sum_1^5 l_i$. The partition $\Pi_1|\ldots|\Pi_6$ of $\Pi \in \alpha^L$ that is consistent with $d(L)$ is denoted $\delta(\Pi)$.

Given a sequence $\Pi \in \alpha^L$, the decomposition, $d(L)$, and the corresponding partition, $\delta(\Pi)$, are used to associate a fixed scheduled process, $S^*(\Pi)$, with Π. $S^*(\Pi)$ is specified recursively by:

1. If $L < 8$ then $S^*(\Pi)$ is the scheduler that minimizes the number of messages sent in a computation of Π.

2. Otherwise, $S^*(\Pi)$ is the minimum extension of S^* for $\delta(\Pi)$.

The fixed scheduling function S^*, is assumed for the remainder of this section. The cost of the continuation of $S^*(\Pi_1|\Pi_2|\ldots|\Pi_k)$ is denoted $CC(\Pi_1|\Pi_2|\ldots|\Pi_k)$. We rely on the simple observation that if $CC(\Pi_1|\Pi_2|\ldots|\Pi_k)$ is sufficiently small (at most one half the length of the smallest sequence in the partition) then $CC(\Pi_i|\Pi_{i+1}) \leq CC(\Pi_1|\Pi_2|\ldots|\Pi_k)$, for $1 \leq i < k$.

Lemma 3.1 bounds the expectation of $CC(\delta(\Pi))$. That is, given that Π is a random L-process partioned consistently with $d(L)$ into $\Pi_1|\cdots|\Pi_6$, Lemma 3.1 bounds the expected number of messages sent in the continuation of S^* for $\Pi_1|\Pi_2|\ldots|\Pi_k$ under the minimum extension. In general, $CC(\Pi_1|\Pi_2|\ldots|\Pi_k)$ is dependent on the choice of Π as well as on the partition of Π. However, the expectation of $CC(\Pi_1|\Pi_2|\ldots|\Pi_k)$ depends only on the decomposition $len(\Pi_1)|\ldots|len(\Pi_k)$. To emphasize this dependence, define $E_{cc}(l_1|\ldots|l_k)$ to be the expected value of $CC(\Pi_1|\Pi_2|\ldots|\Pi_k)$ over all processes Π such that $\Pi_1|\Pi_2|\ldots|\Pi_k$ is a partition of Π consistent with $l_1|\ldots|l_k$.

If $S \subseteq \alpha^l$, then $Pr(S)$ is used to abbreviate $Pr(x \in S | x \in \alpha^l)$.

Lemma 3.1 *For any ϵ-attrition procedure for anonymous bidirectional rings of fixed size n and for all $L \leq n$,*

$$E_{cc}(d(L)) \geq \left(1 - \epsilon^{L/(64n)}\right)^2 \frac{L}{2^{11}} .$$

Proof: Let $d(L) = l_1 | \ldots | l_6$. By definition $l_1 = l_2 = l_3 - 1 = l_4 - 1 = l_5 = l$ and $l_6 > l$ where $l = \left\lfloor \frac{n}{8^{k+1}} \right\rfloor$ for integer k satisfying $\left\lfloor \frac{n}{8^k} \right\rfloor \leq L < \left\lfloor \frac{n}{8^{k-1}} \right\rfloor$. Therefore, $n = 8^{k+1}l + r$ where $r < 8^{k+1}$.

A summary of the central ideas follows. A ring of size n can be partitioned into segments of length l and $l + 1$. If all of $E_{cc}(l|l)$, $E_{cc}(l|l + 1)$, $E_{cc}(l + 1|l)$ and $E_{cc}(l + 1|l + 1)$ are sufficiently small, and adjacent barriers are removed from the partitioned ring, then with high probability, the traces of the message traffic of the continuation will not intersect. Such a situation results in deadlock. So it must be concluded that at least one of $E_{cc}(l|l)$, $E_{cc}(l|l + 1)$, $E_{cc}(l + 1|l)$ or $E_{cc}(l + 1|l + 1)$ is not small. However, all combinations of l and $(l+1)$ are included as adjacent lengths in the decomposition, $d(L)$. Therefore, the expected cost, over all partitions consistent with $d(L)$, of the (minimum) continuation is also not small.

Let α be any ϵ-attrition procedure for anonymous bidirectional rings of fixed size n. Let x and y be two random l-processes from α^l and let w and z be random $(l + 1)$-processes from α^{l+1}. Let $\lambda = 1 - \min\{\Pr(CC(x|y) \leq \frac{l}{2}), \Pr(CC(x|z) \leq \frac{l}{2}), \Pr(CC(z|x) \leq \frac{l}{2}), \Pr(CC(w|z) \leq \frac{l}{2})\}$. Then $\Pr(CC(x|y) \leq \frac{l}{2}) \geq 1 - \lambda$ and $\Pr(CC(x|z) \leq \frac{l}{2}) \geq 1 - \lambda$ and $\Pr(CC(z|x) \leq \frac{l}{2}) \geq 1 - \lambda$ and $\Pr(CC(w|z) \leq \frac{l}{2}) \geq 1 - \lambda$. Let $A_1 = \{s | s \in \alpha^l \wedge \Pr(CC(s|y) \leq \frac{l}{2}) > 1 - \lambda^{1/2}\}$. Then $\Pr(A_1) \geq 1 - \lambda^{1/2}$ since otherwise $\Pr(CC(x|y) \leq \frac{l}{2}) < (1 - \lambda^{1/2}) + \lambda^{1/2}(1 - \lambda^{1/2}) = 1 - \lambda$. Similarly, let:

$$A_2 = \{s | s \in \alpha^l \wedge \Pr(CC(y|s) \leq \frac{l}{2}) > 1 - \lambda^{1/2}\}$$

$$A_3 = \{s | s \in \alpha^l \wedge \Pr(CC(s|z) \leq \frac{l}{2}) > 1 - \lambda^{1/2}\}$$

$$A_4 = \{s | s \in \alpha^{l+1} \wedge \Pr(CC(y|s) \leq \frac{l}{2}) > 1 - \lambda^{1/2}\}$$

$$A_5 = \{s | s \in \alpha^l \wedge \Pr(CC(s|z) \leq \frac{l}{2}) > 1 - \lambda^{1/2}\}$$

$$A_6 = \{s | s \in \alpha^l \wedge \Pr(CC(z|s) \leq \frac{l}{2}) > 1 - \lambda^{1/2}\}$$

Then, in the same way, $\Pr(A_i) \geq 1 - \lambda^{1/2}$ for $i = 2, 3, 4, 5$ and 6. Let $G_{l,l}^l = A_1 \cap A_2$. Let $G_{l,l+1}^l = A_2 \cap A_3$. Let $G_{l,l+1}^{l+1} = A_4 \cap A_5$. Let $G_{l+1,l+1}^{l+1} = A_5 \cap A_6$. Then $\Pr(G) \geq 1 - 2\lambda^{1/2}$ for $G \in \{G_{l,l}^l, G_{l,l+1}^l, G_{l,l+1}^{l+1}, G_{l+1,l+1}^{l+1}\}$.

In summary, with high probability $(1 - \lambda^{1/2})$ an l-process in $G_{l,l}^l$ will combine on either side with a randomly chosen l-process to produce a partition with a small (less than $l/2$) cost of continuation. Similar statements apply for each of $G_{l,l+1}^l, G_{l,l+1}^{l+1}$ and $G_{l+1,l+1}^{l+1}$.

Let $d = 4 \cdot 8^k$ (and hence $n = 2dl + r$). If the remainder r is even, define the class of rings B_{even} with length n by:

$$B_{\text{even}} = \left\{ x_1, \ldots, x_{2d} \mid \; x_i \in G_{l,l}^l \text{ for } i = 1, 3, \ldots, 2d - r - 1, \right.$$
$$x_i \in \alpha^l \text{ for } i = 2, 4, \ldots, 2d - r,$$
$$x_{2d-r+1} \in G_{l,l+1}^{l+1},$$
$$x_i \in G_{l+1,l+1}^{l+1} \text{ for } i = 2d - r + 3, 2d - r + 5, \ldots, 2d - 1,$$
$$x_i \in \alpha^{l+1} \text{ for } i = 2d - r + 2, 2d - r + 4, \ldots, 2d,$$
$$\left. \text{and } CC(x_i|x_{i+1}) \leq \frac{l}{2} \text{ for } 1 \leq i < 2d, \text{ and } CC(x_{2d}|x_1) \leq \frac{l}{2} \right\}$$

Alternatively, if the remainder r is odd define the class of rings B_{odd} with length n by:

$$B_{\text{odd}} = \left\{ x_1, \ldots, x_{2d} \mid \; x_i \in G_{l,l}^l \text{ for } i = 1, 3, \ldots, 2d - r - 2, \right.$$
$$x_i \in \alpha^l \text{ for } i = 2, 4, \ldots, 2d - r - 1,$$
$$x_{2d-r} \in G_{l,l+1}^l,$$
$$x_i \in G_{l+1,l+1}^{l+1} \text{ for } i = 2d - r + 2, 2d - r + 4, \ldots, 2d - 1,$$
$$x_i \in \alpha^{l+1} \text{ for } i = 2d - r + 1, 2d - r + 3, \ldots, 2d,$$
$$\left. \text{and } CC(x_i|x_{i+1}) \leq \frac{l}{2} \text{ for } 1 \leq i < 2d, \text{ and } CC(x_{2d}|x_1) \leq \frac{l}{2} \right\}$$

Then $\Pr(B)$ for $B \in \{B_{\text{even}}, B_{\text{odd}}\}$ in the product space α^n can be bounded as follows. A random l-process $((l+1)$-process$)$ is in the required G set with probability at least $1 - 2\lambda^{1/2}$. Therefore all x_i, for $i = 1, 3, \ldots, 2d - 1$, meet the specification of B with probability at least $(1 - 2\lambda^{1/2})^d$. Given x_i satisfying the specification of B for $i = 1, 3, \ldots, 2d - 1$, $CC(x_i|x_{i+1}) \leq \frac{l}{2}$ with probability at least $1 - \lambda^{1/2}$ and $CC(x_{i+1}|x_{i+2}) \leq \frac{l}{2}$ with probability at least $1 - \lambda^{1/2}$. Hence, for a fixed $i = 2, 4, \ldots, 2d$, the conditions on B are met with probability at least $1 - 2\lambda^{1/2}$. Hence, $\Pr(B) \geq (1 - 2\lambda^{1/2})^d (1 - 2\lambda^{1/2})^d \geq (1 - 2\lambda^{1/2})^{\lfloor n/l \rfloor}$.

For every process sequence $\Pi = x_1, \ldots, x_{2d} \in B$, we have $CC(x_i|x_{i+1}) \leq l/2$, for all i. Since each segment x_i has length at least l, it is impossible for messages generated by the removal of any pair of barriers to interact. Hence, after removal of all barriers there are at most $(l/2)(n/l)$ additional messages before all message traffic ceases. Since these are deadlocking computations, elements of B produce erroneous ϵ-attrition computations under scheduler S'.

Since deadlock occurs with probability at most ϵ, $\Pr(B) \leq \epsilon$, which implies that $(1 - 2\lambda^{1/2})^{\lfloor n/l \rfloor} \leq \epsilon$. Thus, $\lambda \geq (1 - \epsilon^{l/n})^2/4$. But from the definition of λ, either $\Pr(CC(x|y) \leq \frac{l}{2}) = 1 - \lambda$ or $\Pr(CC(x|z) \leq \frac{l}{2}) = 1 - \lambda$ or $\Pr(CC(z|x) \leq \frac{l}{2}) = 1 - \lambda$ or $\Pr(CC(z|w) \leq \frac{l}{2}) = 1 - \lambda$. Hence, either $E_{cc}(l|l) = E(CC(x|y)) \geq \lambda \cdot \frac{l}{2} > (1 - \epsilon^{l/n})^2 \frac{l}{32}$ or $E_{cc}(l|l+1) > (1 - \epsilon^{l/n})^2 \frac{l}{32}$ or $E_{cc}(l+1|l) > (1 - \epsilon^{l/n})^2 \frac{l}{32}$ or $E_{cc}(l+1|l+1) > (1 - \epsilon^{l/n})^2 \frac{l}{32}$.

The decomposition $d(L)$ contains every combination of l and $(l+1)$ as adjacent integers. Furthermore, if a continuation of the whole partition behaves differently than the combination of the continuations on each of the adjacent segments of the partition, then there must have been interaction between these adjacent pieces. This alone would require more than $l/2$ messages. Therefore, the cost of continuation of a partition consistent with $d(L)$ must be at least the maximum of the cost of continuation on the segments composed of adjacent pairs in the partition. Thus, $E_{cc}(l_1|\ldots|l_6) \geq (1 - \epsilon^{l/n})^2 \frac{l}{32}$. But $l = \lfloor \frac{n}{8^{k+1}} \rfloor$ and $L < \lfloor \frac{n}{8^{k-1}} \rfloor$ implying $\frac{L}{64} \leq l$. Hence $E_{cc}(l_1|\ldots|l_6) \geq (1 - \epsilon^{L/(64n)})^2 \frac{L}{2^{11}}$. ∎

Theorem 3.2 *Every ϵ-attrition procedure for anonymous rings of fixed size n has expected message complexity $\Omega\left(n \min\{\log n, \log\log(1/\epsilon)\}\right)$ on rings of size n.*

Proof: Let α be the probability space of processes available to an ϵ-attrition procedure for rings of fixed size n.

Let $\text{cost}_{S^*}(\Pi)$ be the total number of messages sent by scheduled process sequence $S^*(\Pi)$ and define $E_{msg}(L) = E(\text{cost}_{S^*}(\Pi)|\Pi \in \alpha^L)$. Let $\Pi \in \alpha^L$ and recall that $\delta(\Pi)$ is the partition of Π consistent with $d(L)$. Then

$$\text{cost}_{S^*}(\Pi) = CC(\delta(\Pi)) + \sum_{i=1}^{6} \text{cost}_{S^*}(\Pi_i)$$

$$\geq \frac{n}{2^{11}}\frac{1}{4}\min\left\{\log_{64} n, \frac{\log\log{(1/\epsilon)}}{6}\right\}$$

$$= \Omega\left(n\min\left\{\log n, \log\log\frac{1}{\epsilon}\right\}\right).$$

∎

4 Related results

An $O(n\log c)$ expected messages randomized attrition procedure for an anonymous ring with $c \geq 1$ initial contenders is described in [1]. This randomized attrition can be converted to an ϵ-attrition procedure which shows that the $\Omega\left(n\min\left(\log n, \log\log{(1/\epsilon)}\right)\right)$ expected messages bound of Section 3 is tight to within a constant factor. Let $\lambda = \min\left(n, \log{(1/\epsilon)}\right)$. Processors first choose to be contenders with probability λ/n and the contenders run randomized attrition. Since λ contenders are expected, the resulting ϵ-attrition has the desired complexity. The only way the algorithm can err is if $\log{(1/\epsilon)} < n$ and no processor chooses to be a contender. This happens with probability $(1 - \lambda/n)^n \leq e^{-\lambda} = \epsilon$.

Lower bounds on ϵ-attrition imply lower bounds on Monte Carlo leader election. Thus leader election that errs with probability at most ϵ inherits a lower bound of $\Omega(n\min(\log n, \log\log(1/\epsilon)))$ expected messages on rings of fixed size. This lower bound is tight to within a constant factor. In [1] it is shown that a leader election algorithm can be assembled from attrition and solitude verification. This relationship between the three problems extends to the probabilistic case where error with probability at most ϵ is tolerated. But Monte Carlo solitude verification has complexity $\Theta(n\min(\sqrt{\log n}, \sqrt{\log\log(1/\epsilon)} + \log\nu(n), \log\log(1/\epsilon)))$ expected bits on rings of known size, where $\nu(n)$ is the smallest non-divisor of n, [3, 4]. Hence there is a leader election algorithm that errs with probability at most ϵ and has complexity $O\left(n\min\left(\log n, \log\log(1/\epsilon)\right)\right)$ expected bits (and messages).

The preceding discussion illustrates that attrition is an essential and dominant part of leader election in the sense that the complexities of the two problems are equivalent (even in the probabilistic case which permits error with low probability). A number of other common problems have the same relationship to attrition. We show here that the which implies

$$E_{msg}(L) = E\left(CC(\delta(\Pi))\right) + \sum_{i=1}^{6} E_{msg}(l_i)$$

By Lemma 3.1, $E\left(CC(\delta(\Pi))\right)$ is at least $\frac{L}{2^{11}}(1 - \epsilon^{L/(64n)})^2$. Therefore

$$E_{msg}(L) \geq \frac{L}{2^{11}} \cdot (1 - \epsilon^{L/(64n)})^2 + \sum_{i=1}^{6} E_{msg}(l_i)$$

where $\sum_{i=1}^{6} l_i = L$ and $l_i \geq \frac{L}{64}$.

Claim:

$$E_{msg}(L) \geq \frac{L}{2^{11}} \sum_{i=1}^{\log_{64} L} (1 - \epsilon^{L/(64^i n)})^2.$$

Proof of claim: The basis is clear so assume the inductive hypothesis:

$$E_{msg}(l) \geq \frac{l}{2^{11}} \sum_{i=1}^{\log_{64} l} (1 - \epsilon^{l/(64^i n)})^2 \qquad (3.1)$$

for all $l < L$. Then:

$$
\begin{aligned}
E_{msg}(L) &\geq \frac{L}{2^{11}} \cdot (1 - \epsilon^{L/(64n)})^2 + \sum_{i=1}^{6} E_{msg}(l_i) \\
&\geq \frac{L}{2^{11}} \cdot (1 - \epsilon^{L/(64n)})^2 + \sum_{i=1}^{6} \frac{l_i}{2^{11}} \sum_{j=1}^{\log_{64} l_i} (1 - \epsilon^{l_i/(64^j n)})^2 \qquad \text{(by 3.1)} \\
&\geq \frac{L}{2^{11}} \cdot (1 - \epsilon^{L/(64n)})^2 + \sum_{i=1}^{6} \frac{l_i}{2^{11}} \sum_{j=1}^{\log_{64}(L/64)} (1 - \epsilon^{L/(64^{j+1} n)})^2 \\
&\geq \frac{L}{2^{11}} \cdot (1 - \epsilon^{L/(64n)})^2 + \frac{L}{2^{11}} \sum_{j=2}^{\log_{64} L} (1 - \epsilon^{L/(64^j n)})^2 \\
&= \frac{L}{2^{11}} \sum_{j=1}^{\log_{64} L} (1 - \epsilon^{L/(64^j n)})^2 .
\end{aligned}
$$

So the claim holds.

Let $\mathcal{R} = \pi_1, \ldots, \pi_n$ be a random element of α^n. Place a barrier between π_1 and π_n and consider the computation of α on segment $\pi_1, \ldots, \pi_n$ under schedule $\mathcal{S}^*$. Then:

$$E_{msg}(n) \geq \frac{n}{2^{11}} \sum_{j=1}^{\log_{64} n} (1 - \epsilon^{64^{-j}})^2$$

Notice that $(1 - \epsilon^{64^{-x}})^2 \geq \frac{1}{4}$ as long as $x \leq \log_{64} \log(1/\epsilon) = (\log \log(1/\epsilon))/6$. Therefore:

$$E_{msg}(n) \geq \frac{n}{2^{11}} \sum_{i=1}^{\min\{\log_{64} n, \frac{1}{6} \log \log(1/\epsilon)\}} \left(1 - \epsilon^{64^{-i}}\right)^2$$

ϵ-attrition bound extends to computing OR with probability of error at most ϵ on rings of know size.

Theorem 4.1 *The complexity of any algorithm that with probability at least $1 - \epsilon$ computes OR on an anonymous ring of fixed size n is $\Omega\left(n \min\left(\log n, \log\log\left(1/\epsilon\right)\right)\right)$ expected messages.*

Proof: Let α be an algorithm for OR which errs with probability at most ϵ. Let $f(n, \epsilon)$ be the expected message complexity of α. Let γ be any algorithm for attrition on a ring of size n with any non-empty subset of initial contenders such that the expected complexity of γ is $O(n \log c)$ messages, where c is the actual number of contenders (not necessarily known). ([1] describes one such attrition algorithm.) Let $\lambda = \min\left(\log n, \log\log\left(1/\epsilon\right)\right)$. (Throughout this proof "log" refers to the natural logarithm.) Define β to be the distributed algorithm where each processor on an anonymous ring executes the following:

β: 1. generate a random bit, *myflip* $\in \{0, 1\}$ such that $\Pr(1) = \lambda/n$.

   2. if *myflip* $= 1$ become a contender and initiate γ. Henceforth participate in γ and discard all α messages.

   3. if *myflip* $= 0$ initiate α. Participate in α only as long as no γ message arrives. Upon receipt of a γ message, participate in γ as a non-contender and discard all subsequent α messages. If α confirms "all 0's" then restart the algorithm at step 1.

Step 2 performs attrition on an expected small number of contenders. Step 3 alerts the processors to try again in the event that there were no contenders. Thus β is an attrition algorithm that deadlocks with small probability.

Error: The only way that the attrition algorithm, β, deadlocks is if all processors flip 0 and the OR algorithm α fails to confirm all 0's. Therefore the probability of deadlock of β is at most $\epsilon \sum_{i=1}^{\infty}\left(\left(1 - \frac{\lambda}{n}\right)^n\right)^i \leq \frac{\epsilon}{e^{\lambda} - 1} < \epsilon$ as long as $n > 2$ and $\epsilon < 0.135 < 1/e^2$.

Complexity: Let random variable C be the number of processors with *myflip* $= 1$. The expected number of messages sent by β, denoted E(complexity$_\beta$), is given by

$$E\left(\text{complexity}_\beta\right) = E\left(\text{complexity}_\beta | \text{all 0's}\right) \Pr(\text{all 0's}) +$$

$$E\left(\text{complexity}_\beta | \text{at least one 1}\right) \Pr(\text{at least one 1})$$

$$\leq \left(f(n,\epsilon) + E(\text{complexity}_\beta)\right)\left(1 - \frac{\lambda}{n}\right)^n +$$

$$\left(f(n,\epsilon) + E(n \log C)\right)\left(1 - \left(1 - \frac{\lambda}{n}\right)^n\right)$$

$$\leq \frac{f(n,\epsilon) + n \log \lambda \left(1 - \left(1 - \frac{\lambda}{n}\right)^n\right)}{1 - \left(1 - \frac{\lambda}{n}\right)^n}$$

$$< 2f(n,\epsilon) + n \min\left(\log\log n, \log\log\log \frac{1}{\epsilon}\right)$$

$$(\text{if } n > 2 \text{ and } \epsilon < 0.135)$$

Since β is an ϵ-attrition algorithm, β has complexity $\Omega\left(n \min\left(\log n, \log\log\left(1/\epsilon\right)\right)\right)$. Hence $f(n,\epsilon)$ has order $\Omega\left(n \min\left(\log n, \log\log\left(1/\epsilon\right)\right)\right)$. ∎

It is easily verified that ϵ-attrition also reduces to other functions such as AND, PARITY and SUM. Note that this lower bound for OR (and similarly for AND, SUM, and PARITY) is tight to within a constant factor because OR can be computed on a ring of size n by expending an additional $O(n)$ bits after electing a leader.

In an expanded version of this paper we will show how the results presented here can be extended in two directions. First, we show that a variation on the same lower bound argument allows us to extend our bounds to the case where the processors have identities drawn from a space of size $2n$. Second, and more interestingly, we consider the case where only a subset of the processors are contenders for leadership. Together with the results presented here these provide a comprehensive and uniform treatment of randomized leader election on rings of known size.

References

[1] Karl Abrahamson, Andrew Adler, Rachel Gelbart, Lisa Higham, and David Kirkpatrick. The bit complexity of randomized leader election on a ring. *SIAM Journal on Computing*, 18(1):12–29, 1989.

[2] Karl Abrahamson, Andrew Adler, Lisa Higham, and David Kirkpatrick. Randomized function evaluation on a ring. *Distributed Computing*, 3(3):107–117, 1989.

[3] Karl Abrahamson, Andrew Adler, Lisa Higham, and David Kirkpatrick. Optimal algorithms for probabilistic solitude detection on anonymous rings. Technical Report TR 90-3, University of British Columbia, 1990.

[4] Karl Abrahamson, Andrew Adler, Lisa Higham, and David Kirkpatrick. Tight lower bounds for probabilistic solitude verification on anonymous rings. Technical Report TR 90-4, University of British Columbia, 1990.

[5] Hagit Attiya and Mark Snir. Better computing on the anonymous ring. In *Proc. Aegean Workshop on Computing*, pages 329–338, 1988.

[6] Hans L. Bodlaender. *Distributed Algorithms, Structure and Complexity*. PhD thesis, University of Utrecht, 1986.

[7] Hans L. Bodlaender. New lower bound techniques for distributed leader finding and other problems on rings of processors. Technical Report RUU-CS-88-18, Rijksuniversiteit Utrecht, 1988.

[8] J. Burns. A formal model for message passing systems. Technical Report TR-91, Indiana University, 1980.

[9] Danny Dolev, Maria Klawe, and Michael Rodeh. An $O(n \log n)$ unidirectional distributed algorithm for extrema finding in a circle. *J. Algorithms*, 3(3):245–260, 1982.

[10] Pavol Duris and Zvi Galil. Two lower bounds in asynchronous distributed computation (preliminary version). In *Proc. 28nd Annual Symp. on Foundations of Comput. Sci.*, pages 326–330, 1987.

[11] Lisa Higham. *Randomized Distributed Computing on Rings*. PhD thesis, University of British Columbia, Vancouver, Canada, 1988.

[12] Alon Itai and Michael Rodeh. Symmetry breaking in distributed networks. In *Proc. 22nd Annual Symp. on Foundations of Comput. Sci.*, pages 150–158, 1981.

[13] Jan Pachl. A lower bound for probabilistic distributed algorithms. Technical Report CS-85-25, University of Waterloo, Waterloo, Ontario, 1985.

[14] Jan Pachl, E. Korach, and D. Rotem. Lower bounds for distributed maximum finding. *J. Assoc. Comput. Mach.*, 31(4):905–918, 1984.

[15] Gary Peterson. An $O(n \log n)$ algorithm for the circular extrema problem. *ACM Trans. on Prog. Lang. and Systems*, 4(4):758–752, 1982.

AUTHOR INDEX

Lecture Notes in Computer Science

For information about Vols. 1–429
please contact your bookseller or Springer-Verlag